CELLULAR NETWORKS - POSITIONING, PERFORMANCE ANALYSIS, RELIABILITY

Edited by **Agassi Melikov**

Cellular Networks - Positioning, Performance Analysis, Reliability
Edited by Agassi Melikov

Published by InTech
Janeza Trdine 9, 51000 Rijeka, Croatia

Publishing Process Manager Ivana Lorkovic

Technical Editor Teodora Smiljanic

Cover Designer InTech Design Team

Additional hard copies can be obtained from orders@intechopen.com

Cellular Networks - Positioning, Performance Analysis, Reliability,
Edited by Agassi Melikov
p. cm.
ISBN 978-953-307-246-3

For EU product safety concerns:
IN TECH d.o.o., Prolaz Marije Krucifikse Kozulić 3, 51000 Rijeka, Croatia,
info@intechopen.com or visit our website at intechopen.com.

Meet the editor

Agassi Melikov received his degree in Mathematics at Baku State University, Azerbaijan in 1977, PhD in Control Theory from Institute of Automatics, Kiev, in 1984, and Doctor of Sciences in Computer Science from Department of Applied Mathematics at Kiev National Technical University, Ukraine in 1992. He was a Visiting Professor in the Department of Electrical Engineering at the Eastern Mediterranean University of Cyprus from August of 1999 to August of 2000. He was also a Visiting Scholar in the Telecommunication Mathematics Research Centre at the Korea University, Seoul, from April of 2006 to June of 2006. He is currently Head of Department of Aerospace Information Technologies at National Aviation Academy of Azerbaijan and Head of Department of Teletraffic at Institute of Cybernetics, National Academy of Sciences of Azerbaijan. His current research interests are in teletraffic and queuing theory and their applications. He has published around 100 papers in internationally refereed journals such as Automation & Remote Control, Computer Communications, Cybernetics and System Sciences, Journal of Automation and Information Sciences, Automatic Control and Computer Sciences, Engineering Simulation. In 2001 he was selected for Associate Member of National Academy of Sciences of Azerbaijan in the field of Information Technologies. He is a member of Editorial Board of the journals Applied Mathematics (Scientific Research, USA) and Resent Patents on Electrical Engineering (Bentham Science Publisher, USA), Applied and Computational Mathematics (Azerbaijan).

Contents

Preface

Wireless cellular networks are an integral part of modern telecommunication systems. Today it is hard to imagine our life without the use of such networks. Nevertheless, the development, implementation and operation of these networks require engineers and scientists to address a number of interrelated problems. Among them are the problem of choosing the proper geometric shape and dimensions of cells based on geographic location, finding the optimal location of cell base station, selection the scheme dividing the total net bandwidth between its cells, organization of the handover of a call between cells, information security and network reliability, and many others.

This book mainly focuses on three types of problems from the above list - Positioning, Performance Analysis and Reliability. It contains four sections. The first part is devoted to problems of Positioning and contains five chapters. Here, the first three chapters discuss various methods and models to solve these problems. Chapter 1 is a review devoted to a detailed analysis of the main problems regarding Positioning in wireless networks. Chapter 4 is devoted to a comparative analysis of the two most popular choices of the geometric structure of a cell - hexagon and circle. The final chapter 5 of this part discusses some issues on signal processing using Smart Antennas.

Part 2 contains eight Chapters which are devoted to quality of service (QoS) metrics analysis of wireless cellular networks. Chapter 6 is a review of known algorithms to calculate QoS metrics in wireless cellular networks with call admission control based on guard channels. Unified approximate approach to QoS metrics calculations in multi-service wireless cellular networks under two multi-parametric call admission controls is proposed in Chapter 7. The proposed approach provides high accuracy. In Chapter 8, QoS metrics sensitivity to the first three moments of both cell dwell time and unencumbered interruption time in cellular networks is investigated. In Chapter 9, authors propose two channel assignment schemes in multihop cellular networks - asymmetric fixed channel assignment and multihop dynamic channel assignment. For both schemes exact and an approximated multi-dimensional Markov chain models are developed to analyze its QoS metrics. In Chapter 10, call admission control and adaptation (degradation and improvement) issues for elastic calls under restricted resources and bandwidth fluctuation has been considered. In Chapter 11, joint call admission controls algorithms in heterogeneous cellular networks are developed and their performance is investigated through numerical simulations. In Chapter 12, three call admission control strategies (i.e. resource reservation, queuing, and preemption) in public cellular networks with emergency services is proposed. Novel analytical models to evaluate the performance of seamless handover in mobile IPv6-based cellular networks are developed in Chapter 13.

Part 3 contains two Chapters and these Chapters deal with reliability issues of wireless cellular networks. In Chapter 14 Bayesian network model and mobile intelligent agents approaches are combined for automating fault prediction in wireless cellular networks. In Chapter 15, the application of forward error correction in Multimedia Broadcast over Single Frequency Network transmissions over long term evolution wireless cellular networks is examined.

Last Part 4 is a special one and it contains only one Chapter 16 in which basic mechanisms of the metabolic network coordination are proposed and their applications in both primary and complex networks are shown.

The book will be useful to researches in academia and industry and also to post-gradute students in telecommunication specialitiies.

Prof. Dr Agassi Melikov
Institute of Cybernetics, National Academy of Sciences of Azerbaijan,
Azerbaijan

Part 1

Positioning Problems in Cellular Networks

1

Wireless Positioning: Fundamentals, Systems and State of the Art Signal Processing Techniques

Lingwen Zhang[1], Cheng Tao[1] and Gang Yang[2]
[1]*School of Electronics and Information Engineering, Beijing Jiaotong University*
[2]*School of Information Engineering, Communication University of China*
China

1. Introduction

With the astonishing growth of wireless technologies, the requirement of providing universal location services by wireless technologies is growing. The process of obtaining a terminal's location by exploiting wireless network infrastructure and utilizing wireless communication technologies is called wireless positioning (Rappaport, 1996). Location information can be used to enhance public safety and revolutionary products and services. In 1996, the U.S. federal communications commission (FCC) passed a mandate requiring wireless service providers to provide the location of a wireless 911 caller to the nearest public safety answering point (PSAP) (Zagami et al., 1998). The wireless E911 program is divided into two parts- Phase I and Phase II, carriers were required to report the phone number of the wireless E911 caller and the location (Reed, 1998). The accuracy demands of Phase II are rather stringent. Separate accuracy requirements were set forth for network-based and handset-based technologies: For network-based solution: within 100m for 67% of calls, and within 300m for 95% of the calls. For handset-based solutions: within 50m for 67% of calls and within 150m for 95% of calls. Now E911 is widely used in U.S. for providing national security, publish safety and personal emergency location service. Wireless positioning has also been found useful for other applications, such as mobility management, security, asset tracking, intelligent transportation system, radio resource management, etc. As far as the mobile industry is concerned, location based service (LBS) is of utmost importance as it is the key feature that differentiates a mobile device from traditional fixed devices (Vaughan-Nichols, 2009). With this in mind, telecommunications, devices, and software companies throughout the world have invested large amounts of money in developing technologies and acquiring businesses that would let them provide LBS. Numerous companies-such as Garmin, Magellan, and TomTom international-sell dedicated GPS devices, principally for navigation. Several manufactures-including Nokia and Research in Motion-sell mobile phones that provide LBS. Google's My Location service for mobile devices, currently in beta, uses the company's database of cell tower positions to triangulate locations and helps point out the current location on Google map. Various chip makers manufacture processors that provide devices with LBS functionality. These companies' products and services work together to provide location-based services, as Fig. 1. Shows (Vaughan-Nichols, 2009).

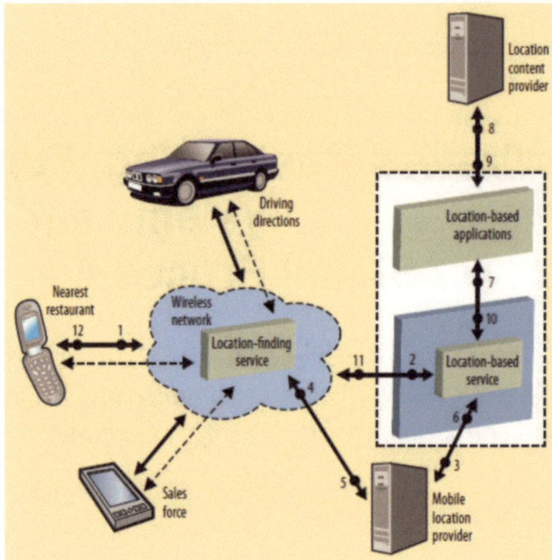

Fig. 1. Diagram shows how various products and services work together to provide location-based services

Thus, location information is extremely important. In order to help the growth of this emerging industry, there is a requirement to develop a scientific framework to lay a foundation for design and performance evaluation of such systems.

1.1 Elements of wireless positioning systems

Fig. 2. illustrates the functional block diagram of a wireless positioning system (Pahlavan, 2002). The main elements of the system are a number of location sensing devices that measure metrics related to the relative position of a mobile terminal (MT) with respect to a known reference point (RP), a positioning algorithm that processes metrics reported by location sensing elements to estimate the location coordinates of MT, and a position computing system that calculate the location coordinates. The location metrics may indicate the approximate arrival direction of the signal or the approximate distance between the MT and RP. The angle of arrival (AOA)/Direction finding (DF) is the common metric used in direction-based systems. The received signal strength (RSS), carrier signal phase of arrival (POA) and time of arrival (TOA), time difference of arrival (TDOA), frequency difference of arrival (FDOA)/Doppler difference (DD) of the received signal are the metrics used for estimation of distance. Which metrics should be measured depends on the positioning

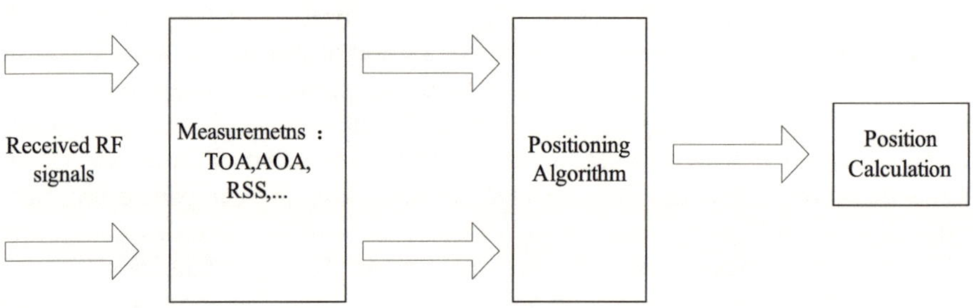

Fig. 2. Basic elements of a wireless positioning system

algorithms. As the measurements of metrics become less reliable, the complexity of the position calculation increased. Some positioning system also has a display system. The display system can simply show the coordinates of the MT or it may identify the relative location of the MT in the layout of an area. This display system could be software residing in a private PC or a mobile locating unit, locally accessible software in a local area network, or a universally accessible service on the web.

1.2 Location measuring techniques

As discussed in section 1.1, received signal strength (RSS), angle of arrival (AOA), time of arrival (TOA), round trip time (RTT), time difference of arrival (TDOA), phase of arrival (POA), and phase difference of arrival (PDOA) can all be used as location measurements (Zhao, 2006).

1.2.1 RSS estimation

RSS is based on predicting the average received signal strength at a given distance from the transmitter (Jian, 2005). Then, the measured RSS can provide ranging information by estimating the distance from the large-scale propagation model. Large-scale propagation model is used to estimate the mean signal strength for an arbitrary transmitter-receiver (T-R) separation distance since they characterize signal strength over large T-R separation distances (several hundreds or thousands of meters). The average large-scale propagation model is expressed as a function of distance by using a path loss exponent, n

$$P_r(d)[dBm] = P_r(d_0)[dBm] - 10n\log(\frac{d}{d_0}) + X_\sigma \tag{1}$$

Where $P_r(d)[dBm]$ is the received power in dBm units which is a function of the T-R distance of d, n is the path loss exponent which indicates the rate at which the path loss increased with distance, d is the T-R separation distance, d_0 is the close-in reference distance, as a known received power reference point. $P_r(d_0)[dBm]$ is the received power at the close-in reference distance. The value $P_r(d_0)[dBm]$ may be predicted or may be measured in the radio environment by the transmitter. For practical system using low-gain antennas in the 1-2GHz region, d_0 is typically chosen to be 1m in indoor environments and 100m or 1km in outdoor environments. X_σ describes the random shadowing effects, and is a zero-mean Gaussian distributed random variable (in dB) with standard deviation σ (also in dB). By measuring $P_r(d)[dBm]$ and $P_r(d_0)[dBm]$, the T-R distance of d may be estimated.

RSS measurement is comparatively simple for analysis and implementation but very sensitive to interference caused by fast multipath fading. The Cramer-Rao lower bound (CRLB) for a distance estimate provides the following inequality (Gezici, 2005):

$$\sqrt{Var(\hat{d})} \geq \frac{\ln 10}{10}\frac{\sigma}{n}d \tag{2}$$

Where d is the distance between the T-R, n is the path loss factor, and σ is the standard deviation of the zero mean Gaussian random variable representing the log-normal channel shadowing effect. It is observed that the best achievable limit depends on the channel parameters and the distance between the transmitter and receiver. It is suitable to use RSS measurements when the target node can be very close to the reference nodes.

1.2.2 TOA and TDOA estimation

TOA can be used to measure distance based on an estimate of signal propagation delay between a transmitter and a receiver since radiowaves travel at the speed of light in free space or air (Alavi,2006). The TOA can be measured by either measuring the phase of received narrowband carrier signal or directly measuring the arrival time of a wideband narrow pulse (Pahlavan, 2002). The ranging techniques of TOA measurement can be classified in three classes: narrowband, wideband and ultra wide band (UWB).

In the narrowband ranging technique, the phase difference between received and transmitted carrier signals is used to measure the distance. The phase of a received carrier signal, ϕ, and the TOA of the signal, τ,are related by $\tau = \phi / \omega_c$,where ω_c is the carrier frequency in radio propagation. However, when a narrowband carrier signal is transmitted in a multipath environment, the composite received carrier signal is the sum of a number of carriers, arriving along different paths, of the same frequency but different amplitude and phase. The frequency of the composite received signal remains unchanged, but the phase will be different form one-path signal. Therefore, using a narrowband carrier signal cannot provide accurate estimate of distance in a heavy multipath environment.

The direct-sequence spread-spectrum (DSSS) wideband signal has been used in ranging systems. In such a system, a signal coded by a known pseudo-noise (PN) sequence is transmitted by a transmitter. Then a receiver cross correlates received signal with a locally generated PN sequence using a sliding correlator or a matched filter. The distance between the transmitter and receiver is determined from the arrival time of the first correlation peak. Because of the processing gain of the correlation process at the receiver, the DSSS ranging systems perform much better than other systems in suppressing interference.

Due to the scarcity of the available bandwidth in practice, the DSSS ranging systems cannot provide adequate accuracy. Inspired by high-resolution spectrum estimation techniques, a number of super-resolution techniques have been studied such as multiple signal classification (MUSIC) (Rieken, 2004).

For a single path additive white Gaussian noise (AWGN) channel, it can be shown that the best achievable accuracy of a distance estimate derived from TOA estimation satisfies the following inequality (Anouar, 2007):

$$\sqrt{Var(\hat{d})} \geq \frac{c}{2\sqrt{2}\pi\sqrt{SNR}\beta} \tag{3}$$

Where c is the speed of light, SNR is the signal-to-noise ratio, and β is the effective signal bandwidth defined by

$$\beta = [\int_{-\infty}^{\infty} f^2 |S(f)|^2 df \,/\, \int_{-\infty}^{\infty} |s(f)|^2 df]^{1/2}$$

and S(*f*) is the Fourier transform of the transmitted signal.

It is observed that the accuracy of a time-based approach can be improved by increasing the SNR or the effective signal bandwidth. Since UWB signals have very large bandwidths exceeding 500MHz, this property allows extremely accurate location estimates using time-based techniques via UWB radios. For example, with a receive UWB pulse of 1.5 GHz bandwidth, an accuracy of less than an inch can be obtained at SNR=0dB.

In general, direct TOA results in two problems. First, TOA requires that all transmitters and receivers in the system have precisely synchronized clocks (e.g.,just 1us of timing error could result in a 300m position location error). Second, the transmitting signal must be labeled with a timestamp in order for the receiver to discern the distance the signal has traveled. For this reason, TDOA measurements are a more practical means of position location for commercial systems. The idea of TDOA is to determine the relative position of the mobile transmitter by examining the difference in time at which the signal arrives at multiple measuring units, rather than the absolute arrival time. Fig.3. is a simulation of a pulse waveform recorded by receivers P0 and P1. The red curve in Fig.3. is the cross correlation function. The cross correlation function slides one curve in time across the other and returns a peak value when the curve shapes match. The peak at time=5 is the TDOA measure of the time shift between the recorded waveforms.

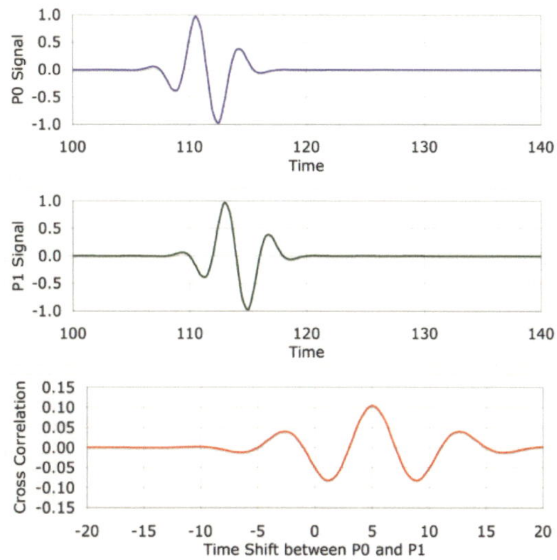

Fig. 3. Cross correlation method for TDOA measurements

1.2.3 AOA estimation

AOA is the measurement of signal direction through the use of antenna arrays. AOA metric has long and widely been studied in many years, especially in radar and sonar technologies for military applications. Using complicated antenna array, high-resolution angle measurement would be obtained.

The advantages of AOA are that a position estimate may be determined with as few as three measuring units for 3-D positioning or two measuring units for 2-D positioning, and that no time synchronization between measuring units is required. The disadvantages include relatively large and complex hardware requirements and location estimate degradation as the mobile target moves farther from the measuring units. For accurate positioning, the angle measurements need to be accurate, but the high accuracy measurements in wireless networks may be limited by shadowing, by multipath reflections arriving from misleading directions, or by the directivity of the measuring aperture. Some literatures also call AOA as direction of arrival (DOA) or direct finding (DF). Classic approaches for AOA estimation include Capon's method (Gershman, 2003; Stoica, 2003). The most popular AOA estimation techniques are based on the signal subspace approach by Schmidt (Swindlehurst, 1992) with

Multiple Signal Classification (MUSIC) algorithm. Subspace algorithms operate by separating a signal subspace from a noise subspace and exploiting the statistical properties of each. Variants of the MUSIC algorithm have been developed to improve its resolution and decrease its computational complexity including Root-MUSIC (Barabell, 1983) and Cyclic MUSIC. Other improved subspace-based AOA estimation techniques include the Estimation of Signal Parameters by Rotational Invariance Techniques (ESPRIT) algorithm and its variants, and a minimum-norm approach.

1.2.4 Joint parameter estimation

Estimators which estimate more than one type of location parameter (e.g., joint AOA/TOA) simultaneously have been developed. These are useful for hybrid location estimation schemes. Most joint estimators are based on ML techniques and signal subspace approaches, such as MUSIC or ESPRIT, and are developed for joint AOA/TOA estimation of a single users multipath signal components at a receiver.

The ML approach in (Wax & Leshem, 1997) for joint AOA/TOA estimation in static channels presents an iterative scheme that transforms a multidimensional ML criterion into two sets of one dimensional problems. Both a deterministic and a stochastic ML algorithm were developed in (Raleigh & Boros, 1998) for joint AOA/TOA estimation in time-varying channels. A novel subspace approach was proposed in (Vanderveen, Papadias & Paulraj, 1997) that jointly estimates the delays and AOAs of multipaths using a collection of space time channel estimates that have constant parameters of interest but different path fade amplitudes. Unlike MUSIC and ESPRIT, this technique has been shown to work when the number of paths exceeds that number of antennas.

1.3 Positioning algorithms

Once the location sensing parameters are estimated using the methods discussed in the previous section, it needs to be considered how to use these measurements to get the required position coordinates. In another words, how to design a geolocation algorithm with these parameters as input and position coordinates as output. In this section, the common methods for determining MT location will be described. It is to be noted that these algorithms assume measurements are made under Line of sight (LOS) conditions.

1.3.1 Geometric location

Geometric location uses the geometric properties to estimate the target location. It has three derivations: trilateration, multilateration and triangulation. Trilateration estimates the position of an object by measuring its distance from multiple reference points. Multilateration locates the object by computing the TDOA from that object to three or more receivers. Triangulation locates an object by computing angles relative to multiple reference points.

A. Trilateration

Trilateration is based on the measurement of distance (i.e. ranges) between MT and RP. The MT lies on the circumference of a circle, with the RP as center and a radius equal to the distance estimate. The desired MT location is determined by the intersection of at least three circle formed by multiple measurements between the MT and several RPs. Common methods for deriving the range measurements include TOA estimation and RSS estimation.

The solution is found by formulating the equations for the three sphere surface and then solving the three equations for the two unknowns: x and y, as shown in Fig.5. It is assumed that the MT located at (x,y), transmits a signal at time t_0, the three RPs located at $(x_1,y_1),(x_2,y_2),(x_3,y_3)$ receive the signal at time t_1,t_2,t_3. The equations for the three spheres are:

$$\sqrt{(x_i-x)^2+(y_i-y)^2}=c(t_i-t_0), i=1,2,3 \tag{4}$$

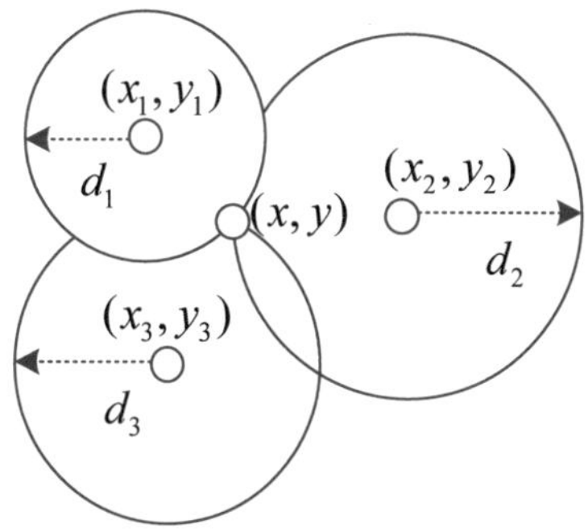

Fig. 5. Trilateration positioning

The next work to do is to find an optimized method to solve these equations under small error conditions. One well-known method is based on cost function. The cost function can be formed by

$$F(x)=\sum_{i=1}^{3}\alpha_i^2 f_i^2(x) \tag{5}$$

Where α_i can be chosen to reflect the reliability of the signal received at the measuring unit i, and $f_i(x)$ is given as follows.

$$f_i(x)=c(t_i-t_0)-\sqrt{(x_i-x)^2+(y_i-y)^2} \tag{6}$$

The location estimate is determined by minimizing the function $F(x)$. There are other algorithms such as closest-neighbor (CN) and residual weighting (RWGH). The CN algorithm estimates the location of the user as the location of the base station or reference point that is located closest to that user. The RWGH algorithm can be viewed as a form of weighted least-square algorithm.

B. Multilateration

Multilateration, also known as hyperbolic positioning, measures the time difference of signals travelled from a MT to a pair of RPs, or vice versa. The MT lies on a hyperbola defined by constant distance difference to the two RPs with the foci at the RPs. The desired location of the MT is determined at the intersection of the hyperbolas produced by multiple measurements as shown in Fig.6. This method requires no timestamp and only the synchronization among the RPs is required.

When the TDOA is measured, a set of equations can be described as follows.

$$R_{i,1} = c(t_i - t_1) = c\Delta\tau_i = R_i - R_1 \tag{7}$$

Where $R_{i,1}$ is the value of range difference from MT to the ith RP and the first RP.
Define

$$R_i = \sqrt{(X_i - x)^2 + (Y_i - y)^2}\,, \quad i = 1, \cdots N$$

(X_i, Y_i) is the RP coordinate, (x, y) is the MT location, R_i is the distance between the RP and MT, N is the number of BS, c is the light speed, $\Delta\tau_i$ is the TDOA between the service RP and the ith RP_i. In the geometric point of view, each equation presents a hyperbolic curve. Eq. (7) is a set of nonlinear equations. Fang (Fang, 1990) gave an exact solution when the number of equations is equal to the number of unknown coordinates. This solution, however, cannot make use of extra measurements, available when there are extra sensors, to improve position accuracy. In reality, the surfaces rarely intersect, because of various errors. In this case, the location problem can be posed as an optimization problem and solved using, for example, a least square method. The more general situation based on least square algorithm with extra measurements was considered by Friendlander (Friendlander, 1987). Although closed-form solution has been developed, the estimators are not optimum. Chen gave a closed-form, non-iterative solution utilizing the least square algorithm two times which performs well when the TDOA estimation errors are small. However, as the estimation errors increase, the performance declines quickly. Taylor-series method (Foy, 1976) is an iterative method which starts with an initial guess which is in the condition of close to the true solution to avoid local minima.

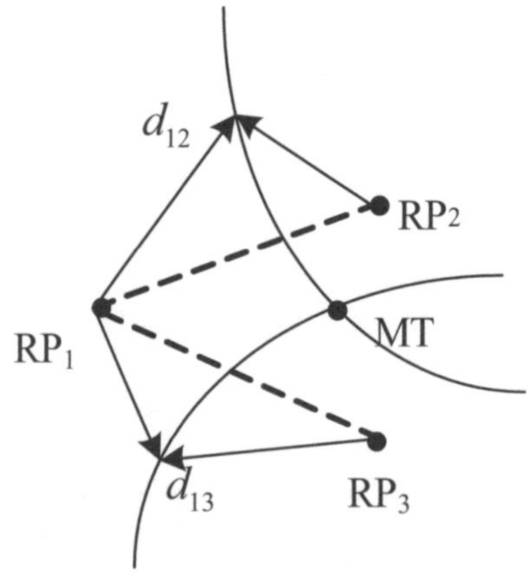

Fig. 6. Hyperbolic positioning

C. Triangulation positioning

When the AOA is measured, the location of the desired target can be found by the intersection of several pairs of angle direction lines. As shown in Fig. 7., at least two known RP and two measured angles are used to derive the 2-D location of the MT. The advantages of triangulation are that a position estimate may be determined with as few as three measuring units for 3-D positioning or two measuring units for 2-D positioning, and that no time synchronization between measuring units is required. In cellular systems, the deployment of smart antenna makes AOA practical. However, the drawback of this method includes complexity and cost for the deployment of antennas at the RP side and impractical implementation at the MT side; susceptibility to linear orientation of RPs; accuracy deterioration with the increase in distance between the MT and the RP owing to fundamental limitations of the devices used to measure the arrival angles. The accuracy is limited by shadowing, multipath reflections arriving from misleading directions, or the directivity of the measuring aperture.

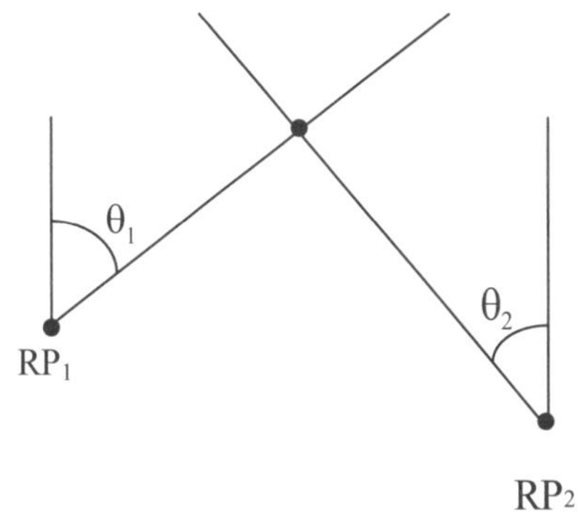

Fig. 7. Triangulation positioning

1.3.2 Hybrid positioning

Since the above reviewed location methods complement each other, hybrid techniques, which use a combination of available range, range-difference or angle measurements, or other methods to solve for locations, have been extensively investigated (see for example). Hybrid techniques are also studied to combat the problems, e.g. hearability (Zhao, 2006), accuracy , NLOS problems which will be discussed in the next section. Hybrid methods are especially useful in hearability conditions when the number of available BSs in cellular networks is limited. Most typical hybrid method combines TOA (TDOA) location with AOA location (Thomas, 2001). The scheme proposed in (Catovic & Sahinoglu, 2004) combines TDOA with RSS measurements.

1.3.3 Fingerprinting

Fingerprinting refers to techniques that match the fingerprint of some characteristic of a signal that is location dependent. There are two stages for location fingerprinting: offline stage and online stage. During the offline stage, a site survey is performed in an environment. The location coordinates and respective signal strengths from nearby RPs are collected. During the online stage, a fingerprinting algorithm is used to identify the most likely recorded fingerprinting to the measured one and to infer the target location. The main challenges to the techniques based on location fingerprinting is that the received signal strength could be affected by diffraction, reflection, and scattering in the propagation environments. There are at least five location fingerprinting-based positioning algorithm using pattern recognition technique so far: probabilistic methods, k-nearest-neighbor, neural networks, support vector machine, and smallest M-vertex polygon.

In urban areas, when the multipath problem is quite severe, both AOA and TOA/TDOA may encounter difficulties. To solve this problem, the multipath characteristics can be considered as the fingerprinting of mobile phones, as shown in Fig. 8. The design involves a location server with a database that includes measured and predicted signal characteristics for a specific area. When an E911 call is made, the location of the mobile phone can be computed by comparing signals received by the mobile with the signal values stored in the database. Various signal characteristics, including received signal levels and time delays may be utilized.

Using a multipath delay profile to locate a mobile terminal is possible with fingerprinting. This avoids many of the problems that multipath propagation posed for conventional location methods. This method could obtain high accurate location as long as offline stage collects adequate and update information. However, the high cost for deployment and maintenance is obvious and unavoidable. As a result, it is a promising technique but not a mainstream option for the time being.

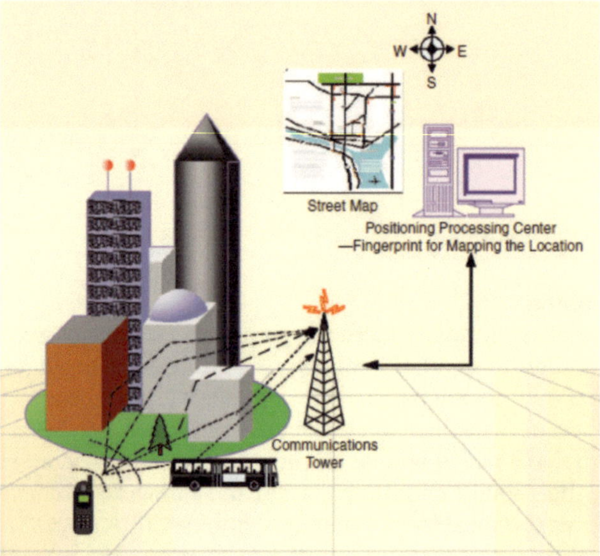

Fig. 8. Fingerprinting of mobile phones

2. Current location systems

Network-aided positioning has attracted much research attention in recent years. Different network topologies pose various technical challenges to design faster, more robust and more accurate positioning systems. There are numerous methods for obtaining the location information, depending on different location systems.

Location systems can be grouped in many different ways, including indoor versus outdoor systems or cellular versus sensor network positioning, as shown in Fig.9 (Guolin, 2005). Global positioning systems and cellular based location system can be used for outdoor positioning while indoor location used existing wireless local access network (WLAN) infrastructures for positioning. An overview of indoor positioning versus outdoor positioning by satellite is shown in Table 1. Sensor networks vary significantly from traditional cellular networks, where access nodes are assumed to be small, inexpensive, cooperative, homogeneous and often relatively autonomous. A number of location-aware protocols have been proposed for "ad hoc" routing and networking. Sensor networks have also been widely used for intrusion detection in battlefields as well as for monitoring wildlife.

Different network topologies, physical layer characteristics, media access control layer characteristics, devices and environment require remarkably different positioning system solutions. In this section, an overview of positioning solutions applied in GPS, cellular networks and WLAN will be provided.

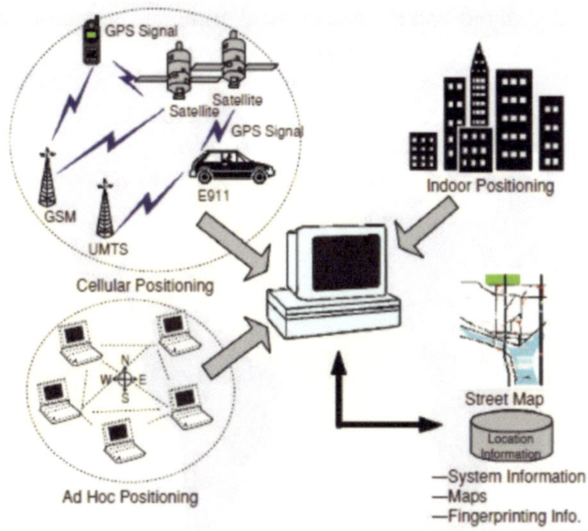

Fig. 9. Overview of indoor versus outdoor positioning systems

INDOOR	OUTDOOR (POSITIONING BY SATELLITE)
WLAN —CLIENT-BASED SYSTEM DESIGN —CLIENT-ASSISTED SYSTEM DESIGN SENSOR NETWORK —LOCALIZATION WITH BEACONS —LOCALIZATION WITH MOVING BEACONS —BEACON-FREE LOCALIZATION UWB —A PROMISING APPROACH FOR INDOOR GEOLOCATION —CAN ACHIEVE VERY ACCURATE SHORT DISTANCE ESTIMATION	GPS —REQUIRES MINIMAL OBSTRUCTIONS —LONG ACQUISITION TIMES (30 s–15 min) —HAS TO BE SYNCHRONOUS —HIGH POWER CONSUMPTION AND HIGH UNIT COST A-GPS —MUCH MORE ACCURATE: ACCURACY OF 10–50 m CAN BE USED EVEN FOR INDOOR POSITIONING —IMPROVES ACQUISITION TIME (< 10 s) —SYNCHRONOUS OR ASYNCHRONOUS —MORE COST EFFECTIVE THAN GPS —LITTLE/NO HARDWARE CHANGES REQUIRED IN BASE STATIONS

Table. 1. Overview of indoor positioning versus outdoor positioning by satellite

2.1 GPS

The Global Positioning System (GPS) is a satellite-based positioning system that can provide 3-D position and time information to users in all weather and at all times and anywhere on or near the earth when and where there is an unobstructed line of sight to four or more GPS satellites. It is maintained by the United States government and is freely accessible by anyone with a GPS receiver. GPS was created by U.S. Department of Defense and was originally run with 24 satellites. It was established in 1973.

2.1.1 GPS structure

GPS consists of three parts: the space segment, the control segment and the user segment. The space segment is composed of 24 to 32 satellites in medium earth orbit and also includes the boosters required to launch them into orbit. As of March 2008, there are 31 active broadcasting satellites in the GPS constellation shown in Fig.10., and two older, retired from active service satellites kept in the constellation as orbital spares. The additional satellites improve the precision of GPS receiver calculations by providing redundant measurements. The control segment is composed of a master control station, an alternate master control station, and a host of dedicated and shared ground antennas and monitor stations. The user segment is composed of hundreds of thousands of U.S. and allied military users of the secure GPS precise positioning service, and tens of millions of civil, commercial, and scientific users of the standard positioning service. In general, GPS receivers are composed of an antenna, receiver-processors and a highly stable clock.

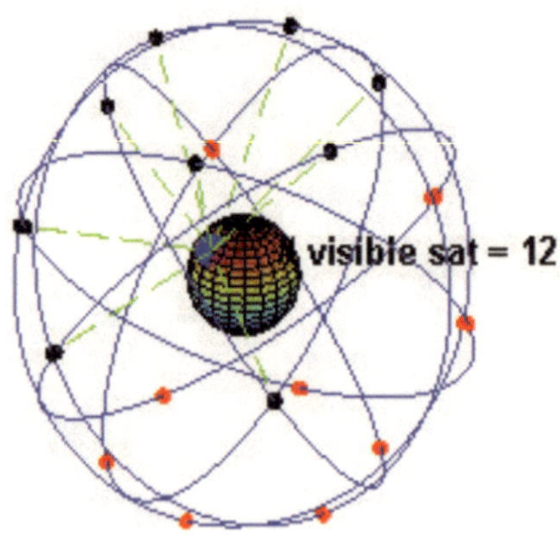

Fig. 10. GPS constellation

2.1.2 GPS signals

Each GPS satellite continuously broadcasts a navigation message at a rate of 50 bits per second. Each complete message is composed of 30 second frames shown in Fig. 11. All satellites broadcast at the same two frequencies, 1.57542GHz (L1 signal) and 1.2276 GHz (L2 signal). The satellite network uses a CDMA spread-spectrum technique where the low bit rate message data is encoded with a high rate pseudo random (PN) sequence that is different for each satellite as shown in Fig. 12. The receiver must be aware of the PN codes for each satellite to reconstruct the actual message data. The C/A code, for civilian use, transmits data at 1.023 million chips per second, whereas the P code, for U.S. military use, transmits at 10.23 million chips per second. The L1 carrier is modulated by both the C/A and P codes, while the L2 carrier is only modulated by the P code. The P code can be encrypted as a so-called P(Y) code which is only available to military equipment with a proper decryption key.
Since all of the satellite signals are modulated onto the same L1 carrier frequency, there is a need to separate the signals after demodulation. This is done by assigning each satellite a unique binary sequence known as a Gold code. The signals are decoded after demodulation, using addition of the Gold codes corresponding to the satellite monitored by the receiver as shown in Fig. 13.

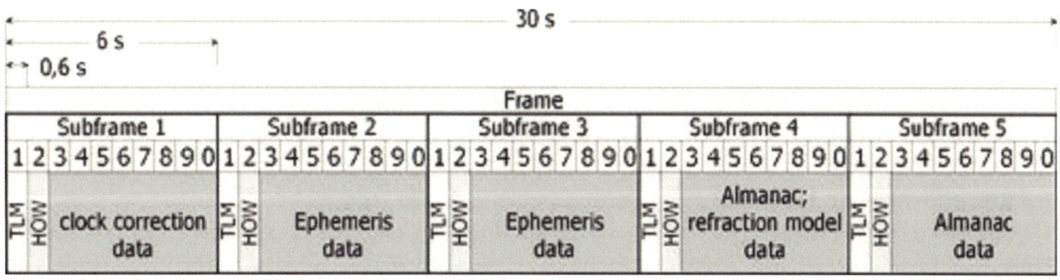

Fig. 11. GPS message frame

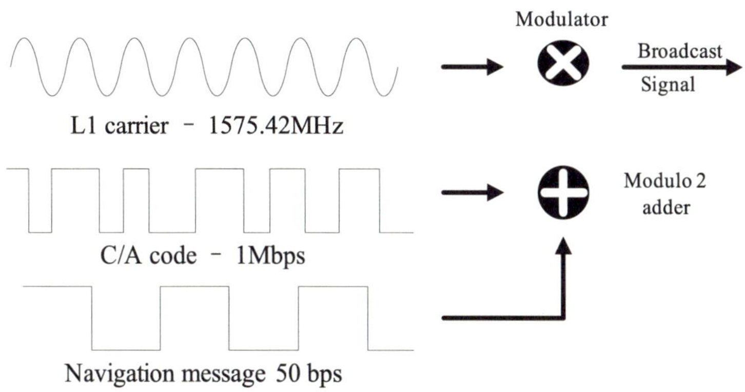

Fig. 12. Modulating and encoding GPS satellite signal using C/A code

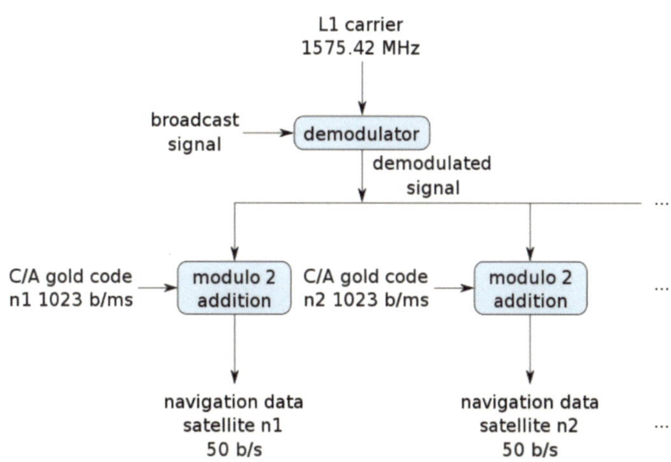

Fig. 13. Demodulating and decoding GPS satellite signal using C/A code

When the receiver uses messages to obtain the time of transmission and the satellite position, trilateration method is used to form equations and optimized algorithm is used to solve the equations as mentioned above.

The main advantages of GPS are its global coverage and high accuracy within 50 meters. And GPS receivers are not required to transmit anything to satellites, so there is no limit to the number of users that can use the system simultaneously. However, there also exist several issues that affect the effectiveness of GPS, especially in dealing with emergency services: response time, the time to first fix (TTFF) which may be greater than 30 seconds. Besides, GPS cannot provide accurate location under the obstructed signal case (e.g. in the unban city area, inside buildings). Taking these drawbacks into consideration, GPS are not suitable for some location services such as emergency call.

2.2 Standardization methods for positioning in cellular networks

There are various location techniques that are used in cellular-based positioning system. They can be classified by three types: mobile-based solution in which the positioning is carried out in handset and sent back to the network, mobile-assisted solution in which handset makes the measurements, reports these to the network where the serving mobile location center node calculated the position, and network-based solution in which the measuring and positioning are done by network.

In 1997, TIA led the standardization activities for the positioning in GSM. Four positioning methods were included. They are cell identity and timing advance, uplink time of arrival, enhanced observed time difference (E-OTD) and Assisted GPS (A-GPS). There are two stages of standardizations, the first version specification supports circuit-switch connections and the second version specification provides the same support in the packet-switch domain.

Cell-ID is a simple positioning method based on knowing which cell sector the target belongs to. The sector is known only during an active voice or data call. With this method, no air interface resources are required to obtain cell sector information (if the user is active), and no modifications to handset hardware are required. The disadvantage obviously is that the location is roarse.

Time advance (TA) represents the round trip delay between the mobile and serving base station (BS), it is represented by a 6-bit integer number in the GSM frame. In addition, RXLEV is the measurement of the strength of signals received by a mobile, therefore, with suitable propagation models, the distance between a mobile and BS can be estimated.

Since Cell-ID is not accurate, Cell-ID+TA and Cell-ID+TA+RXLEV such hybrid positioning methods are used as shown in Fig.14.

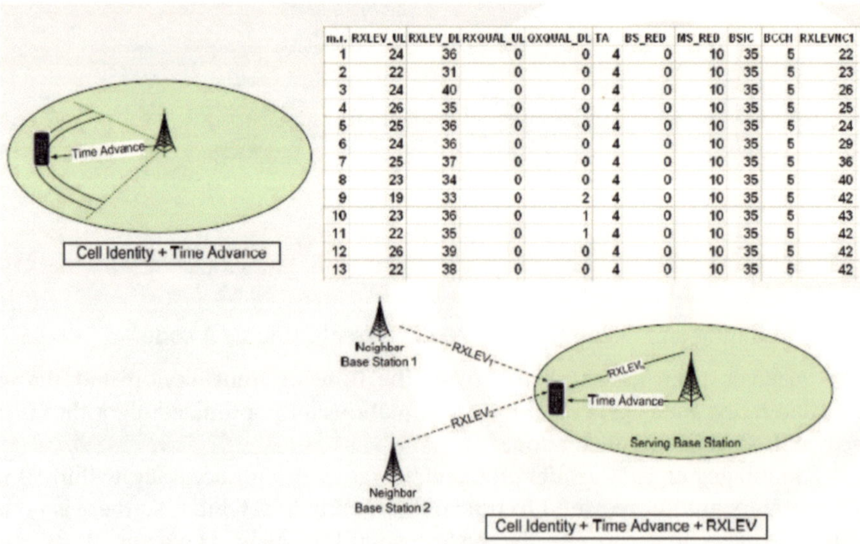

m.r.	RXLEV_UL	RXLEV_DL	RXQUAL_UL	OXQUAL_DL	TA	BS_RED	MS_RED	BSIC	BCCH	RXLEVNC1
1	24	36	0	0	4	0	10	35	5	22
2	22	31	0	0	4	0	10	35	5	23
3	24	40	0	0	4	0	10	35	5	26
4	26	35	0	0	4	0	10	35	5	25
5	25	36	0	0	4	0	10	35	5	24
6	24	36	0	0	4	0	10	35	5	29
7	25	37	0	0	4	0	10	35	5	36
8	23	34	0	0	4	0	10	35	5	40
9	19	33	0	2	4	0	10	35	5	42
10	23	36	0	1	4	0	10	35	5	43
11	22	35	0	1	4	0	10	35	5	42
12	26	39	0	0	4	0	10	35	5	42
13	22	38	0	0	4	0	10	35	5	42

Fig. 14. Cell-ID with TA for GSM

E-OTD is based on TDOA measured by the mobile between the receptions of bursts transmitted from the reference BS and each neighboring BS which value is called geometric time difference (GTD), requiring a synchronous network. However, GSM is not synchronous. Location measurement unit (LMU) devices are therefore required to compute the synchronization difference between two BSs which is called real time difference (RTD). The GTD can be obtained by GTD=OTD-RTD. Fig. 15 illustrates the solution of E-ODT.

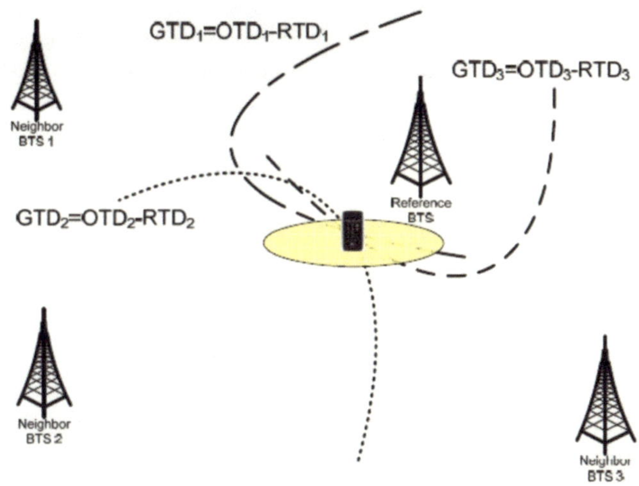

Fig. 15. E-OTD positioning for GSM

In the A-GPS for GSM, the GSM network informs the mobile about the data that GPS satellites are sending.

The standard positioning methods supported within UTRAN are:

- Cell-ID based method;
- OTDOA method that may be assisted by network configurable idle periods;
- Network-assisted GNSS methods;
- U-TDOA.

In the cell ID based (i.e. cell coverage) method, the position of an UE is estimated with the knowledge of its serving Node B. The information about the serving Node B and cell may be obtained by paging, locating area update, cell update, URA update, or routing area update. The cell coverage based positioning information can be indicated as the Cell Identity of the used cell, the Service Area Identity or as the geographical co-ordinates of a position related to the serving cell. The position information shall include a QoS estimate (e.g. regarding achieved accuracy) and, if available, the positioning method (or the list of the methods) used to obtain the position estimate. When geographical co-ordinates are used as the position information, the estimated position of the UE can be a fixed geographical position within the serving cell (e.g. position of the serving Node B), the geographical centre of the serving cell coverage area, or some other fixed position within the cell coverage area. The geographical position can also be obtained by combining information on the cell specific fixed geographical position with some other available information, such as the signal RTT in FDD or Rx Timing deviation measurement and knowledge of the UE timing advance, in TDD.

In OTDOA-IPDL method, the Node B may provide idle periods in the downlink, in order to potentially improve the hearability of neighbouring Node Bs. The support of these idle periods in the UE is optional. Support of idle periods in the UE means that its OTDOA

performance will improve when idle periods are available. Alternatively, the UE may perform the calculation of the position using measurements and assistance data.
Global Navigation Satellite System (GNSS) methods make use of UEs, which are equipped with radio receivers capable of receiving GNSS signals. Examples of GNSS include GPS, Modernized GPS, Galileo, GLONASS, Satellite Based Augmentation Systems (SBAS), and Quasi Zenith Satellite System (QZSS).In this concept, different GNSS (e.g. GPS, Galileo, etc.) can be used separately or in combination to perform the location of a UE.
The U-TDOA positioning method is based on network measurements of the Time of Arrival (TOA) of a known signal sent from the UE and received at four or more LMUs. The method requires LMUs in the geographic vicinity of the UE to be positioned to accurately measure the TOA of the bursts. Since the geographical coordinates of the measurement units are known, the UE position can be calculated via hyperbolic trilateration. This method will work with existing UE without any modification.
The standard positioning methods supported for E-UTRAN access are:

- network-assisted GNSS methods;
- downlink positioning;
- hanced cell ID method.

Hybrid positioning using multiple methods from the list of positioning methods above is also supported.
These positioning methods may be supported in UE-based, UE-assisted/E-SMLC-based, or eNB-assisted versions. Table 2 indicates which of these versions are supported in this version of the specification for the standardized positioning methods.

Method	UE-based	UE-assisted, E-SMLC-based	eNB-assisted	SUPL
A-GNSS	Yes	Yes	No	Yes (UE-based and UE-assisted
Downlink	No	Yes	No	Yes (UE-assisted)
E-CID	No	Yes	Yes	Yes (UE-assisted)

Table 2. Supported versions of UE positioning methods

The downlink (OTDOA) positioning method makes use of the measured timing of downlink signals received from multiple eNode Bs at the UE. The UE measures the timing of the received signals using assistance data received from the positioning server, and the resulting measurements are used to locate the UE in relation to the neighbouring eNode Bs.
Enhanced Cell ID (E-CID) positioning refers to techniques which use additional UE and/or E-UTRAN radio resource and other measurements to improve the UE location estimate. Although E-CID positioning may utilize some of the same measurements as the measurement control system in the RRC protocol, the UE generally is not expected to make additional measurements for the sole purpose of positioning; i.e., the positioning procedures do not supply a measurement configuration or measurement control message, and the UE reports the measurements that it has available rather than being required to take additional measurement actions. In cases with a requirement for close time coupling between UE and eNode B measurements (e.g., TADV type 1 and UE Tx-Rx time difference), the eNode B

configures the appropriate RRC measurements and is responsible for maintaining the required coupling between the measurements.

2.3 Indoor location system

Since cellular-based positioning methods or GPS cannot provide accurate indoor geolocation, which has its own independent applications and unique technical challenges, this section focuses on positioning based on wireless local area network (WLAN) radio signals as an inexpensive solution for indoor environments.

2.3.1 IEEE 802.11

What is commonly known as IEEE 802.11 actually refers to the family of standards that includes the original IEEE 802.11 itself, 802.11a, 802.11b, 802.11g and 802.11n. Other common names by which the IEEE standard is known include Wi-Fi and the more generic wireless local area network (WLAN). IEEE 802.11 has become the dominant wireless computer networking standard worked at 2.4GHz with a typical gross bit rate of 11,54,108 Mbps and a range of 50-100m.

Using an existing WLAN infrastructure for indoor location can be accomplished by adding a location server. The basic components of an infrastructure-based location system are shown in Fig.16. The mobile device measures the RSS of signals from the access points (APs) and transmits them to a location server which calculates the location.

There are several approaches for location estimation. The simpler method which is to provide an approximate guess on AP that receives the strongest signal. The mobile node is assumed to be in the vicinity of that particular AP. This method has poor resolution and poor accuracy. The more complex method is to use a radio map. The radio map technique typically utilizes empirical measurements obtained via a site survey, often called the offline phase. Given the RSS measurements, various algorithms have been used to do the match such as *k*-nearest neighbor (*k*-NN), statistical method like the hidden Markov model (HMM). While some systems based on WLAN using RSS requires to receive signals at least three APs and use TDOA algorithm to determine the location.

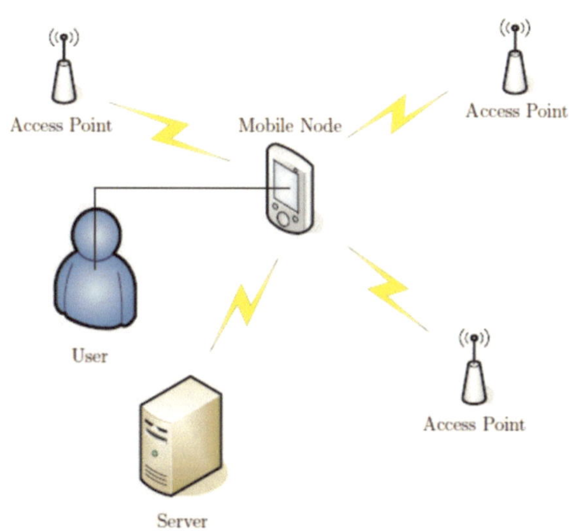

Fig. 16. Typical architecture of WLAN location system

3. Advanced signal processing techniques for wireless positioning

Although many positioning devices and services are currently available, some important problems still remain unsolved. This chapter gives some new ideas, results and advanced signal processing techniques to improve the performance of positioning.

3.1 Computational algorithms of TDOA equations

When TDOA measurements are employed, a set of nonlinear hyperbolic equations has been set up, the next step is to solve these equations and derive the location estimate. Usually, these equations can be solved after being linearized. These algorithms can be grouped into two types: non-iterative methods and iterative methods.

3.1.1 Non-iterative methods

A variety of non-iterative methods for position estimation have been investigated. The most common ones are direct method (DM), least-square (LS) method, Chan method.
When the TDOA is measured, a set of equations can be described as follows.

$$R_{i,1} = c(t_i - t_1) = c\Delta\tau_i = R_i - R_1$$

Where $R_{i,1}$ is the value of range difference from MT to the ith RP and the first RP.

Define

$$R_i = \sqrt{(X_i - x)^2 + (Y_i - y)^2}\ , \quad i = 1,\cdots N \tag{8}$$

(X_i, Y_i) is the RP coordinate, (x, y) is the MT location, R_i is the distance between the RP and MT, N is the number of BS, c is the light speed, $\Delta\tau_i$ is the TDOA between the service RP and the ith RP_i.
Squaring both sides of (8)

$$R_i^{\ 2} = (X_i - x)^2 + (Y_i - y)^2\ , \quad i = 1,\cdots N \tag{9}$$

Substracting (9) for i=2,…N by (8) for i=1

$$X_{i,1}x + Y_{i,1}y = d_{i,1}, i = 2,...N \tag{10}$$

Where $X_{i,1} = X_i - X_1; Y_{i,1} = Y_i - Y_1$ and $d_{i,1} = ((X_i^2 + Y_i^2) - (X_1^2 + Y_1^2) + R_1^2 - R_i^2)/2$

3.1.1.1 Direct method

It assumes that three RPs are used. The solution to (10) gives:

$$\hat{x} = \frac{Y_{2,1}d_{3,1} - Y_{3,1}d_{2,1}}{X_{3,1}Y_{2,1} - X_{2,1}Y_{3,1}}; \hat{y} = \frac{X_{3,1}d_{2,1} - X_{2,1}d_{3,1}}{X_{3,1}Y_{2,1} - X_{2,1}Y_{3,1}} \tag{11}$$

The solution shows that there are two possible locations. Using a priori information, one of the value is chosen and is used to find out the coordinates.

3.1.1.2 Least square methods

Reordering (10) the terms gives a proper system of linear equations in the form $A\theta = B$, where

$$A = \begin{pmatrix} X_{21} & Y_{21} \\ X_{31} & Y_{31} \end{pmatrix}; \theta = \begin{bmatrix} x \\ y \end{bmatrix}; B = \begin{bmatrix} d_{2,1} \\ d_{3,1} \end{bmatrix}$$

The system is solved using a standard least-square approach:

$$\hat{\theta} = (A^T A)^{-1} A^T B. \tag{12}$$

3.1.1.3 Chan's method

Chan's method (Chan, 1994) is capable of achieving optimum performance. If we take the case of three RPs, the solution of (10) is given by the following relation:

$$\begin{bmatrix} x \\ y \end{bmatrix} = -\begin{pmatrix} X_{21} & Y_{21} \\ X_{31} & Y_{31} \end{pmatrix}^{-1} \left\{ \begin{bmatrix} d_{21} \\ d_{31} \end{bmatrix} \times R_1 + 0.5 \times \begin{bmatrix} d_{21}^2 - K_2 + K_1 \\ d_{31}^2 - K_3 + K_1 \end{bmatrix} \right\} \tag{13}$$

Where
$K_i = X_i^2 + Y_i^2, i = 1,2,3$

3.1.2 Iterative method

Taylor series expansion method is an iterative method which starts with an initial guess which is in the condition of close to the true solution to avoid local minima and improves the estimate at each step by determining the local linear least-squares.
Eq. (10) can be rewritten as a function

$$f_i(x,y) = \sqrt{(x - X_{i+1})^2 + (y - Y_{i+1})^2} - \sqrt{(x - X_1)^2 + (y - Y_1)^2} \qquad i = 1, \cdots N\text{-}1 \tag{14}$$

Let $\hat{t}_i$ be the corresponding time of arrival at BS_i. Then,

$$f_i(x,y) = \hat{d}_{i+1,1} + \varepsilon_{i+1,1} \qquad i = 1, \cdots N\text{-}1 \tag{15}$$

Where

$$\hat{d}_{i+1,1} = c(\hat{t}_{i+1} - \hat{t}_1) \tag{16}$$

$\varepsilon_{i,1}$ is the corresponding range differences estimation error with covariance R.
If (x_0, y_0) is the initial guess of the MS coordinates, then

$$x = x_0 + \delta_x, \quad y = y_0 + \delta_y \tag{17}$$

Expanding Eq. (15) in Taylor series and retaining the first two terms produce

$$f_{i,0} + a_{i,1}\delta_x + a_{i,2}\delta_y \approx \hat{d}_{i+1,1} + \varepsilon_{i+1,1} \qquad i = 1, \cdots N - 1 \tag{18}$$

Where

$$\begin{aligned}
&f_{i,0} = f_i(x_0, y_0)\\
&a_{i,1} = \left.\frac{\partial f_i}{\partial x}\right|_{x_0, y_0} = \frac{X_1 - x_0}{\hat{d}_1} - \frac{X_{i+1} - x_0}{\hat{d}_{i+1}}\\
&\hat{d}_i = \sqrt{(x_0 - X_i)^2 + (y_0 - Y_i)^2}\\
&a_{i,2} = \left.\frac{\partial f_i}{\partial y}\right|_{x_0, y_0} = \frac{Y_1 - y_0}{\hat{d}_1} - \frac{Y_{i+1} - y_0}{\hat{d}_{i+1}}
\end{aligned} \tag{19}$$

Eq. (18) can be rewritten as

$$\mathrm{A}\delta = \mathrm{D} + \mathrm{e} \tag{20}$$

Where

$$\mathrm{A} = \begin{bmatrix} a_{1,1} & a_{1,2} \\ a_{2,1} & a_{2,2} \\ \vdots & \vdots \\ a_{N-1,1} & a_{N-1,2} \end{bmatrix}, \quad \delta = \begin{bmatrix} \delta_x \\ \delta_y \end{bmatrix}$$

$$\mathrm{D} = \begin{bmatrix} \hat{d}_{2,1} - f_{1,0} \\ \hat{d}_{3,1} - f_{2,0} \\ \vdots \\ \hat{d}_{N,1} - f_{N-1,0} \end{bmatrix}, \quad \mathrm{e} = \begin{bmatrix} \varepsilon_{2,1} \\ \varepsilon_{3,1} \\ \vdots \\ \varepsilon_{N,1} \end{bmatrix}$$

The weighted least square estimator for (20) produces

$$\delta = \left[\mathrm{A}^{\mathrm{T}}\mathrm{R}^{-1}\mathrm{A}\right]^{-1}\mathrm{A}^{\mathrm{T}}\mathrm{R}^{-1}\mathrm{D} \tag{21}$$

R is the covariance matrix of the estimated TDOAs.
Taylor series method starts with an initial guess (x_0, y_0), in the next iteration, (x_0, y_0) are set to $(x_0 + \delta_x, y_0 + \delta_y)$ respectively. The whole process is repeated until (δ_x, δ_y) are sufficiently small. The Taylor series method can provide accurate results, however the convergence of the iterative process depends on the initial value selection. The recursive computation is intensive since least square computation is required in each iteration.

3.1.3 Steepest decent method

From the above analysis, the convergence of Taylor series expansion method and the convergence speed directly depends on the choice of the MT initial coordinates. This iterative method must start with an initial guess which is in the condition of close to the true solution to avoid local minima. Selection of such a starting point is not simple in practice.
To solve this problem, steepest decent method with the properties of fast convergence at the initial iteration and small computation complexity is applied at the first several iterations to

get a corrected MT coordinates which are satisfied to Taylor series expansion method. The algorithm is described as follows.
Eq. (18) can be rewritten as

$$\varphi_i(x,y) = f_i(x,y) - \hat{d}_{i+1,1} + \varepsilon_{i+1,1} \qquad i = 1,\cdots N\text{-}1 \tag{19}$$

Construct a set of module functions from Eq. (18)

$$\Phi(x,y) = \sum_{i=1}^{N-1} \left[\varphi_i(x,y)\right]^2 \tag{20}$$

The solution to Eq. (18) is translated to compute the point of minimum Φ. In geometry, $\Phi(x,y)$ is a three-dimension curve, the minimum point equals to the tangent point between $\Phi(x,y)$ and xOy. In the region D of $\Phi(x,y)$, any point is passed through by an equal high line. If starting with an initial guess (x_0,y_0) in the region D, declining $\Phi(x,y)$ in the direction of steepest descent until $\Phi(x,y)$ declines to minimum, and then we can get the solution.
Usually, the normal direction of an equal high line is the direction of the gradient vector of $\Phi(x,y)$ which is denoted by

$$G = \left(\frac{\partial\Phi}{\partial x}, \frac{\partial\Phi}{\partial y}\right)^{\mathrm{T}} \tag{21}$$

The opposite direction to the gradient vector is the steepest descent direction.
Given (x_0,y_0) is an approximate solution, compute the gradient vector at this point

$$G_0 = (g_{10}, g_{20})^{\mathrm{T}}$$

Where

$$\begin{cases} g_{10} = \left.\dfrac{\partial\Phi}{\partial x}\right|_{(x_0,y_0)} = 2\left[\sum_{i=1}^{N-1}\left(\dfrac{\partial\varphi_i}{\partial x}\right)\varphi_i\right]_{(x_0,y_0)} \\ g_{20} = \left.\dfrac{\partial\Phi}{\partial y}\right|_{(x_0,y_0)} = 2\left[\sum_{i=1}^{N-1}\dfrac{\partial\varphi_i}{\partial y}\right]\varphi_i\,]_{(x_0,y_0)} \end{cases} \tag{22}$$

Then, start from (x_0,y_0), cross an appropriate step-size in the direction of $-G_0$, λ is the step-size parameter, get a new point (x_1,y_1)

$$\begin{cases} x_1 = x_0 - \lambda g_{10} \\ y_1 = y_0 - \lambda g_{20} \end{cases} \tag{23}$$

Choose an appropriate λ in order to let (x_1,y_1) be the relative minimum in $-G_0$,

$$\Phi(x_1,y_1) \approx \min\{\Phi(x_0 - \lambda g_{10}, y_0 - \lambda g_{20})\}$$

In order to fix on another approximation close to (x_0, y_0), expand $\varphi_i(x_0 - \lambda g_{10}, y_0 - \lambda g_{20})$ at (x_0, y_0), omit λ^2 high order terms, get the approximation of Φ

$$\Phi(x_0 - \lambda g_{10}, y_0 - \lambda g_{20}) = \sum_{i=1}^{N-1} [\varphi_i(x_0 - \lambda g_{10}, y_0 - \lambda g_{20})]^2$$

$$\approx \{\sum_{i=1}^{N-1} (\varphi_i)^2 - 2\lambda[\sum_{i=1}^{N-1} \varphi_i(g_{10}\frac{\partial \varphi_i}{\partial x} + g_{20}\frac{\partial \varphi_i}{\partial y})] + \lambda^2[\sum_{i=1}^{N-1} (g_{10}\frac{\partial \varphi_i}{\partial x} + g_{20}\frac{\partial \varphi_i}{\partial y})^2]\}_{(x_0, y_0)}$$

Let $\partial\Phi / \partial\lambda = 0$,

$$\lambda = \left[\frac{\sum_{i=1}^{N-1} \varphi_i(g_{10}\frac{\partial \varphi_i}{\partial x} + g_{20}\frac{\partial \varphi_i}{\partial y})}{\sum_{i=1}^{N-1} (g_{10}\frac{\partial \varphi_i}{\partial x} + g_{20}\frac{\partial \varphi_i}{\partial y})^2} \right]_{(x_0 \cdot y_0)} \tag{24}$$

Subtract Eq. (24) from Eq. (23), we obtain a new (x_1, y_1), and regard this as a relative minimum point of Φ in the direction of $-G_0$, then start at this new point (x_1, y_1), update the position estimate according to the above steps until Φ is sufficiently small.

In general, the convergence of steepest descent method is fast when the initial guess is far from the true solution, vice versa. Taylor series expansion method has been widely used in solving nonlinear equations for its high accuracy and good robustness. However, this method performs well under the condition of close to the true solution, vice versa. Therefore, hybrid optimizing algorithm (HOA) is proposed combining both Taylor series expansion method and steepest descent method, taking great advantages of both methods, optimizing the whole iterative process, improving positioning accuracy and efficiency.

In HOA, at the beginning of iteration, steepest descent method is applied to let the rough initial guess close to the true solution. Then, a further precise adjustment is implemented by Taylor series expansion method to make sure that the final estimator is close enough to the true solution. HOA has the properties of good convergence and improved efficiency. The specific flow is

1. Give a free initial guess (x_0, y_0), compute $i = 1 \cdots N-1$, $\frac{\partial \varphi_i}{\partial x}, \frac{\partial \varphi_i}{\partial y}$
2. Compute the gradient vector g_{10}, g_{20} at the point (x_0, y_0) from Eq. (22)
3. Compute λ from Eq. (24)
4. Compute (x_1, y_1) from Eq. (23)
5. If $\Phi \approx 0$, stop; otherwise, substitute (x_1, y_1) for (x_0, y_0), iterate (2)(3)(4)(5)
6. Compute $\hat{d}_{i+1,1}$ when $i = 1 \cdots N-1$ from Eq. (16)
7. Compute $\hat{d}_1, \hat{d}_{i+1}, f_{i,0}, a_{i,1}, a_{i,2}$ when $i = 1 \cdots N-1$ from Eq. (19)
8. Compute δ from Eq. (21)
9. Continually refine the position estimate from (7)(8)(9) until δ satisfies the accuracy

According to the above flow, the performance of the proposed HOA is evaluated via Matlab simulation software. In the simulation, we model a cellular system with one central BS and two other adjacent BS. More assistant BS can be utilized for more accuracy, however, in cellular communication systems, one of the Main design philosophies is to make the link loss between the target mobile and the home BS as small as possible, while the other link loss as large as possible to reduce the interference and to increase signal-to-interference ratio for the desired communication link. This design philosophy is not favorable to position location (PL), and leads to the main problems in the current PL technologies, i.e. hearability and accuracy. Considering the balance between communication link and position accuracy, two assistant BS is chose. We assume that the coordinates of central BS is (x1=0m ; y1=0m), the two assistant BS coordinates is (x2=2500m ; y2=0m); (x3=0m ; y3=2500m) respectively, MS coordinates is (x=300;y=400). A comparison of HOA and Taylor series expansion method is presented.

A lot of simulation computation demonstrates: there are 3 situations. The first one is that HOA is more accuracy and efficiency under the precondition of the same initial guess and the same measured time. In the second situation, HOA is more convergence to any initial guess than Taylor series expansion method under the precondition of the same initial guess and the same measured time. In the third situation, at the prediction of inaccurate measurements, the same initial guess, HOA is proved to be more accuracy and efficiency. The simulation results are given in Tables 3,4,5 respectively.

As shown in Table 3, the steepest decent method performs much better at the convergence speed. Indeed, the location error is smaller than Taylor series expansion method for 10^3. Meanwhile, the computation efficiency is improved by 23.35%. The result is that HOA is more accuracy and efficiency.

As shown in Table 4, when the initial guess is far from the true location, Taylor series expansion method is not convergent while HOA is still convergent which declines the constraints of the initial guess.

As shown in Table 5, when the measurements are inaccurate, the HOA location error is smaller than Taylor series expansion method for 10 times. Meanwhile, the computation efficiency is improved by 23.14%.

algorithms	Iterative results(m)	errors(m)	time(ms)
HOA	x =299.9985 y =400.0006	xx =-0.0015 yy =0.0006	0.374530
Taylor	x=301.1 y=400.4482	xx=1.1000 yy=0.4482	0.488590

Table 3. Comparison of HOA and Taylor series expansion method when the initial guess is close to the true solution and the measured time is accurate

algorithms	Iterative results(m)	errors(m)	time(ms)
HOA	x =299.9985 y =400.0006	xx =-0.0015 yy =0.0006	1.025930
Taylor	x =+∞ y =+∞	Not convergent	

Table 4. Comparison of HOA and Taylor series expansion method when the initial guess is far from the true solution and the measured time is accurate

algorithms	Iterative results(m)	errors(m)	time(ms)
HOA	x= 301.1297 y= 400.4492	xx=1.1297 yy=0.4492	0.376400
Taylor	x =317.8 y =396.0549	xx=17.8000 yy=-3.9451	0.489680

Table 5. Comparison of HOA and Taylor series expansion method when the initial guess is the same and the measured time is inaccurate

3.2 Data fusion techniques

Date fusion techniques include system fusion and measurement data fusion (Sayed, 2005). For example, a combination of GPS and cellular networks can provide greater location accuracy, and that is one kind of system fusion. Measurement data fusion combines different signal measurements to improve accuracy and coverage. This section mainly concerns how to use measurement data fusion techniques to solve problems in cellular-based positioning system.

3.2.1 Technical Challenges in cellular-based positioning

The most popular cellular-based positioning method is multi-lateral localization. In such positioning system, there are two major challenges, non-line-of-sight (NLOS) propagation problem and hearability.

A. Hearability problem

In cellular communication systems, one of the main design philosophies is to make the link loss between the target mobile and the home BS as small as possible, while the other link loss as large as possible to reduce the interference and to increase signal to noise ratio for the desired communication link. In multi-lateral localization, the ability of multiple base stations to hear the target mobile is required to design the localization system, which deviates from the design of wireless communication system , and this phenomenon is referred as hearability (Prretta, 2004).

B. The non-line-of-sight propagation problem

Most location systems require line-of-sight radio links. However, such direct links do not always exist in reality because the link is always attenuated or blocked by obstacles. This phenomenon, which refers as the NOLS error, ultimately translates into a biased estimate of the mobile's location (Cong, 2001).

As illustrated by the signal transmission between BS7 and MS in Fig.17. A NLOS error results from the block of direct signal and the reflection of multipath signals. It is the extra distance that a signal travels from transmitter to receiver and as such always has a nonnegative value. Normally, NLOS error can be described as a deterministic error, a Gaussian error, or an exponentially distributed error.

In order to demonstrate the performance degradation of a time-based positioning algorithm due to NLOS errors, taking the TOA method as an example. The least square estimator used for MS location is of the following form

$$\hat{\mathbf{x}} = \arg\min \sum_{i \in S} (r_i - \|\mathbf{x} - \mathbf{X}_i\|)^2 \tag{25}$$

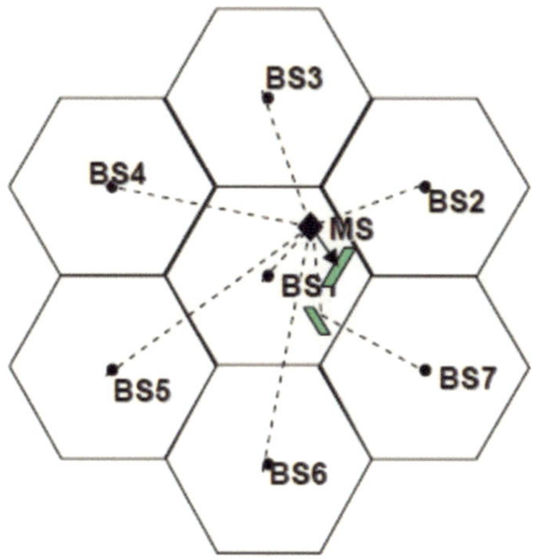

Fig. 17. NLOS error

$$r_i = L_i + n_i + e_i, \quad i = 1,2,...N \tag{26}$$

Where r is the range observation, L is LOS range, n is receiver noise, e is NLOS error.

$$\mathbf{r} = \mathbf{L} + \mathbf{n} + \mathbf{e} \tag{27}$$

If the true MS location is used as the initial point in the least square solution, the range measurements can be expressed via a Taylor series expansion as

$$\mathbf{r} \approx \mathbf{L} + \mathbf{G}\begin{bmatrix} \Delta x \\ \Delta y \end{bmatrix} \tag{28}$$

$$\begin{bmatrix} \Delta x \\ \Delta y \end{bmatrix} = (\mathbf{G}^T\mathbf{G})^{-1}\mathbf{G}.\mathbf{n} + (\mathbf{G}^T\mathbf{G})^{-1}\mathbf{G}.\mathbf{e} \tag{29}$$

Where G is the design matrix, and $[\Delta x, \Delta y]$ is the MS location error. Because NLOS errors are much larger than the measurement noise, the positioning errors result mainly from NLOS errors if NLOS errors exist.

3.2.2 Data fusion architecture

The underlying idea of data fusion is the combination of disparate data in order to obtain a new estimate that is more accurate than any of the individual estimates. This fusion can be accomplished either with raw data or with processed estimates. One promising approach to the general data fusion problem is represented by an architecture that was developed in 1992 by the data fusion working group of the joint directors of laboratories (JDL) (Kleine-Ostmann, 2001). This architecture is comprised of a preprocessing stage, four levels of fusion and data management functions. As a refinement of this architecture, Hall proposed a hybrid approach to data fusion of location information based on the combination of level one and level two fusion (Kleine-Ostmann, 2001).

Based on the JDL model and its specialization to first and second level hybrid data fusion, an architecture for the position estimation problem in cellular networks is constructed. Fig. 18. shows the data fusion model that uses four level data fusion.

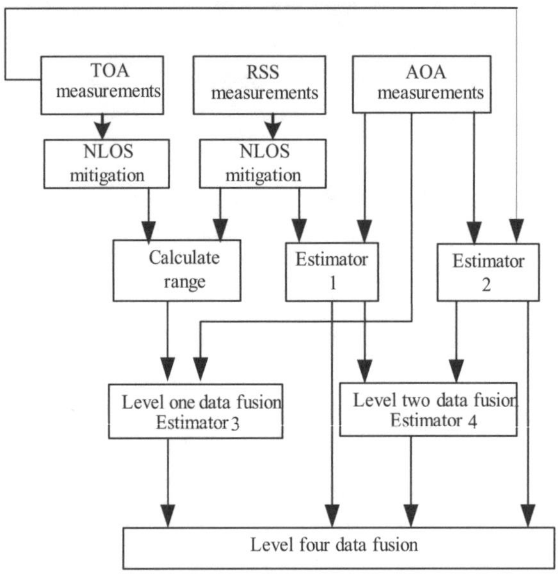

Fig.18. Data fusion model

Position estimates are obtained by four different approaches in this model. The first approach uses TOA/AOA hybrid method. The second position estimate is based on RSS /AOA hybrid method. The other two estimates are obtained by level one and level 2 data fusion methods.

A. Level one fusion

Firstly, we use the method shown in (Wylie, 1996) to mitigate TOA NLOS error and calculate the LOS distance d_{TOA}. As the same way, we mitigate RSS NLOS error and calculate the LOS distance d_{RSS}. Then, the independent d_{TOA} and d_{RSS} are fused into d. The derivation of d is below.
Let

$$\begin{cases} \text{Var}(d_{\text{TOA}}) = \sigma_{\text{TOA}}^2 \\ \text{Var}(d_{\text{RSS}}) = \sigma_{\text{RSS}}^2 \end{cases}$$

Define,

$$d = f(d_{\text{TOA}}, d_{\text{RSS}}) = ad_{\text{TOA}} + bd_{\text{RSS}} \tag{30}$$

The constrained minimization problem is described as (31)

$$\begin{aligned} &\arg_{\min}[\text{Var}(d)] = \arg_{\min}[\text{E}(d - \bar{d})^2] \\ &a + b = 1 \end{aligned} \tag{31}$$

By using Lagrange Multipliers, the solution of (31) is obtained as (32)

$$a = \frac{\sigma_{\text{RSS}}^2}{\sigma_{\text{RSS}}^2 + \sigma_{\text{TOA}}^2}, \quad b = \frac{\sigma_{\text{TOA}}^2}{\sigma_{\text{TOA}}^2 + \sigma_{\text{RSS}}^2} \tag{32}$$

The data fusion result is given by (33)

$$d = \frac{\sigma_{\text{RSS}}^2 d_{\text{TOA}} + \sigma_{\text{TOA}}^2 d_{\text{RSS}}}{\sigma_{\text{TOA}}^2 + \sigma_{\text{RSS}}^2} \tag{33}$$

Using (32)(33), the variance of d is

$$\text{Var}(d) = \left(\frac{1}{\sigma_{\text{TOA}}^2} + \frac{1}{\sigma_{\text{RSS}}^2}\right)^{-1} \tag{34}$$

Therefore,

$$\begin{aligned} \text{Var}(d) &\le \left(\frac{1}{\sigma_{\text{TOA}}^2}\right)^{-1} = \text{Var}(d_{\text{TOA}}) \\ \text{Var}(d) &\le \left(\frac{1}{\sigma_{\text{RSS}}^2}\right)^{-1} = \text{Var}(d_{\text{RSS}}) \end{aligned} \tag{35}$$

So, the data fusion estimator is more accurate than estimator 1 or 2.

B. Level two fusion

By utilizing the result proved in (32)(33)(34), the estimator 4 fused solution and its variance are of the following equations.

$$x_{\text{C}} = \frac{\sigma_{\text{TOA/AOA}}^2 x_{\text{RSS/AOA}} + \sigma_{\text{RSS/AOA}}^2 x_{\text{TOA/AOA}}}{\sigma_{\text{TOA/AOA}}^2 + \sigma_{\text{RSS/AOA}}^2} \tag{36}$$

$$\sigma_{\text{C}}^2 = \left(\frac{1}{\sigma_{\text{TOA/AOA}}^2} + \frac{1}{\sigma_{\text{RSS/AOA}}^2}\right)^{-1} \tag{37}$$

Where $x_{\text{RSS/AOA}}$ and variance $\sigma_{\text{RSS/AOA}}^2$ are the mean and variance of estimator 1, $x_{\text{TOA/AOA}}$ and $\sigma_{\text{TOA/AOA}}^2$ are the mean and variance of estimator 2, x_{C} and σ_{C}^2 are the mean and variance of estimator 4.

C. Level three fusion

In general, the estimate that exhibits the smallest variance is considered to be the most reliable estimate. However, the choice cannot be based solely on variance. In a poor signal propagation situation when the MS is far from BSs, the RSS estimate becomes mistrust.

3.2.3 Single base station positioning algorithm based on data fusion model

To solve the problem, a single home BS localization method is proposed in this paper. In (Wylie, 1996), a time-history-based method is proposed to mitigate NLOS error. Based on

this method, a novel single base station positioning algorithm based on data fusion model is established to improve the accuracy and stability of localization.
Fig.19. illustrates the geometry fundamental of this method. The MT coordinates (x, y) is simply calculated by (38)

$$\begin{aligned} x &= d\cos\alpha \\ y &= d\sin\alpha \end{aligned} \quad (38)$$

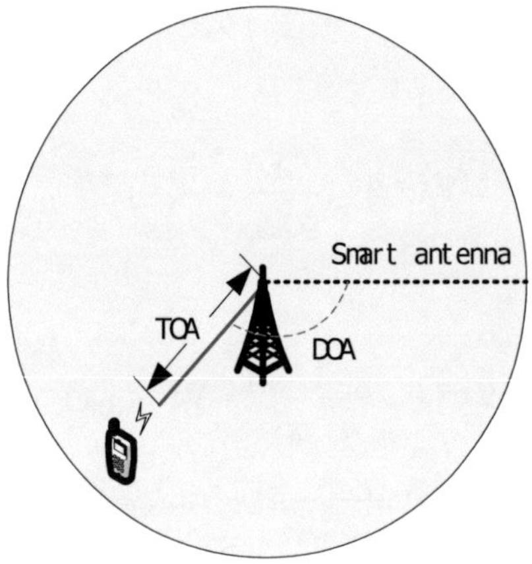

Fig. 19. Geometry of target coordinates (x, y)

The MT localization is determined by d and α where d denotes the line-of-sight (LOS) distance between the MT and the home BS, α denotes the signal direction from the home BS to the TM. The above two parameters are important for localization accuracy. Data fusion model discussed above can be utilized to get a more accurate localization.
In this section, we present some examples to demonstrate the performance of the proposed method. We suppose the MT's trajectory is x=126.9+9.7t，y=286.6+16.8t, sampling period is 0.05s, 200 samples are taken, 50 random tests are taken in one sample. The velocity is constant at $v_x = 9.7\text{m/s}$, $v_y = 16.8\text{m/s}$. The TOA measurements error is Gaussian random variable with zero mean and standard variance 20, NLOS error is exponential distribution with mean 100. RSS medium-scale path loss is a zero mean Gaussian distribution with standard deviation 20 and small-scale path loss is a Rayleigh distribution with $\sigma_{ss}^2 = 79.7885$. The home BS is located at (0,0).
Simulation 1, when the NLOS and measurements error are added to the TOA, we utilize (Wylie, 1996) to reconstruct LOS. Fig.20. shows the results. From the results, we can see that NLOS error is the major effect to bias the true range up to 900m. Due to NLOS, at most of the time, the measurements are much larger than the true range. After the reconstruction, the corrected range is near the true range and float around the true range.
Simulation 2, when the medium-scale path loss and small-scale path loss are added to the RSS, we utilize (Wylie, 1996) to reconstruct LOS. Fig.21. shows the results. From the results, we can see that the small-scale error (NLOS error) is the major effect to bias the true range up to 700m. Due to the NLOS, at most of the time, the measurements are much larger than the true range. After the reconstruction, the corrected range is near the true range and float around the true range.

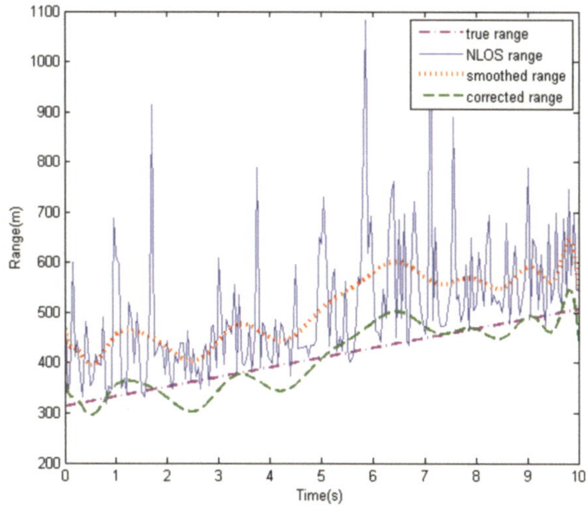

Fig. 20. TOA LOS reconstruction from NLOS measurements

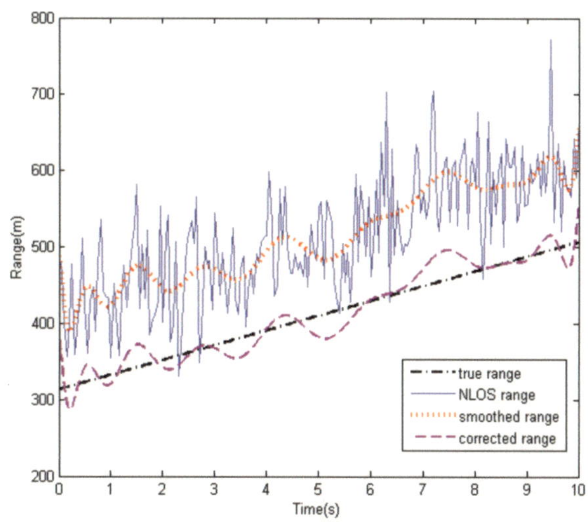

Fig. 21. RSS LOS reconstruction from NLOS measurements

Simulation 3 is about the localization improvement. The results are shown in Fig.22. It indicates that the standard variance of the proposed method is smaller than any of TOA or RSS. HLMR technique is able to significantly reduce the estimation bias when compared to the classic NLOS mitigation method shown by (Wylie, 1996). By statistical calculation, the mean of TOA standard variance by (Wylie, 1996) is 37.382m, while the data fusion aided method is 17.695m. The stability is more than one time higher. Fig.23. demonstrates the Euclidean distance between the true range and estimation range by data fusion based method, TOA and AOA. The mathematical expressions are given in (39)(40)(41). By statistical calculation, the Euclidean distance of TOA is 37.44, the proposed method is 3.1318 which is ten times more accurate.

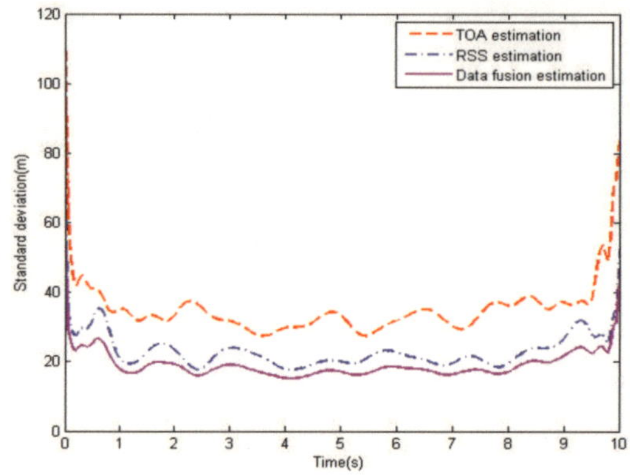

Fig. 22. Standard variance of estimation range

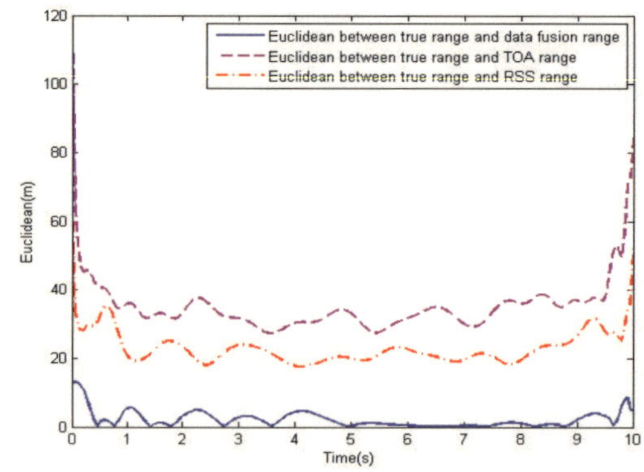

Fig. 23. Euclidean distance between true range and estimation range

$$\|\mathbf{r}-\mathbf{d}\| = \sqrt{\sum_{i=1}^{N}(r_i - d_i)^2} \tag{39}$$

$$\|\mathbf{r}-\mathbf{d}_{\mathbf{TOA}}\| = \sqrt{\sum_{i=1}^{N}(r_i - d_{\mathrm{TOA}_i})^2} \tag{40}$$

$$\|\mathbf{r}-\mathbf{d}_{\mathbf{RSS}}\| = \sqrt{\sum_{i=1}^{N}(r_i - d_{\mathrm{RSS}_i})^2} \tag{41}$$

3.3 UWB precise real time location system

Reliable and accurate indoor positioning for moving users requires a local replacement for satellite navigation. Ultra WideBand (UWB) technology is particularly suitable for such local systems, for its good radio penetration through structures, the rapid set-up of a stand-alone system, tolerance of high levels of reflection, and high accuracy even in the presence of severe multipath (Porcino, 2003).

3.3.1 UWB localization challenges

UWB technology is defined by the Federal Communications Commission (FCC) as any wireless transmission scheme that occupies a fractional bandwidth $W / f_c \geq 20\%$ where W is the transmission bandwidth and fc is the band center frequency, or more than 500 MHz of absolute bandwidth. FCC approved the deployment of UWB on an unlicensed basis in the 3.1-10.6GHz band with limited power spectral density as shown in Fig.24.

UWB signal is a kind of signals which occupies several GHz of bandwidth by modulating an impulse-like waveform. A typical baseband UWB signal is Gaussian monocycle obtained by differentiation of the standard Gaussian waveform (Roy, 2004). A second derivative of Gaussian pulse is given by

$$p(t) = A[1 - 4\pi(\frac{t}{T_d})^2]e^{-2\pi(\frac{t}{T_d})^2} \tag{42}$$

Where the amplitude A can be used to normalize the pulse energy. Fig.25 shows the time domain waveform of (42). From Fig.25, we see that the duty cycle (the pulse duration divided by the pulse period) is really small. In other aspect of view, UWB signal is sparse in time domain. The Fourier transform (Fig.26) is occupied from near dc up to the system bandwidth $B_S \approx 1/T_d$.

A. CRLB for time delay estimation

The CRLB defines the best estimation performance, defined as the minimum achievable error variance, which can be achieved by using an ideal unbiased estimator. It is a valuable tool in evaluating the potential of UWB signals for TOA estimation. In this section, we will derive the expression of the CRLB of TOA estimation for UWB signals.

Consider the signal in (42) is sampled with a sampling period T_s. The sequence of the samples is written as

$$r_n = s_n(\tau) + w_n \tag{43}$$

The joint probability of r_n conditioned to the knowledge of delay τ:

$$p(r_n|\tau) = (2\pi\sigma^2)^{-\frac{N}{2}} \exp(-\frac{1}{2\sigma^2}\sum_{n=1}^{N}(r_n - s_n(\tau)^2)) \tag{44}$$

Where N is the number of samples, σ^2 is the variance of r_n.

In order to get the continuous probability of (44)

$$\begin{aligned} p(r|\tau) &= \lim_{N\to+\infty} p(r_n|\tau) \\ &= (2\pi\sigma^2)^{-\frac{N}{2}} \exp(-\frac{1}{2\sigma^2}\int_0^T (r(t) - s(t;\tau))^2 dt) \end{aligned} \tag{45}$$

The log-likelihood function of (45)

$$\ln p = \ln(2\pi\sigma^2)^{-\frac{N}{2}} - \frac{1}{2\sigma^2}\int_0^T (r(t) - s(t;\tau))^2 dt \tag{46}$$

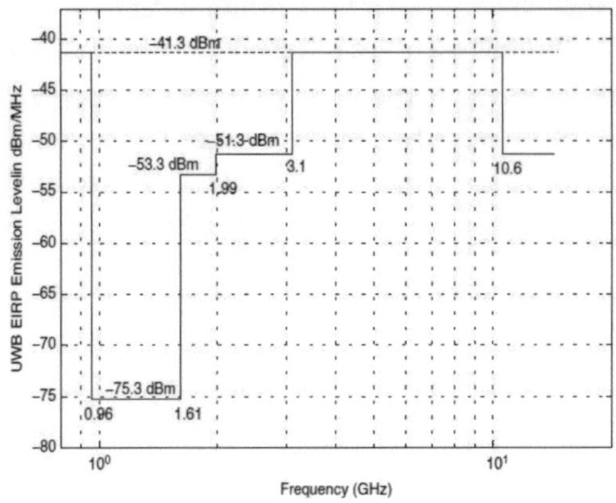

Fig. 24. UWB spectral mask

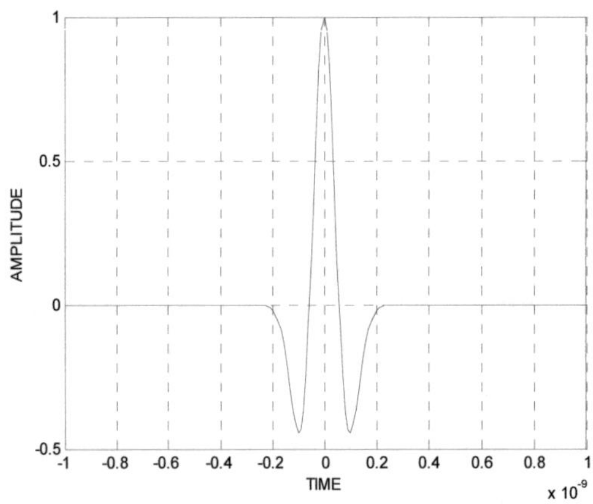

Fig. 25. UWB signal in time domain

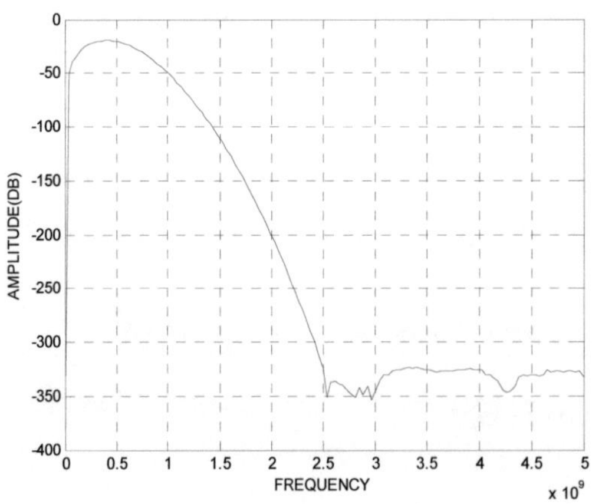

Fig. 26. UWB signal in frequency domain

The second differentiation of (46) is

$$\frac{\partial^2 \ln p}{\partial^2 \tau} = \frac{1}{\sigma^2}\left(\int_0^T s''(t;\tau)(r(t) - s(t;\tau))dt + \int_0^T -(s'(t;\tau))^2 dt\right) \quad (47)$$

The average value of (47)

$$E\left(\frac{\partial^2 \ln p}{\partial^2 \tau}\right) = -\frac{1}{\sigma^2}\int_0^T (s'(t;\tau))^2 dt$$

The minimal achievable variance for any unbias estimation (CRLB) is thus:

$$\begin{aligned}\sigma_t^2 &= -\frac{1}{E\left(\frac{\partial^2 \ln p}{\partial^2 \tau}\right)} = \frac{\sigma^2}{\int_0^T (s'(t;\tau))^2 dt} \\ &= \frac{\sigma^2}{\int_0^T s^2(t;\tau)dt} \cdot \frac{\int_0^T s^2(t;\tau)dt}{\int_0^T (s'(t;\tau))^2 dt} \\ &= \left(\frac{E}{N}\right)^{-1} \cdot \beta^2\end{aligned} \quad (48)$$

Where

$$\begin{aligned}\beta^2 &= \frac{\int_0^T s^2(t;\tau)dt}{\int_0^T (s'(t;\tau))^2 dt} = \frac{\int S^2(f;\tau)df}{\int f^2 S^2(f;\tau)df} \\ &\geq \frac{\int S^2(f;\tau)df}{\int f^2 df . \int S^2(f;\tau)df} = \frac{1}{\int f^2 df}\end{aligned} \quad (49)$$

The equality holds if $f = kS(f;\tau)$, where k is an arbitrary constant. E/N is the signal to noise ratio, and $S(f)$ is the fourier transform of the transmitted signal.
Inequation (49) shows that the accuracy of TOA is inversely proportional to the signal bandwidth. Since UWB signals have very large bandwidth, this property allows extremely accurate TOA estimation. UWB signal is very suitable for TOA estimation. However, there are many challenges in developing such a real time indoor UWB positioning system due to the difficulty of large bandwidth sampling technique and other challenges.

3.3.2 Compressive sensing based UWB sampling method

As discussed above, due to a large bandwidth of a UWB signal, it can't be sampled at receiver directly, how to compress and reconstruct the signal is a problem. To solve this problem, this section gives a new perspective on UWB signal sampling method based on Compressive Sensing (CS) signal processing theory (Candès, 2006; Candès,2008; Richard, 2007).

CS theory indicates that certain digital signals can be recovered from far fewer samples than traditional methods. To make this possible, CS relies on two principles: sparsity and incoherence.

Sparsity expresses the idea that the number of freedom degrees of a discrete time signal may be much smaller than its length. For example, in the equation $x=\psi\alpha$, by K-sparse we mean that only $K\leq N$ of the expansion coefficients α representing $x=\psi\alpha$ are nonzero. By compressible we mean that the entries of a, when sorted from largest to smallest, decay rapidly to zero. Such a signal is well approximated using a K-term representation.

Incoherent is talking about the coherence between the measurement matrix ψ and the sensing matrix Φ. The sensing matrix is used to convert the original signal to fewer samples by using the transform $y=\Phi x=\Phi\Psi\alpha$ as shown in Fig.27. The definition of coherence is $\mu(\Phi,\Psi)=\sqrt{n}.\max_{1\leq k,j\leq n}\left|\left\langle \phi_k,\varphi_j\right\rangle\right|$. It follows from linear algebra that is $\mu(\Phi,\Psi)\in[1,\sqrt{n}]$. In CS, it concerns about low coherence pairs. The results show that random matrices are largely incoherent with any fixed basis Ψ. Gaussian or ±1 binaries will also exhibit a very low coherence with any fixed representation Ψ.

Since $M<N$, recovery of the signal x from the measurements y is ill-posed; however the additional assumption of signal sparsity in the basis Ψ makes recovery possible and practical.

The signal can be recovered by solving the following convex program as shown in Fig. 27.

$\alpha = \arg\min ||\alpha||_1$ s.t. $y = \Phi\Psi\alpha$.

And M should obey $M\geq C.\mu^2(\Phi,\Psi).K.\log N$, where C is a small constant, K is the number of non-zero elements, N is the length of the original signal.

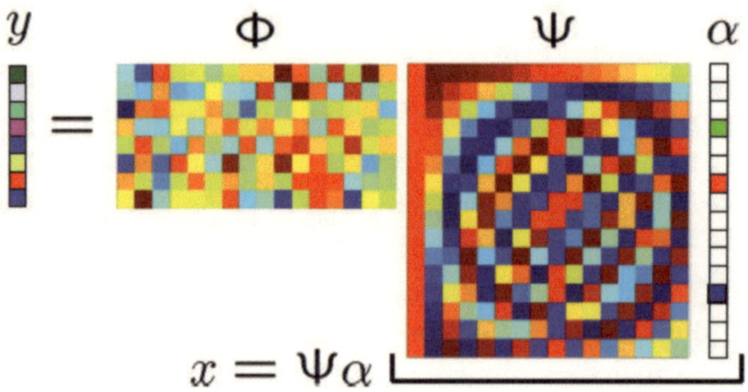

Fig. 27. Compressive sensing transform

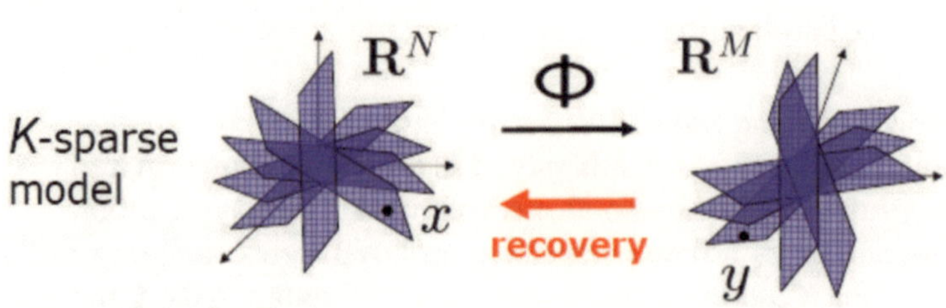

Fig. 28. Signal recovery algorithm

we utilize the temporal sparsity property of UWB signals and CS technique. There are three key elements needed to be addressed in the use of CS theory into UWB signal sampling: 1) How to find a space in which UWB signals have sparse representation 2) How to choose random measurements as samples of sparse signal 3) How to reconstruct the signal.
CS is mainly concerned with low coherent pairs. How to find a good pair of Φ and Ψ in which UWB signals have sparse representation is the problem we need to solve. Since UWB signal is sparse in time domain, we choose Ψ is spike basis $\phi_k(t) = \delta(t-k)$ and Φ is random Gaussian matrix.
The mathematical principle can be formulated as

$$\mathbf{s} = \mathbf{G}.\mathbf{E}.p(t)\big|_{t=kT} + \mathbf{n} \quad k{=}1\ldots N \tag{50}$$

Where $\mathbf{s}$ is the sensing vector, $\mathbf{G}$ is random Gaussian matrix, $\mathbf{E}$ is spike matrix, $p(t)\big|_{t=kT}$ is the Nyquist samples with sample period T, total samples N. $\mathbf{n}$ is the additive noise vector with bounded energy $\|\mathbf{n}\|_2 \leq \varepsilon$.
The coherence between measurement matrix **E** and sensing matrix **G** is near 1. **G** matrix is largely incoherent with **E**. Therefore, in our method, the precondition of sparsity and incoherent are satisfied.

$$u(\mathbf{G},\mathbf{E}) = \sqrt{n}.\max_{1\leq k,j\leq n}\left|\left\langle g_k, e_j\right\rangle\right| \tag{51}$$

Since **E** is spike matrix, **G.E=G.**
(50) can be simplified by

$$\mathbf{s} = \mathbf{G}.p(t)\big|_{t=kT} + \mathbf{n} \quad k{=}1\ldots N \tag{52}$$

In (52), the CS method is simplified, and the multiply complexity is reduced by MN^2. Therefore, the UWB signal is suitable for CS, moreover it makes CS simpler and reduces the computation complexity.
The recovery algorithm is

$$\arg\min\left\|p(t)\big|_{t=kT}\right\|_1 \text{ such that } \left\|\mathbf{G}.p(t)\big|_{t=kT} - \mathbf{s}\right\|_2 \leq \varepsilon \tag{53}$$

The recovery multiply complexity is reduced by N^2.
Theorem:
Fix $p(t)\big|_{t=kT} \in \mathbb{R}^N$, and it is K sparse on a certain basis Ψ. Select M measurements in the Φ domain uniformly at random. Then if

$$M \geq c.\mu^2(\Phi,\Psi).K.\log N \tag{54}$$

For some positive constant c, the solution to (10) is success with high probability. From (54), we see that M is proportional to three factors: μ, K and N. If μ and N are fixed, the sparser K can reduce the measurements needed to reconstruct the signal. From Fig.29, we see that the spike basis can recover the signal.

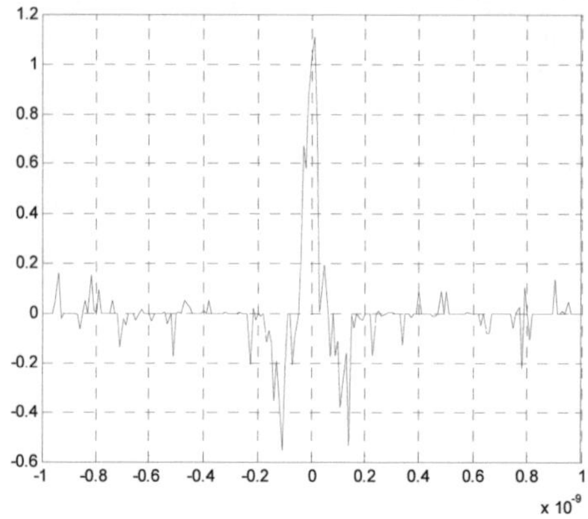

Fig. 29. Reconstructed signal from measurementsat 20% of the Nyquist rate

In this part, examples are done to show the comparison of the proposed method with traditional methods.
In all of the examples, the transmitted signal is expressed as

$$s(t) = (1 - 4\pi(\frac{t}{0.2\times 10^{-9}})^2)\times \exp(-2\pi(\frac{t}{0.2\times 10^{-9}})^2).$$

The bandwidth of the signal is 2.5GHz, and the traditional sampling frequency is 5GHz.

A. Example 1 (in LOS environment)

In the first example, we assume that the signal is passed through Rician channel and the number of multipath is six. In the first simulation (see Fig.30), we set the observed time is 0.2um. Fig.30(a) shows the UWB signal without channel interference. Fig.30(b) shows the

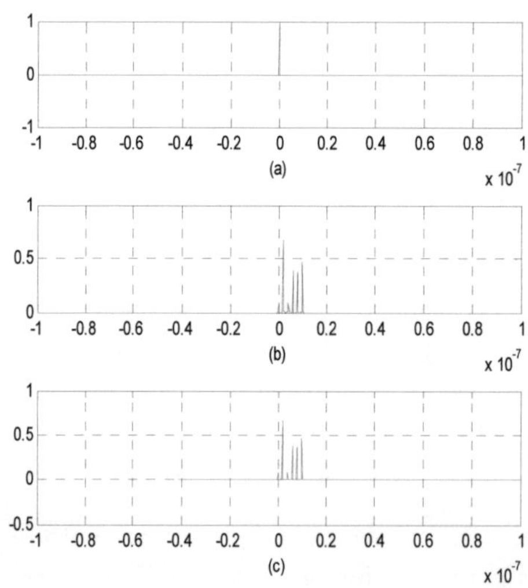

Fig. 30. (a) Ideal reconstructed UWB signal (b) Reconstructed UWB signal with Nyquist rate (c) Reconstructed UWB signal with 10% of the Nyquist rate

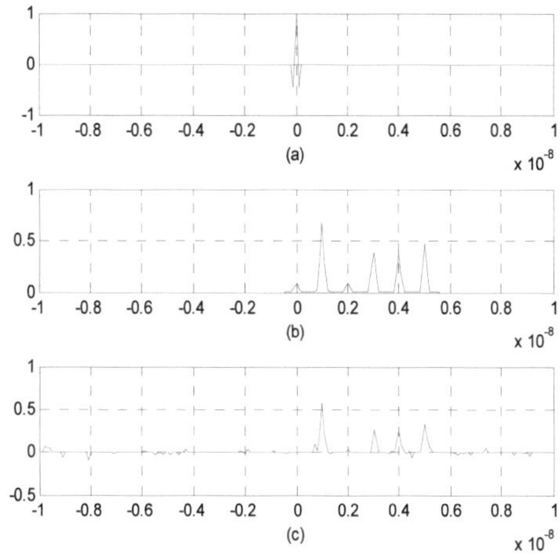

Fig. 31. (a) Ideal reconstructed UWB signal (b) Reconstructed UWB signal with Nyquist rate (c) Reconstructed UWB signal with 30% of Nyquist rate

reconstructed UWB signal at Nyquist sampling rate. Fig30(c) shows the reconstructed signal by 10% of the Nyquist sampling rate. The time delay error of both methods is about 1nm.

In the second simulation (see Fig.31), we shorten the observed time to 0.02um, all of the other parameters are the same. Fig.31(c) shows the measurement we need to reconstruct the signals is up to 30%. And much more details of the signal can be seen. And the time delay error is 1nm.

Comparing these two simulations, the conclusion is that 1) by using 10% of Nyquist sampling rate, the accuracy of TOA estimation is the same as that by using full Nyquist rate. 2) By enlarging the sampling rate by 30%, more detail information of the signal can be recovered. However, for TOA estimation, we do not need to recover the full signal but the peak location of the signal which makes the use of 10% Nyquist sampling rate possible.

B. Example 2 (in NLOS environment)

In the second example, we simulate the TOA estimation of UWB signal in NLOS environment (the number of multipath is set to six).

At the first simulation (see Fig.32), we set the observed time is 0.2um. Fig.32(a) shows the ideal received UWB signal without Rayleigh channel interference. Fig.32(b) shows the detected UWB signal at Nyquist sampling rate. Fig.32(c) shows the detected UWB signal at 11% Nyquist sampling rate by using our method. We can see that Fig.32(c) can recover the signal (in Fig.32(b)) well, although lose some detail information. And the time delay errors of them are both 1nm.

At the second simulation (see Fig.33), we shorten the observed time to 0.02um, all of the other parameters are the same. It is shown in Fig.33(c) that the measurement we need to reconstruct the signals is 35% and much more details of the signal can be seen compared with Fig.33(c). And the time delay error is 1nm.

Comparing these two simulations, the conclusion is that 1) the accuracy of TOA estimation achieved by 11% Nyquist sampling rate is the same as that by full Nyquist sampling rate. 2) When more sampling rate is used, more detail information can be recovered. However, in TOA estimation, we do not need to recover the whole signal but the peak location of the signal. Finally, we can get the TOA estimation by 11% Nyquist sampling rate and the drawback is that some detail information of the signal is lost.

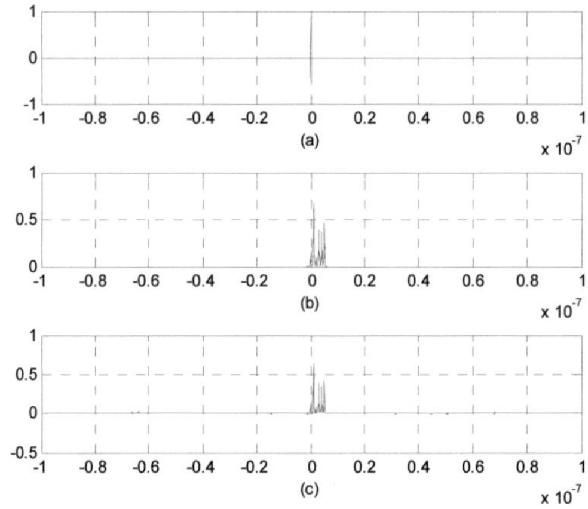

Fig. 32. (a)Ideal reconstructed UWB signal (b) Reconstructed UWB signal with Nyquist rate (c) Reconstructed UWB signal with 11% of the Nyquist rate

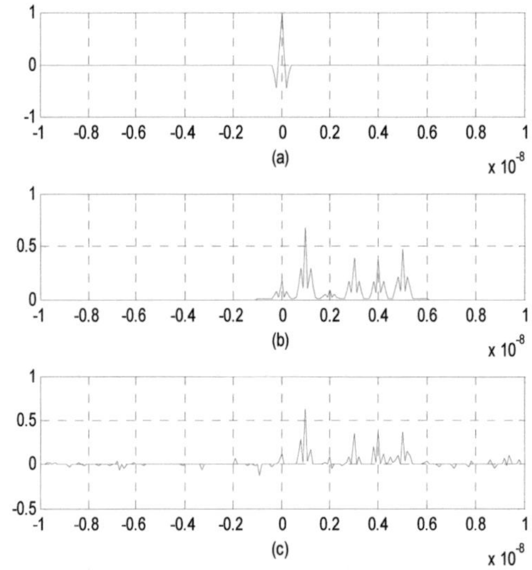

Fig. 33. (a)Ideal reconstructed UWB signal (b)Reconstructed UWB signal with Nyquist rate (c) Reconstructed UWB signal with 35% of Nyquist rate

3.3.3 Tracking system

Fig.34. is Ubisense precise real time location system, tracking unlimited number of people and objects in any space of any size with 15cm 3D tracking accuracy and high reliability. In this system, Ubisense UWB hardware is the platform and Ethernet (wire/wireless) is used as a transmission network. The UWB sensors are deployed around the room, generally on the wall. The target is attached with a UWB tag. When the target come into the area where is covered by UWB sensors, the sensors locate the target and provide location and speed information to the user.

UWB tracking systems have inherent advantages over other technologies:

a. Exceptional performance —Performs in high multi-path environments
b. Excellent real-time location accuracy — Better than 30cm (1 foot)
c. Long tag battery life —Up to 7 years at 1 Hz blink rate
d. Long Range — Up to 200 meters (650 feet) with line of sight
e. Unmatched real-time location tag throughput — Up to2700 tags/hub
f. Fast tag transmission rates —Up to 25 times/second
g. Fast intuitive setup — typical single location set-up in one day

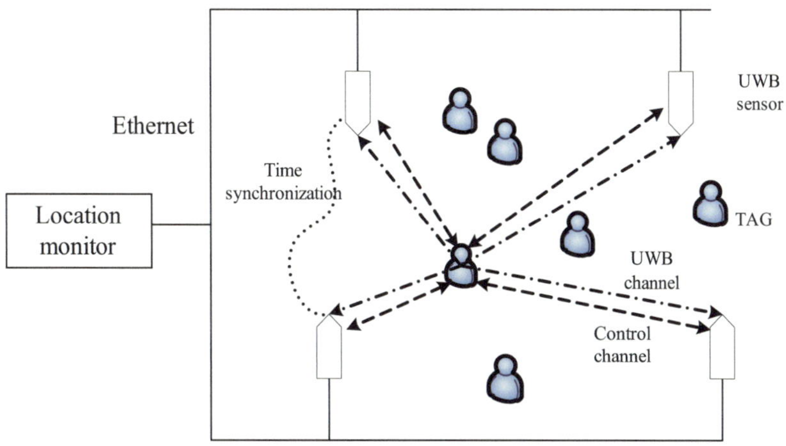

Fig. 34. UWB real time location system

3.4 Smart antennas technique

Smart antennas are often used for providing accurate AOA estimation. The commonly used methods for AOA estimation are beam forming (BF) (Van, 1998), minimum variance distortionless response (MVDR), multiple signal classification (MUSIC) (Vaidyanathan, 1995), maximum likelihood (ML) (Stoica, 1990).

3.4.1 Array signal processing

Before we describe the conventional methods of AOA estimation, it is necessary to present the array signal processing issues by smart antennas. In the array signal process, there are four issues of interest:

a. Array configuration
b. Spatial and temporal characteristics of the signal
c. Spatial and temporal characteristics of the interference
d. Objective of the array processing

Here, we consider the smart antenna as a uniform linear array (ULA). For the second issue, we set the signal structure as a known plane-wave signal from unknown directions. The interference is white Gaussian noise that is statistically independent in time and space domain. The objective is to estimate the AOA of multiple plane-wave signals in the presence of noise.

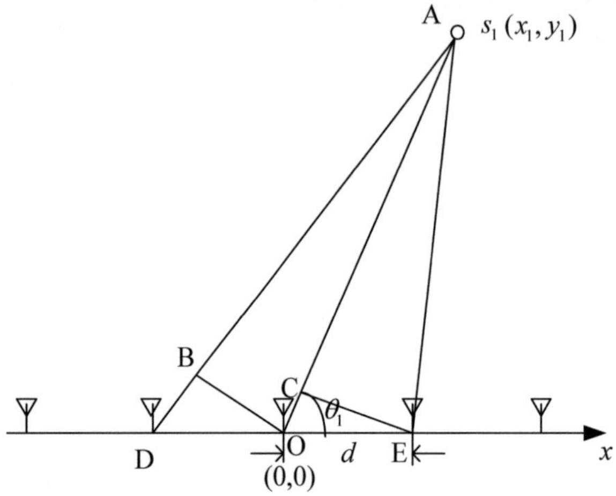

Fig. 35. Array processing observation model

3.4.2 AOA estimation methods

Before we describe the conventional methods of AOA estimation, it is necessary to present the mathematical model for the problem. Consider the basic case, the narrowband sources in the farfield of a uniform linear arrays (ULA) as shown in Fig.35. ULA consists of M omni-directional sensors with equal spacing d, residing on the x-coordinate axis. Taking the phase center of the array at the origin, the position of the m-th sensor is

$$p_m=(m-(M+1)/2)d,\ m\in\{1,\ldots M\}$$

The modulated signal in narrowband case can be expressed as $u(t)\exp(j\omega_0 t)$, where $u(t)$ is the baseband signal.

The output of the sensor at origin is

$$y_0(t)=u(t-\tau_{center})\exp(j\omega_0(t-\tau_{center})) \tag{55}$$

τ_{center} is the delay from the source to the phase-center. After demodulating, it can be represented as

$$y_0(t)=u(t-\tau_{center})\exp(-j\omega_0\tau_{center}) \tag{56}$$

Assume that the time delay relative to the origin sensor is τ_m

The output of sensor m is

$$y_m(t)=u(t-\tau_{center}-\tau_m)\exp(j\omega_0(-\tau_{center}-\tau_m)) \tag{57}$$

Since the signal is narrowband, it is able to ignore the delay between the sensors.

$$y_m(t)=u(t-\tau_{center})\exp(j\omega_0(-\tau_{center}-\tau_m)) \tag{58}$$

By measuring time relative to the phase center, the dependence on τ_{center} can be dropped. Thus, the output of sensor m is

tfor a single source, the complex envelope of the sensor outputs has the following form:

$$y_m(t) = u(t)\exp(-j\omega_0\tau_m) \quad (59)$$

where $\tau_m = -\frac{(m-(M+1)/2)}{c} d\cos\theta_1$

$$y_m(t) = u(t)\exp(j\omega_0(\frac{(m-(M+1)/2)}{c} d\cos\theta_1)) \quad (60)$$

Define $k^T p_m = -(\omega_0 / c)(m-(M+1)/2)d\cos\theta_1$, $k = \omega_0 / c$, therefore,

$$y_m(t) = u(t)\exp(-jk^T p_m) \quad (61)$$

Define angle vector $\mathbf{a}(\theta_1) = \exp(-jk^T \mathbf{p})$

Thus, for a single source, the complex envelope of the sensor outputs has the following form:

$$\mathbf{y}(t) = \mathbf{a}(\theta_1)u(t) + \mathbf{n}(t) \quad (62)$$

Define angle matrix $A(\theta) = [\mathbf{a}(\theta_1)^T, \mathbf{a}(\theta_2)^T ..., \mathbf{a}(\theta_K)^T]$, where $\boldsymbol{\theta} = [\theta_1,...,\theta_K]$ is the vector of unknown emitters' AOAs, The (m,k) element represents the kth source AOA information to the mth sensor. $\mathbf{u}(t) = [s_1(t), s_2(t),..., s_k(t)]'$ is the signals from K emitters. Taking noise into account, the final version of the model takes the following form:

$$\mathbf{y}(t) = \mathbf{A}(\boldsymbol{\theta})\mathbf{u}(t) + \mathbf{n}(t) \quad (63)$$

In order to characterize the arriving signal, several time samples are required, this is the Snapshot Model

$$\mathbf{y}(t) = \mathbf{A}(\boldsymbol{\theta})\mathbf{u}(t) + \mathbf{n}(t), \quad t = 1,2,...N \quad (64)$$

For simplicity, the noise is assumed to be spatially and temporally stationary and white, uncorrelated with the source. The covariance matrix takes the following form:

$$\mathrm{E}[\mathbf{n}(t)\mathbf{n}^H(t)] = \sigma^2\mathbf{I}$$

Where **I** is an identity matrix.

The covariance of the baseband signal $u(t)$ is given by

$$\mathbf{P} = \mathrm{E}[\mathbf{u}(t)\mathbf{u}^H(t)] = \lim_{N\to\infty} \frac{1}{N}\sum_{t=1}^{N} \mathbf{u}(t)\mathbf{u}^H(t) \quad (65)$$

N times snapshots approximation is computed by

$$\widehat{P} = \frac{1}{N}\sum_{t=1}^{N} \mathbf{u}(t)\mathbf{u}^H(t) \quad (66)$$

The covariance of the sensor output signal $y(t)$ is given by

$$\mathbf{R} = \mathrm{E}[\mathbf{y}(t)\mathbf{y}^H(t)] = \mathrm{A}(\theta)\mathrm{P}\mathrm{A}^H(\theta) + \sigma^2\mathbf{I} \quad (67)$$

N times snapshots approximation is used:

$$\hat{\mathbf{R}} = \frac{1}{N}\sum_{t=1}^{N}\mathbf{y}(t)\mathbf{y}^H(t) \tag{68}$$

A. Beamforming

The beamforming method uses complex weights **w** on the sensors output to achieve maximum power. The array output thus becomes

$$z(t) = \mathbf{w}^H\mathbf{y}(t) \tag{69}$$

$$P_{bf}(\theta) = \mathrm{E}[z(t)z^H(t)] = \mathbf{w}^H\mathrm{E}[\mathbf{y}(t)\mathbf{y}^H(t)]\mathbf{w} = \mathbf{w}^H\mathbf{R}\mathbf{w} \tag{70}$$

For simplicity, we assume that the source comes from the direction of θ_1, then the output power is given by

$$\begin{aligned} P &= \mathrm{E}[\mathbf{w}^H\mathbf{y}(t)\mathbf{y}^H(t)\mathbf{w}] = \mathbf{w}^H\mathrm{E}[\mathbf{y}(t)\mathbf{y}^H(t)]\mathbf{w} \\ &= \mathbf{w}^H\mathrm{E}[(\mathbf{a}(\theta_1)u_1(t) + \mathbf{n}(t))(u_1(t)^H\mathbf{a}^H(\theta_1) + \mathbf{n}^H(t))]\mathbf{w} \\ &= \mathbf{w}^H[\mathbf{a}(\theta_1)\mathrm{E}[u_1(t)u_1(t)^H]\mathbf{a}^H(\theta_1) + \sigma^2\mathbf{I}]\mathbf{w} \\ &= \mathbf{w}^H\mathbf{a}(\theta_1)\mathrm{E}[u_1(t)u_1(t)^H]\mathbf{a}^H(\theta_1)\mathbf{w} + \mathbf{w}^H\sigma^2\mathbf{I}\mathbf{w} \\ &= \left\|\mathbf{w}^H\mathbf{a}(\theta_1)\right\|_2^2\mathrm{E}[u_1(t)u_1(t)^H] + \mathbf{w}^H\mathbf{w}\sigma^2\mathbf{I} \end{aligned} \tag{71}$$

From the above equation, it is observed that when $\mathbf{w} = \mathbf{a}(\theta_1)$, the power is maximum. From the view of physical concept, the maximum power is achieved by steering at the direction from which the waves are arriving.
The normalized form of **w** is given by

$$\mathbf{w} = \frac{\mathbf{a}(\theta)}{\|\mathbf{a}(\theta)\|} \tag{72}$$

Thus, the output power takes the form:

$$P_{bf}(\theta) = \frac{\mathbf{a}^H(\theta)\mathbf{R}\mathbf{a}(\theta)}{\|\mathbf{a}(\theta)\|^2} \tag{73}$$

In practice, the N times snapshots are used to compute the power

$$\hat{P}_{bf}(\theta) = \frac{\mathbf{a}^H(\theta)\hat{\mathbf{R}}\mathbf{a}(\theta)}{\|\mathbf{a}(\theta)\|^2} \tag{74}$$

Where $\hat{\mathbf{R}} = \frac{1}{N}\sum_{t=1}^{N}\mathbf{y}(t)\mathbf{y}^H(t)$

Beamforming is a very simple and robust approach, which is widely used in practice. However, the method performance cannot be improved by increasing SNR or observation time.

B. Minimum variance distortionless response (MVDR)

The classical beamforming method has weights which are independent of the signals and noise.

The idea of MVDR is to use the estimated signal and noise parameters to improve the performance. It attempts to minimize the variance due to noise, while keeping the gain in the direction of steering equal to unity:

$$\mathbf{w}(\theta) = \arg\min_w(\mathbf{w}^H\mathbf{R}\mathbf{w}), \text{ subject to } \mathbf{w}^H\mathbf{a}(\theta) = 1$$

The solution of this optimization problem can be shown to have the following form:

$$\mathbf{w} = \frac{\mathbf{R}^{-1}\mathbf{a}(\theta)}{\mathbf{a}^H(\theta)\mathbf{R}^{-1}\mathbf{a}(\theta)} \tag{75}$$

The resulting spectrum has an expression as:

$$P(\theta) = \mathbf{w}_{opt}\mathbf{R}\mathbf{w}_{opt} = \frac{\mathbf{a}^H(\theta)\mathbf{R}^{-1}\mathbf{R}\mathbf{R}^{-1}\mathbf{a}(\theta)}{[\mathbf{a}^H(\theta)\mathbf{R}^{-1}\mathbf{a}(\theta)]^2} = \frac{1}{\mathbf{a}^H(\theta)\mathbf{R}^{-1}\mathbf{a}(\theta)} \tag{76}$$

The main benefit of this method is a substantial increase in resolution compared with beamforming. The resolution increases without limit as SNR or the observation time are increased. Shortcomings include an increase in the amount of computation compared to beamforming, poor performance with small amounts of time-samples and inability to handle strongly correlated or coherent sources.

C. Multiple signal classification (MUSIC)

The MUSIC method is the most prominent member of the family of eigen-expansion based source location estimators. The underlying idea is to separate the eigenspace of the covariance matrix of sensor outputs into the signal and noise components using the knowledge about the covariance of the noise. The sensor output correlation matrix admits the following decomposition:

$$\mathbf{R} = \mathrm{A}(\theta)\mathrm{P}\mathrm{A}^H(\theta) + \sigma^2\mathbf{I} = \mathbf{U}\mathbf{\Lambda}\mathbf{U}^H = \mathbf{U}_s\mathbf{\Lambda}_s\mathbf{U}_s{}^H + \mathbf{U}_n\mathbf{\Lambda}_n\mathbf{U}_n{}^H \tag{77}$$

$\mathbf{U}$ and $\mathbf{\Lambda}$ form the eigenvalue decomposition of $\mathbf{R}$, and $\mathbf{U}_s, \mathbf{\Lambda}_s$ are the partitions of signal subspace, $\mathbf{U}_n, \mathbf{\Lambda}_n$ are the partitions of noise subspace, $\mathbf{\Lambda}_n$ equals to σ^2 。Provided that $\mathrm{A}(\theta)\mathrm{P}\mathrm{A}^H(\theta)$ has rank *K*. The number of sources, *K* has to be strictly less than the number of sensors *M*. **R** has *K* eigenvalues which are due to the combined signal plus noise subspace, and *M-K* eigenvalues due to the noise subspace. Due to the orthogonality of eigensubspaces corresponding to different eigenvalues for Hermitian matrices, the noise subspace is orthogonal to the direction vector of signals, thus

$$\mathbf{U}_n^H\mathbf{a}(\theta) = 0 \tag{78}$$

MUSIC spectrum is obtained by putting the squared norm of this term into the denominator, which leads to very sharp estimates of the positions of the sources

$$P_{MUS}(\theta) = \frac{1}{\mathbf{a}^H(\theta)\mathbf{U}_n\mathbf{U}_n^H\mathbf{a}(\theta)} \tag{79}$$

In contrast with the previously discussed techniques, MUSIC spectrum has no direct relation to power, also it cannot be used as a beamformer, since the spectrum is not obtained by steering the array. MUSIC provides a consistent estimate of the locations of the sources, as SNR and the number of sensors go to infinity. Despite the dramatic improvement in resolution, MUSIC suffers from a high sensitivity to model errors, such as sensor position uncertainty. Also, the resolution capabilities decrease when the signals are correlated. When some of the signals are coherent, the method fails to work. The computational complexity is dominated by the computation of the eigenexpansion of the covariance matrix.

D. Sparsity angle sensing (SAS)

Many algorithm design challenges arise when the sources are close, Signal to Noise Ratio is low, correlated and coherent sources, less number of time samples. To improve the estimation performance and robustness, sparsity-based signal processing techniques for AOA estimation have been popularity.

To cast this problem into a sparse representation problem, the basic steps are:

1. Construct a known vector $\tilde{\boldsymbol{\theta}}$ which is the expansion of vector $\boldsymbol{\theta}$ considering all possible source locations. Let $\tilde{\boldsymbol{\theta}}$ be filled with N vectors $\{\theta_1,...,\theta_N\}$ which are the possible locations of unknown emitters. $\Delta(\tilde{\theta})$ is the ideal spatial resolution ability.
2. Fill each column of $\tilde{\mathbf{A}}(\tilde{\theta})$ with each potential emitter location: $\tilde{\mathbf{A}} = [\mathbf{a}(\tilde{\theta}_1),\mathbf{a}(\tilde{\theta}_2),...,\mathbf{a}(\tilde{\theta}_N)]$. Suppose the number of sensor arrays is M, the number of all possible emitters is N, the number of real unknown emitters is K. $\tilde{\mathbf{A}}$ is a $M\times N$ matrix. The relationship among M, N and K is $K<M<N$.
3. Reconstruct the output signal model as

$$\mathbf{y}(t) = \hat{\mathbf{A}}(\boldsymbol{\theta})\hat{\mathbf{u}}(t) + \mathbf{n}(t)$$

   Where $\hat{\mathbf{u}}(t) = [u_1(t),u_2(t),...u_N(t)]$ represents a N virtual transmitted signal vector in which only the signal corresponding to the true angle directions are non-zero, other directional signals are zero, which means $\|\hat{\mathbf{u}}(t)\|_0 = K$

4. The nonzero elements of $\hat{\mathbf{u}}(t) = [u_1(t),u_2(t),...u_N(t)]$ corresponds to the estimated AOAs. Thus, the AOA estimation could be transfered to the estimation o $\hat{\mathbf{u}}(t)$. First to solve $\hat{\mathbf{u}}(t)$ and then get the AOAs.
5. $\hat{\mathbf{u}}(t)$ can be solved by l_1-denoise optimization algorithm with quadratic constraints.

$$\min\|\hat{\mathbf{u}}(t)\|_1 \text{ subject to } \|\mathbf{y}(t) - \hat{\mathbf{A}}(\boldsymbol{\theta})\hat{\mathbf{u}}(t)\|_2 \le \varepsilon \tag{80}$$

   where $\|\hat{\mathbf{u}}(t)\|_1 = \sum_{i=1}^{K}|u_i(t)|$, ε is the error threshold.

Several experimental results are shown by comparing four AOA estimation approaches: BF, MVDR, MUSIC and SAS.

In Fig.36., the SNR is -7dB and the distance between two sources are 20°, all the methods are able to solve the two sources. In Fig.37., the SNR is still -7dB, but the distance between two sources are close to 10° separation, the BF method begins to merge the two peaks while the other three methods are able to solve two sources. In Fig.38., when the two sources are close to 4° separation, SNR is -1dB, BF, MVDR and MUSIC all emerge except SAS. Four approaches are compared and the results demonstrate that SAS outperforms the other three approaches in terms of robustness and spatial resolution.

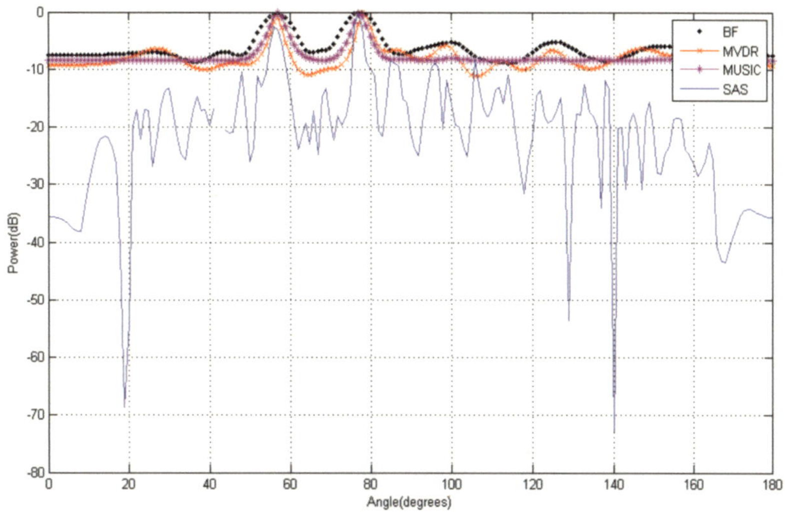

Fig. 36. Spatial spectra for BF, MVDR, MUSIC and SAS. SOAs: 57°and 77°.SNR=-7dB

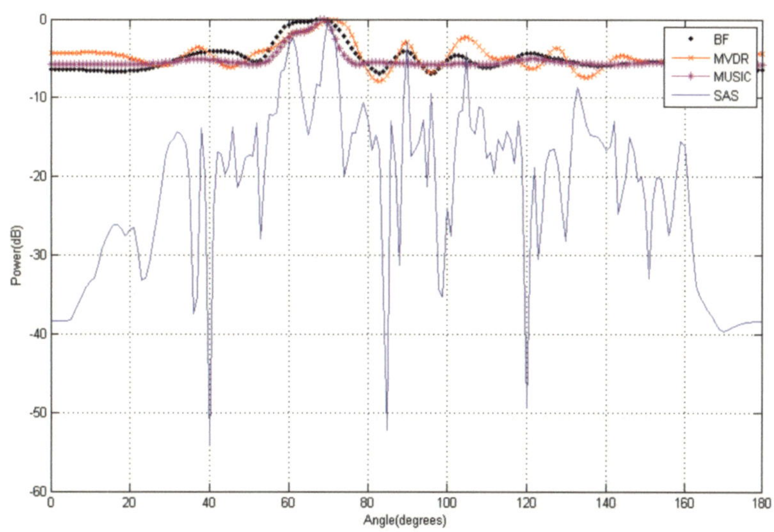

Fig. 37. Spatial spectra for BF, MVDR, MUSIC and SAS. AOAs: 60°and 70°.SNR=-7dB

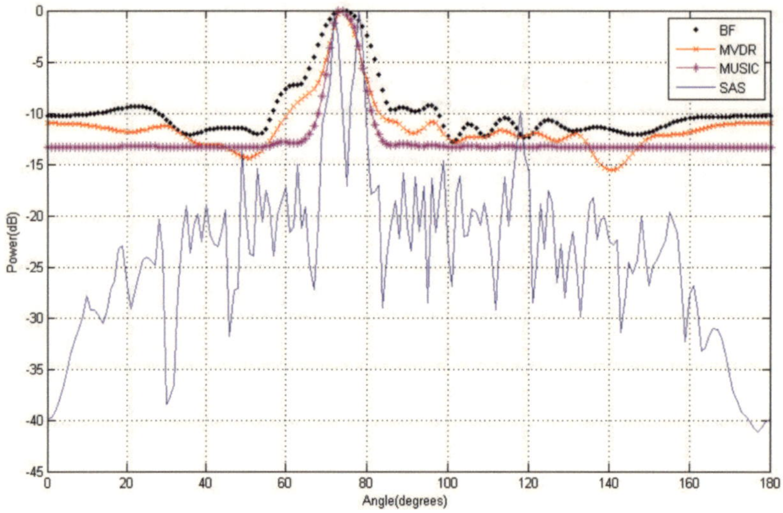

Fig. 38. Spatial spectra for BF, MVDR, MUSIC and SAS. AOAs: 73°and 77°,SNR=-1dB

4. Conclusion

This chapter is focused on the design and analysis of wireless positioning systems. An overview of basic principles, latest developed systems and state of the art signal processing techniques for wireless positioning are presented. This chapter aims to provide the concepts related to localization systems as well as the methods to localize terminals in different wireless networks. As an important part of the chapter, potential challenges and new techniques for wireless positioning are provided to the readers. The authors hope that this chapter will help readers identify the key technical challenges in wireless positioning and be interested in this emerging area.

5. Acknowledgement

This work was supported by Campus Talent Grant No. W10RC00040, the National Science and Technology Major Project Grant No.2009ZX03003-008-02 and National Science Foundation No.61071075.

6. References

Alavi, B. & Pahlavan, K. (2006). Modeling of the TOA-based distance measurement error using UWB indoor radio measurements. *Communications Letters, IEEE*, Vol.10, No.4, pp.275-277, ISSN 1089-7798

Barabell A.J. (1983). Improving the resolution performance of eigenstructure based direction-finding algorithms. *IEEE international conference on acoustics, speech, and signal processing*, pp. 336-339

Candès, E. J. & Wakin, M. B. (2008). An introduction to compressive sampling. *IEEE Signal Processing Magazine*, Vol.25, No.2, pp. 21-30

Catovic, A.Sahinoglu, Z. (2004). The Cramer-Rao bounds of hybrid TOA/RSS and TDOA/RSS location estimation schemes. *Communications Letters, IEEE,* Vol.8, No.10, pp.626-628, ISSN 1089-7798

Chan Y.T.& Ho K.C. (1994). A simple and efficient estimator for hyperbolic location. *IEEE Trans on Signal Processing,* Vol. 42, No.8, pp. 1905-1915

Cong, L. & Weihua Z. (2001). Non-line-of-sight error mitigation in TDOA mobile location. *GLOBECOM '01. IEEE* , Vol.1, No.pp.680-684

Candès, E. J.; Romberg J.& Tao T. (2006). Robust uncertainty principles: Exact signal reconstruction from highly incomplete frequency information. *IEEE Trans. Inform. Theory,* Vol.2,No.2, pp.489-509

Fang B.T. (1990). Simple solutions for hyperbolic and related position fixes. *IEEE Trans on AES,* Vol. 26, No.9, pp.748-753

Friedlander B. (1987). A passive localization algorithm and its accuracy analysis. *IEEE J. Ocean. Eng.,* Vol.12, No.1, pp. 234-245

Foy W. H., (1976) Position-location solutions by Taylor-series estimation . *IEEE Trans on AES,* Vol.12, No.3, pp. 187-194

Gezici, S. et al. (2005). Localization via ultra-wideband radios: a look at positioning aspects for future sensor networks. *Signal Processing Magazine, IEEE,* Vol.22, No.4, pp.70-84, ISSN 1053-5888

Gershman, A.B. (2003).Robust adaptive beamforming: an overview of recent trends and advances in the field. *International Conference on Antenna Theory and Techniques,* pp.30-35, Sevastopol, Ukraine

Guolin, S. (2005). Signal processing techniques in network-aided positioning: a survey of state-of-the-art positioning designs. *Signal Processing Magazine, IEEE,* Vol.22, No.4, pp.12-23, ISSN 1053-5888

Jian Z.; Durgin, G.D. (2005). Indoor/outdoor location of cellular handsets based on received signal strength. *Vehicular Technology Conference,* Vol. 1, pp. 92 - 96

Kleine-Ostmann T. & Bell A.E. (2001). A data fusion architecture for enhanced position estimation in wireless networks. *IEEE Communications Letters,* Vol.5, No.8, pp. 343-345

Porretta, M. et al. (2004). A novel single base station location technique for microcellular wireless networks: description and validation by a deterministic propagation model. Vehicular Technology, *IEEE Transactions on,* Vol. 53, No.5, pp. 1502-1514, ISSN 0018-9545

Pahlavan, K.; Xinrong, Li & Makela, J. P. (2002). Indoor geolocation science and technology. *Communications Magazine, IEEE,* Vol.40, No.2, pp. 112-118, ISSN 0163-6804

Porcino, D. & Hirt, W. (2003). Ultra-wideband radio technology: potential and challenges ahead. *Communications Magazine, IEEE,* Vol.41, No., pp.66-74, ISSN 0163-6804

Rappaport, T.S.; Reed J.H. (1996). Position Location Using Wireless Communications on Highways of the Future. *IEEE Communications Magazine,* Vol.34, No.10, pp.33-41

Reed, J. H. et al. (1998). An overview of the challenges and progress in meeting the E-911 requirement for location service. *Communications Magazine, IEEE,* Vol.36, No.4, pp.30-37, ISSN0163-6804

Rieken, D. W. & Fuhrmann, D. R. (2004). Generalizing MUSIC and MVDR for multiple noncoherent arrays. *Signal Processing, IEEE Transactions on,* Vol. 52, No. 9, pp. 2396-2406, ISSN 1053-587X

Roy S., et al. (2004). Ultrawideband radio design: the promise of high-speed, short-range wireless connectivity. Proceedings of the IEEE, Vol.92, No. 2, pp. 295-311

Richard B. (2007). Compressive sensing. *IEEE Signal Processing Magzine,* Vol.24, No.4, pp.118-121.

Raleigh G. & Boros T. (1998). Joint space-time parameter estimation for wireless communication channels. *IEEE trans. on signal processing,* Vol. 46, pp.1333-1343

Stoica, Z. Wang, J. (2003). Robust capon beamforming. *IEEE Signal Processing Letters,* Vol.10, June 2003

Swindlehurst, A. Kailath, T. (1992).A performance analysis of subspace-based methods in the presence of model Errors-Part 1: the MUSIC algorithm. *IEEE Trans. Signal Processing,* Vol.40, No.7,pp.1758-1774,July

Sayed, A. H.; Tarighat, A.& Khajehnouri, N. (2005). Network-based wireless location: challenges faced in developing techniques for accurate wireless location information. *Signal Processing Magazine, IEEE,* Vol.22, No.4, pp.24-40, ISSN1053-5888

Thomas N.; Cruickshank D. & Laurenson D. (2001). Performance of a TDOA/AOA hybrid mobile location system. *International conference on 3G mobile communication technologies,* pp. 216-220

Vaughan-Nichols, S. J.(2009). Will Mobile Computing's Future Be Location, Location, Location? *IEEE Computer,*Vol.42, No.2, pp. 14-17, ISSN 0018-9162

Vanderveen M.; Papadias C. & Paulraj A. (1997). Joint angle and delay estimation for multipath signals arriving at an antenna array. *IEEE communication letters,* Vol. 1, pp. 12-14

Van, B.D. & Buckley, K.M. (1998). Beamforming: a versatile approach to spatial filtering. *IEEE ASSP Magzine,* Vol.5, pp.4-24

Vaidyanathan C.& Buckley, K.M. (1995). Comparative studies of MUSIC and MVDR location estimators for model perturbations. *IEEE International Conference on Acoustics, Speech and Signal Processing,* Vol.3, pp.9-12,1995

Wax M.; Leshem A. (1997). Joint estimation of time delays and directions of arrival of multiple relections of a known signal. *IEEE trans. on signal processing,* Vol. 45, PP.2477-2484

Wylie, M.P. & Holtzman, J. (1996). The non-line of sight problem in mobile location estimation. *5th IEEE International Conference on Universal Personal Communications,* Vol.2, No.29, pp. 827-831

Zagami, J. M. (1998). Providing universal location services using a wireless E911 location network. *Communications Magazine, IEEE,* Vol.36, No.4, pp. 66-71, ISSN 0163-6804

Zhao, L.; Yao, G.&Mark, J.W. (2006). Mobile positioning based on relaying capability of mobile stations in hybrid wireless networks. *IEE Proceedings on Communications,* Vol. 153, No.5, pp.762-770

2

Positioning in Cellular Networks

Mirjana Simić and Predrag Pejović
University of Belgrade
Serbia

1. Introduction

Cellular networks are primarily designed to provide communication to mobile users. Besides the main application, determining location of mobile users (stations) within the cellular networks like Global System for Mobile Communications (GSM) and Universal Mobile Telecommunications System (UMTS) became an interesting additional feature. To provide the location based services (LBS), radio communication parameters already available in the network are preferably used, while some methods require investment in additional hardware to improve precision of the positioning. Positioning methods applied in cellular networks are characterized by tradeoff between the positioning precision and the requirements for additional hardware.

The idea to determine user location in cellular networks originated in the USA to support 911 service for emergency calls. The Federal Communications Commission (FCC) in 1996 initiated a program in which mobile operators are required to provide automatic location determination with specified accuracy for the users that make emergency calls. The new service is named Enhanced 911 (E-911). Similar service was initiated in Europe somewhat later, and it is called E-112. Besides the security related applications, availability of the user location information in cellular networks opened significant commercial opportunities to mobile operators.

In this chapter, methods to determine the mobile station position according to the radio communication parameters are presented. Position related radio communication parameters and their modeling are discussed, and algorithms to process collected data in order to determine the mobile station position are presented. Finally, standardized positioning methods are briefly reviewed.

2. Position related parameters

2.1 Received signal strength

According to wave propagation models (Rappaport, 2001), the received signal power may be related to distance r between the mobile station and the corresponding base station by

$$P(r) = P(r_0) \left(\frac{r_0}{r}\right)^m \qquad (1)$$

where m is the path loss exponent, r_0 is the distance to a reference point, and $P(r_0)$ is the power at the reference distance, i.e. the reference power, obtained either by field measurements at r_0

or using the free space equation

$$P(r_0) = \frac{\lambda^2}{(4\pi)^2\, r_0^2}\, P_t\, G_t\, G_r \tag{2}$$

where λ is the wavelength, P_t is the transmitted power, G_t is the transmitter antenna gain, and G_r is the receiver antenna gain. According to (2), the received signal power depends on the transmitter antenna gain, which is dependent on the mobile station relative angular position to the transmitter antenna. Also, (2) assumes direct wave propagation. In the case the wave propagation is direct, and the antenna gain is known, (1) may be used to determine the distance between the mobile station and the base station from

$$r = r_0 \left(\frac{P(r_0)}{P(r)} \right)^{\frac{1}{m}} \tag{3}$$

which constitutes deterministic model of the received signal strength as a position related parameter.

The information provided by (3) may be unreliable in the case the antennas have pronounced directional properties and/or the propagation is not line-of-sight. In that case, an assumption that the received signal power cannot be larger than in the case the antennas are oriented to achieve the maximal gain and the wave propagation is direct may be used. Received signal power under this assumption locates the mobile station within a circle centered at the base station, with the radius specified by (3). This results in a probability density function

$$p_P(x, y) = \begin{cases} \frac{1}{\pi r^2} & \text{for } (x - x_{BS})^2 + (y - y_{BS})^2 \le r^2 \\ 0 & \text{elsewhere} \end{cases} \tag{4}$$

where x_{BS} and y_{BS} are coordinates of the base station, and r is given by (3). This constitutes probabilistic model of the received signal power as a position related parameter.

2.2 Time of arrival

Another parameter related to mobile station location is the time of arrival, i.e. the signal propagation time. This parameter might be extracted from some parameters already measured in cellular networks to support communication, like the timing advance (TA) parameter in GSM and the round trip time (RTT) parameter in UMTS. Advantage of the time of arrival parameter when used to determine the distance between the mobile station and the base station is that it is not dependent on the whether conditions, nor on the angular position of the mobile station within the radiation pattern of the base station antenna, neither the angular position of the mobile station to the incident electromagnetic wave. However, the parameter suffers from non-line-of-sight propagation, providing false information of the distance being larger than it actually is. In fact, the time of arrival provides information about the distance wave traveled, which corresponds to the distance between the mobile station and base station only in the case of the line-of-sight propagation.

To illustrate both deterministic and probabilistic modeling of the information provided by the time of arrival, let us consider TA parameter of GSM systems. Assuming direct wave propagation, information about the coordinates of the base station (x_{BS}, y_{BS}) and the corresponding TA parameter value TA localize the mobile station MS, (x_{MS}, y_{MS}), within an annulus centered at the base station specified by

$$TA\, R_q \le r \le (TA + 1)\, R_q \tag{5}$$

where r is the distance between the base station and the mobile station

$$r = \sqrt{\left(x_{MS} - x_{BS}\right)^2 + \left(y_{MS} - y_{BS}\right)^2} \tag{6}$$

and $R_q = 553.46$ m is the TA parameter distance resolution quantum, frequently rounded to 550 m. The annulus is for $TA = 2$ shown in Fig. 1.

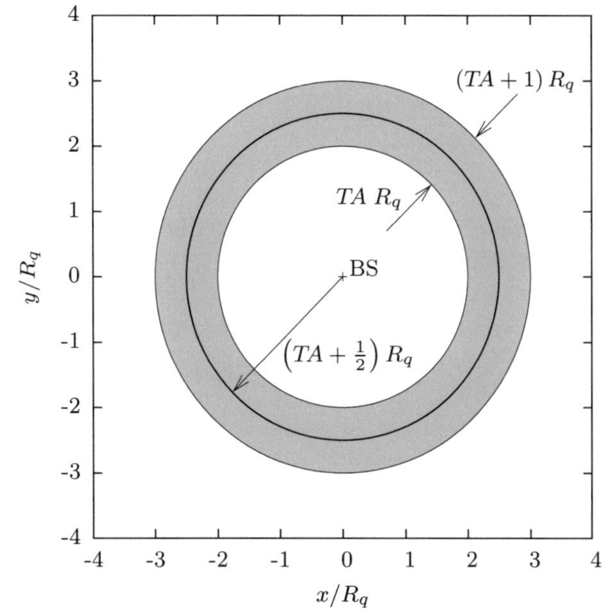

Fig. 1. Position related information derived from TA parameter, $TA = 2$.

In a probabilistic model, the mobile station localization within the annulus is represented by the probability density function

$$p_{T1}(x, y) = \begin{cases} \dfrac{1}{\pi\,(2\,TA + 1)\,R_q^2} & \text{for } TA\,R_q \leq r \leq (TA + 1)\,R_q \\ 0 & \text{elsewhere.} \end{cases} \tag{7}$$

The probability density function of (7) assumes direct wave propagation, which is a reasonable assumption in some rural environments. However, in urban environments, as well as in some rural environments, indirect propagation of waves might be expected. As detailed in (Simić & Pejović, 2009), in environments where indirect wave propagation might be expected the TA parameter value guarantees only that the mobile station is located within a circle

$$r \leq (TA + 1)\,R_q. \tag{8}$$

In absence of a better model, uniform distribution within the circle might be assumed, resulting in the probability density function

$$p_{T2}(x, y) = \begin{cases} \dfrac{1}{\pi\left((TA + 1)\,R_q\right)^2} & \text{for } r \leq (TA + 1)\,R_q \\ 0 & \text{elsewhere.} \end{cases} \tag{9}$$

For both of the probability density functions, the area where the probability density functions might take nonzero value is limited to a square

$$x_{BS} - (TA + 1)\, R_q \leq x \leq x_{BS} + (TA + 1)\, R_q \tag{10}$$

and

$$y_{BS} - (TA + 1)\, R_q \leq y \leq y_{BS} + (TA + 1)\, R_q \tag{11}$$

which will be used to join the data collected from various information sources.
Choice of the probability density function that represents the information about the base station coordinates and the TA parameter value depends on the environment. The probability density function (7) should be used where the line-of-sight propagation is expected, while (9) should be used otherwise.
Deterministic model of the mobile station position information contained in the TA parameter value is much simpler. Assuming uniform distribution within the mobile station distance limits, and assuming the line-of-sight propagation, the distance between the mobile station and the base station is estimated as

$$r = \left(TA + \frac{1}{2}\right) R_q. \tag{12}$$

The circle of possible mobile station location that results from $TA = 2$ is shown in Fig. 1.

2.3 Time difference of arrival

To measure wave propagation time, clocks involved in the measurement should be synchronized. Term "synchronization" when used in this context means that information about a common reference point in time is available for all of the synchronized units. The requirement may be circumvented in time of arrival measurements if the round trip time is measured, which requires only one clock. Also, the requirement for mobile stations to be synchronized is avoided when the time difference of signal propagation from two base stations to the mobile station is measured. In this case, offset in the mobile station clock is canceled out, and only the base stations are required to be synchronized. The time difference of arrival might be extracted from time measurements on the Broadcast Control Channel (BCCH) or Traffic Channel (TCH) in GSM, or from SFN-SFN (System Frame Number) observed time difference measurements on the Common Pilot Channel (CPICH) in UMTS. Measured difference of the time of signal propagation results in information about the difference in distances between the mobile station and the two participating base stations. Value of the information provided by the time difference of arrival is not sensitive on the signal propagation loss, neither on the mobile station angular position, but suffers from non-line-of-sight wave propagation.

2.4 Angle of arrival

Historically, angle of arrival was the first parameter exploited to determine position of radio transmitters, as utilized in goniometric methods. The angle of signal arrival might be determined applying direction sensitive antenna systems. Application of specific antenna systems is the main drawback for application in cellular networks, since specific additional hardware is required. Besides, the information of the angle of arrival is not included in standardized measurement reports in cellular networks like GSM and UMTS. To extract useful information from the angle of arrival, line-of-sight propagation is required, again. Due to the drawbacks mentioned, positioning methods that utilize this parameter are not standardized for positioning applications in cellular networks yet.

3. Position estimation

After the position related parameters are collected, position of the mobile station is determined by joining of the collected data applying some of the available methods. The methods can be classified as deterministic, probabilistic, and fingerprinting.

3.1 Deterministic methods

Deterministic methods apply geometric relations to determine position of the mobile station according to known coordinates of the base stations and distances and/or angles extracted from the radio parameters. The extracted distances and/or angles are treated as known, and uncertainty and/or inconsistency of the data are observed only when redundant measurements are available. In this section, geometric parameters extracted from the radio measurements and known coordinates of the base stations are related to the coordinates of the mobile station. It is assumed that the base stations, as well as the mobile station, are located in the same plane, i.e. that the problem is two-dimensional. All of the equations are derived for the two-dimensional case, and generalization to the three-dimensional case is outlined.

3.1.1 Angulation

To determine coordinates (x_{MS}, y_{MS}) of a mobile station (MS) applying angulation method, at least two base stations are needed, BS1 and BS2, and their coordinates $(x_{BSk}, y_{BSk}), k \in \{1, 2\}$ should be known. The only information base stations provide are the angles $\varphi_k, k \in \{1, 2\}$ the rays (half-lines) that start from the base station BS*k* and point towards the mobile station form with the positive ray of the x-axis, $y = 0$, $x > 0$. The angles are essentially azimuth angles, except the azimuth angles are referred to the north, and the positive ray of the x-axis points to the east. The choice is made to comply with common notation of analytical geometry. The angle measurement is illustrated in Fig. 2, where the mobile station located at $(x_{MS}, y_{MS}) = (5, 5)$ is observed from three base stations, $(x_{BS1}, y_{BS1}) = (3, 5)$ with $\varphi_1 = 0$, $(x_{BS2}, y_{BS2}) = (5, 2)$ with $\varphi_2 = 90°$, and $(x_{BS3}, y_{BS3}) = (9, 8)$ with $\varphi_3 = -143.13°$.

Coordinates of the base stations and the mobile station observation angles locate the mobile station on a line

$$\frac{y_{MS} - y_{BSk}}{x_{MS} - x_{BSk}} = \tan \varphi_k \tag{13}$$

which can be transformed to

$$y_{MS} - x_{MS} \tan \varphi_k = y_{BSk} - x_{BSk} \tan \varphi_k \tag{14}$$

if $\varphi_k \neq \pi/2 + n\pi, n \in Z$, i.e. $x_{BSk} \neq x_{MS}$. In the case $\varphi_k = \pi/2 + n\pi$ the equation degenerates to

$$x_{MS} = x_{BSk}. \tag{15}$$

The observation angle provides more information than contained in (13), locating the mobile station on the ray given by (14) and $x_{MS} > x_{BS_k}$ for $-\pi/2 < \varphi_k < \pi/2$, or on the ray $x_{MS} < x_{BS_k}$ for $-\pi < \varphi_k < -\pi/2$ or $\pi/2 < \varphi_k < \pi$. This might be used as a rough test of the solution consistency in the case of ill-conditioned equation systems.

To determine the mobile station coordinates, at least two base stations are needed. In general, two base stations $k \in \{1, 2\}$ form the equation system

$$\begin{bmatrix} \tan \varphi_1 & -1 \\ \tan \varphi_2 & -1 \end{bmatrix} \begin{bmatrix} x_{MS} \\ y_{MS} \end{bmatrix} = \begin{bmatrix} x_{BS1} \tan \varphi_1 - y_{BS1} \\ x_{BS2} \tan \varphi_2 - y_{BS2} \end{bmatrix} \tag{16}$$

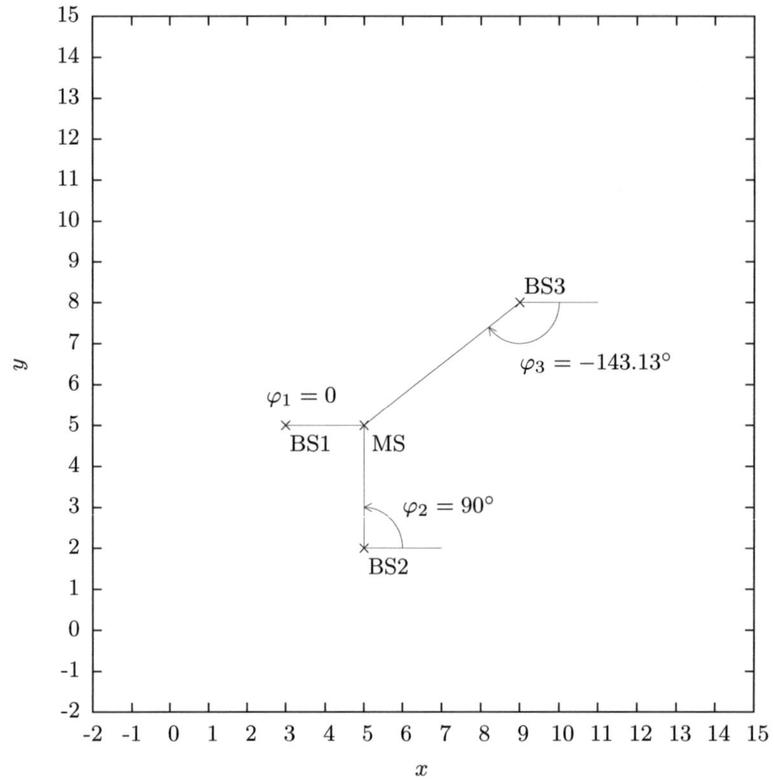

Fig. 2. Angulation.

assuming finite values for $\tan\varphi_k$. In the opposite case, corresponding equation should be replaced by an equation of the form (15).

In the case $\tan\varphi_1 = \tan\varphi_2$, the base stations and the mobile station are located on the same line, and the equation system (16) is singular. An additional base station is needed to determine the mobile station coordinates, but it should not be located on the same line as the two base stations initially used. Furthermore, mobile station positions close to the line defined by the two base stations result in ill-conditioned equation system (16). This motivates introduction of additional base stations, and positions of three or more base stations on the same line, or close to a line, should be avoided.

In the example of Fig. 2, three base stations are available, and taking any two of the base stations to form (16) correct coordinates of the mobile station are obtained, since the systems are well-conditioned and the data are free from measurement error. If BS2 is involved, equation of the form (15) should be used.

In practice, more than two base stations might be available, and an overdetermined equation system might be formed,

$$\begin{bmatrix} \tan\varphi_1 & -1 \\ \vdots & \vdots \\ \tan\varphi_n & -1 \end{bmatrix} \begin{bmatrix} x_{MS} \\ y_{MS} \end{bmatrix} = \begin{bmatrix} x_{BS1}\tan\varphi_1 - y_{BS1} \\ \vdots \\ x_{BSn}\tan\varphi_n - y_{BSn} \end{bmatrix} \qquad (17)$$

where $n \in N$ and $n \geq 2$. The system (17) may be written in a matrix form

$$\mathbf{A}\begin{bmatrix} x_{MS} \\ y_{MS} \end{bmatrix} = \mathbf{b}. \tag{18}$$

The system (18) can be solved in a least-squares sense (Bronshtein et al., 2007), (Press et al., 1992) forming the square system

$$\mathbf{A}^T\mathbf{A}\begin{bmatrix} x_{MS} \\ y_{MS} \end{bmatrix} = \mathbf{A}^T\mathbf{b}. \tag{19}$$

To determine the mobile station location in three dimensions, coordinates of at least two base stations should be available in three-dimensional space, as well as two observation angles, the azimuth and the elevation angle. With the minimum of two base stations, two measured angles result in an overdetermined equation system over three mobile station coordinates. In practice, the two rays defined by their azimuth and elevation angles would hardly provide an intersection, due to the presence of measurement errors. Thus, linear least-squares solution (19) should be used even in the case only two base stations are considered. Let us underline that in three-dimensional case two base stations are still sufficient to determine the mobile station position.

A similar technique is applied in surveying, frequently referred to as "triangulation", since the object position is located in a triangle vertex, while the remaining two vertexes of the triangle are the base stations. Two angles are measured in order to determine the object position. The angles are frequently measured relative to the position of the other base station. Having the coordinates of the base stations known, the angles can be recalculated and expressed in the terms used here.

3.1.2 Circular lateration

Circular lateration is a method based on information about the distance r_k of the mobile station (MS) from at least three base stations BSk, $k \in \{1, \ldots n\}$, $n \geq 3$. Coordinates (x_{BSk}, y_{BSk}) of the base stations are known. An example for circular lateration using the same coordinates of the base stations and the mobile station as in the angulation example is presented in Fig. 3, where information about the mobile station position is contained in distances r_k instead of the angles φ_k.

Let us consider a minimal system of equations for circular lateration

$$(x_{MS} - x_{BSk})^2 + (y_{MS} - y_{BSk})^2 = r_k^2 \tag{20}$$

for $k \in \{1, 2, 3\}$. The equation system is nonlinear. According to the geometrical interpretation depicted in Fig. 3, each of the equations represents a circle, centered at the corresponding base station, hence the name of the method—circular lateration. If the system is consistent, each pair of the circles provides two intersection points, and location of the mobile station is determined using the information provided by the third base station, indicating which of the intersection points corresponds to the mobile station location. The problem becomes more complicated in the presence of measurement uncertainties, making exact intersection of three circles virtually impossible. An exception from this situation is the case where the two base stations and the mobile station are located on a line, resulting in tangent circles. In this case, the third base station won't be needed to determine the mobile station

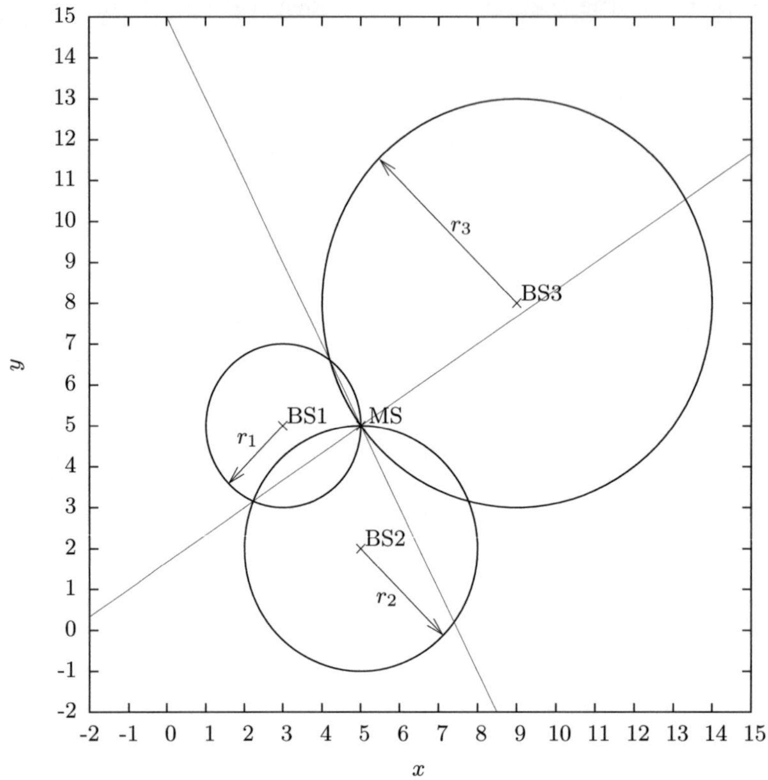

Fig. 3. Circular lateration.

position. Due to measurement uncertainty, it is also possible that measured distances result in circles that do not intersect.

The nonlinear system of equations (20) could be transformed to a linear system of equations (Bensky, 2008) applying algebraic transformations. This removes problems associated with solution methods for nonlinear equations and ambiguities about the mobile station location in the case the circle intersections that do not match. The first step in algebraic transformations is to expand the squared binomial terms

$$x_{MS}^2 - 2\,x_{MS}\,x_{BSk} + x_{BSk}^2 + y_{MS}^2 - 2\,y_{MS}\,y_{BSk} + y_{BSk}^2 = r_k^2. \tag{21}$$

Next, all squared terms are moved to the right-hand side

$$-\,2\,x_{MS}\,x_{BSk} - 2\,y_{MS}\,y_{BSk} = r_k^2 - x_{BSk}^2 - y_{BSk}^2 - x_{MS}^2 - y_{MS}^2. \tag{22}$$

Up to this point, equations of the form (20) were subjected to transformation separately. Now, let us add the equation for $k = 1$ multiplied by -1

$$2\,x_{MS}\,x_{BS1} + 2\,y_{MS}\,y_{BS1} = -r_1^2 + x_{BS1}^2 + y_{BS1}^2 + x_{MS}^2 + y_{MS}^2 \tag{23}$$

to the remaining two equations. Terms x_{MS}^2 and y_{MS}^2 on the right hand side are cancelled out, resulting in a linear system of two equations in the form

$$2\,(x_{BS1} - x_{BSk})\,x_{MS} + 2\,(y_{BS1} - y_{BSk})\,y_{MS} = r_k^2 - r_1^2 + x_{BS1}^2 - x_{BSk}^2 + y_{BS1}^2 - y_{BSk}^2 \tag{24}$$

for $k \in \{2, 3\}$. Each of the equations of the form (24) determines a line in the (x, y) plane. The line equation is formed manipulating the equations of the circles centered at BS1 and BSk, and it is satisfied at both intersections of the circles, if the intersections exist. Thus, the line obtained from the two circle equations passes through the circle intersections. This geometrical interpretation is illustrated in Fig. 3 for both of the line equations (24).

The system of equations (24) could be expressed in a matrix form

$$\begin{bmatrix} x_{BS1} - x_{BS2} & y_{BS1} - y_{BS2} \\ x_{BS1} - x_{BS3} & y_{BS1} - y_{BS3} \end{bmatrix} \begin{bmatrix} x_{MS} \\ y_{MS} \end{bmatrix} = \frac{1}{2} \begin{bmatrix} r_2^2 - r_1^2 + x_{BS1}^2 - x_{BS2}^2 + y_{BS1}^2 - y_{BS2}^2 \\ r_3^2 - r_1^2 + x_{BS1}^2 - x_{BS3}^2 + y_{BS1}^2 - y_{BS3}^2 \end{bmatrix}. \quad (25)$$

In the case all three of the base stations are located on the same line, their coordinates satisfy

$$\frac{y_{BS1} - y_{BS2}}{x_{BS1} - x_{BS2}} = \frac{y_{BS1} - y_{BS3}}{x_{BS1} - x_{BS3}} \quad (26)$$

which results in a singular system (25) since the determinant of the system matrix is zero. In that case, measurements from additional base stations should be used to determine the mobile station location, similar to the case the angulation method is applied and a singular system is reached. Also, base stations located in a close to a line arrangement result in ill-conditioned system (25) and huge sensitivity on the distance measurement error. Again, to avoid such situation additional base stations are needed, and their close to a line arrangement should be avoided as much as possible.

In the case measurements from n base stations, $n \geq 3$, are available, the system of equations is overdetermined and it takes a matrix form

$$\begin{bmatrix} x_{BS1} - x_{BS2} & y_{BS1} - y_{BS2} \\ \vdots & \vdots \\ x_{BS1} - x_{BSn} & y_{BS1} - y_{BSn} \end{bmatrix} \begin{bmatrix} x_{MS} \\ y_{MS} \end{bmatrix} = \frac{1}{2} \begin{bmatrix} r_2^2 - r_1^2 + x_{BS1}^2 - x_{BS2}^2 + y_{BS1}^2 - y_{BS2}^2 \\ \vdots \\ r_n^2 - r_1^2 + x_{BS1}^2 - x_{BSn}^2 + y_{BS1}^2 - y_{BSn}^2 \end{bmatrix}. \quad (27)$$

This system takes the same compact form as (18), and should be solved as (19).

Presented analysis is performed under the assumption of consistent data, resulting in intersection of all circles in a single point. However, even in the case when only three base stations are considered (25) it is likely that the three circles do not intersect in the same point. The mobile station location is determined as an intersection of lines defined by intersections of pairs of circles. Obtained solution may be checked for compliance with the starting circle equations (20) to verify the solution and to estimate the error margins. This also applies in the case the linear least-squares method of (19) is applied to solve (27).

Presented method of linearization of the equations for the circular lateration method could be generalized to three-dimensional case in a straightforward manner. An additional base station would be needed to provide enough data to determine three unknown coordinates of the mobile station. The equation system would take the form similar to (25) and (27).

3.1.3 Hyperbolic lateration

Hyperbolic lateration is a method to determine the mobile station location applying information about differences in distance of the mobile station to a number of pairs of base stations with known coordinates. To determine the mobile station position this would require at least four base stations and information about at least three differences in distance.

To get acquainted with hyperbolic lateration, let us consider a case when the first base station (BS1) is located at $(x_{BS1}, y_{BS1}) = (c, 0)$, while the second base station (BS2) is located at

$(x_{BS2}, y_{BS2}) = (-c, 0)$. It is assumed that $c > 0$. Arbitrary locations of two base stations could be transformed to this case applying translation and rotation of the coordinates, as it will be discussed in detail later. The distance between the base stations is $D = 2c$. Let us also assume that a mobile station (MS) located at (x_{MS}, y_{MS}) is for d more distant from BS2 than from BS1, i.e.

$$r_2 - r_1 = d \tag{28}$$

where

$$r_1 = \sqrt{(x_{MS} - c)^2 + y_{MS}^2} \tag{29}$$

and

$$r_2 = \sqrt{(x_{MS} + c)^2 + y_{MS}^2} \tag{30}$$

are the distances between the mobile station and base stations BS1 and BS2, respectively. It should be noted that d might either be positive or negative, being positive in cases when BS1 is closer to MS, and negative in the opposite case. According to the triangle inequality

$$r_2 + 2c > r_1 \tag{31}$$

and

$$r_1 + 2c > r_2 \tag{32}$$

which after algebraic manipulations limits d to the interval

$$-2c < d < 2c. \tag{33}$$

To provide convenient notation, let us introduce

$$a = \frac{d}{2} \tag{34}$$

which is according to (33) limited to

$$-c < a < c. \tag{35}$$

After the notation is introduced, the equation for the distance difference (28) becomes

$$\sqrt{(x_{MS} + c)^2 + y_{MS}^2} - \sqrt{(x_{MS} - c)^2 + y_{MS}^2} = 2\,a. \tag{36}$$

To remove the radicals, (36) has to be squared twice. After the squaring and after some algebraic manipulation, (36) reduces to

$$\left(c^2 - a^2\right) x_{MS}^2 - a^2\, y_{MS}^2 - a^2 \left(c^2 - a^2\right) = 0. \tag{37}$$

At this point it is convenient to introduce parameter b as

$$b^2 = c^2 - a^2 \tag{38}$$

which reduces (37) to

$$b^2\, x_{MS}^2 - a^2\, y_{MS}^2 - a^2\, b^2 = 0. \tag{39}$$

For $a \neq 0$ and $b \neq 0$, (39) might be expressed in a standard form

$$\frac{x_{MS}^2}{a^2} - \frac{y_{MS}^2}{b^2} = 1 \tag{40}$$

which is the equation that defines a pair of hyperbolas (Bronshtein et al., 2007).
Out of the two hyperbolas defined by (40), one is located in the right half-plane, $x > 0$, which corresponds to the positive distance difference, $d > 0$, while the one located in the left half-plane corresponds to $d < 0$. In the case $d = 0$, which implies $a = 0$, according to (39) the hyperbolas degenerate to a line

$$x_{MS} = 0 \tag{41}$$

which specifies the set of points equally distant from BS1 and BS2.
Parametric description of a hyperbola that satisfies (39) is given by

$$x_{MS} = a\ \cosh t \tag{42}$$

and

$$y_{MS} = b\ \sinh t \tag{43}$$

where t is a dummy variable. For $d > 0$, i.e. $a > 0$, the parametrically specified hyperbola is located in the right half-plane, while for $d < 0$ it is located in the left half-plane, which meets the requirements of the physical model. Besides, the parametric description covers the degenerate case $d = 0$, which is of interest in practice. In this manner, the set of points that are for d more distant from BS2 than from BS1 is specified by

$$x_{MS} = \frac{d}{2}\ \cosh t \tag{44}$$

and

$$y_{MS} = \frac{1}{2}\sqrt{D^2 - d^2}\ \sinh t \tag{45}$$

where D is the distance between the base stations

$$D = \sqrt{(x_{BS1} - x_{BS2})^2 + (y_{BS1} - y_{BS2})^2}. \tag{46}$$

This notation is convenient to plot the hyperbolas.
Far from the hyperbola center, which in the considered case is located at the origin $(0, 0)$, in non-degenerate case $d \neq 0$ the hyperbola is approximated by its asymptotic rays

$$y = \pm\frac{b}{a}\ x = \pm x\sqrt{\left(\frac{D}{d}\right)^2 - 1} \tag{47}$$

which applies for $x > 0$ if $d > 0$ or for $x < 0$ if $d < 0$. Approximation of the hyperbola with these two asymptotic rays is helpful in estimating the number of intersections of two hyperbolas. Besides, it states that in the area far from the base stations, $r_1, r_2 \gg D$, the information about the distance difference reduces to the angle of signal arrival, with some ambiguity regarding the asymptote that corresponds to the incoming signal, i.e. the sign of (47) which applies.
For application in hyperbolic lateration, arbitrary coordinates of the base stations must be allowed. This is achieved by rotation and translation of the hyperbola already analyzed. Let us assume that coordinates of the base stations are (x_{BS1}, y_{BS1}) and (x_{BS2}, y_{BS2}). The line segment that connects BS1 and BS2 is inclined to the ray $x = 0$, $x > 0$ for angle θ whose sine and cosine are

$$\sin\theta = \frac{y_{BS1} - y_{BS2}}{D} \tag{48}$$

and

$$\cos\theta = \frac{x_{BS1} - x_{BS2}}{D}. \tag{49}$$

Counterclockwise rotation for angle θ of the line segment that connects the base stations is achieved multiplying the vector of coordinates by the rotation matrix

$$\begin{bmatrix} x_{new} \\ y_{new} \end{bmatrix} = \begin{bmatrix} \cos\theta & -\sin\theta \\ \sin\theta & \cos\theta \end{bmatrix} \begin{bmatrix} x_{old} \\ y_{old} \end{bmatrix}. \tag{50}$$

Center of the hyperbola is located at the midpoint between the base stations

$$\begin{bmatrix} x_C \\ y_C \end{bmatrix} = \frac{1}{2} \begin{bmatrix} x_{BS1} + x_{BS2} \\ y_{BS1} + y_{BS2} \end{bmatrix}. \tag{51}$$

After the rotation and the translation, the hyperbola is given by

$$\begin{bmatrix} x \\ y \end{bmatrix} = \begin{bmatrix} x_C \\ y_C \end{bmatrix} + \begin{bmatrix} \cos\theta & -\sin\theta \\ \sin\theta & \cos\theta \end{bmatrix} \begin{bmatrix} a \cosh t \\ b \sinh t \end{bmatrix} \tag{52}$$

which is after substitution expressed in positioning related terms as

$$\begin{bmatrix} x \\ y \end{bmatrix} = \frac{1}{2} \begin{bmatrix} x_{BS1} + x_{BS2} \\ y_{BS1} + y_{BS2} \end{bmatrix} + \frac{1}{2} \begin{bmatrix} x_{BS1} - x_{BS2} & -y_{BS1} + y_{BS2} \\ y_{BS1} - y_{BS2} & x_{BS1} - x_{BS2} \end{bmatrix} \begin{bmatrix} \frac{d}{D} \cosh t \\ \sqrt{1 - \left(\frac{d}{D}\right)^2} \sinh t \end{bmatrix}. \tag{53}$$

In some situations, it might be convenient to express the hyperbola by a single quadratic form over x and y, instead of the two parametric equations. This can be achieved by substituting

$$\cosh t = \frac{p^2 + 1}{2p} \tag{54}$$

and

$$\sinh t = \frac{p^2 - 1}{2p} \tag{55}$$

and eliminating the dummy variable $p = e^t$ (thus, $p > 0$) and one of the equations applying algebraic manipulations.

To illustrate hyperbolic lateration, the set of base stations BS1: $(3, 5)$, BS2: $(5, 2)$, BS3: $(9, 8)$ and the mobile station positioned at $(5, 5)$ are used, as shown in Fig. 4. The same setup is already used to illustrate angulation and circular lateration. Differences in distances between the base stations and the mobile station are $d_2 = d_{2,1} = r_2 - r_1 = 1$ and $d_3 = d_{3,1} = r_3 - r_1 = 3$. These two differences in distance define two hyperbolas with focal points in BS1 and BS2, as well as BS1 and BS3, respectively, specified by

$$-3\,x^2 + 12\,x\,y - 8\,y^2 - 18\,x + 8\,y + 25 = 0 \tag{56}$$

and

$$-27\,x^2 - 36\,x\,y + 558\,x + 216\,y - 2295 = 0. \tag{57}$$

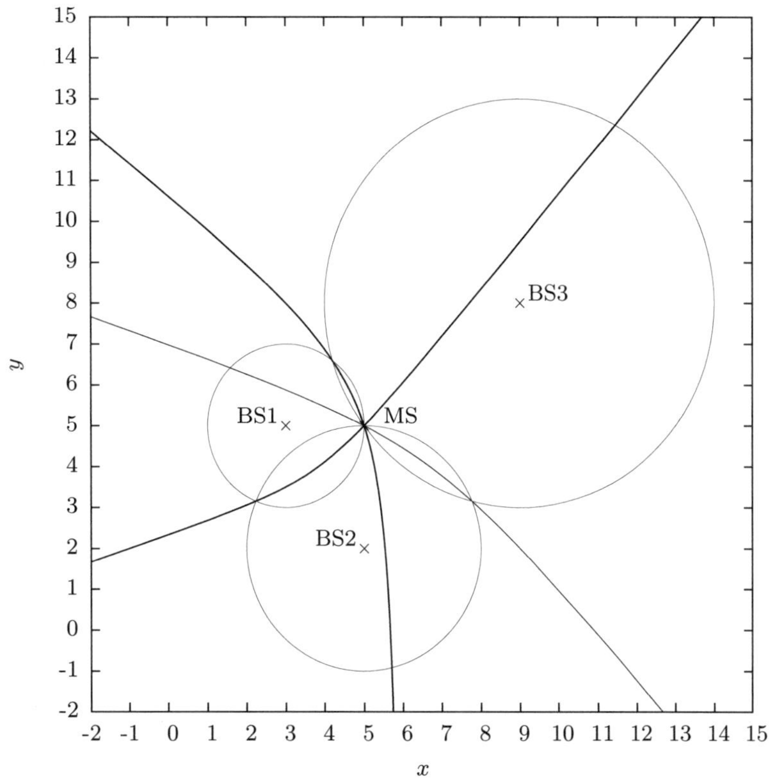

Fig. 4. Hyperbolic lateration, unique solution.

These hyperbolas pass through the intersections of circles defined by fixed distance between the base stations and the mobile station, similarly to the lines in the case of linearized system of equations for circular lateration (27). An additional difference in distance $d_{3,2}$ defined as

$$d_{3,2} = r_3 - r_2 = (r_3 - r_1) - (r_2 - r_1) = d_3 - d_2 \tag{58}$$

results in an another hyperbola

$$-3\,x^2 - 12\,x\,y - 8\,y^2 + 102\,x + 164\,y - 755 = 0 \tag{59}$$

having focal points in BS3 and BS2, shown in Fig. 4 in thin line, but $d_{3,2}$ is linearly dependent on d_2 and d_3 and does not add any new information about the mobile station location.

It is possible to determine the mobile station location as an intersection of hyperbolas directly solving the nonlinear system of equations that arises from (28), like the system formed of (56) and (57). This approach requires iterative solution and raises convergence issues related to the numerical methods for nonlinear equation systems. However, it is possible to reduce the problem to a single quadratic equation or even to a system of linear equations applying appropriate algebraic transformations.

Let us start with the equation (24) derived for the method of circular lateration

$$2\,(x_{BS1} - x_{BSk})\,x_{MS} + 2\,(y_{BS1} - y_{BSk})\,y_{MS} = r_k^2 - r_1^2 + x_{BS1}^2 - x_{BSk}^2 + y_{BS1}^2 - y_{BSk}^2. \tag{60}$$

Instead of having distances r_k, distance differences

$$d_k = d_{k,1} = r_k - r_1 \tag{61}$$

are available. To eliminate the terms that involve r_k consider

$$r_k^2 - r_1^2 = (r_1 + d_k)^2 - r_1^2 = 2\, d_k\, r_1 + d_k^2. \tag{62}$$

In this manner, r_k is eliminated, while r_1 remains present in a linear term. After the transformation, the set of equations (60) becomes

$$2\,(x_{BS1} - x_{BSk})\, x_{MS} + 2\,(y_{BS1} - y_{BSk})\, y_{MS} = 2\, d_k\, r_1 + d_k^2 + x_{BS1}^2 - x_{BSk}^2 + y_{BS1}^2 - y_{BSk}^2 \tag{63}$$

which for $k \in \{2, 3\}$ results in the equation system expressed in a matrix form

$$\mathbf{A} \begin{bmatrix} x_{MS} \\ y_{MS} \end{bmatrix} = r_1 \mathbf{b}_1 + \mathbf{b}_0 \tag{64}$$

where

$$\mathbf{A} = \begin{bmatrix} x_{BS1} - x_{BS2} & y_{BS1} - y_{BS2} \\ x_{BS1} - x_{BS3} & y_{BS1} - y_{BS3} \end{bmatrix} \tag{65}$$

$$\mathbf{b}_1 = \begin{bmatrix} d_2 \\ d_3 \end{bmatrix} \tag{66}$$

and

$$\mathbf{b}_0 = \frac{1}{2} \begin{bmatrix} d_2^2 + x_{BS1}^2 - x_{BS2}^2 + y_{BS1}^2 - y_{BS2}^2 \\ d_3^2 + x_{BS1}^2 - x_{BS3}^2 + y_{BS1}^2 - y_{BS3}^2 \end{bmatrix}. \tag{67}$$

Solution of the linear system is

$$\begin{bmatrix} x_{MS} \\ y_{MS} \end{bmatrix} = \mathbf{A}^{-1} \mathbf{b}_1\, r_1 + \mathbf{A}^{-1} \mathbf{b}_0 \tag{68}$$

where r_1 is not known yet, while coordinates of the mobile station are provided as linear functions of r_1

$$x_{MS} = k_x\, r_1 + n_x \tag{69}$$

and

$$y_{MS} = k_y\, r_1 + n_y. \tag{70}$$

The value of r_1 is computed from

$$(x_{MS} - x_{BS1})^2 + (y_{MS} - y_{BS1})^2 = r_1^2 \tag{71}$$

which after substitution of (69) and (70) results in a quadratic equation

$$\left(k_x^2 + k_y^2 - 1\right) r_1^2 + 2\left(k_x\,(n_x - x_{BS1}) + k_y\,(n_y - y_{BS1})\right) r_1 + (n_x - x_{BS1})^2 + (n_y - y_{BS1})^2 = 0. \tag{72}$$

In general, the quadratic equation provides two solutions. If the equation corresponds to the physical model, at least one of the solutions should be positive. Possible negative solution for r_1 should be rejected for the lack of physical meaning. However, it is possible to obtain two positive solutions for r_1, which faces us with a dilemma where the mobile station is located. Such situation is illustrated in Fig. 5, where in comparison to Fig. 4 BS1 is moved from (3, 5) to $(6, 5)$, resulting in $r_1 = 1$ for MS at $(5, 5)$. Two solutions for r_1 are obtained, $r_{1a} = 1$ corresponding to MS at $(5, 5)$, and $r_{1b} = 0.62887$ corresponding to MS at (6.1134, 4.3814). Circles that correspond to r_{1b} are shown in dashed lines. In (Bensky, 2008), a priori knowledge about the mobile station location is advised as a tool to overcame the situation. In the example of Fig. 5, it is shown that the two solutions can be close one to another, which would require significant amount of a priori knowledge to determine the location of the mobile station. The situation where two solutions are close might be expected where one of the intersecting hyperbolas is highly curved, which is caused by $|d/D|$ approaching 1. To determine actual position of the mobile station, an additional source of information is needed, i.e. another base station that provides information about the difference in distances. However, in the case an additional source of information is available, solving of the quadratic equation is not required, and the problem could be transformed to linear (Gillette & Silverman, 2008).

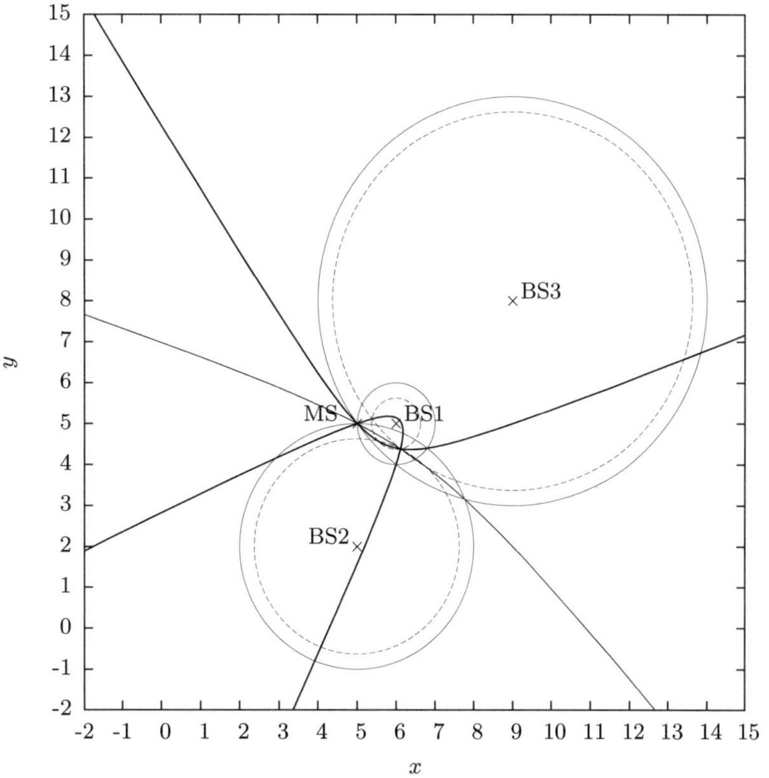

Fig. 5. Hyperbolic lateration, pair of solutions.

Let us consider (63). Variable r_1 is unknown, and it is represented by a linear term on the right hand side. Simple transfer of the term that involves r_1 to the left hand side of (63) results in

$$2\,(x_{BS1} - x_{BSk})\,x_{MS} + 2\,(y_{BS1} - y_{BSk})\,y_{MS} - 2\,d_k\,r_1 = d_k^2 + x_{BS1}^2 - x_{BSk}^2 + y_{BS1}^2 - y_{BSk}^2. \quad (73)$$

Adding the information that originates from the fourth base station, BS4, the system of linear equations over x_{MS}, y_{MS}, and r_1 is obtained as

$$\begin{bmatrix} x_{BS1}-x_{BS2} & y_{BS1}-y_{BS2} & -d_2 \\ x_{BS1}-x_{BS3} & y_{BS1}-y_{BS3} & -d_3 \\ x_{BS1}-x_{BS4} & y_{BS1}-y_{BS4} & -d_4 \end{bmatrix} \begin{bmatrix} x_{MS} \\ y_{MS} \\ r_1 \end{bmatrix} = \frac{1}{2} \begin{bmatrix} d_2^2 + x_{BS1}^2 - x_{BS2}^2 + y_{BS1}^2 - y_{BS2}^2 \\ d_3^2 + x_{BS1}^2 - x_{BS3}^2 + y_{BS1}^2 - y_{BS3}^2 \\ d_4^2 + x_{BS1}^2 - x_{BS4}^2 + y_{BS1}^2 - y_{BS4}^2 \end{bmatrix}. \tag{74}$$

Solution of the system provides unique information about the mobile station coordinates expressed in a closed-form. In the case more than four base stations provided information about the distance difference, an overdetermined system of equations is obtained as

$$\begin{bmatrix} x_{BS1}-x_{BS2} & y_{BS1}-y_{BS2} & -d_2 \\ \vdots & \vdots & \vdots \\ x_{BS1}-x_{BSn} & y_{BS1}-y_{BSn} & -d_n \end{bmatrix} \begin{bmatrix} x_{MS} \\ y_{MS} \\ r_1 \end{bmatrix} = \frac{1}{2} \begin{bmatrix} d_2^2 + x_{BS1}^2 - x_{BS2}^2 + y_{BS1}^2 - y_{BS2}^2 \\ \vdots \\ d_n^2 + x_{BS1}^2 - x_{BSn}^2 + y_{BS1}^2 - y_{BSn}^2 \end{bmatrix} \tag{75}$$

and it can be solved in the least-squares sense (19). The method that involves the square equation over r_1 (72) should be applied in cases where information from only three base stations are available, which results in uniquely determined position of the mobile station in some cases, and in two possibilities for the mobile station location in the remaining cases.

3.2 Probabilistic methods

In contrast to deterministic methods that apply geometric relations to estimate the mobile station position assuming fixed distances and/or angles extracted from radio propagation parameters, probabilistic methods treat available data about the mobile station location as spatial probability density functions. This approach is suitable when accuracy of available data is poor, which frequently is the case in mobile station positioning problems in cellular networks. After the available information about the position related parameters are collected, corresponding probability density functions are joined to a single probability density function that describes position of the mobile station within the cellular network. Coordinates of the mobile station are estimated as expectation of the random variable that corresponds to the resulting probability density function. In comparison to the deterministic methods, probabilistic methods are computationally more intensive.

3.2.1 Joining of the probability density functions

Let us assume that an information source indexed i provides information in the form of a two-dimensional probability density function $p_i(x, y)$ stating that probability that the mobile station is located in a rectangle specified by $x_1 < x < x_2$ and $y_1 < y < y_2$ is

$$P_i(x_1 < x < x_2, y_1 < y < y_2) = \int_{y_1}^{y_2} \int_{x_1}^{x_2} p_i(x, y)\, dx\, dy. \tag{76}$$

Probabilistic methods join the probability density functions collected from various sources of information in order to improve the information about the mobile station location. At first, let us consider two probability density functions $p_i(x, y)$ and $p_j(x, y)$. Multiplying the probability density functions provides a four-dimensional probability density function

$$q_{ij}(x_1, y_1, x_2, y_2) = p_i(x_1, y_1)\, p_j(x_2, y_2). \tag{77}$$

The probability density function of interest applies under constraints $x_1 = x_2 = x$ and $y_1 = y_2 = y$, i.e. that both of the sources of information provide the same answer about the mobile station location. Joined probability density function is then

$$p_{ij}(x, y) = \frac{p_i(x, y)\ p_j(x, y)}{\int_{-\infty}^{+\infty}\int_{-\infty}^{+\infty} p_i(x, y)\ p_j(x, y)\ dx\,dy}. \tag{78}$$

This result generalized for n probability density functions, from $p_1(x, y)$ to $p_n(x, y)$, is

$$p(x, y) = \frac{\prod_{i=1}^{n} p_i(x, y)}{\int_{-\infty}^{+\infty}\int_{-\infty}^{+\infty} \prod_{i=1}^{n} p_i(x, y)\ dx\,dy}. \tag{79}$$

Coordinates (x_{MS}, y_{MS}) that most likely reveal the mobile station position are obtained as an expected value of the two-dimensional random variable (x, y) with the probability density function $p(x, y)$

$$x_{MS} = \int_{-\infty}^{+\infty}\int_{-\infty}^{+\infty} x\,p(x, y)\ dx\,dy = \int_{-\infty}^{+\infty}\left(\int_{-\infty}^{+\infty} p(x, y)\ dy\right) x\,dx \tag{80}$$

and

$$y_{MS} = \int_{-\infty}^{+\infty}\int_{-\infty}^{+\infty} y\,p(x, y)\ dx\,dy = \int_{-\infty}^{+\infty}\left(\int_{-\infty}^{+\infty} p(x, y)\ dx\right) y\,dy. \tag{81}$$

As a measure of precision the mobile station coordinates are determined, standard deviation may be used,

$$\sigma = \sqrt{\int_{-\infty}^{+\infty}\int_{-\infty}^{+\infty} \left((x - x_{MS})^2 + (y - y_{MS})^2\right)\ p(x, y)\ dx\,dy} \tag{82}$$

being lower for better precision.

Coordinates of the mobile station provided computing the random variable expected value sometimes might provide an absurd result, since they may identify location of the mobile station at a point where the probability density is equal to zero, and where the mobile station cannot be located. However, the expected value provides the best guess of the mobile station location in the sense the standard deviation is minimized.

3.2.2 Limits of the mobile station possible location

To determine the mobile station location numerically while minimizing the computational burden, the space where the mobile station might be located should be reduced as much as possible according to the available information. Let us assume that $p_i(x, y) = 0$ everywhere outside a rectangular region specified by

$$x_{i\,min} \leq x \leq x_{i\,max} \tag{83}$$

and

$$y_{i\,min} \leq y \leq y_{i\,max}. \tag{84}$$

Inside the region, there might be points and even sub-regions where $p_i(x, y) = 0$, but outside the rectangular region there should not be any point with $p_i(x, y) \neq 0$. Values of $x_{i\,min}$ and $y_{i\,min}$ should be the highest, while $x_{i\,max}$ and $y_{i\,max}$ should be the lowest values that provide $p_i(x, y) = 0$ outside the region specified by (83) and (84). This results in the smallest rectangle that encloses nonzero values of the probability density function. Some probability density functions, including widely used normal distribution, take nonzero value in the entire region they are defined. In these cases, reasonable approximations should be used to neglect low nonzero values of the probability density.

According to (79), the joined probability density function takes nonzero value in points where all of the probability density functions are nonzero. Thus, the region where the joined probability density function may take nonzero value is an intersection of the rectangles specified by each of the probability density functions, given by

$$x_{min} \leq x \leq x_{max} \tag{85}$$

and

$$y_{min} \leq y \leq y_{max} \tag{86}$$

where

$$x_{min} = \max_{1 \leq i \leq n} (x_{i\ min}) \tag{87}$$

$$x_{max} = \min_{1 \leq i \leq n} (x_{i\ max}) \tag{88}$$

$$y_{min} = \max_{1 \leq i \leq n} (y_{i\ min}) \tag{89}$$

and

$$y_{max} = \min_{1 \leq i \leq n} (y_{i\ max}). \tag{90}$$

In (Simić & Pejović, 2009), a probabilistic algorithm that stops here is proposed to estimate the mobile station location, being named the method of squares. Probability density function that assumes uniform distribution within the rectangle defined by (85) and (86) is assumed, and center of the resulting rectangle

$$x_{MS} = \frac{x_{min} + x_{max}}{2} \tag{91}$$

and

$$y_{MS} = \frac{y_{min} + y_{max}}{2} \tag{92}$$

is proposed as the mobile station position estimate. This results in standard deviation of the mobile station coordinates treated as the two-dimensional random variable given by

$$\sigma = \sqrt{\frac{(x_{max} - x_{min})^2 + (y_{max} - y_{min})^2}{12}}. \tag{93}$$

This method is computationally efficient, but suffers from poor precision, i.e. significant standard deviation (93). To improve the precision, numerical methods are applied to refine the region defined by (85) and (86).

3.2.3 Discretization of space

To facilitate numerical computation, the obtained region of interest specified by (85) and (86), is discretized into a grid of n_X segments over x coordinate resulting in the segment width

$$\Delta x = \frac{x_{max} - x_{min}}{n_X} \tag{94}$$

and n_Y segments over y coordinate

$$\Delta y = \frac{y_{max} - y_{min}}{n_Y} \tag{95}$$

wide. It is common to choose n_X and n_Y to provide $\Delta x = \Delta y$. In this manner, the space of the mobile station possible location is discretized into the grid of $n_X \times n_Y$ segments.
After the segmentation of coordinate axes is performed, let us discretize x coordinate in the region of interest (85) into a vector of discrete values

$$x_k = x_{min} + \left(k - \frac{1}{2}\right) \Delta x \tag{96}$$

for $k = 1, \ldots n_X$. Discrete coordinate values of x_k correspond to coordinates of central points of the segments. In the same manner, the discretization is performed over y coordinate in the region (86),

$$y_l = y_{min} + \left(l - \frac{1}{2}\right) \Delta y \tag{97}$$

for $l = 1, \ldots n_Y$. Discretized coordinates would represent positions of grid elements in subsequent computations.
According to the space discretization and (76), the probability density functions should be integrated over each grid element to obtain the probability that the mobile station is located in that element

$$P_{i,k,l} = \int_{y_{min}+(l-1)\,\Delta y}^{y_{min}+l\,\Delta y} \int_{x_{min}+(k-1)\,\Delta x}^{x_{min}+k\,\Delta x} p_i\left(x, y\right)\, dx\, dy \tag{98}$$

for $i \in \{1, \ldots n\}$, which might be approximated as

$$P_{i,k,l} \approx \Delta x\, \Delta y\, p_i\left(x_k, y_l\right). \tag{99}$$

In this manner, the probability density functions $p_i\left(x, y\right)$ are for $i \in \{1, \ldots n\}$ discretized into matrices of probabilities $P_{i,k,l}$ for $k = 1, \ldots n_X$ and $l = 1, \ldots n_Y$.
In terms of discretized space, elements of joined probability matrix are obtained as

$$P_{k,l} = \frac{\prod_{i=1}^{n} P_{i,k,l}}{\sum_{k=1}^{n_X} \sum_{l=1}^{n_Y} \prod_{i=1}^{n} P_{i,k,l}}. \tag{100}$$

The mobile station coordinates are obtained computing the expected value in discretized terms applying

$$x_{MS} = \sum_{k=1}^{n_X} \left(\sum_{l=1}^{n_Y} P_{k,l} \right) x_k \tag{101}$$

and

$$y_{MS} = \sum_{l=1}^{n_Y} \left(\sum_{k=1}^{n_X} P_{k,l} \right) y_l. \tag{102}$$

The standard deviation is obtained as

$$\sigma = \sqrt{\sum_{k=1}^{n_X} \sum_{l=1}^{n_Y} \left(\left(x_k - x_{MS}\right)^2 + \left(y_l - y_{MS}\right)^2 \right) P_{k,l}}. \tag{103}$$

By (101)–(103), computation of integrals is replaced by summations. Reducing the space of interest to the rectangular area specified by (87)–(90), the sums are made finite and with fixed limits.

3.2.4 Implementation of the algorithm for probability density functions of the exclusion type

The method in its discretized form requires the probability matrix that contains $n_X \times n_Y$ entries. Handling of this matrix might be computationally inefficient in some cases, and may require significant storage space. To simplify the computation, property of some probability density functions to provide information where the mobile station cannot be located, and uniform probability density in the areas where the mobile station can be located may be utilized. These probability density functions are named probability density functions of the exclusion type (Simić & Pejović, 2009). Typical examples are the probability density functions obtained from parameters significantly discretized in the mobile communication system, such as the timing advance (TA) parameter in GSM, and the round trip time (RTT) parameter in UMTS. After discretization, these probability density functions result in a set S_{NZ} of n_{NZ} grid elements, $n_{NZ} \leq n_X \times n_Y$, determined by their indices $\left(k_j, l_j\right)$ for $j = 1, \ldots n_{NZ}$, where probability that the mobile station is located within the grid element takes nonzero value equal to $1/n_{NZ}$, while it takes zero value elsewhere, i.e.

$$P_{i,k,l} = \begin{cases} 1/n_{NZ} & \text{for } (k, l) \in S_{NZ} \\ 0 & \text{for } (k, l) \notin S_{NZ}. \end{cases} \tag{104}$$

Actual nonzero magnitude of the probability density function in the case of the probability density functions of the exclusion type is of low importance, since it can easily be computed if needed according to the normalization criterion, which after discretization for the joined probability takes form

$$\sum_{k=1}^{n_X} \sum_{l=1}^{n_Y} P_{k,l} = 1. \tag{105}$$

Thus, to store probabilities that arise from the probability density functions of the exclusion type it is enough to use one binary digit per grid element. Sources of information about the mobile station location are represented by their bitmaps $b_{i,k,l}$ that take value 1 for nonzero probability, indicating that it is possible that the mobile station is located within the corresponding grid element, and 0 for the probability equal to zero, indicating that it is not possible that the mobile station is located within the considered grid element. Besides the significant reduction in storage requirements, this simplifies joining of the probabilities obtained from different sources of information. Multiplication of probability values reduces to logical AND operation over the bits that show is it possible for the mobile station to be located within the considered grid element or not. After the joined probability bitmap is obtained as

$$b_{k,l} = \bigwedge_{i=1}^{n} b_{i,k,l} \tag{106}$$

normalization to corresponding joined probability matrix can be performed applying

$$P_{k,l} = \frac{1}{n_1} b_{k,l} \tag{107}$$

where n_1 is the number of elements in $b_{k,l}$ that take value 1,

$$n_1 = \sum_{k=1}^{n_X} \sum_{l=1}^{n_Y} b_{k,l}. \tag{108}$$

The normalization should not be performed at the matrix level, since it is more convenient to perform it on the level of coordinate and standard deviation computation,

$$x_{MS} = \frac{1}{n_1} \sum_{k=1}^{n_X} \left(\sum_{l=1}^{n_Y} b_{k,l} \right) x_k \tag{109}$$

$$y_{MS} = \frac{1}{n_1} \sum_{l=1}^{n_Y} \left(\sum_{k=1}^{n_X} b_{k,l} \right) y_l \tag{110}$$

and

$$\sigma = \sqrt{\frac{1}{n_1} \sum_{k=1}^{n_X} \sum_{l=1}^{n_Y} \left((x_k - x_{MS})^2 + (y_l - y_{MS})^2 \right) b_{k,l}}. \tag{111}$$

Sometimes it might be useful to approximate the probability density functions that are not of the exclusion type by the exclusion type ones, sacrificing some of the precision in order to improve the computational efficiency. Besides, the bitmap of (106) provides useful visual information that can be presented to the user as a map of the mobile station possible location.

3.2.5 An example

To illustrate application of probabilistic methods, an example that includes two base stations of a GSM network located at $(x_{BS1}, y_{BS1}) = (0, 0)$ with $TA = 1$ and $(x_{BS2}, y_{BS2}) = (1.8\,R_q, 1.6\,R_q)$ with $TA = 0$ is created, as shown in Fig. 6. All of the distances considered in this example are expressed in terms of the GSM spatial resolution quantum $R_q = 553.46$ m $\approx$ 550 m. Line-of-sight wave propagation is assumed, thus probabilistic model of (7) is applied. Under these assumptions, according to (10), (11), and (87)–(90), the space where the mobile station might be located is limited to $0.8\,R_q \leq x_{MS} \leq 2\,R_q$ and $0.6\,R_q \leq y_{MS} \leq 2\,R_q$. To perform discretization of space, the same discretization quantum is applied for both of the axes, $\Delta x = \Delta y = 0.1\,R_q$. In Fig. 6, in the region of the mobile station possible location spatial grid of $n_X \times n_Y = 12 \times 14 = 168$ elements is drawn, and center points of the grid elements are indicated by dots. The probability density function of (7) is of the exclusion type, and applying the algorithm for this class of functions the grid elements are classified regarding possible position of the mobile station. To perform this task, distance between two points had to be determined $2 \times 168 = 336$ times. Grid elements where the mobile station might be located are shaded in Fig. 6. Applying (109) and (110), the mobile station coordinates are estimated as $x_{MS} = 1.2935\,R_q$ and $y_{MS} = 1.1471\,R_q$, while according to (111) standard deviation of the location estimation is obtained as $\sigma = 0.40556\,R_q$. For comparison, the method of squares (91)–(93) provides $x_{MS} = 1.4\,R_q$ and $y_{MS} = 1.3\,R_q$ with $\sigma = 0.53229\,R_q$, without any need to compute distances mentioned 336 times.

To apply deterministic methods in order to provide a comparison, coordinates of two base stations and distances to the mobile station estimated according to (12) as $r_1 = 1.5\,R_q$ and $r_2 = 0.5\,R_q$ are available, as well as the distance difference $d_2 = r_2 - r_1 = R_q$. The available set of data is not sufficient to determine position of the mobile station. However, some information about the mobile station position might be extracted.

For circular lateration, the distance between the base stations $D = 2.40832\,R_q$ is larger than $r_1 + r_2 = 2\,R_q$, thus the circles of possible mobile station location do not intersect. However, equation of the type (24) locates the mobile station on a line $18x + 16y = 39$, which is shown

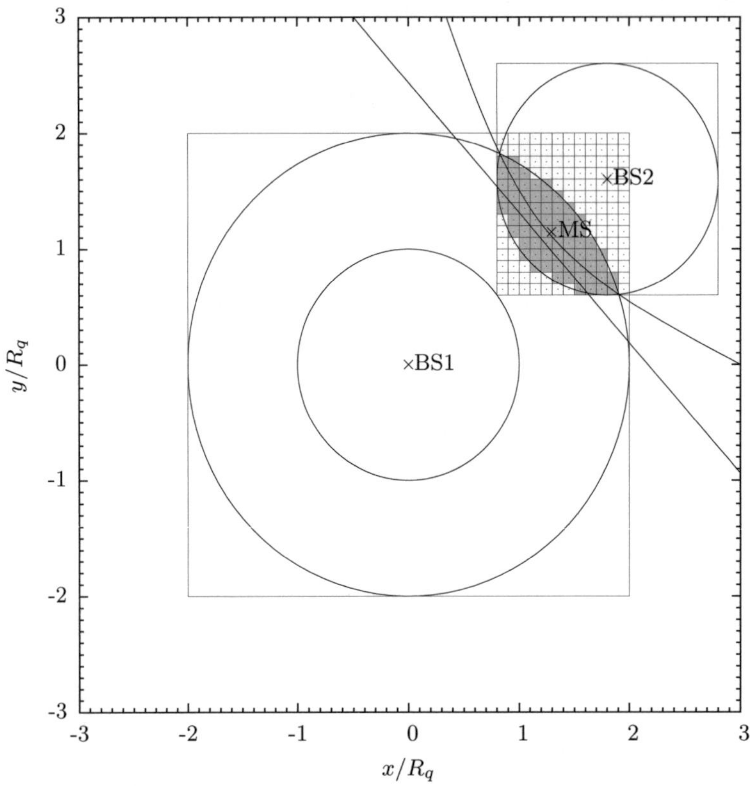

Fig. 6. Probabilistic methods, an example.

in Fig. 6. On the other hand, hyperbolic lateration locates the mobile station on a hyperbola obtained applying (53) as

$$x = 0.9 + 0.3737 \cosh t - 0.7278 \sinh t \tag{112}$$

and

$$y = 0.8 + 0.3322 \cosh t + 0.8187 \sinh t. \tag{113}$$

The hyperbola passes through the intersections of the outer boundary circles of the regions defined by (7), which have the radii for $\frac{1}{2}R_q$ higher than the estimated distances of the mobile station from corresponding base stations. In the distance difference, these offsets cancel out, thus the hyperbola passes through the intersections of circles, as it would pass through the intersections of circles that represent estimated distance to the mobile station if they had intersected.

As illustrated in this example, probabilistic approach is able to provide an estimate of the mobile station coordinates even in cases when available data is insufficient for deterministic methods. However, probabilistic methods are computationally more intensive.

3.3 Fingerprinting methods

Fingerprinting methods (Küpper, 2005) treat a vector of position related parameters that mobile station observes as a fingerprint of the mobile station position. The position is determined comparing the observed vector to vectors stored in a predetermined database of fingerprints. The database is obtained either by field measurements or precomputed applying appropriate wave propagation models. Point in the database having the vector of position related parameters the closest to the vector observed by the mobile station is assumed as the mobile station position.

Radio propagation parameter convenient to be used in a vector of position related parameters is the received signal power. This is particularly convenient for application in WLAN networks (Küpper, 2005). Let us assume that applying some information about the mobile station location (the user is in a building, ID of the serving cell is known, etc.) the set of base stations that might be observed by the mobile station is reduced to a set of n base stations. Observed vector of position related parameters is

$$\vec{P} = \left[P_{O\,1}, \dots P_{O\,n}\right]. \tag{114}$$

Let us also assume that a database of M entries is predetermined, consisting of vectors

$$\vec{P}_m = \left[P_{DB\,m,1}, \dots P_{DB\,m,n}\right] \tag{115}$$

accompanied by corresponding coordinates (x_m, y_m) for $m \in \{1, \dots M\}$. As a measure of difference between the observed vector of position related parameters, Euclidean distance

$$\delta_m^2 = \left|\vec{P} - \vec{P}_m\right|^2 = \sum_{k=1}^{n} \left(P_{O\,k} - P_{DB\,m,k}\right)^2 \tag{116}$$

might be used. Index m_{MS} of the database entry that corresponds to the mobile station position is obtained as

$$\delta_{m_{MS}}^2 = \min_{1 \le m \le M} \delta_m^2 \tag{117}$$

and the mobile station coordinates are obtained as $(x_{m_{MS}}, y_{m_{MS}})$. The problem that might appear here is that (117) might provide multiple solutions for m_{MS}, likelihood of which is increased by rougher quantization of the received signal power. The situation would not be frequent in well designed systems, with adequate choice of the vector of position related parameters. In the case the vector of position related parameters contains variables with different physical dimensions, to enable determination of Euclidean distance normalization, i.e. nondimensionalization of parameters, should be performed. This introduces weighting coefficients for the elements of the vector of position related parameters, which can be introduced even in the case the elements of the vector are of the same physical dimension.

Described fingerprinting method for each positioning request requires Euclidean distance computation and minimum search over the entire database. This might be computationally intensive and time consuming. To simplify the database search, in (Simić & Pejović, 2008) a set of n_{CBS} base stations with the strongest received signal power is proposed as a fingerprint for GSM networks. The position fingerprint contains only indexes of the base stations, not the information about the received signal power. Since the information about received signal power of up to seven base stations observed by the mobile station is included in standard measurement report in GSM, it is convenient that $1 \le n_{CBS} \le 7$. According to the set of base stations with the highest received signal power, the space is divided into segments characterized by the same fingerprint. Within the segment, the probabilistic approach is applied assuming uniform probability density function. The information each fingerprint provides are estimated coordinates of the mobile station, precomputed as mathematical expectation of the coordinates treated as a random variable of uniform probability density within the segment, corresponding standard deviation, and the map of the segment. In (Simić & Pejović, 2008), a simple propagation model that assumes line-of-sight propagation and monotonous decrease of signal power with distance is applied to illustrate the method. However, more complex propagation models might be applied, as well as field measurements. Search for the fingerprint in the database might be performed as a binary search, and it is computationally efficient. The method could be generalized to other cellular networks. Furthermore, the method could be extended, and instead of the set of base stations with the strongest received signal power, the fingerprint might consist of indexes of the base stations ordered according to the received signal power. Again, information about the actual received signal power is not included in the fingerprint. This approach would increase the number of available fingerprints and reduce the segment size.

4. Standardized positioning methods in cellular networks

Implementation of the location based services required special standards to be developed. The standardization is performed by The Third Generation Partnership Project (3GPP).
For GSM cellular networks, the following positioning methods have been standardized (3GPP TS 43.059, 2007):

- Cell-ID+TA (Cell Identification + Timing Advance),
- E-OTD (Enhanced Observed Time Difference),
- U-TDOA (Uplink Time Difference of Arrival), and
- A-GNSS (Assisted Global Navigation Satellite System).

For UMTS cellular networks, the following positioning methods have been standardized (3GPP TS 25.305, 2007):

- Cell-ID (Cell Identification),
- OTDOA-IPDL (Observed Time Difference Of Arrival-Idle Period Downlink),
- A-GNSS (Assisted Global Navigation Satellite System), and
- U-TDOA (Uplink Time Difference of Arrival).

Cell-ID is the easiest and most widespread method to obtain the location of the mobile station in cellular networks. This method is based on proximity sensing principle, i.e. estimated coordinates of the mobile station are geographic coordinates of the base station (Node B) currently serving the mobile station. A drawback of this method is poor accuracy, dependent on the serving cell size.
To achieve location estimation in GSM networks with higher accuracy, Cell-ID is combined with the timing advance value resulting in Cell-ID+TA positioning method. The TA parameter value is related to the wave propagation round trip time, which is proportional to the distance from the serving base station to the mobile station, and thus the location of the mobile station can be constrained to a ring centered at the base station.
E-OTD method is based on measurements of the signal propagation time on the downlink, and applies hyperbolic lateration. In E-OTD, the mobile station measures the time difference of arrival (TDOA) of the signals from the neighboring base stations, which are not synchronized in GSM. Hence, it is necessary to install additional components in the network to deal with the synchronization issues. These components are called Location Measurement Units (LMU).

OTDOA is an equivalent to E-OTD positioning method in GSM. The differences are caused by different structure of UMTS radio interface. Timing measurements at the mobile station in the systems based on CDMA principle suffer from so-called hearability problem. This problem occurs when the mobile station is located near the service base station, which may block signals from other base stations that operate on the same frequency. In order to overcome this problem, each base station must interrupt its transmission in a short period of time to allow the mobile station to detect signals from neighboring base stations and to perform all the measurements necessary for lateration. These periods are called idle periods, while the mechanism that controls them is called IPDL (Idle Period Downlink). OTDOA positioning method which applies IPDL is called OTDOA-IPDL positioning method.

U-TDOA is a method standardized for positioning both in GSM and UMTS. Similarly to E-OTD and OTDOA, U-TDOA positioning method is based on measurements of the signal propagation time and applies hyperbolic lateration. However, in U-TDOA the measurements are made on the uplink, i.e. the time of arrival of the signal emitted from the mobile station is observed by the serving base station and a number of LMUs. Thus, this method also requires installation of additional hardware components, LMUs, and in larger number than in the case of E-OTD or OTDOA method.

A-GNSS is a method standardized for positioning both in GSM and UMTS cellular networks. Unlike the positioning methods previously described, completely based on the cellular network infrastructure, A-GNSS positioning method is based on the satellite infrastructure of the Global Positioning System, GPS. This method requires mobile stations to be equipped with a GPS receiver, and they are supplied by some additional assistance data from the network, which allow to reduce the acquisition time and to increase the accuracy.

5. Conclusions

Methods that utilize propagation parameters of radio waves to determine mobile station location are analyzed in this chapter. Propagation parameters of radio waves are briefly analyzed, and it is shown that distance can be estimated from the propagation path loss and from the wave propagation time. To measure the wave propagation time, clocks involved in the measurement should be synchronized. In the case the distance between two points is measured, to avoid synchronization the round trip time might be measured, which involves only one clock. In the case the base stations are synchronized, the time difference of the signal propagation between the mobile station and two base stations can be measured without the requirement for mobile station to be synchronized. Besides distances and differences in distance, angles of signal arrival might be used to perform positioning.

After the wave propagation data are collected, they should be processed to provide estimate of the mobile station location. Estimation algorithms are based either on deterministic or probabilistic principles. Deterministic algorithms that process the angle of signal arrival, the path loss or the wave propagation time, and the time difference of signal propagation are analyzed, named after their geometrical interpretations as angulation, circular lateration, and hyperbolic lateration, respectively. It is shown that in all three of the cases coordinates of the mobile station could be computed as a solution of a system of linear equations. To provide unique solution for the mobile station coordinates in two dimensions, it is shown that angulation requires data from at least two base stations, circular lateration from three, and hyperbolic lateration from four base stations. Additional data might be incorporated to refine the estimate of the mobile station coordinates applying linear least-squares method to solve the resulting overdetermined system of linear equations.

Probabilistic methods model information about the mobile station location extracted from the wave propagation parameters as probability density functions. A method to process the probability density functions in order to increase the positioning precision is presented. An emphasis to computational efficiency is made. In order to reduce the memory requirement and to simplify the computation, probability density functions of the exclusion type were introduced. Application of the proposed probabilistic method results in estimated coordinates, corresponding standard deviation, and a map of the mobile station possible location. Probabilistic methods are convenient in cases where available information about the mobile station location is poor.

The third group of the methods to process the position related data analyzed in this chapter are fingerprinting methods. These methods reduce the problem to database search, treating collected set of radio propagation position related parameters as a position fingerprint. The database might be formed either applying deterministic or probabilistic methods, or even applying field measurements. Variants of the method that use the fingerprint formed as a vector of received signal power from various base stations and the fingerprint formed as a set of base station indexes formed according to the received signal power are analyzed.

A brief review of standardized positioning methods in GSM and UMTS systems is given.

6. References

3GPP TS 43.059, release 8 (v8.0.0), */EDGE Radio Access Network, Functional stage 2 description of Location Services (LCS) in GERAN*, 2007.

3GPP TS 25.305, release 8 (v8.0.0), *Stage 2 Functional specification of UE positioning in UTRAN*, 2007.

Bensky, A. (2008). *Wireless Positioning Technologies and Applications*, Artech House, ISBN 978-1-59693-130-5, Norwood.

Bronshtein, I. N.; Semendyayev, K. A.; Musiol, G. & Muehlig, H. (2007). *Handbook of Mathematics*, Springer, ISBN 978-3-540-72121-5, Berlin, Heidelberg, New York.

Gillette, M. D. & Silverman, H. F. (2008). A linear closed-form algorithm for source localization from time-differences of arrival. *IEEE Signal Processing Letters*, Vol. 15, No. 1, Jan. 2008, 1–4, ISSN 1070-9908

Küpper, A. (2005). *Location-based Services*, John Wiley & Sons, ISBN 978-0-470-09231-6, Chichester.

Press, W. H.; Teukolsky S. A.; Vetterling, W. T. & Flannery, B. P. (1992). *Numerical Recipes in C*, Cambridge University Press, ISBN 0-521-43108-5, Cambridge.

Rappaport, T. S. (2001). *Wireless communications principles and practice*, Prentice Hall, ISBN 0130422320, Englewood Cliffs.

Simić, M. & Pejović, P. (2008). An algorithm for determining mobile station location based on space segmentation. *IEEE Communications Letters*, Vol. 12, No. 7, July 2008, 499–501, ISSN 1089-7798

Simić, M. & Pejović, P. (2009). A probabilistic approach to determine mobile station location with application in cellular networks. *Annals of Telecommunications*, Vol. 64, No. 9–10, Oct. 2009, 639–649, ISSN 0003-4347 (Print), 1958-9395 (Online)

Simić, M. & Pejović, P. (2009). A comparison of three methods to determine mobile station location in cellular communication systems. *European Transactions on Telecommunications*, Vol. 20, No. 8, Dec. 2009, 711–721, ISSN 1124-318X (Print), 1541-8251 (Online)

3

Middleware for Positioning in Cellular Networks

Israel Martin-Escalona, Francisco Barcelo-Arroyo and Marc Ciurana
Universitat Politècnica de Catalunya (UPC)
Spain

1. Introduction

Middleware is defined in (Mahmoud, 2004) as, "a distributed software layer that sits above the network operating system and below the application layer and abstracts the heterogeneity of the underlying environment". According to this definition, the purpose of a middleware is to isolate technology. This task requires that a new layer responsible for handling data between two systems be defined, so that the technology used in each system is transparent to the other system. This situation is illustrated in Fig. 1, in which a middleware is used for communication between the two systems (i.e., A and B). Accordingly, two interfaces are defined, one for the data exchange between system A and the middleware and another for the link between the middleware and system B. Every time a system begins a communication, the middleware handles the data, so that they are adapted to the technology available in each system.

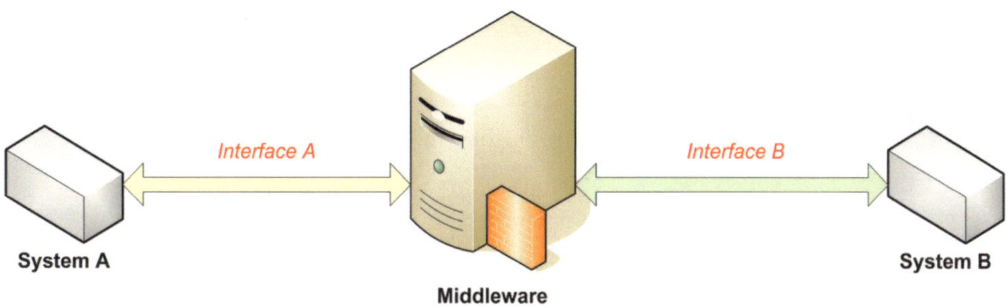

Fig. 1. Middleware interconnecting two systems

The scenario in Fig. 1, in which the middleware splits the communication between two systems, means that several abilities can be expected of a middleware even though these abilities can be grouped into three categories as shown in Fig. 2.As previously mentioned, the main purpose of a middleware is to isolate technology, i.e., making the technology of one system transparent to other systems so that any technology upgrade in one system does not affect any other one. This scenario is especially interesting in the positioning field, in which there are several technologies for fixing the position of customers. Location Based Services (LBS) cannot deal with all those technologies, and the role middleware plays, seems the natural solution to managing the location data. For instance, the Location API for Java 2 Micro Edition (Java Community Process) is a simple solution for writing LBS applications for constrained devices and uses this technology-isolation approach. A distributed approach for technology isolation was proposed by 3GPP in their location platform (3GPP, 2002), which gave the role of middleware to the Gateway Mobile Location Center (GMLC) that acts as mediation for LBS providers.

In the past, the roles of location and LBS providers were combined in a single entity: the network operator. However, the ability of network operators to provide new LBS applications is limited. Several LBS providers appeared in the market, with a large number of applications ready to be released. Although the first approaches for location middleware provided technology isolation, they did not account for quick development, deployment and maintenance of LBS by third parties. This led to middleware proposals that included a framework, i.e., an additional application layer with the purpose of allowing third parties to develop, deploy and maintain their applications in the middleware context. There were several proposals. The Place Lab platform (LaMarca et al., 2005) defines a location middleware based on gathering beacon data for several technologies and storing them in a central database. The main purpose was to provide richer cell-identification (cell-ID) position fixes because dense cells can overlap. Furthermore, the availability of the location solution was improved because several techniques were taken into account. The Place Lab also provides a framework to build clients (i.e., applications) that can exchange data with the platform. A more generic solution is NEXUS (Fritsch & Volz, 2003), which proposes a distributed middleware for a Geographical Information Systems (GIS) platform based on a three-tier architecture: application, federation and service. Application and service can be represented as the location client and server, and the federation is a middle layer that manages the data exchange in a distributed fashion. The main purpose of this solution is the provision of a distributed data model for GIS representation (Augmented World Query Language), which does not depend on the technology used for positioning and allowing spatial overlapping. Other solutions following the same approach are the MiddleWhere (Raganathan et al., 2004), the OpenLS (OGC) and the POLOS platform (Spanoudakis et al., 2004).

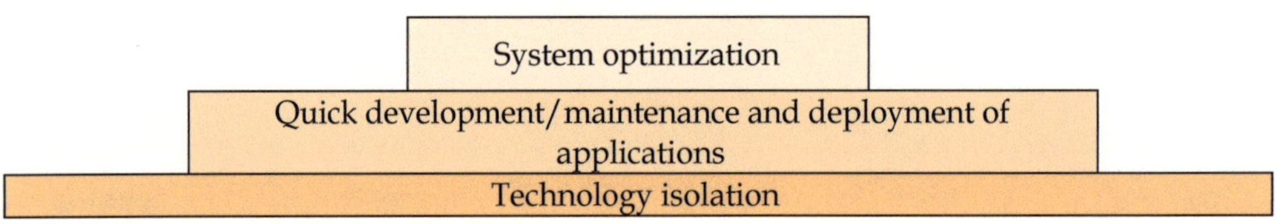

Fig. 2. Main purposes of a middleware

Most of the features associated with the middleware concept are addressed to isolating technology and framework provisioning. However, middleware can also include a functional layer, the purpose of which is to define how the duties are made in order to optimize the system. This optimization can involve several parameters according to the sort of middleware being implemented. In the case of a positioning middleware, optimization can involve economic-aspect enhancement or even tuning parameters directly involved in the operation and maintenance tasks of the communication networks supporting the positioning system. This chapter presents a middleware that has a twofold purpose: 1) optimizing the performance of the system and 2) fulfilling the Quality of Service (QoS) requested. There are few studies of middleware that address optimization. The closest to the proposal presented in this chapter is called TraX (Küpper et al, 2006), which proposes an intermediary device-centric architecture as presented in Fig. 3. The TraX platform distributes the middleware among an LBS provider, a location provider, a content provider and the target device. The *LBS provider* makes the LBS accessible to the external clients (i.e., users) and communicates with *content providers* to add value to the target's position, which is supplied by the *location provider*. The target device is responsible for gathering the measurements necessary to compute the position and, depending on the technique, fixing the position.

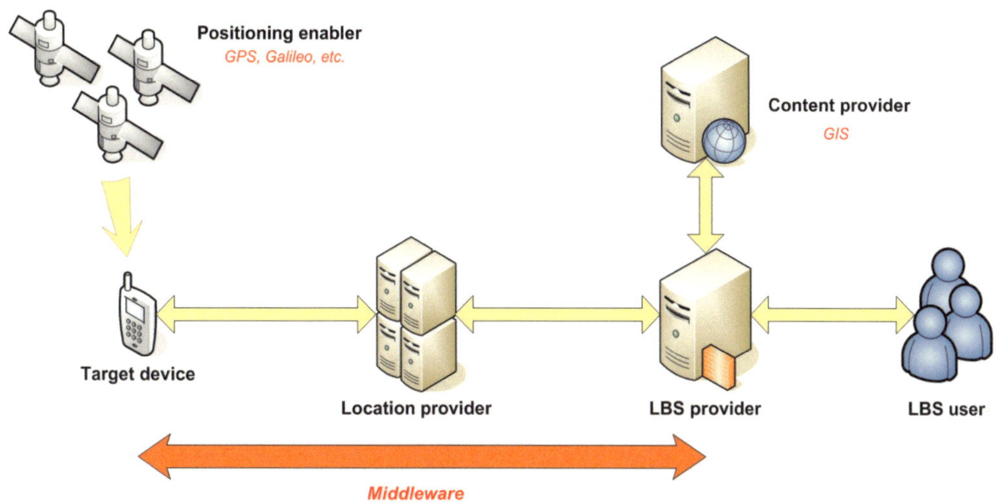

Fig. 3. Intermediary device-centric middleware architecture

The TraX platform consists of three layers: positioning (e.g., A-GPS, WLAN, RFID, OTDOA, etc.), position management (e.g., polling, periodic position updating context awareness, etc.) and application layers (e.g., emergency services, navigation LBS, goods tracking, ...). In TraX, the most suitable location technique is selected and then activated to perform the Location Service (LCS) request.

The solution proposed in this document is a middleware for system optimization. It is designed to fulfill the requirements of LBS, and, at the same time, optimize the positioning procedure so that the performance of the whole location solution is improved. Further details of the architecture and performance assessment of this middleware location platform are provided in the following sections.

2. Middleware for location cost optimization

2.1 Middleware architecture

The resources consumed by location systems generally belong to the underlying networks, on which the location solution runs. It means that LBS share the resources with the regular services provided by the network. Thus, allocating resources for LBS involves reducing the carried traffic for these regular services. The solution proposed in this chapter is a middleware that addresses optimizing the use of resources in location systems. This middleware, which is named MILCO, i.e., Middleware for Location Cost Optimization, has been developed in the frame of (Ministerio de Ciencia y Educacion, 2009). The performance of MILCO consist of analyzing the QoS of the LBS requests, filtering out those location techniques not suitable for a specific request and selecting the optimum technique among the remaining ones according to the resources that are expected they use. MILCO accounts for other factors that constrain the performance location system, such as the location techniques implemented in both, user terminal and core network, the environment where the user is, etc. This approach differs from those taken in standard location middleware solutions because they are usually focused on providing technology-independence or the rapid development of LBS to third-parties, rather than on resource use efficiency.

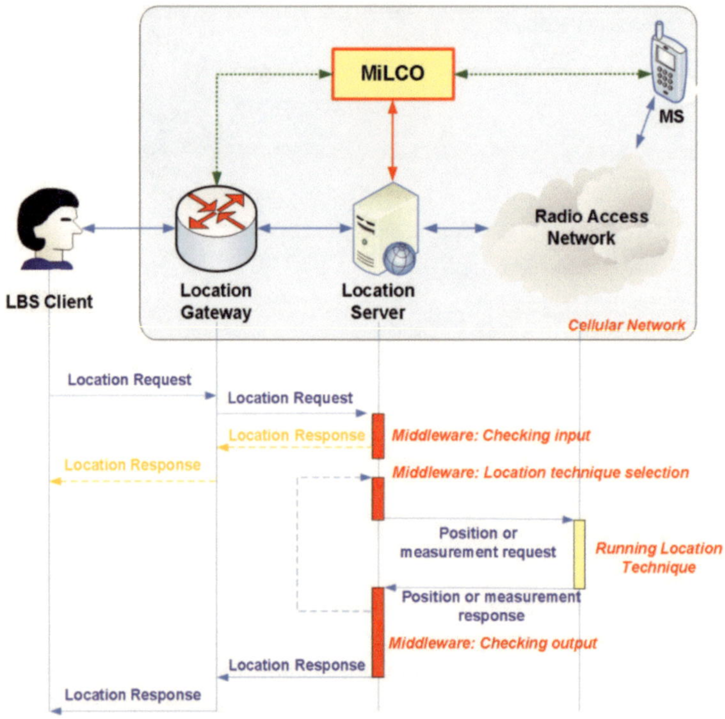

Fig. 4. MILCO system architecture

MILCO is designed to be implemented in terminals, location providers and LBS providers or in a subset of them. However, the usual implementation for MILCO is as a new piece of software inside location providers, e.g., inside Serving Mobile Location Centers (SMLCs) in the case of the ETSI/3GPP notation (3GPP, 2004). Fig. 4 shows a location system architecture that incorporates MILCO in the location provider. Nevertheless, the mobile station (MS) and LBS providers can include certain MILCO functionalities, which are illustrated as green dotted lines in Fig. 4. Under this architecture, each time a LBS request reaches the location system via the location gateway (e.g. GMLC in the case of ETSI/3GPP notation), it is delivered to a location server (e.g. the SMLC in the case of ETSI/3GPP notation). The location server handles the request and forwards it to the MILCO entity, which is placed in the topmost layer of the protocol stack. MILCO then run several input modules to assess whether the request requires executing a location technique. If it is not the case, the input modules will return an estimated position to the LBS client. Otherwise, MILCO selects the optimum location technique for the request, i.e., the one that is expected to provide the requested QoS at the minimum cost. Once it is selected, MILCO uses the network facilities provided by the location server to run the technique and fix the user's position. Finally, if the position fulfills the requested QoS, it will be forwarded to the LBS client. Otherwise, MILCO will iterate again using another location technique.

It must be noted that the MILCO architecture can easily be extended to any cellular system (e.g., 4G PLMNs, WLAN, etc.) as they only need to include MILCO as the topmost application layer in the location stack of any or several devices in the LBS supply chain.

MILCO requires several data in order to carry out its tasks. Most of these data is included in the LBS request or can be easily achieved. These data are detailed below:

- *Location request data.* This information is composed of all the data related to the LCS request, such as the LCS client identifier, the sort of position (i.e. 2D/3D), the periodicity, etc.
- *QoS requirements*. The QoS is required by the LBS client. This QoS can involve several parameters, but it is mainly measured in terms of the minimum accuracy and the maximum delay required by the service as stated in previous chapters.
- *Cell identity*. These data indicate the cell to which the target user is linked. This information is used to compute a coarse position for the target user as well as to optimize the performance of MILCO.
- *Network and handset capabilities*. This information feeds MILCO with the location techniques available in both the network and the target user's handset. MILCO uses these data to filter all the location techniques that are not available in both network and handset simultaneously.

MILCO's procedure is depicted in Fig. 5. This procedure comprises three stages:

- *Pre-filtering* is the process by which any location technique not suitable for the request is filtered out. Location techniques may be marked as unsuitable for three reasons:
  - *Missing technique*, i.e., the location technique is not implemented in either the network or the user terminal.
  - *Poor QoS*, i.e., the location technique is unable, even in the best case, to perform the QoS requested.
  - *Off-line estimation*, i.e., MILCO is able to attend the request and achieve the requested QoS without running any of the location techniques.
- *Selection* is the second stage, and it involves the selection of the best location technique for the request being performed. At this stage, MILCO ranks the remaining set of location techniques (i.e., those available after filtering) according to the optimality for attending the request. This step is achieved by means of a cost function, which quantifies the resource consumption of each of the location techniques. Further details on the cost function are provided in the next section.
- The *Post-processing* stage is responsible for managing the results. The procedure followed in the case of location failures, i.e. QoS offered by the system lower than the requested, is to execute the next location technique in the MILCO's ranking, provided the response-time required has not run out. Notice that this behavior can be modified adding as many output modules as necessary.

2.2 Input modules

Input modules are used at the pre-filtering stage to extend the functionalities of MILCO and to improve its performance. The reference implementation for MILCO accounts for two input modules: a location cache and a concurrence manager. These two modules help reduce the number of requests reaching the cost function. As a consequence, the overall

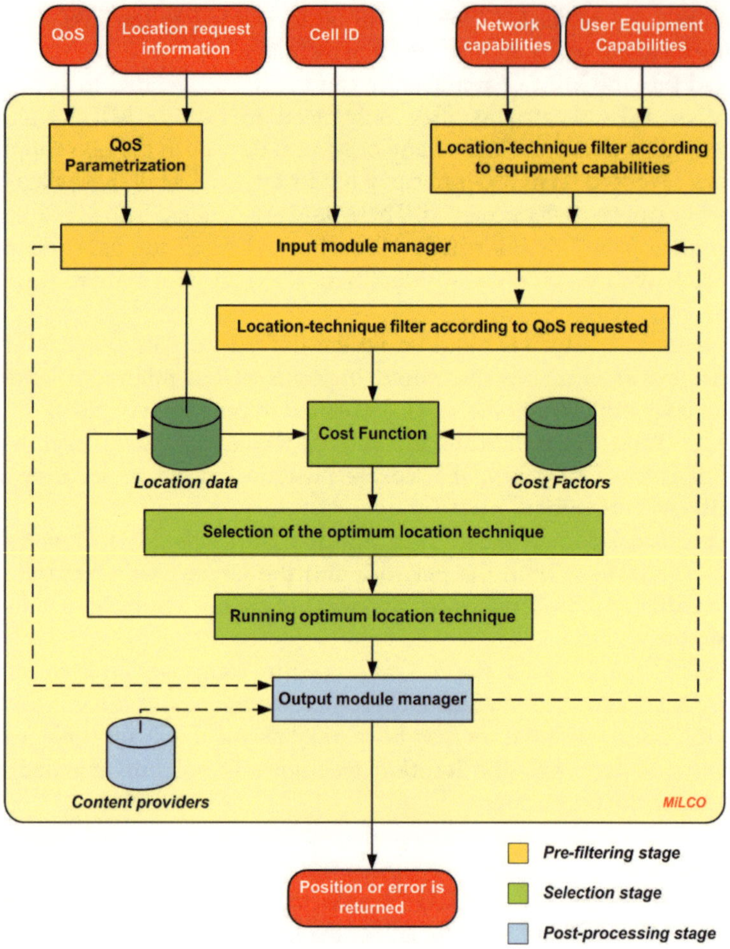

Fig. 5. Block diagram of MILCO

amount of resources used for the location is reduced, since no location technique is run to attend the request, and heavier traffic conditions can be handled.

The location cache saves positions reported in the past to estimate new positions in a near future. The main assumption taken by this module is the user being close enough to those past positions. It means that this module is addressed to users with a slow and pretty constant speed mobility pattern. There are several approaches to verify that the terminal position is close enough to the last stored position (Biswas et al., 2002). The one taken by MILCO consists of building a database with the positions fixed, the QoS achieved and the time at which positions where returned, using this latter information to compute the age of the stored positions and assess if cache module can be run. If the positions stored in the database are close enough to the current time, the cache modules computes the average speed and direction of the user terminal and uses these data to estimate the current position of the mobile station. Subsequently, this estimation may be sufficient, depending on the QoS required, for the task at hand; hence, fewer resources are required for positioning. Accordingly, the performance of the module depends on how old are the positions stored in the database, the mobility pattern of users and the level of QoS requested.

Concurrence aims at avoiding collisions at request level, i.e., a location request is received while another requiring better or equal QoSis still in progress, both asking for the position of the same user. Under such situations, the concurrence manager removes unnecessary traffic in the network, blocking the last received request until the ongoing one finishes. Then, the resulting position is shared by the two requests, even though this may result in a situation where some of the positions returned provide better QoS than necessary. Consequently, the concurrence manager is required to store the input data related to the request (i.e., those data feeding MILCO) to match the QoS obtained by the current technique to the one required by the blocked request. All these data are necessary to handle those cases in which concurrence fails and other input modules or the cost function must be run.

2.3 Cost function

The cost function can be considered the MILCO's core It ranks the location techniques suitable for the request (i.e., those available after filtering) according to each technique's resource consumption. Therefore, the more resources the technique consumes the lower it is ranked

The use of resources can be computed based on several factors. All these factors would be subsequentlycombined to obtain an overall cost so that location techniques are ranked. The way in which these factors are combined is defined by the cost function, as shown below:

$$Z_i(t) = f\left(\left\{\alpha_1(t), z_1^i(t)\right\}, \ldots, \left\{\alpha_n(t), z_n^i(t)\right\}\right), \tag{1}$$

where $Z_i(t)$ represents the resources consumed by the ith location technique at a specific time t, f stands for a given function, and a_j and $z_j^i(t)$ are the weight and the value of the jth factor applied to the ith location technique, respectively. Several functions f may be used to calculate the use of resources. The reference implementation for MILCO uses a simple additive function with m, defined as

$$Z_i(t) = \sum_{j=1}^{m} \alpha_j(t) z_j^i(t). \tag{2}$$

It must be noted that Equation (2) is a first approach to the cost function. It has been formulated on the premise of simplicity and its main purpose is to evaluate the performance of MILCO under low-requirement conditions. Better results could be expected when using more complex functions, but the impact of such complexity on the response time of location requests needs to be quantified and could involve a serious constraint. Furthermore, the actual response time would depend on the hardware and software implementation, which is beyond the scope of this chapter.

2.4 Output modules

Output modules are responsible for managing the result of the positioning. The purpose of output modules is twofold: to help recover from location errors and to optimize the computed position. The basic output module deals with location errors and its performance consists in retrying the MILCO procedure as long as it is expected to conclude before reaching the QoS-imposed deadline.

Additional output modules are expected to work with MILCO, such as those related to content providers, which can greatly enhance the QoS of the position reported especially in terms of accuracy.

4. Performance assessment

The middleware has been analyzed through simulation. The simulator wraps the simulation area to minimize the impact of the edge effects on the results. The simulation area is turned into a torus (Zander & Kim, 2001) thus becoming a virtually infinite surface with regard to mobility and propagation patterns. This tool is used in upcoming sections to evaluate the middleware under several architectures, networks, location techniques and scenarios.

4.1 Network-based implementation

This section explores the performance of the middleware when it is implemented in the core elements of a UMTS network.

4.1.1 Cost factors

4.1.1.1 Signaling volume

This cost factor accounts for the amount of information exchanged by each technique. This factor is aimed at favoring lighter techniques, i.e., those requiring less traffic on the network to compute the target position.

In the computation of the signaling volume, the following assumptions are made:

- Only the topmost protocol in the stack (e.g. RANAP, NBAP, etc.) is taken into account.
- A-GPS does not include acquisition assistance information.
- OTDOA and A-GPS can be run with and without assistance data.
- A-GPS running without assistance data means not including the *Almanac* information.
- Hybrid OTDOA/A-GPS includes acquisition assistance information.

Table 1 summarizes the quantification of the signaling volume cost factor for the location techniques allowed by 3GPP in UMTS networks N_{NB} and N_{SAT} in Table 1 stand for the amount of Node-B and satellites involved in the positioning, respectively.

Technique	***Assistance***	***Cost***
Cell-ID	No	0
OTDOA	Yes	$375+134 \cdot N_{NB}$
OTDOA	No	268
A-GPS	Yes	$473+1199 \cdot N_{SAT}$
A-GPS	No	$461+647 \cdot N_{SAT}$
Hybrid	Yes	$653+134 \cdot N_{NB} + 1254 \cdot N_{SAT}$

Table 1. Quantification of the signaling volume

4.1.1.2 Use of wideband interfaces

This cost factor favors those techniques that use wideband channels. Accordingly, it favors those techniques operating in the core network (i.e., network-based techniques). The cost associated with this factor is computed as

$$z_1 = \sum_i r_i^{-1} \quad [ns / bit], \tag{3}$$

where r stands for the throughput of a given channel i and z_1 accounts for the cost of all the channels involved in the location process. The Cell-ID is assumed to be delivered to MILCO, and hence, the cost for this factor is 0. On the other hand, the other techniques (i.e. OTDOA, A-GPS and hybrid) are mobile-based and involve the same amount of messages and channels. Under the assumption of I_{ub} and U_u channels having a throughput of 155 Mbps and 384 Kbps respectively, z_1 for mobile-based techniques is

$$z_1 = 2\left(\frac{1}{155Mbps} + \frac{1}{384Kbps}\right)10^9 \quad [ns / bit], \tag{4}$$

4.1.1.3 Energy consumption

The last cost factor proposed for UMTS networks accounts for the amount of energy required by each technique to fix the position. This factor aims to maximize the lifetime of the terminal. Power consumption largely depends on the user terminal performance. Here, a simple approach for quantifying power consumption is proposed, which is based on the amount of sources involved in the positioning. The cost of this factor for the location techniques in UMTS is summarized in the Table 2. It must be highlighted that this approach is meant to qualitatively compare the battery consumption of the various techniques, not to set up differences of actual consumptions.

Technique	*Cost*
Cell-ID	0
OTDOA	N_{NB}
A-GPS	N_{SAT}
Hybrid	$N_{NB} + N_{SAT}$

Table 2. Quantification of the energy consumed by each location technique

4.1.2 Scenarios simulated

The first scenario in which MILCO is evaluated corresponds to a UMTS network (Martin-Escalona & Barcelo-Arroyo, 2006). The call admission control (CAC) used in the simulator was proposed in (Capone & Redana, 2001) and it is based on the impact of new users on the Signal to Interference Ratio (SIR) of ongoing services. It accepts new users whenever the actual SIR (SIR_2) of all of ongoing calls in the cell does not drop below the target SIR by more than 1 dB. Otherwise, the service request is blocked.

The power control algorithm was borrowed from (Nuyami, Lagrange, & Godlewski, 2002). Its performance is illustrated in Fig. 6, where P_{tx} and P_{rx} stand for the signal strength transmitted by the mobile station and received in serving node-B, respectively. The algorithm checks whether the transmitting power of the MS should be increased or decreased Δ dB according to the target SIR and sensibility measured in serving node-B. Table 3 shows the values used in the simulator for all the parameters required by the power control algorithm.

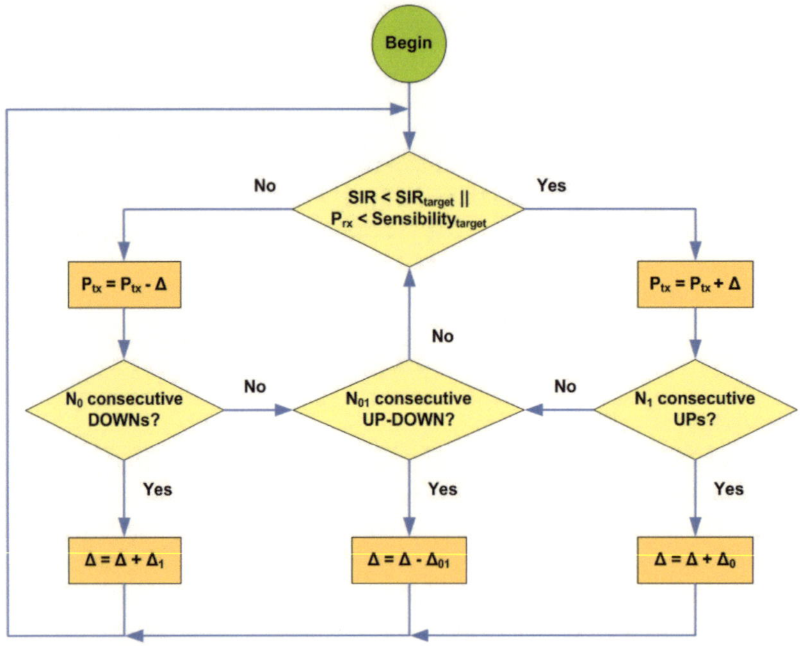

Fig. 6. Power control algorithm

A basic scenario simulates several location loads ranging from 0.01 to 1 request per second. This basic scenario was composed of 100 Node Bs (NBs), which were uniformly placed in a square-shaped simulation area. Each node-B, which involves a cell with a theoretical coverage of 1135-meters, is placed in the center of a square-shaped building. Fig. 7 shows the simulation layout. It must be noted that an important share of the whole area is an overlapping region, i.e., covered by more than one Node-B. This feature puts the simulation closer to reality and at the same time allows OTDOA, which is not possible in areas covered by only one or two Node-Bs.

Parameter	*Value*
$\eta_0 = \eta_1 = \eta_{01}$	2
$\Delta_0 = \Delta_1 = \Delta_{01}$	10 dB
Δ (maximum)	10 dB
Δ (minimum)	-20 dB
Δ (initial)	0 dB
Power updates between movements	20

Table 3. Parameters of the power control algorithm

Buildings simulate indoor conditions and as consequence, the signal reception inside them is limited. Users move freely within the simulation boundaries and are able to enter the buildings. It must be noted that MILCO makes decisions according to the location request features, the ultimate target of which is a specific mobile station. Consequently, no matter how many users are in the network to carry out the performance assessment. The mobility pattern follows a random-walk approach (Atsan & Özkasap, 2006), in which the user's speed is updated once per second and velocity in both directions, x and y, are modeled as normal random variables. Pedestrian users are taken into account and therefore, a mean and a standard deviation of 0.6 m/s and 0.18 m/s respectively are set for the user's speed random variables.

The propagation pattern is based on the Okumura-Hata model. According to (Holma & Toskala, 2000), the path-loss slope and zero-meter losses for the pretended scenario were set to 4 and 23dB, respectively. The SIR is calculated according to (3GPP, 2004), which accounts for a spreading factor of 10 dB and an orthogonality factor of 0.4, respectively. Handoffs are requested each time the received power or SIR in a Node B or MS fall below a given threshold, which is known as the handoff threshold. The handoff request is held until either a new channel becomes free and the handoff is then achieved or the SIR or the received power falls below the sensitivity level for more than 15 seconds, which produces a handoff failure and the service disruption. Successful handoffs drop all ongoing location requests carried by the mobile station and unsuccessful handoffs shut down the user terminal for a mean exponential time of 5 seconds. The main propagation pattern parameters have been taken from (Holma & Toskala, 2000) and (3GPP, 2004) and are displayed in Table 4.

Parameter	*Value*
Minimum SIR	-9 dB
Sensitivity of the stations	-109.2 dBm
Maximum MS transmission power	21 dBm
Minimum MS transmission power	-44 dBm
Node B transmission power	43 dBm
Handoff threshold for received power	-106.2 dBm
Handoff threshold for the SIR at reception	SIR_{min} - 6 dB

Table 4. Propagation pattern parameters

The cell-ID, OTDOA and A-GPS location techniques were taken into account, in addition to a hybrid tight-synchronized OTDOA/A-GPS location technique (Barcelo & Martin-Escalona, 2004). The QoS provided by such techniques, in terms of the expected accuracy and response time, is shown in Table 5, where the *mean* and the *std* stands for the average and standard deviation respectively and the *range* indicates the set of values that the variable may take. The availability of the OTDOA depends on the radio propagation pattern and it is computed on execution time depending on the received power and the SIR. In the case of satellite-based techniques, availability is accounted for differently. The default number of satellites at a sight is set to 5. This number drops to a uniformly distributed value from 0 to 2 satellites inside buildings. It must be noted that the QoS provided by the coupling technique is worse than that achieved by the A-GPS as standalone. This result is due to the greater availability of the hybrid technique, which is favored instead of the accuracy. Furthermore, the lifecycle of the assistance data for the OTDOA and A-GPS is set to 30 seconds, i.e., the assistance information expires 30 seconds after it has been received.

Four LBS generate requests for the station (Martin-Escalona & Barcelo-Arroyo, 2007): emergency, tracking, push and tracing. Table 6 shows the QoS requested by these services and their cadence, i.e., the time between consecutive requests. This later is exponentially distributed in all services. Tracing service differs from the rest in the fact request are received as a burst, i.e., each LBS request involve several LCS requests. The number of LCS requests in the burst is uniformly distributed from 1 to 5, each of the requests separated 20

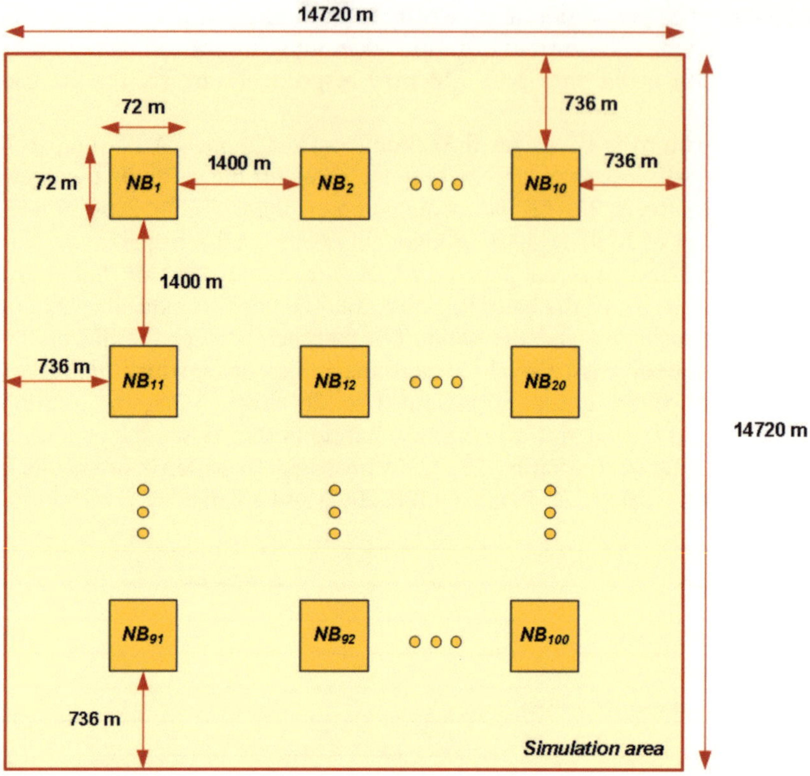

Fig. 7. Simulation layout

	Accuracy (meters)				Delay (seconds)	
	Distribution	*Mean*	*Std*	*Range*	*Distribution*	*Mean*
Cell-ID	Deterministic	1135	0	1135	Deterministic	0
OTDOA	Uniform	100	28.7	[50,150]	Exponential	7
A-GPS	Gaussian	3	0.9	[0,+∞)	Exponential	11
Hybrid	Gaussian	50	15	[0,+∞)	Exponential	27

Table 5. QoS achieved by the location techniques

seconds. Not satisfying either the accuracy or the delay involves not fulfilling the QoS requested. Other QoS approaches are allowed, with more parameters and different constraints, but the most restrictive definition (according to 3GPP) is used in this performance assessment.

With respect to the input modules, location cache stores the positions for 2 seconds and then they are removed from the database. The maximum value of a weighted factor was set to 1, which means that the importance of all the cost factors is the same. The maximum value of the cost function is then 3. Weights for the cost factors are assumed to be deterministic and are computed according to that equality assumption. Tuning the weights of the factors is beyond the scope of this work because it is assumed that the setting of these weights would be a task for network operators, thus allowing them to focus their attention on the factors they consider more important at the time.

Service	*Average time between requests*	*Accuracy*	*Response time*
Emergency	30 min	50 m	10 s
Tracking	2 min	150 m	15 s
Push service	300 min	1500 m	15 s
Tracing	10 min	50 m	15 s

Table 6. Main parameters for services simulated

4.1.3 Simulation results

The performance of plain MILCO systems, i.e., systems based only on the cost function and that discard all the input modules, must be analyzed first. Table 7 shows the percentage of successful LCS, the average number of location techniques used in successful LBS (i.e., the requested QoS was finally delivered) and the cost of delivering the LBS. The latter applies not only to those LCS successfully attended, but also accounts for all the LCS run until the QoS requested for the LBS is achieved. Thus, the cost per LBS can exceed the maximum per LCS, i.e., 3. The results in Table 7 correspond to the scenario based on the data in Table 6. Figures for scenarios with a heavier load are not included since they are statistically the same in all the scenarios (i.e., they are not sensitive to the load). Location techniques used as standalone are included for the sake of comparison.

Location Technique	*Average number of techniques*	*Percentage of successful LCS*	*Overall cost*
MILCO	1.36	64.84 %	2.06
CI	1.00	00.39 %	0.00
OTDOA	1.58	52.17 %	3.02
A-GPS	1.00	64.11 %	2.70
HYBRID	1.04	16.01 %	2.81

Table 7. Performance of MILCO based on the cost function

According to data in Table 7, MILCO achieves the best performance in terms of successful LBS, with figures very close to those achieved by A-GPS. Statistically, it can be stated that there are no differences between them. However, MILCO provides all these LBS with the lowest cost. The performance of MILCO is noticeable better if compared with the OTDOA, both in terms of technique executions and cost. It must be noted that MILCO runs more than one technique per LBS to achieve these figures. However, the cost function compensates this increase in the amount of techniques run does not impact the overall cost because *cheaper* techniques are run first. The poor availability and high cost of the hybrid technique constrains its results when used as standalone. Finally, Cell-ID is the more available and least costly technique, but it yields the least successful LBS rate. According to the results, Cell-ID and hybrid solutions are not suitable for being used as standalone; OTDOA and A-GPS can be understood as a trade-off solution, while MILCO provides the best results.

Performance was expected to be improved by input modules. Hereafter, all the results account for the cost function, and the location cache and the concurrence manager input modules are already enabled. Fig. 8 displays the evolution of the successful LCS requests with the load. The request rates in Fig. 8 start from the rate of services in Table 6 up to 100 times these rates. Therefore, simulations ranging from 0.01 requests per second (light-load profile) to 1.45 requests per second (heavy-load profile) were performed, which is assumed to be is sufficient for assess MILCO under the most demanding applications (e.g., tracking, tracing, etc.). Fig. 8 shows that at medium/low request rates (e.g., 0.05 requests per second), MILCO gives a successful LCS rate of around 65%. These poor figures are due to the QoS definition used in this performance assessment: an LCS is successful only if the accuracy and response time requirements are met. Furthermore, it must be noted that the higher the load, the higher the successful LCS rate. This behavior is due to the fact that the cache and concurrence modules are more likely to be used when less time is spent between consecutive location requests, i.e., more intense is the location traffic. This result proves that the scalability of the proposed approach is guaranteed. In the case of heavier loads, input modules enhance the percentage of successful requests.

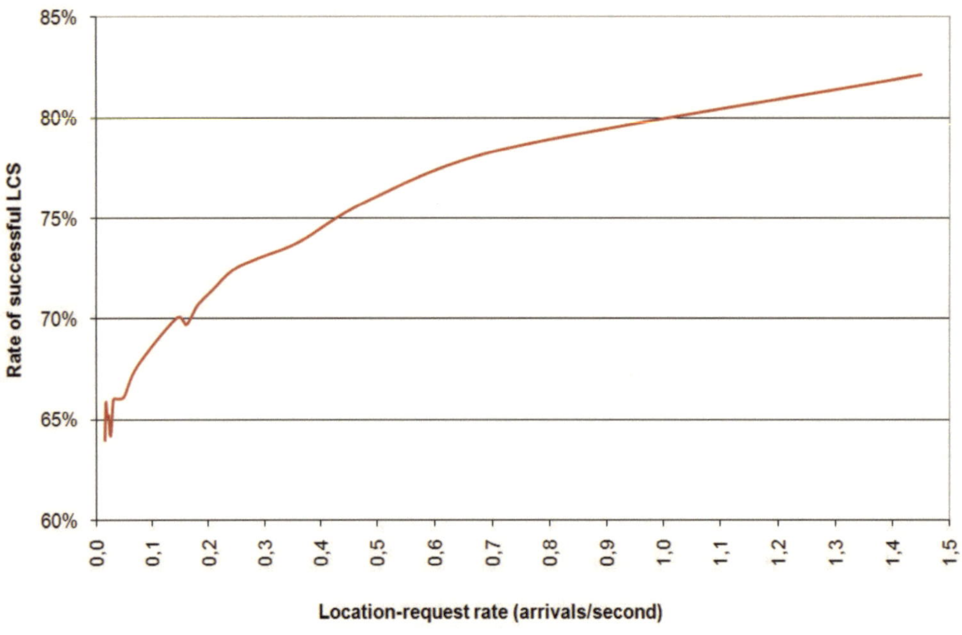

Fig. 8. Evolution of LCS successfully attended

Reducing the use of resources is another strong point of MILCO. Fig. 9 shows the average resources consumed by LBS successfully performed by MILCO. The maximum resources consumed by a single technique is set to 3 under the assumption that all cost factors are weighted the same. This cost is achieved by the hybrid approach, which usually consumes more resources to fix a position. This threshold is depicted in Fig. 9 with a green line. The resource consumption for successful LBS is always below the threshold. Furthermore, the consumption of resources drops as the load increases. This improvement is due to the increasing use of input modules (i.e., the location cache and concurrence manager) because these modules deal with location requests at no cost. Consequently, MILCO is a good approach for reducing the consumption of resources in location systems.

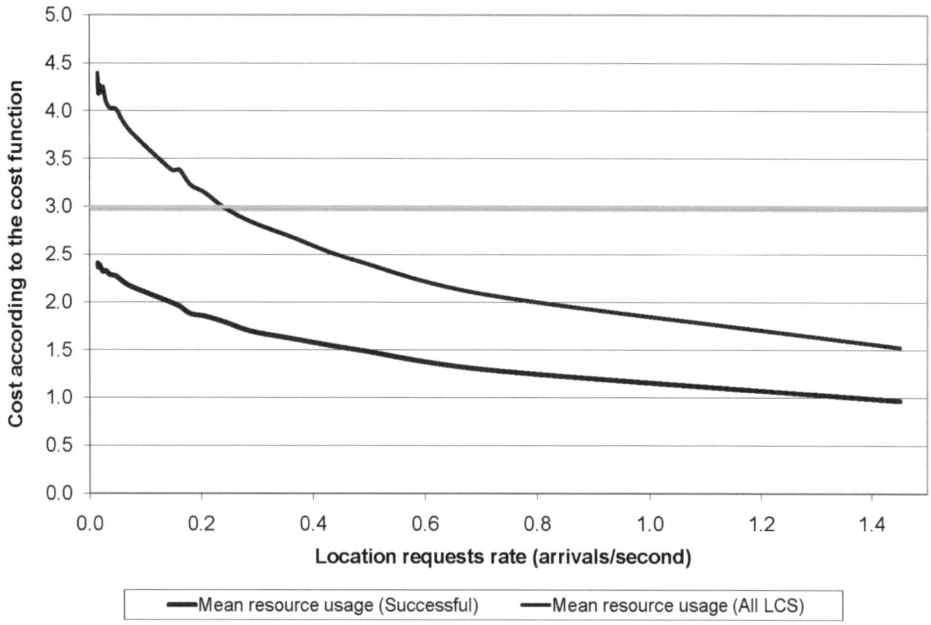

Fig. 9. Average cost of providing LCS in MILCO

Fig. 9 also shows the resources required by MILCO for attending all the LBS, i.e., those successful and unsuccessful. With lighter loads, the cost of providing the LBS is higher than the threshold. This result is due to the fact that unsuccessful LBS usually involve several techniques and hence a cost that is likely to be higher than 3. However, as the successful LBS rate increases with the load, unsuccessful LBS have less impact on the total amount of resources required for LBS delivery. Therefore, the advantages of using MILCO are more noticeable for heavier loads. The resources used are reduced by up to 50.88% in the scenario loaded with 1.45 requests per second and up to 32.2% in the same scenario if only successful LBS are taken into account.

Fig. 10 shows the performance of the cache input module as well as the average number of techniques run per successful LBS. The impact of the cache is stronger as the load becomes heavier. This result was expected because the system is more likely to receive several requests involving short displacements and consequently use the cache feature. In fact, in the scenario involving the heaviest load, the cache handled 52.57% of the requests. The intensive use of the cache results in a reduction of the average number of location techniques used per LBS, and consequently, there is a drop in the amount of resources consumed to attend the location traffic. Furthermore, because the cache is only valid for 2 seconds, 100% of the positions fixed through the cache fulfilled the QoS requirements. The lifetime of cache data could be extended according to the mobility pattern of users at the cost of more complexity in the MILCO implementations. Moreover, Fig. 10 demonstrates the scalability of MILCO, which reduces a 53.67% the average number of techniques used if compared with figures reported in the lightest load scenario.

Simulations show that the impact of the concurrence manager is negligible if compared with the cache module. Fig. 11 displays the percentage of LBS in which the concurrence manager

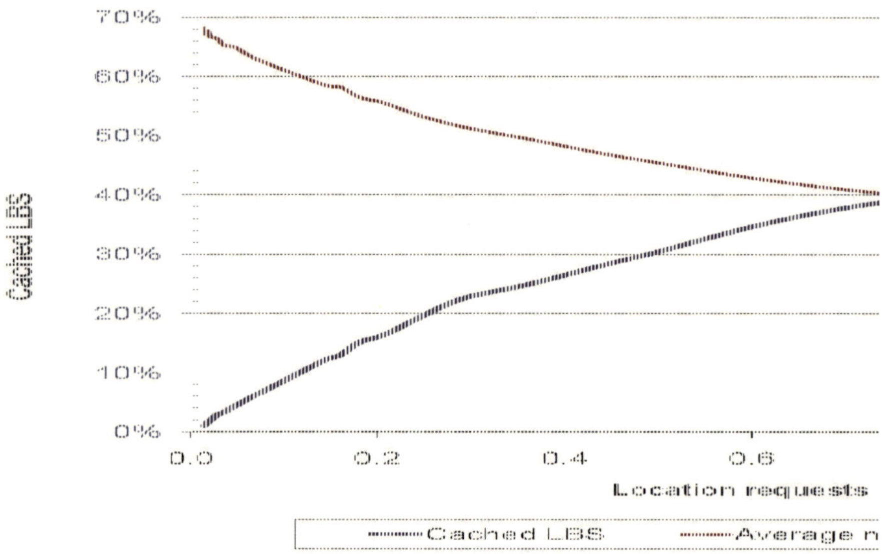

Fig. 10. Performance of the cache module

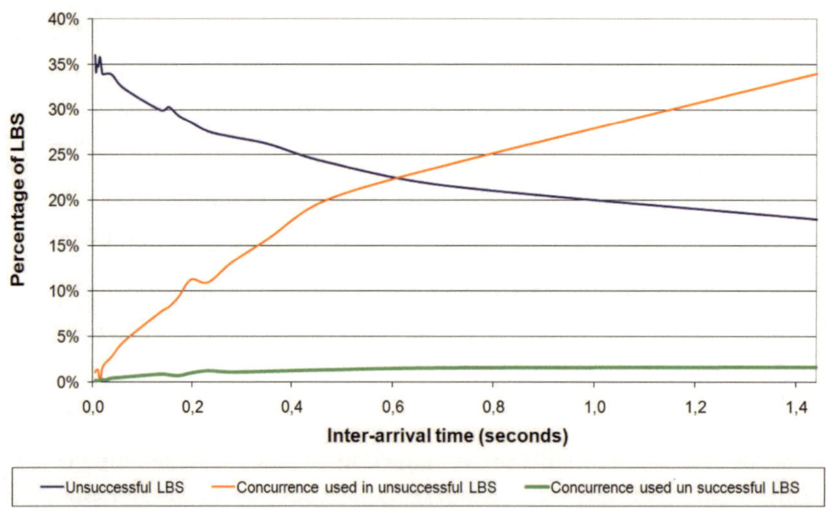

Fig. 11. Performance of the concurrence module

is involved. The rate of unsuccessful LBS is included as a reference. In the best case, only 1.3% of the successful LBS are handled by the concurrence manager. This behavior does not depend on the load. Although every improvement on the successful LBS is welcome, the performance of the concurrence manager for the location is far from being optimum. As long as the load increases, the percentage of unsuccessful LBS decreases and concurrence management appears to be the main reason for LBS to fail. This behavior is due to the fact that a higher load involves more blocked requests and therefore a greater impact on positioning failures. It is expected that running multiple location techniques instead of blocking them will slightly improve the percentage of successful LBS but at the cost of a noticeable increase in the resources consumed. Furthermore, location devices are usually small and computationally restricted, which means that several techniques can rarely run simultaneously.

4.2 Handset-based implementation

MILCO can be implemented either in the core network or the user terminal as a new piece of software. The latter approach has been followed to evaluate the performance of MILCO in wireless LAN (WLAN) networks (Martin-Escalona & Barcelo-Arroyo, 2008). Handset-based implementations allow the middleware to use any information available in the user terminal with minimal delay because all data exchange to the middleware is done locally. However, these operations are performed at the cost of reducing the grade of optimization that can be achieved. Only optimizations local to the user terminal can be done because full system optimizations require middleware components to be distributed along the entire LBS supply chain.

The location system architecture is similar to the one presented in Fig. 4. Each time a location request reaches the location system, it is delivered to the user terminal, where the request is finally handled by the middleware. Once there, the middleware analyzes all the requirements included in the location request (e.g., the QoS demanded) and gathers all the facilities provided by the user terminal (e.g., the location techniques implemented). Then, the middleware selects the location technique that best fits the request, i.e., the one expected to achieve the requested QoS with the minimum amount of resources. Finally, the middleware uses the user-terminal facilities to fix the user's position and forward the result to the location service (LBS) client that requested it.

In this implementation of MILCO, input modules are not accounted for even though the application of those modules to MS-based MILCO is obviously feasible. This was ignored because the main purpose of this study was to evaluate the performance of the cost function because similar results for cache and concurrence manager modules are expected independent of the device in which MILCO is implemented.

4.2.1 Cost factors

4.2.1.1 Success probability

This cost factor computes the probability of a location technique reaching the QoS requested by means of two histograms, one for the accuracy and one for the response time. Then, the success probability is calculated as:

$$z_2 = Pr\left[A\left(LT_i\right) \leq Accuracy_{requested}\right] \cdot Pr\left[\tau\left(LT_i\right) \leq \tau_{requested}\right], \qquad (5)$$

where z_2 stands for the success probability, and A and τ are the estimates for the accuracy and the response time of location technique LT_i. Histograms are built locally to a certain area (SP_CELL), usually smaller than the simulation area, to increase the precision of the success-probability estimation. The smaller the *SP_CELLs* are, the more accurate. The drawback of this cellular-fashioned approach is the memory requirement, which increases according to the number of *SP_CELLs* (i.e., the number of histograms computed). Therefore, there is a trade-off between accuracy and memory consumption. The *SP_CELL* matches the coverage area of an access point (AP), i.e., of a cell. Therefore, two histograms are built for each access point available in the network. The mobile equipment uses the pair corresponding to the access point that it is associated with or the one corresponding to the access point with highest RSSI.

WLAN networks are usually deployed indoors, and consequently, the location solution is expected to work under constrained conditions. This behavior means that signal conditions and consequently QoS offered by location techniques may change drastically. Consequently, the histogram computation follows a non-linear approach. Thus, not all the samples in the histogram are weighted the same. Recent samples are favored because they are more likely to be correlated with future positions than the older samples stored in the histogram. The weight of each sample is computed as

$$\alpha_2(n) = \begin{cases} g_{min} + B \cdot \log(n), & 1 \le n \le M \\ g_{max} & , \quad M \le n \le N \end{cases}, \tag{6}$$

where g_{min} and g_{max} are the minimum and maximum gains, respectively, M stands for the number of weighted samples and N is the maximum number of samples used to compute the histogram. B is a scale factor that is based on the g_{min}, g_{max} and M parameters. A sliding windows of N samples is run to build the histogram, i.e., if a new sample is added to a histogram with N samples, the oldest sample is removed to make room for the new one and the rest of samples are shifted one position. This approach allows memory in the user terminal to be saved. Notice that with larger values of N, more accurate results are expected.

4.2.1.2 Energy consumption

Energy consumption is one of the most common issues in user terminals implementing location techniques because running such techniques usually demands much more energy than simple communications tasks. This drain is much more noticeable as the number of techniques implemented in the terminal increases. As in case of UMTS networks, this factor constrains the use of the techniques according to the energy consumption and the remaining battery in the terminal. The values proposed for this cost factor, which are displayed in Table 8, are only provided as a proof-of-concept of the location middleware according to the authors' experience. N_{AP} and N_{SAT} in Table 8 stand for the number of access points and satellites that are involved in the positioning process, respectively.

The user terminal consumes energy for several reasons:

- *Attending to incoming services*. These tasks involve an energy drop due to signal demodulation and packet building and interpretation. This process is quantified as one unit of energy dropping.
- *Location technique execution*. The station consumes energy each time a location technique is run.

Table 8 shows, comparatively, the energy drop expected from each location technique. The quantification of this factor should depend on the remaining battery of the terminal because highly demanding location techniques could deplete the battery in a short time making further positioning impossible. The middleware weights this factor as

$$\alpha_3(t,t_0)=\begin{cases}\alpha_3(t_0)\left(1-log\left(\dfrac{Battery(t)}{Battery(t_0)}\right)\right)^{3/2} & ,\alpha_3(t,t_0)<\beta \\ \beta & ,\ \alpha_3(t,t_0)\geq\beta\end{cases} \tag{7}$$

where t_0 is the time at which the battery is completely charged, and *Battery(t)* is a function that calculates the remaining battery in the terminal at a certain time *t*. The maximum weight for this factor needs also to be set to limit the impact of this factor in the cost function.

Technique	*Cost*
WLAN Fingerprinting	$10 + N_{AP}$
MEMS	1
Assisted GPS	$10 + N_{SAT}$

Table 8. Energy consumption according to the location technique

4.2.1.3 Expected accuracy

Although location QoS includes several parameters, it is often reduced to a couple of metrics describing the accuracy or delay. An examination of user requirements reveals that accuracy is more restrictive than delay, i.e., users are willing to wait longer for more accurate results. This cost factor aims at giving less weight to those techniques that are not likely tofulfill the accuracy requirements.

The expected accuracy is computed as the average accuracy of each location technique. This should be a static cost factor because the expected accuracy comes from a previous analysis of the performance of each location technique. However, the accuracy of some techniques is dependent on time. It is the case of inertial solutions (MEMS), which depends on the distance travelled since the last reference positioning (e.g., a GPS position). Consequently, this cost factor updates its value over time for these time-dependent techniques while for other techniques, such as WLAN fingerprinting (WLAN-FP) or A-GPS, the cost factor value is constant. Further details on the values of this factor can be found in Table 11, in which average accuracies for the simulated location techniques are provided.

4.2.2 Simulated scenario

The simulator was used to model an indoor WLAN network. The proposed scenario consists of a single service and station. The simulation layout models a square-shaped corridor. The user moves freely through corridors, which are 4 m wide. However, they cannot cross the forbidden area (simulation area outside the corridors). Access points placed in the forbidden area simulate instances in different rooms/floors than the user. The propagation pattern follows the Okumura-Hata model for indoor scenarios, with path-loss slope and zero-meter losses set to 3.5 and 40 dB, respectively. Handoffs are handled as explained for the UMTS network simulation. If a new channel is not found during the handoff, the service is interrupted and the user terminal backs off for an exponential time with a mean of 5 seconds. Table 9 reports the main parameters of the propagation pattern, which are based on current industry equipment and the authors' experience.

Parameter	*Value*
Minimum SIR	-9 dB
Sensitivity of the stations	-65 dBm
Maximum MS transmission power	17 dBm
Minimum MS transmission power	0 dBm
AP transmission power	17 dBm
Handoff threshold for received power	-62 dBm
Handoff threshold for SIR at reception	-6 dB

Table 9. Main parameters of the propagation pattern

This basic scenario is simulated with 9, 16, 25 and 36 access points. These access points are uniformly spread along a square-shaped simulation area, which gives each of them a minimum of 63 meters of coverage at minimum throughput according to the data in Table 9. Table 10 shows the coverage expected in terms of access points available in each scenario. *Scenario_3* models regular network deployments, where the stations receive signal from 2 to 4 APs. More than 4 APs under coverage are not considered because it is unlikely that such a network plan exists in actual WLAN deployments. Hence, *Scenario_4* is included as an example of an over-coverage network, whereas *Scenario_1* and *Scenario_2* are examples of constrained scenarios. These latter scenarios represent realistic situations with only a partially working infrastructure. The minimum coverage is computed according to analytical propagation models. However, the simulations involve factors not included in the analytical calculation.

Scenario name	*Number of APs*	*Minimum coverage*	*Maximum coverage*
Scenario_1	9	0 AP	1 AP
Scenario_2	16	0 AP	2 AP
Scenario_3	25	2 AP	4 AP
Scenario_4	36	4 AP	4 AP

Table 10. The scenarios simulated

Four location techniques have been taken into account in these scenarios: WLAN fingerprinting (FP) and A-GPS as standalone techniques and A-GPS/MEMS and FP/MEMS couplings. Table 11 shows the average accuracy expected for each standalone technique, in which d stands for the distance travelled since the last positioning was calculated with WLAN-FP or A-GPS. All these data (along with other data related to the capabilities of the location techniques) have been borrowed from (Thales Selena Space et al., 2007).

Fig. 12 displays the distribution of the positioning error of the WLAN-FP technique according to the data supplied in (Thales Selena Space et al., 2007). The first and second rows in Fig. 12 stand for the error module in the x and y coordinates, respectively. Only 2D positioning is considered. The column in Fig. 12 represents the number of access points involved in the positioning, which range from 1 (top left) to 4 (top right). Fig. 13 shows the accuracy expected from MEMS in a light indoor scenario according to the data provided in (Thales Selena Space et al., 2007). The simulator couples MEMS with another technique as long as the position provided by such a technique has better accuracy than 4 meters. MEMS keeps working in coupled mode until a position is expected to provide an error beyond 6 meters. Consequently, the results are expected to be slightly conservative because in real scenarios MEMS could be used in more positioning processes.

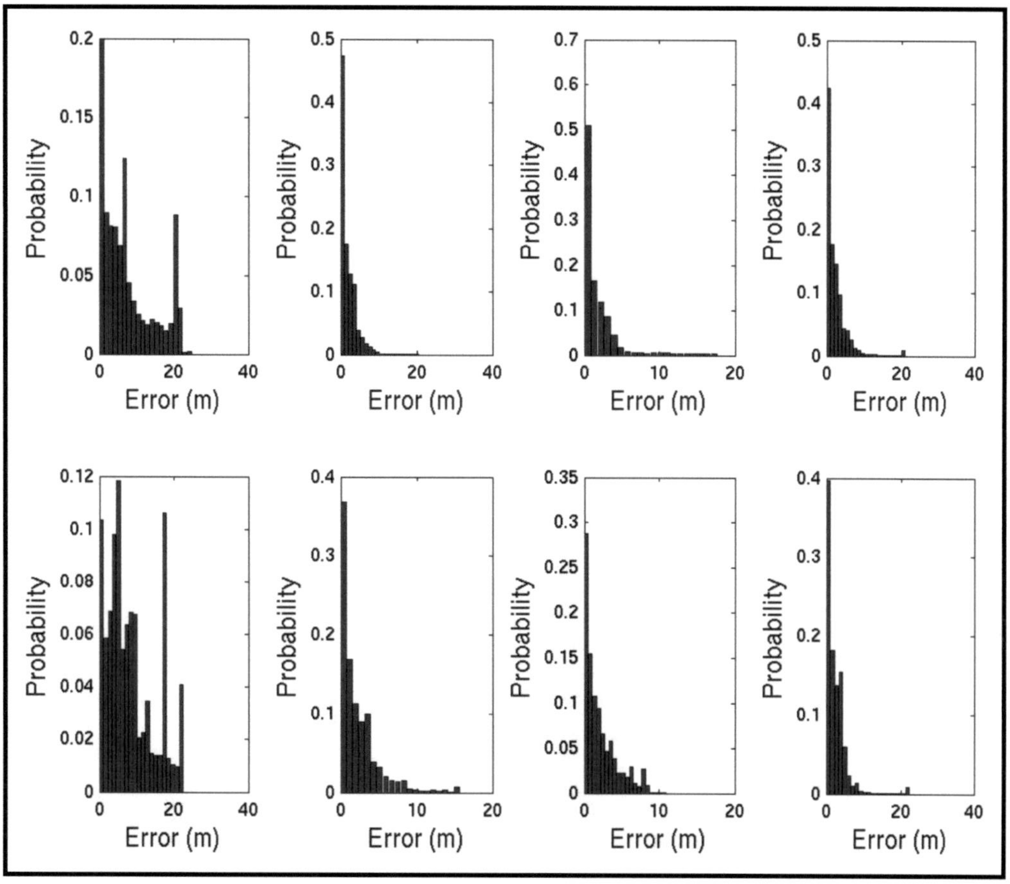

Fig. 12. The accuracy of x (first row) and y (second row) coordinates provided by the WLAN-FP technique with 1 AP (leftmost column) to 4 AP (rightmost column) in sight.

To reduce the complexity of implementing the whole satellite map and estimate the signal availability, the simulator computes the availability of GPS similarly to the UMTS case. the simulator provides the availability for A-GPS satellites uniformly distributed from 2 to 4 satellites if the user is at most 1 meter away from the simulation area edges. These emplacements are considered as light indoor scenarios (i.e., close to windows) and thus A-GPS would be able to receive weak signals from few satellites. Other locations are assumed to be in deep indoor conditions and thus no position at all is provided by A-GPS. The expected values for the accuracy of all techniques are presented in Table 11. In the case of A-GPS, the positioning error is Gaussian distributed with a square coefficient of variation of 0.3.

The cost function includes all the cost factors presented: success probability, energy consumption and expected accuracy. The N and M parameters in Equation (6) are set to 256 and 512 samples, respectively, and the minimum (g_{min}) and maximum (g_{max}) gains for those samples are 1 and 8, respectively.

The weights of the factors in the cost function are 1 and 0.0017 for the successful probability and expected accuracy, respectively. These figures are used to provide a cost of 1 under the worst conditions. The weight of the energy-consumption factor is provided by Equation (7), in which $\alpha_3(t_0)$ is set to 1 and the maximum value (β) is limited to 3. Consequently, the cost

Technique	*Expected accuracy*
WLAN Fingerprinting	12.2766 m (1 access point) 3.4058 m (2 access points) 3.1982 m (3 access points) 3.9329 m (4 access points)
MEMS	$1.65 + 0.2825 \cdot d$ meters
Assisted GPS	3 meters (only very light indoor scenarios)

Table 11. Expected accuracy of location techniques as standalone

function can produce values from 0 to 5. These values allow the optimum technique to be used as long as the battery in the user equipment has enough charge and smoothly switches to a power-saving technique as long as the energy is going to run out. Once the battery runs out, the station switches off for 5 seconds and then turns on completely recharged. The time the station spends between switching off and on simulates the network re-association process.

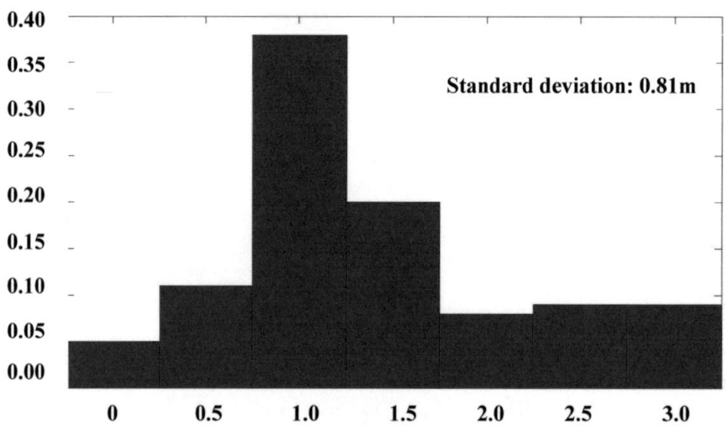

Fig. 13. The accuracy of MEMS in light indoor scenario

One single LBS is simulated, generating one request each 5 seconds, and requesting an accuracy of 6 meters. Simulations do not account for the response-time in the QoS requirements. This approach was taken because in indoors, customers perceive more degradation in the QoS when the required accuracy is not achieved. Furthermore, the response-time in mobile-based techniques is expected to be mostly the same, more if it is taken into account that most of the time used by the LCS is spent communicating with the network, not on executing the technique. The cost function is run twice at most to avoid infinite looping and save resources in the terminal.

4.2.1 Simulation results

This section presents the performance results obtained by MILCO and compares them with those achieved using WLAN-FP and A-GPS as standalone. MEMS is not evaluated on its own because this technique positions relatively to a previous location provided by WLAN-FP or A-GPS. Therefore, the positioning error in MEMS drifts with the distance covered, and as consequence MEMS needs correction updates from other location techniques periodically.

Table 12 presents the QoS achieved by means of the location techniques when used as standalone and the results obtained when middleware is run. The first parameter taken into account is the location traffic carried. Two situations may lead to a location request not being performed: the station being in a position without radio network coverage, and the station being in a *recharging* condition. The best performance was achieved by WLAN-FP and MILCO, which are able to carry more than the 80% of the traffic. Only slight differences can be found between the performances of these two techniques. However, MILCO's performance is more stable due to the better management of resources (i.e. less *recharging* situations), and the support of MEMS (i.e. coverage improvement). Although A-GPS performs poorly as a standalone system, because it only works under light indoor conditions, it must be noted that a single A-GPS position can enable the MEMS techniques for a long time. In all scenarios, the traffic carried by MILCO is at least as good as that carried by WLAN-FP used as a standalone system and usually better than the latter. It must be noted that A-GPS as a standalone system provides excellent figures for the rest of the variables in this study, but they only apply to less than 0.5% of the traffic (i.e., location traffic already carried). Accordingly, A-GPS results cannot be considered suitable to deliver any kind of LBS as a standalone system, and consequently, related results will not be commented on hereafter.

The percentage of traffic successfully handled measures the amount of carried traffic that yields a successfully attended request (i.e., with the required QoS already achieved). Under excellent coverage conditions (i.e., *Scenario_4*), MILCO and regular location techniques provide almost the same ratio of successfully handled LBS. However, reducing the number of access points in sight drastically impacts the figures provided by the WLAN-FP solution. The same does not apply to MILCO, which is not as sensitive to the number of access points in sight. This behavior is due to the fact that MILCO is able to use MEMS when positions provided by WLAN fingerprinting become noisy. As previously observed, under the worst conditions (i.e. *Scenario_1*), MILCO successfully handles 91% of carried traffic versus 49.1% achieved by WLAN fingerprinting used as a standalone system. It is because MILCO takes also benefit from A-GPS positions. According to these results, MILCO is more robust in front of integrity failures, since it manages several location techniques and modulates their use according to the resources in the network.

Data in Table 12 shows that in the first three scenarios MILCO outperforms WLAN-FP in terms of average accuracy, whereas WLAN-FP provides better accuracy in *Scenario_4*. However, the LBS client demands positioning errors lower than 6 meters, and in this scenario, both MILCO and WLAN-FP provide figures for positioning error below this threshold. The less accurate positions for MILCO in this scenario are a consequence of the noisier positions provided by the MEMS technique. On the other hand, the use of MEMS reduces battery consumption. MILCO looks for the optimum technique for each request and thus modulates the use of MEMS to fulfill the QoS requirements and at the same time save network resources.

Similar results are reported in terms of resource consumption (according to the cost function). The maximum cost expected according to the simulation parameters is 10, which is twice the achievable maximum cost. According to the data in Table 12, MILCO reduces the cost of providing LBS by more than 46% in all scenarios. As expected, the cost increases with the lack of available access points because unsuccessful LBS involve several techniques being run. Better results are observed only if successful LBS are taken into account, which achieve a reduction in the consumption of resources higher than 50% in all the scenarios.

Parameter	*Scenario*	*A-GPS*	*WLAN-FP*	*MILCO*
Carried location traffic	Scenario 1	0,39%	85,70%	90,56%
	Scenario 2	0,42%	80,14%	99,93%
	Scenario 3	0,39%	99,73%	99,94%
	Scenario 4	0,42%	92,29%	99,97%
Successful LBS (only carried traffic)	Scenario 1	100,00%	49,11%	91,01%
	Scenario 2	100,00%	51,17%	92,70%
	Scenario 3	100,00%	61,96%	94,81%
	Scenario 4	100,00%	97,45%	99,60%
Accuracy	Scenario 1	2,82 m	9,72 m	5.23 m
	Scenario 2	3,13 m	9,15 m	5.06 m
	Scenario 3	3,09 m	7,69 m	4.73 m
	Scenario 4	2,96 m	2,60 m	3.92 m
Average cost	Scenario 1	2,22	5,40	2,88
	Scenario 2	2,21	5,35	2,79
	Scenario 3	2,05	5,11	2,61
	Scenario 4	2,18	4,13	1,97
Amount of location techniques per LBS	Scenario 1	1.00	1,71	1,56
	Scenario 2	1.00	1,67	1,51
	Scenario 3	1.00	1,55	1,41
	Scenario 4	1.00	1,10	1,12

Table 12. The QoS achieved by techniques as standalone and MILCO

According to the results shown, extended battery lifetime and improved performance is expected when using MILCO. Furthermore, the average number of techniques required by LBS is reduced as the availability conditions improve (as can be expected). MILCO uses fewer techniques to attend LBS on average, except for the best scenario, in which the excellent success percentages cause MILCO to achieve the same performance as WLAN-FP. Even though the results are statistically similar, the techniques used by MILCO involve less resource consumption than in the case of WLAN-FP as a standalone technique.

5. Concluding remarks

This chapter offers a brief overview of middleware for positioning. A new middleware for optimizing the cost of LBS provisioning was presented. This novel approach has not been examined closely by the research community even though a great demand for LBS is expected. Different implementations of the middleware (handset-based and network-based) were presented and evaluated, and in all of them, the middleware provides a way to reduce the resources consumed to provide LBS and, at the same time, optimize several parameters of the LBS provisioning chain, such as the stability of the accuracy or the scalability of the location system.

6. References

3GPP. (2004). Functional stage 2 description of Location Services (LCS). Release 6. *3GPP TS 23.271 (v6.8.0)* . 3GPP.

3GPP. (2002). Location Services (LCS); Functional description; Stage 2. *3GPP TS 03.71 (v8.7.0)* . 3GPP.

3GPP. (2004). Universal Mobile Telecommunications System (UMTS); RF system scenarios. *3GPP TR 25.942 (ETSI TR 125 942) , 6.3.0* . 3GPP.

Atsan, E., & Özkasap, Ö. (2006). A Classification and Performance Comparison of Mobility Models for Ad Hoc Networks. *Ad-Hoc, Mobile, and Wireless Networks (LNCS) , 4104/2006*, 444-457.

Barcelo, F., & Martin-Escalona, I. (2004). Coverage of Hybrid Terrestrial-Satellite Location in Mobile Communications. *European Wireless Conference: Mobile and Wireless Systems beyond 3G*, (pp. 475-479).

Biswas, P., Han, S., & Wu, J. (2002). Location Caching in the Mobile Middleware Platform. *Third International Conference on Mobile Data Management*, (pp. 172-174).

Capone, A., & Redana, S. (2001). Call Admission Control Techniques for UMTS. *Vehicular Technology Conference (VTC-Fall). 2*, pp. 925-929. IEEE.

EMILY system trials report. (2005). *EMILY project (IST-2000-26040)* . European Comission.

Fritsch, D., & Volz, S. (2003). Nexus - the mobile GIS-environment. *Workshop on Mobile Future and Symposium on Trends in Communications (SympoTIC)*, (p. 1.4).

Holma, H., & Toskala, A. (Eds.). (2000). *WCDMA for UMTS: Radio Access for Third Generation Mobile Communications.* Wiley Technology Publishing.

Java Community Process. (n.d.). *JSR 179: Location API for J2METM.* Retrieved 2009 йил May from Community Development of Java Technology Specifications: http://www.jcp.org/en/jsr/detail?id=179

Küpper, A., Treu, G., & Linnhoff-Popien, C. (2006). TraX: a device-centric middleware framework for location-based services. *Communications Magazine , 44* (9), 114-120.

LaMarca, A., Chawathe, Y., Consolvo, S., Hightower, J., Smith, I., Scott, J., et al. (2005). Place Lab: Device Positioning Using Radio Beacons in the Wild. In *Pervasive Computing (Lecture Notes in Computer Science)* (Vol. 3468/2005, pp. 116-133). Springer Berlin / Heidelberg.

Mahmoud, Q. H. (Ed.). (2004). *Middleware for Communications.* John Wiley & Sons, Ltd.

Martin-Escalona, I., & Barcelo-Arroyo, F. (2006). A middleware approach for reducing the network cost of location traffic in cellular networks. *Journal of Communications Software and Systems (JCOMSS) , 2* (4), 305-312.

Martin-Escalona, I., & Barcelo-Arroyo, F. (2008). An approach to increase the scalability of Location Systems in WLAN networks. *Conference on MOBILe Wireless MiddleWARE, Operating Systems, and Applications (MOBILWARE). 278*, pp. 1-6. Innsbruck (Austria): ACM.

Martin-Escalona, I., & Barcelo-Arroyo, F. (2007). Performance evaluation of Middleware for provisioning LBS in cellular networks. *International Conference on Communications (ICC)* (pp. 5537 - 5544). Glasgow: IEEE.

Ministerio de Ciencia y Educacion. (2009). Encaminamiento en redes ad-hoc basado en localizaci6n de terminales. Aplicaciones y sinergias con localización pasiva. *TEC2009-08198* .

Nuyami, L., Lagrange, X., & Godlewski, P. (2002). A Power Control Algorithm for 3G WCDMA System. *European Wireless.* IEEE.

OGC. (n.d.). *Location Service (OpenLS).* Retrieved 2009 йил May from Open Geospatial Consortium, Inc.: http://www.opengeospatial.org/standards/ols

Ranganathan, A., Al-Muhtadi, J., Chetan, S., Campbell, R., & Mickunas, M. D. (2004). MiddleWhere: a middleware for location awareness in ubiquitous computing applications. *ACM/IFIP/USENIX international conference on Middleware, 78,* pp. 397-416.

Spanoudakis, M., Batistakis, A., Priggouris, I., Ioannidis, A., Hadjiefthymiades, S., & Merakos, L. (2004). Extensible Platform for Location Based Services Provisioning. *International Conference on Web Information Systems Engineering Workshops (WISEW)* (pp. 1-8). IEEE.

Thales Selena Space; EPFL; UPC; France Telecom. (2007). *Deliverable D077,* . 6th FP IST Liaison project: LIAISON.

Zander, J., & Kim, S. L. (2001). Radio Resource Management for Wireless Networks. Artech House.

4

Hexagonal vs Circular Cell Shape: A Comparative Analysis and Evaluation of the Two Popular Modeling Approximations

Konstantinos B. Baltzis
Aristotle University of Thessaloniki
Greece

1. Introduction

Recent years have witnessed an explosion in wireless communications. In the last decades, the development of wireless communication systems and networks is taking us from a world where communications were mostly carried over PSTN, packet-switched and high speed LAN networks to one where the wireless transmission dominates. Nowadays, high data rates carry multimedia communications, real-time services for delay-sensitive applications are added and networks are asked to deal with a traffic mix of voice, data and video. Next generation mobile systems will further include a variety of heterogeneous access technologies, support multimedia applications and provide end-to-end IP connectivity (Bolton et al., 2007; Xylomenos et al., 2008; Demestichas et al., 2010). Undoubtedly, new possibilities are created for both telcos and users and important design and traffic issues emerge. This revolution has spurred scientists toward the development of reliable and computationally efficient models for evaluating the performance of wireless networks.

A crucial parameter in the modeling of a cellular communication system is the shape of the cells. In real life, cells are irregular and complex shapes influenced by terrain features and artificial structures. However, for the sake of conceptual and computational simplicity, we often adopt approximate approaches for their design and modeling. In the published literature, cells are usually assumed hexagons or circles. The hexagonal approximation is frequently employed in planning and analysis of wireless networks due to its flexibility and convenience (Jan et al., 2004; Goldsmith, 2005; Pirinen, 2006; Chan & Liew, 2007; Hoymann et al., 2007; Baltzis, 2008, 2010a; Choi & You, 2008; Dou et al., 2008; Xiao et al., 2008; Baltzis & Sahalos, 2010). However, since this geometry is only an idealization of the irregular cell shape, simpler models are often used. In particular, the circular–cell approximation is very popular due to its low computational complexity (Petrus et al., 1998; Baltzis & Sahalos, 2005, 2009b; Goldsmith, 2005; Pirinen, 2006; Bharucha & Haas, 2008; Xiao et al., 2008; Baltzis, 2010b).

Among various performance degradation factors, co-channel interference (CCI) is quite significant since the cells in cellular networks tend to become denser in order to increase system capacity (Stavroulakis, 2003). The development of models that describe CCI generates great interest at the moment. Several reliable models can be found in the published literature (Butterworth et al., 2000; Cho et al., 2000; Cardieri & Rappaport, 2001; Zhou et al., 2003; Grant and Cavers, 2004; Masmoudi and Tabbane, 2006). However, their practical application is restricted by their algorithmic complexity and computational cost, which results in the development of simpler models. A simplified approach is the geometrical-based modeling that allows an approximate but adequate evaluation of system performance (Petrus et al., 1998; Au et al. 2001; Zhang et al., 2004, 2006; Baltzis & Sahalos, 2005, 2009b, 2010; Panagopoulos et al., 2007; Baltzis, 2008).

Another important issue in the study of a wireless system is the prediction of signal attenuation (Parsons, 2000; Fryziel et al. 2002; Baltzis, 2009). In general, path loss models approximate signal attenuation as a function of the distance between communicating nodes. The majority of the developed models are based on empirical measurements over a given distance in a specific frequency range and a particular environment (Ghassemzadeh, 2004; Moraitis and Constantinou, 2004; Holis and Pechac, 2008). However, the application of these models is not recommended in general environments because they are closely related to network and environmental parameters. Therefore, several other methods have been further developed for the description of path loss. For example, Haenggi introduced a unified framework that allowed the geometric characterization of fading (Haenggi, 2008). Proposals based on neural network techniques are also of great interest (Cerri et al. 2004, Popescu et al., 2006; Östlin et al., 2010). Approximate analytical methods that reduce significantly the computational requirements and the cost of the simulations can also be found in the published literature (Bharucha & Haas, 2008; Baltzis, 2010a,b).

In this chapter, we discuss and evaluate the hexagonal and the circular cell shape modeling approximations in a cellular network. The analysis focuses on the impact of the shape of the cells on co-channel interference and path loss attenuation. First, we present three two-dimensional geometrical-based models for the Angle-of-Arrival (AoA) of the uplink interfering signals in a cellular system and their application in the study of co-channel interference. Next, we present analytical and approximate closed-form expressions for the path loss statistics in cellular networks with hexagonal and circular cells; the role of cell shape in signal level is explored by representative examples.

The chapter proceeds as follows: Section 2 highlights some theoretical background on the hexagonal and the circular cell shape approximations. Then we study the impact of cell shape on performance degradation due to co-channel interference and on RF signal attenuation. In particular, Section 3 provides and evaluates some geometrical-based models for the description of co-channel interference in cellular networks with circular or hexagonal cells and Section 4 discusses the statistics of path loss in these systems. Finally, some research ideas are conclusions are pointed out.

2. The approximations of the hexagonal and the circular cell

Cellular networks are infrastructure-based networks deployed throughout a given area. One of their features is the efficient utilization of spectrum resources due to frequency reuse. In practice, frequency reuse is a defining characteristic of cellular systems. It exploits the fact that signal power falls of with distance to reuse the same frequency spectrum at spatially separated locations (cells).

In a cellular communication system, cell shape varies depending on geographic, environmental and network parameters such as terrain and artificial structures properties, base station location and transmission power, accessing techniques, etc. Nevertheless, for representation and analytical simplicity, cells are usually approximated as regular shapes such as hexagons and circles. In the following paragraphs, we will discuss the main features of the two approximations. The interested reader can consult the published literature for further information (MacDonald, 1979; Kim, 1993; Lee and Miller, 1998).

A hexagon is a tessellating cell shape in that cells can be laid next to each other with no overlap; therefore, they can cover the entire geographical region without any gaps. This approximation is frequently employed in planning and analysis of cellular networks. In practice, hexagons are commonly used to approximate cell shapes in macrocellular systems with base stations placed at the top of buildings (Goldsmith, 2005). Figure 1 shows a typical hexagonal cell layout with frequency reuse factor seven. Cells are numbered according to the frequency band they use. In this figure, we further depict the desired cell (red-colored), the six equidistant nearest interfering cells, i.e. the first ring of cells that use the same spectrum resources (grey-colored), and the second tier of interferers (black-colored).

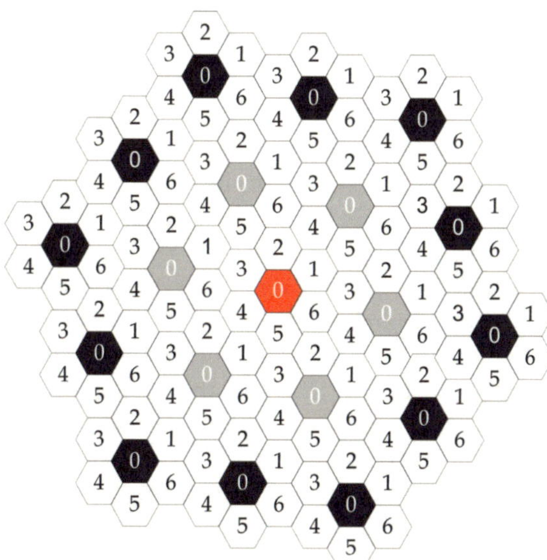

Fig. 1. Hexagonal cell layout.

A simpler assumption, the circular-shaped cell, is also common in the literature. A reasonable approximation of this assumption is provided when signal propagation follows path loss models that consider constant signal power level along a circle around the base station (Goldsmith, 2005). In fact, an omni-directional base station antenna may cover a circular area that is defined as the area for which the propagating downlink signal exceeds a certain threshold; however, even in this case, this is only an approximation due to the impact of the environment. The major drawback of the circular approximation is that circular cells must partially overlap in order to avoid gaps. Examples of simple layouts are depicted in Fig. 2. Figure 2a illustrates the hexagonal cell layout. The inradius and the circumradius of the hexagonal cell are r and a, respectively. In Fig. 2b, the radius of the circular cell R equals to r; as a result, the cells do not cover the whole network coverage area. In Fig. 2c, cells are partially overlapped because R equals to the hexagon's circumradius. In this case, the model considers nodes not belonging to the cell of interest. In the next sections, the impact of cell shape on the estimation of specific performance metrics is discussed.

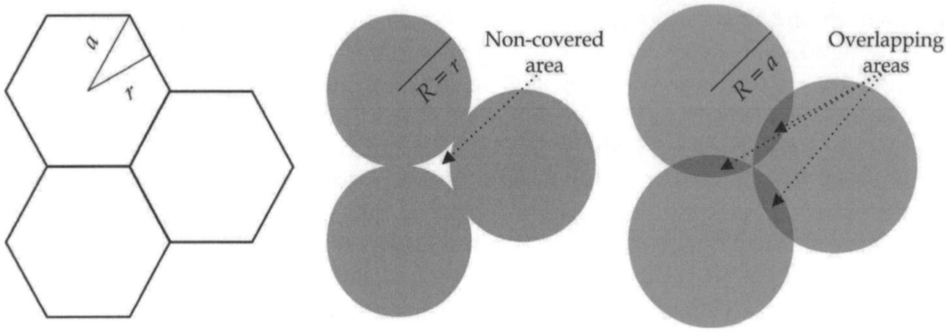

Fig. 2. Hexagonal cell layout (a) and idealized circular coverage areas (b), (c).

3. The impact of the shape of the cells on co-channel interference analysis

In this section, we explore the relation between co-channel interference and cell shape. A common measure of performance degradation due to CCI is the probability an interferer from another cell is causing interference to the cell of interest. We present three 2-D geometrical-based models for the description of the probability of uplink interference that assume circular and hexagonal cells. At this point, we have to notice that two-dimensional modeling is adequate for the description of the azimuth angular spreading of the propagating signals but fails to describe any signal variations in the elevation plane (Fuhl et al., 1997; Kuchar et al., 2000; Baltzis & Sahalos, 2009a, b; Nawaz et al., 2010).

In the development of the models, certain assumptions have been made. The base stations (BSs) are located at the centers of the cells and the mobile users (MUs) are uniformly distributed within each cell. The models assume a single line-of-sight signal path between the interferers and the BS and they consider only the first ring of co-channel interferers. They also assume that the interference power is much larger than the thermal noise power (interference-limited environment), the BS main beam points to the direction of the desired user and the transmitted BS power is maximized according to the desired user with no concern of the others.

In the circular-cell approach (Petrus et al., 1998), see Fig. 3, the cells are circles with radius R. BS_0 and BS_i are the base stations at the desired and an interfering cell, respectively. MUs are uniformly distributed within the cell; therefore, the area within the shaded region is proportional to the probability of the AoA of the uplink interfering signals.

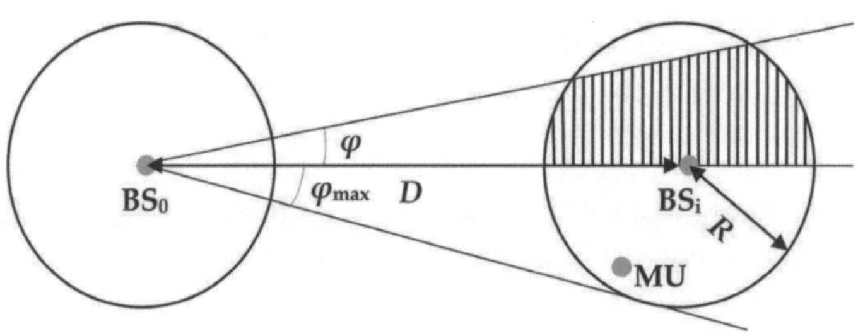

Fig. 3. The circular model.

In this case, the cdf of the AoA of the uplink interfering signals equals to the area of the shaded region multiplied by the user density. Its differentiation with respect to φ gives the corresponding pdf

$$f(\phi)=\frac{2D\cos\phi}{\pi R^2}\sqrt{R^2-D^2\sin^2\phi}\,\mathrm{U}\left(\sin^{-1}\left(\frac{R}{D}\right)-|\phi|\right) \tag{1}$$

where $\mathrm{U}(\cdot)$ is the unit step function.

A method for hexagonal-shaped cells was provided by Baltzis (Baltzis, 2008). In that proposal, the probability of the AoA of the uplink interfering signals was proportional to the area of the polygon defined from the line that connected the BSs of the cell of interest and the interfering cell, the line segment determined from the angle ϕ, and the boundaries of the interfering cell, see Fig. 4.

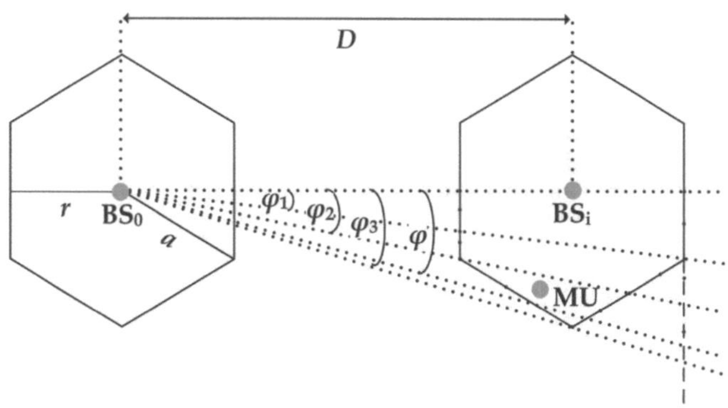

Fig. 4. The single cluster hexagonal model.

The pdf of the AoA of the uplink interfering signals is given by the expression

$$f(\phi)=\begin{cases}\dfrac{D}{\sqrt{3}r\cos^2\phi} & |\phi|\le\phi_1\\[2ex] \dfrac{1}{\cos^2\phi}\left(\dfrac{D}{\sqrt{3}r}-\dfrac{(\tan|\phi|-\tan\phi_1)(D+r)^2\left[2+\sqrt{3}(\tan|\phi|+\tan\phi_1)\right]}{4r^2\left(1+\sqrt{3}\tan|\phi|\right)^2}\right), & \phi_1\le|\phi|\le\phi_2\\[2ex] \dfrac{\left[r+\sqrt{3}(D+r)\tan\phi_1-\sqrt{3}D\tan|\phi|\right]\left\{D-\sqrt{3}\left[r+\sqrt{3}(D+r)\tan\phi_1\right]\tan|\phi|\right\}}{\sqrt{3}r^2\left(1-3\tan^2\phi\right)^2\cos^2\phi}, & \phi_2\le|\phi|\le\phi_3\\[2ex] 0, & |\phi|>\phi_3\end{cases} \tag{2}$$

where

$$\phi_i=\cot^{-1}\left(\frac{\sqrt{3}D}{\mu_i r}+\nu_i\right),\quad i=1,2,3 \tag{3}$$

with $\mu_{1..3}=\{1,1,2\}$ and $\nu_{1..3}=\{\sqrt{3},-\sqrt{3},0\}$.

The model is valid for single cluster networks or topologies in which the closest edges of the two cells are properly aligned, e.g. CDMA and WCDMA networks, OFDMA systems or special cases of ad hoc networks (Almers et al., 2007; Andrews et al., 2007; Webb, 2007; Lu et al., 2008).

Recently (Baltzis and Sahalos, 2010), the previous model was extended and included networks with hexagonal cells of arbitrary cluster size and orientation. In that proposal, the boundaries of the hexagonal cell were expressed in polar coordinates with origin the position of the desired BS, see Fig. 5, (with no loss of generality, the circumradius of the hexagon was normalized to unity) as

$$\begin{aligned} \rho_{\substack{(1)\\(4)}}(\phi) &= \frac{D\cos\phi_0 - \sqrt{3}\left(D\sin\phi_0 \mp 1\right)}{\cos\phi - \sqrt{3}\sin\phi} \\ \rho_{\substack{(2)\\(5)}}(\phi) &= \frac{D\cos\phi_0 \pm \sqrt{3}/2}{\cos\phi} \\ \rho_{\substack{(3)\\(6)}}(\phi) &= \frac{D\cos\phi_0 + \sqrt{3}\left(D\sin\phi_0 \pm 1\right)}{\cos\phi + \sqrt{3}\sin\phi} \end{aligned} \tag{4}$$

under the constraint

$$\left|\rho_{(i)}(\phi) - D\cos(\phi-\phi_0)\right| \le \sqrt{1 - D^2\sin^2(\phi-\phi_0)},\quad i = 1..6 \tag{5}$$

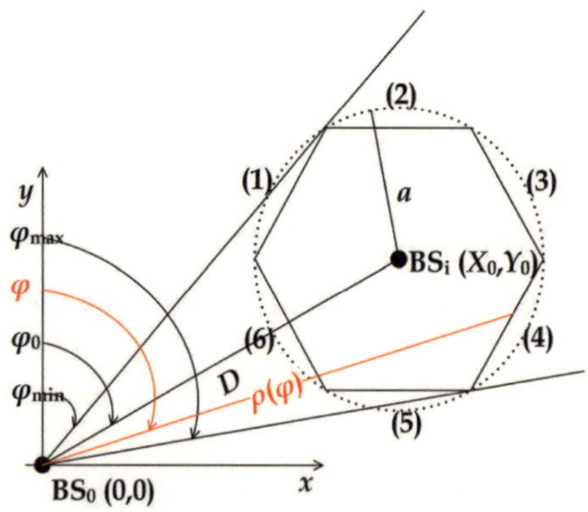

Fig. 5. Geometry of the generalized hexagonal model.

The pdf of the AoA of the uplink interfering signals is

$$f(\phi) = \frac{\rho_2^2(\phi) - \rho_1^2(\phi)}{3\sqrt{3}} \mathrm{U}\left(\phi + |\phi_{\min}|\right)\mathrm{U}\left(\phi_{\max} - \phi\right) \tag{6}$$

where the angles $\phi_{\min}$ and $\phi_{\max}$ depend on system geometry and

$$\rho_1(\phi) = \max_{i=5,6}\left\{\rho_{(i)}(\phi)\right\} \tag{7}$$

$$\rho_2(\phi) = \min_{i=1..4}\left\{\rho_{(i)}(\phi)\right\} \tag{8}$$

Next, we provide representative examples to demonstrate the efficacy of the aforementioned models. Our discussion is mainly focused on the analysis of the impact of cell shape on system performance. Figures 6 and 7 illustrate the pdfs and cdfs of the AoA of the uplink interfering signals for cellular systems with frequency reuse factor, *K*, one, three and seven.

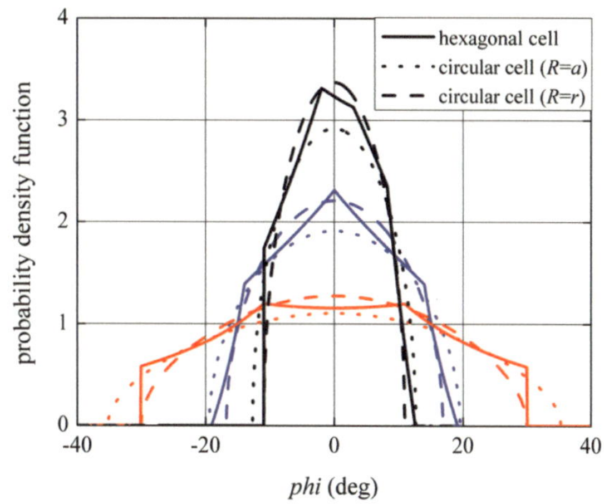

Fig. 6. Pdf of the AoA of the uplink interfering signals; the frequency reuse factor is seven (black curves), three (blue curves) and one (red curves).

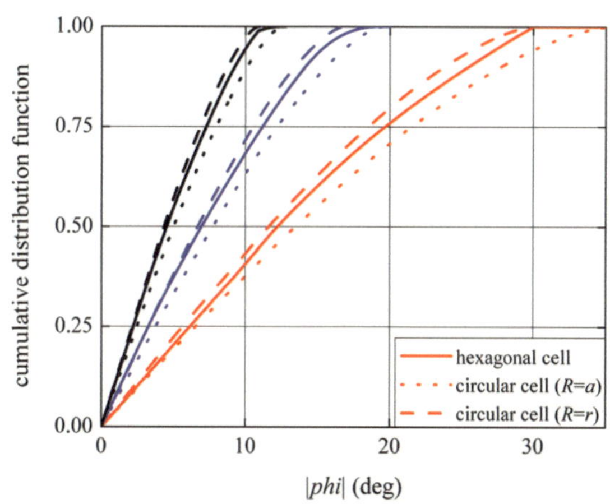

Fig. 7. Cdf of the AoA of the uplink interfering signals; the frequency reuse factor is seven (black curves), three (blue curves) and one (red curves).

Figure 6 shows that the circular and the hexagonal cell pdfs differ for small values of ϕ. In the first case, the pdfs are even functions maximized at $\phi = 0$. On the other hand, the hexagonal model for frequency reuse factor one or seven estimates the maxima of the pdfs at $\phi \neq 0$; moreover, when $K = 7$ the pdf curve is no longer symmetric with respect to $\phi = 0$. Differences are also observed at large values of ϕ. These differences are related to the different size of the cells (obviously, a circular cell with radius equal to the hexagon's inradius (circumradius) has a smaller (greater) coverage area compared to the hexagonal cell) and their relative positions in the cluster. Noticeable differences are also observed between the cdf curves, see Fig. 7. In comparison with the hexagonal approach, the inradius (circumradius) approximation overestimates (underestimates) the amount of interference at small angles. In general, the inradius approximation gives results closer to the hexagonal solution compared with the circumradius one.

For a given azimuth angle, the probability that the users of another cell interfere with the desired uplink signal is given by the convolution of the desired BS antenna radiation pattern with the pdf of the AoA of the incoming interfering signals. The summation of all the possible products of the probability that n cells are interfering by the probability that the remaining $N - n$ do not gives the probability that n out of the possible N interfering cells are causing interference over ϕ (Petrus et al., 1998; Baltzis & Sahalos, 2005, 2009b).

Let us assume a single cluster WCDMA network with a narrow beam BS antenna radiation pattern and a three- and six-sectored configuration. The BS antenna radiation patterns are cosine-like with side lobe level -15 dB and half-power beamwidth 10, 65, and 120 degrees, respectively (Czylwik & Dekorsy, 2004; Niemelä et al., 2005). Figure 8 depicts the probability that an interfering cell causes interference over $|\phi|$ in the network (in a single cluster system, this probability is even function). We observe differences between the hexagonal and the circular approaches for small angles and angles that point at the boundaries of the interfering cell. Increase in half-power beamwidth reduces the difference between the models but increases significantly the probability of interference.

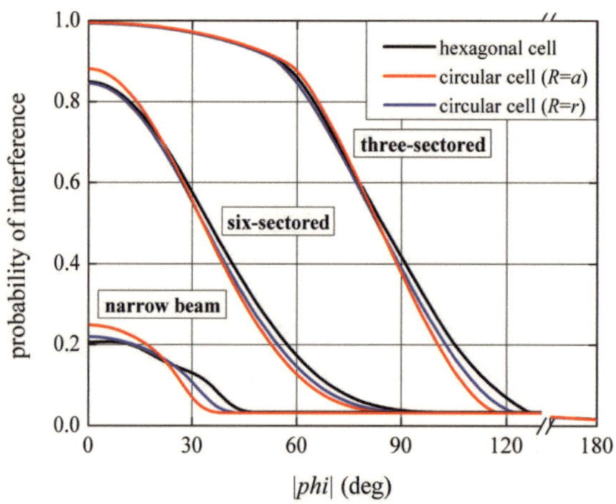

Fig. 8. Probability that an interfering cell is causing interference over $|\phi|$.

The validation of the previous models using simulation follows. The pdfs in (1) and (6) are calculated for a single cluster size WCDMA network. The users are uniformly distributed within the hexagonal cells; therefore, user density is (Jordan et al., 2007)

$$p(x,y)=\frac{1}{3ar}\mathrm{U}(a-|x|)\mathrm{U}(r-|y|)\mathrm{U}\left(2r-\left|\sqrt{3}x+y\right|\right) \tag{9}$$

considering that the center of the cell is at (0,0). System parameters are as in Aldmour et al. (Aldmour et al., 2007). In order to generate the random samples, we employ the DX-120-4 pseudorandom number generator (Deng & Xu, 2003) and apply the rejection sampling method (Raeside, 1976). The simulation results are calculated by carrying out 1000 Monte Carlo trials. Table I presents the mean absolute, e_p, and mean relative, ε_p, difference between the theoretical pdf values and the simulation results (estimation errors). The simulation results closely match the theoretical pdf of (6); however, they differ significantly from the circular-cell densities. A comparison between the inradius and the circumradius approximations shows the improved accuracy of the first.

Hexagonal model		Circular model			
		$R=r$		$R=a$	
e_p	ε_p	e_p	ε_p	e_p	ε_p
1.48%	1.87%	8.17%	10.40%	9.76%	20.36%

Table 1. Probability density function: Estimation errors.

Among the measures of performance degradation due to CCI, a common one is the probability an interferer is causing interference at the desired cell. Table 2 lists the mean absolute, e_P, and mean relative, ε_P, difference between theoretical values and simulations results, i.e. the estimation error, of this probability. We consider a six-sectored and a narrow-beam system architecture. The rest of the system parameters are set as before. In the six-sectored system, we observe a good agreement between the theoretical values and the simulation results for all models. However, in the narrow-beam case, noticeable differences are observed. Again, the circumradius approximation gives the worst results.

System architecture	Hexagonal model		Circular model			
			$R=r$		$R=a$	
	e_P	ε_P	e_P	ε_P	e_P	ε_P
six-sectored system	0.59%	1.18%	1.34%	2.03%	1.86%	2.54%
narrow-beam system	0.47%	2.51%	1.19%	5.77%	3.99%	19.35%

Table 2. Probability of interference: Estimation errors.

Use of the previous models allows the approximate calculation of the co-channel interference in a cellular network. By setting *CIR* the Carrier-to-Interference Ratio, *Q* the Protection ratio, Z_d the Carrier-to-Interference plus Protection Ratio (*CIRP*), $P(n)$ the average probability that n out of the possible N interfering cells are causing interference over ϕ and $P(Z_d<0\,|\,n)$ the conditional probability of outage given n interferers, this term depends on fading conditions (Muammar & Gupta, 1982; Petrus et al., 1998; Au et al., 2001; Baltzis & Sahalos, 2009b), we can express the average probability of outage of CCI as

$$P_{out} \underset{def}{=} P(CIR<Q) = \sum_{n=1}^{N} P(Z_d<0\,|\,n)P(n) \tag{10}$$

As an example, Fig. 9 illustrates the outage curves of a WCDMA cellular system for different BS antenna half-power beamwidths. The antennas are flat-top beamformers; an example of an omni-directional one is also shown. In the simulations, the protection ratio is 8 dB and the activity level of the users equals to 0.4. Decrease in the beamformer's beamwidth up to a point reduces significantly the outage probability of co-channel interference indicating the significance of sectorization and/or the use of narrow-beam base station transmission antennas.

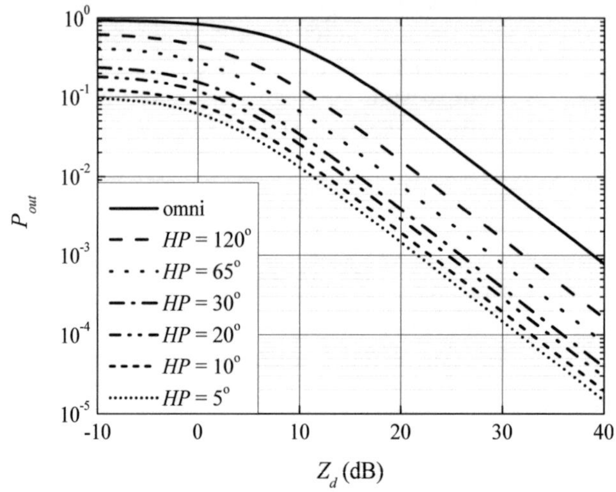

Fig. 9. Plot of outage curves as a function of *CIRP*.

In the calculation of co-channel interference, the inradius approximation considers part of the cell coverage area; on the contrary, the circumradius approach takes into account nodes not belonging to the cell, see Fig. 2. In both cases, an initial network planning that employs hexagonal cells but applies a circular model for the description of co-channel interference does not utilize network resources effectively. A hexagonal model is more accurate when network planning and design consider hexagonal-shaped cells. The comparisons we performed show that the inradius approximation compared with the circumradius one gives results closer to the hexagonal approach. In fact, it has been found that circles with radius that range between $1.05r$ and $1.1r$ give results closer to the hexagonal solution (Baltzis & Sahalos, 2010). Similar results are drawn for several other performance metrics (Oh & Li, 2001).

4. Cell shape and path loss statistics

In system-level simulations of wireless networks, path loss is usually estimated by distributing the nodes according to a known distribution and calculating the node-to-node distances. Thereafter, the application of a propagation model gives the losses. In order to increase the solution accuracy, we repeat the procedure many times but at the cost of simulation time. Therefore, the analytical description of path loss reduces significantly the computational requirements and may provide a good trade-off between accuracy and computational cost.

In the wireless environment, path loss increases exponentially with distance. The path loss at a distance d greater than the reference distance of the antenna far-field d_0 may be expressed in the log-domain (Parsons, 2000; Ghassemzadeh, 2004; Baltzis, 2009) as

$$L = L_0 + 10\gamma \log(d/d_0) + X_S + Y, \quad d > d_0 \qquad (11)$$

where L_0 is the path loss at d_0, γ is the path loss exponent, X_S is the shadowing term and Y is the small-scale fading variation. Shadowing is caused by terrain configuration or obstacles

between the communicating nodes that attenuate signal power through absorption, reflection, scattering and diffraction and occurs over distances proportional to the size of the objects. Usually, it is modeled as a lognormal random process with logarithmic mean and standard deviation μ and σ, respectively (Alouini and Goldsmith, 1999; Simon and Alouini, 2005). Small-scale fading is due to constructive and destructive addition from multiple signal replicas (multipaths) and happens over distances on the order of the signal wavelength when the channel coherence time is small relative to its delay spread or the duration of the transmitted symbols. A common approach in the literature, is its modeling by the Nakagami-*m* distribution (Alouini and Goldsmith, 1999; Simon and Alouini, 2005; Rubio et al., 2007).
The combined effect of shadowing and small-scale fading can be modeled with the composite Nakagami-lognormal distribution. In this case, the path loss pdf between a node distributed uniformly within a circular cell with radius R and the center of the cell is (Baltzis, 2010b)

$$f_L(l)=\frac{1}{\xi\gamma R^2}\exp\left(2\left[\frac{l-L_0-\mu_C}{\xi\gamma}+\left(\frac{\sigma_C}{\xi\gamma}\right)^2\right]\right)\operatorname{erfc}\left(\frac{l-L_0-10\gamma\log R-\mu_C}{\sqrt{2}\sigma_C}+\frac{\sqrt{2}\sigma_C}{\xi\gamma}\right) \tag{12}$$

with $\xi=10/\ln 10\approx 4.343$, m the Nakagami fading parameter and

$$\begin{aligned}\mu_C&=\xi[\Psi(m)-\ln m]\\ \sigma_C^2&=\sigma^2+\xi^2\zeta(2,m)\end{aligned} \tag{13}$$

where $\Psi(\cdot)$ is the Euler's psi function and $\zeta(\cdot,\cdot)$ is the generalized Reimann's zeta function (Gradshteyn & Ryzhik, 1994). In the absence of small-scale fading, (12) is simplified (Bharucha & Haas, 2008) into

$$f_L(l)=\frac{\ln 10}{bR^2}\exp\left(\frac{2b(l-L_0)\ln 10+2(\sigma\ln 10)^2}{b^2}\right)\operatorname{erfc}\left(\frac{l-L_0-b\log R+\dfrac{2\sigma^2\ln 10}{b}}{\sqrt{2}\sigma}\right) \tag{14}$$

where $b=10\gamma$.
In the case of hexagonal instead of circular cells, the path loss pdf (in the absence of small-scale fading; the incorporation of this factor is a topic for a potential next stage of future work extension) is (Baltzis, 2010a)

$$f(l)=\frac{\pi\ln 10}{2\sqrt{3}br^2}\left(\begin{array}{l}100^{(l-L_0)/b}\exp(N^2)\left[\operatorname{erfc}\left(\dfrac{l}{\sqrt{2}\sigma}-M+N-S\right)-2\left(\begin{array}{l}\operatorname{erf}\left(\dfrac{l}{\sqrt{2}\sigma}-M+N-S\right)\\-\operatorname{erf}\left(\dfrac{l}{\sqrt{2}\sigma}-M+N-T\right)\end{array}\right)\right]\\ -\dfrac{6}{\pi}\displaystyle\sum_{j=0}^{+\infty}\left(\begin{array}{l}\dfrac{|P_{2j}(0)|}{2j+1}r^{2j+1}\exp\left(\left[(j-1/2)N\right]^2\right)10^{-(2j-1)(l-L_0)/b}\\ \times\left(\begin{array}{l}\operatorname{erf}\left(-\dfrac{l}{\sqrt{2}\sigma}+M+S+(j-1/2)N\right)\\-\operatorname{erf}\left(-\dfrac{l}{\sqrt{2}\sigma}+M+T+(j-1/2)N\right)\end{array}\right)\end{array}\right)\end{array}\right) \tag{15}$$

with $P_{2j}(x), j \in \mathbf{N}$ the Legendre polynomials of order 2*j*, $M = 2^{-1/2}\sigma^{-1}L_0$, $N = \sqrt{2}\sigma b^{-1}\ln 10$, $S = 2^{-1/2}\sigma^{-1}b\log r$ and $T = 2^{-1/2}\sigma^{-1}b\log\alpha$. A closed-form approximation of this expression is

$$f(l) = \frac{\sqrt{3}\ln 10}{2br^2}\left(100^{(l-L_0)/b}\exp(N^2)\left(\operatorname{erfc}\left(\frac{l}{\sqrt{2}\sigma} - M + N - S\right) + \frac{\sqrt{3}}{\sqrt{3}-2}\left(\operatorname{erf}\left(\frac{l}{\sqrt{2}\sigma} - M + N - S\right) - \operatorname{erf}\left(\frac{l}{\sqrt{2}\sigma} - M + N - T\right)\right)\right) - \frac{2r\exp\left(N^2/4\right)}{\sqrt{3}-2}10^{(l-L_0)/b}\left(\operatorname{erf}\left(\frac{l}{\sqrt{2}\sigma} - M - S + \frac{N}{2}\right) - \operatorname{erf}\left(\frac{l}{\sqrt{2}\sigma} - M - T + \frac{N}{2}\right)\right)\right) \quad (16)$$

A significant difference between the circular and the hexagonal cell models appears in the link distance statistics. The link distance pdf from the center of a circular cell with radius *R* to a spatially uniformly distributed node within it is (Omiyi et al., 2006)

$$f(d) = \frac{2d}{R^2}\mathrm{U}(d)\mathrm{U}(R-d) \quad (17)$$

The link distance pdf within a centralised hexagonal cell with inradius *r* and circumradius *a* is (Pirinen, 2006)

$$f(d) = \begin{cases} \dfrac{\pi d}{\sqrt{3}r^2}, & 0 \le d \le r \\ \dfrac{2\sqrt{3}d}{r^2}\left[\dfrac{\pi}{6} - \cos^{-1}\left(\dfrac{r}{d}\right)\right], & r \le d \le a \\ 0, & d \ge a \end{cases} \quad (18)$$

Figure 10 shows the link distance pdf and cdf curves for centralized hexagonal and circular cells. Notice the differences between the hexagonal and the circular approach. We further see that the inradius circular pdf and cdf are closer to the hexagonal ones compared with the circumradius curves.

Let us now consider a cellular system with typical UMTS air interface parameters (Bharucha & Haas, 2008). In particular, we set $\gamma = 3$ and L_0 = 37dB while shadowing deviation equals to 6dB or 12dB. The cells are hexagons with inradius 50m or 100m. Figure 11 shows the path loss pdf curves derived from (14)-(16). The corresponding cdfs, see Fig. 12, are generated by integrating the pdfs over the whole range of path losses. A series of simulations have also been performed for the cases we studied. For each snapshot, a single node was positioned inside the hexagonal cell according to (9). Then, the distance between the generated node and the center of the hexagon was calculated and a different value of shadowing was computed. After one path loss estimation using (11) (recall that small-scale fading was not considered), another snapshot continued. For each set of σ and *r*, 100,000 independent simulation runs were performed. In Fig. 11, the simulation values were averaged over a path loss step-size of one decibel.

Figure 11 shows a good agreement between theory and simulation. We also notice that increase in σ flattens the pdf curve; as cell size increases the curve shifts to the right. The inradius approximation considers part of the network coverage area; as a result the pdf curve shifts to the left. The situation is reversed in the circumradius approximation because it considers nodes not belonging to the cell of interest. In practice, the first assigns higher probability to lower path loss values overestimating system performance. In this case, initial network planning may not satisfy users' demands and quality of service requirements. On the other hand, the circumradius approach assigns lower probability to low path loss values and underestimates system performance. As a result, network resources are not utilized efficiently. Again, the inradius approximation gives result closer to the hexagonal model.

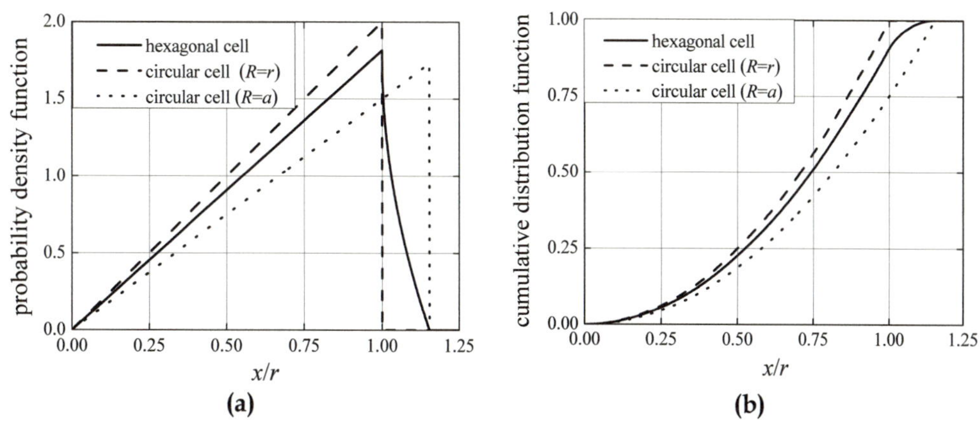

Fig. 10. Probability density function (a) and cumulative distribution function (b) curves.

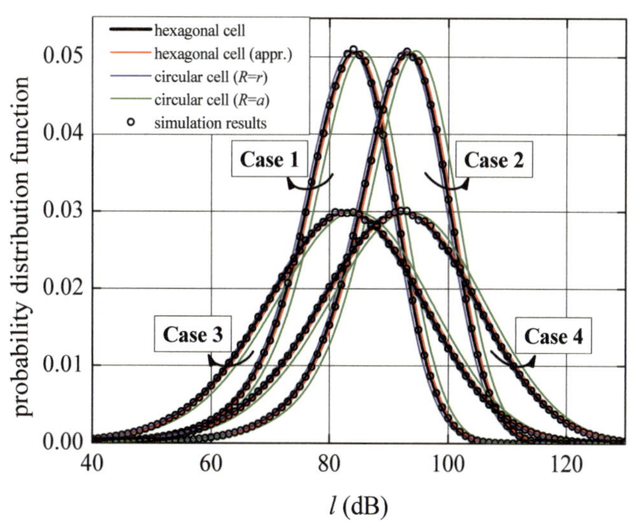

Fig. 11. Path loss pdf curves and simulation results; Case 1: $\sigma = 6\text{dB}$ and $r = 50\text{m}$; Case 2: $\sigma = 6\text{dB}$ and $r = 100\text{m}$; Case 3: $\sigma = 12\text{dB}$ and $r = 50\text{m}$; Case 4: $\sigma = 12\text{dB}$ and $r = 100\text{m}$.

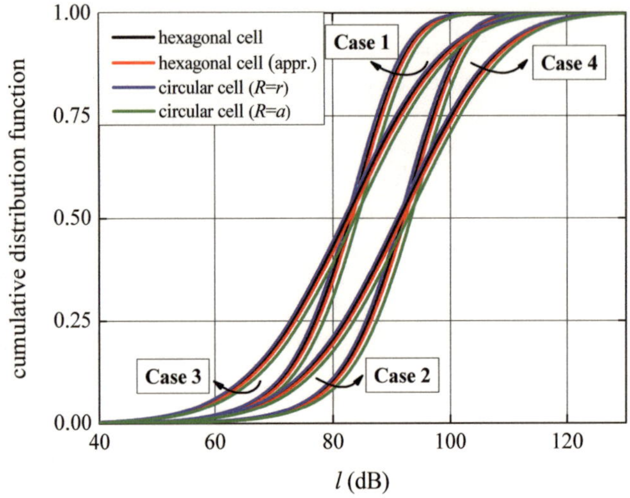

Fig. 12. Path loss cdf curves. (Cases 1 to 4 are defined as in Fig. 11).

Similar to before, we observe a good agreement between the hexagonal and the inradius circular approximation in Fig. 12. As it was expected, the curves shift to the right with cell size. However, the impact of shadowing is more complicated. Increase in σ, shifts the cdf curves to the left for path loss values up to a point; on the contrary, when shadowing deviation decreases the curves shift to the left with l. Moreover, Figs. 11 and 12 point out the negligible difference between the exact and the approximate hexagonal solutions.
Finally, Table 3 presents the predicted mean path loss values for the previous examples. The results show that the difference between the cell types is rather insignificant with respect to mean path loss. Notice also that the last does not depend on shadowing.

		Mean path loss (dB)			
σ(dB)	r (m)	hexagonal (15)	hexagonal (16)	inradius appr.	circumradius appr.
6	50	82.1	82.4	81.5	83.3
12	50	82.1	82.4	81.5	83.3
6	100	91.1	91.4	90.5	92.4
12	100	91.1	91.4	90.5	92.4

Table 3. Predicted mean path loss values.

A comparison between the proposed models and measured data (Thiele & Jungnickel, 2006; Thiele et al.; 2006) can be found in the literature (Baltzis, 2010a). In that case, the experimental results referred to data obtained from 5.2GHz broadband time-variant channel measurements in urban macro-cell environments; in the experiments, the communicating nodes were moving toward distant locations at low speed. It has been shown that the results derived from (15) and (16) were in good agreement with the measured data. The interested reader can also consult the published literature (Baltzis, 2010b) for an analysis of the impact of small-scale fading on path loss statistics using (12).

5. Research ideas

As we have stated in the beginning of this chapter, cells are irregular and complex shapes influenced by natural terrain features, man-made structures and network parameters. In most of the cases, the complexity of their shape leads to the adoption of approximate but simple models for its description. The most common modeling approximations are the circular and the hexagonal cell shape. However, alternative approaches can also be followed. For example, an adequate approximation for microcellular systems comprises square- or triangular-shaped cells (Goldsmith & Greenstein, 1993; Tripathi et al., 1998). Nowadays, the consideration of more complex shapes for the description of cells in emerging cellular technologies is of significant importance. An extension of the ideas discussed in this chapter in networks with different cell shape may be of great interest.

Moreover, in the models we discussed, several assumptions have been made. Further topics that illustrate future research trends include, but are not limited to, the consideration of non-uniform nodal distribution (e.g. Gaussian), the modeling of multipath uplink interfering signal, the use of directional antennas, the modeling of fading with distributions such as the generalized Suzuki, the *G*-distribution and the generalized *K*-distribution (Shankar; 2004; Laourine et al., 2009; Withers & Nadarajah, 2010), etc.

6. Conclusion

This chapter discussed, evaluated and compared two common assumptions in the modeling of the shape of the cells in a wireless cellular network, the hexagonal and the circular cell shape approximations. The difference in results indicated the significance of the proper choice of cell shape, a choice that is mainly based on system characteristics. In practice, use of the hexagonal instead of the circular–cell approximation gives results more suitable for the simulations and planning of wireless networks when hexagonal–shaped cells are employed. Moreover, it was concluded that the inradius circular approximation gives results closer the hexagonal approach compared to the circumradius one.

The chapter also provided a review of some analytical models for co-channel interference analysis and path loss estimation. The derived formulation allows the determination of the impact of cell shape on system performance. It further offers the capability of determining optimum network parameters and assists in the estimation of network performance metrics and in network planning reducing the computational complexity.

7. References

Aldmour, I. A.; Al-Begain, K. & Zreikat, A. I. (2007). Uplink capacity/coverage analysis of WCDMA with switched beam smart antennae. *Wireless Personal Communications*, Vol. 43, no. 4, Dec. 2007, 1705-1715

Almers, P. et al. (2007). Survey of channel and radio propagation models for wireless MIMO systems. *EURASIP Journal on Wireless Communications and Networking*, Vol. 2007, 19 pages, doi:10.1155/2007/19070

Alouini, M.-S. & Goldsmith, A. J. (1999). Area spectral efficiency of cellular mobile radio systems. *IEEE Transactions on Vehicular Technology*, Vol. 48, no. 4, July 1999, 1047-1066

Andrews, J. G.; Weber, S. & Haenggi, M. (2007). Ad hoc networks: To spread or not to spread?. *IEEE Communications Magazine*, Vol. 45, no. 12, Dec. 2007, 84-91

Au, W. S.; Murch, R. D. & Lea, C. T. (2001). Comparison between the spectral efficiency of SDMA systems and sectorized systems. *Wireless Personal Communications*, Vol. 16, no. 1, Jan. 2001, 51-67

Baltzis, K. B. (2008). A geometrical-based model for cochannel interference analysis and capacity estimation of CDMA cellular systems. *EURASIP Journal on Wireless Communications and Networking*, Vol. 2008, 7 pages, doi:10.1155/2008/791374

Baltzis, K. B. (2009). Current issues and trends in wireless channel modeling and simulation. *Recent Patents on Computer Science*, Vol. 2, no. 3, Nov. 2009, 166-177

Baltzis, K. B. (2010a). Analytical and closed-form expressions for the distribution of path loss in hexagonal cellular networks. *Wireless Personal Communications*, Mar. 2010, 12 pages, doi:10.1007/s11277-010-9962-2

Baltzis, K. B. (2010b). Closed-form description of microwave signal attenuation in cellular systems. *Radioengineering*, Vol. 19, no. 1, Apr. 2010, 11-16

Baltzis, K. B. & Sahalos, J. N. (2005). A 3-D model for measuring of the interference degradation of wireless systems, *Proceedings of Mediterranean Microwave Symposium 2005 (MMS'05)*, pp. 85-90, Athens, Sept. 2005

Baltzis, K. B. & Sahalos, J. N. (2009a). A simple 3-D geometric channel model for macrocell mobile communications. *Wireless Personal Communications*, Vol. 51, no. 2, Oct. 2009, 329-347

Baltzis, K. B. & Sahalos, J. N. (2009b). A low-complexity 3-D geometric model for the description of CCI in cellular systems. *Electrical Engineering (Archiv für Elektrotechnik)*, Vol. 91, no. 4-5, Dec. 2009, 211-219

Baltzis, K. B. & Sahalos, J. N. (2010). On the statistical description of the AoA of the uplink interfering signals in a cellular communications system. *European Transactions on Telecommunications*, Vol. 21, no. 2, Mar. 2010, 187-194

Bharucha, Z. & Haas, H. (2008). The distribution of path losses for uniformly distributed nodes in a circle. *Research Letters in Communications*, Vol. 2008, 4 pages, doi:10.1155/2008/376895

Bolton, W.; Xiao, Y. & Guizani, M. (2007). IEEE 802.20: mobile broadband wireless access. *IEEE Wireless Communications*, Vol. 14, no. 1, Feb. 2007, 84-95

Butterworth, K. S.; Sowerby, K. W. & Williamson, A. G. (2000). Base station placement for in-building mobile communication systems to yield high capacity and efficiency. *IEEE Transactions on Communications*, Vol. 48, no. 4, Apr. 2000, 658-669

Cardieri, P. & Rappaport, T. S. (2001). Application of narrow-beam antennas and fractional loading factors in cellular communication systems. *IEEE Transactions on Vehicular Technology*, Vol. 50, no. 2, Mar. 2001, 430-440

Cerri, G.; Cinalli, M.; Michetti, F. & Russo, P. (2004). Feed forward neural networks for path loss prediction in urban environment. *IEEE Transactions on Antennas and Propagation*, Vol. 52, no. 11, Nov. 2004, 3137-3139

Chan, A. & Liew, S. C. (2007). VoIP capacity over multiple IEEE 802.11 WLANs, *Proceedings of 2007 IEEE International Conference on Communications (ICC'07)*, pp. 3251-3258, Glasgow, June 2007

Cho, H.-S.; Kwon, J. K. & Sung, D. K. (2000). High reuse efficiency of radio resources in urban microcellular systems. *IEEE Transactions on Vehicular Technology*, Vol. 49, no. 5, Sep. 2000, 1669-1667

Choi, S.-O. & You, K.-H. (2008). Channel adaptive power control in the uplink of CDMA systems. *Wireless Personal Communications*, Vol. 47, no. 3, Nov. 2008, 441-448

Czylwik, A. & Dekorsy, A. (2004). System-level performance of antenna arrays in CDMA-based cellular mobile radio systems. *EURASIP Journal on Applied Signal Processing*, Vol. 2004, no. 9, 1308-1320

Demestichas, P.; Katidiotis, A.; Petromanolakis, D. & Stavroulaki, V. (2010). Management system for terminals in the wireless B3G world. *Wireless Personal Communications*, Vol. 53, no. 1, Mar. 2010, 81-109

Deng, L.-Y & Xu, H. (2003). A system of high-dimensional, efficient, long-cycle and portable uniform random generators. *ACM Transactions on Modeling and Computer Simulation*, Vol. 13, no. 4, Oct. 2003, 299-309

Dou, J.; Guo, Z.; Cao, J. & Zhang, G. (2008). Lifetime prolonging algorithms for wireless sensor networks, *Proceedings of 4th IEEE International Conference on Circuits and Systems for Communications (ICCSC 2008)*, pp. 833-837, Shanghai, May 2008

Fryziel, M.; Loyez, C.; Clavier, L.; Rolland, N. & Rolland, P. A. (2002). Path-loss model of the 60-GHz indoor radio channel. *Microwave and Optical Technology Letters*, Vol. 34, no. 3, Aug. 2002, 158-162

Fuhl, J.; Rossi, J.-P. & Bonek, E. (1997). High-resolution 3-D direction-of-arrival determination for urban mobile radio. *IEEE Transactions on Antennas and Propagation*, Vol. 45, no. 4, Apr. 1997, 672-682

Ghassemzadeh, S. S.; Jana, R.; Rice, C. W.; Turin, W. & Tarokh, V. (2004). Measurement and modeling of an ultra-wide bandwidth indoor channel. *IEEE Transactions on Communications*, Vol. 52, no. 10, Oct. 2004, 1786-1796

Gilhousen, K. S.; Jacobs, I. M.; Padovani, R.; Viterbi, A. J.; Weaver, L. A. Jr. & Wheatley, C. E. III (1991). On the capacity of a cellular CDMA system. *IEEE Transactions on Vehicular Technology*, Vol. 40, no. 2, May 1991, 303-312

Goldsmith, A. (2005). *Wireless Communications*, Cambridge University Press, New York

Goldsmith, A. J. & Greenstein, L. J. (1993). A measurement-based model for predicting coverage areas of urban microcells. *IEEE Journal on Selected Areas in Communications*, Vol. 11, no. 7, Sep. 1993, 1013-1023

Gradshteyn, I. S. & Ryzhik, I. M. (1994). *Table of Integrals, Series, and Products*, 5th ed., Academic Press, London

Grant, S. J. & Cavers, J. K. (2004). System-wide capacity increase for narrowband cellular systems through multiuser detection and base station diversity arrays. *IEEE Transactions on Wireless Communications*, Vol. 3, no. 6, Nov. 2004, 2072-2082

Haenggi, M. (2008). A geometric interpretation of fading in wireless networks: Theory and applications. *IEEE Transactions on Information Theory*, Vol. 54, no. 12, Dec. 2008, 5500-5510

Holis, J. & Pechac, P. (2008). Elevation dependent shadowing model for mobile communications via high altitude platforms in built-up areas. *IEEE Transactions on Antennas and Propagation*, Vol. 56, no. 4, Apr. 2008, 1078-1084

Hoymann, C.; Dittrich, M. & Goebbels, S. (2007). Dimensioning cellular multihop WiMAX networks, *Proceedings of 2007 IEEE Mobile WiMAX Symposium*, pp. 150-157, Orlando, Mar. 2007

Jan, R.-H.; Chu, H.-C. & Lee, Y.-F. (2004). Improving the accuracy of cell-based positioning for wireless systems, *Computer Networks*, Vol. 46, no. 6, Dec. 2004, 817-827

Jordan, M.; Senst, M.; Yang, C.; Ascheid, G. & Meyr, H. (2007). Downlink based intercell time synchronization using maximum likelihood estimation, *Proceedings of 16th IST Mobile and Wireless Communications Summit*, pp. 1-5, Budapest, July 2007, doi:10.1109/ISTMWC.2007.4299174

Kim, K. I. (1993). CDMA cellular engineering issues. *IEEE Transactions on Vehicular Technology*, Vol. 42, no. 3, Aug. 1993, 345-350

Kuchar, A.; Rossi, J.-P. & Bonek, E. (2000). Directional macro-cell channel characterization from urban measurements. *IEEE Transactions on Antennas and Propagation*, Vol. 48, no. 2, Feb. 2000, 137-146

Laourine, A.; Alouini, M.-S.; Affes, S. & Stéphenne, A. (2009). On the performance analysis of composite multipath/shadowing channels using the G-distribution. *IEEE Transactions on Communications*, Vol. 57, no. 4, Apr. 2009, 1162-1170.

Lee, J. S. & Miller, L. E. (1998). *CDMA Systems Engineering Handbook*, Artech House, Boston

Lu, K.; Qian, Y.; Chen, H.-H. & Fu, S. (2008). WiMAX networks: From access to service platform. *IEEE Network*, Vol. 22, no. 3, May-Jun. 2008, 38-45

MacDonald, V. H. (1979). The cellular concept. *The Bell System Technical Journal*, Vol. 58, no. 1, Jan. 1979, 15-41

Masmoudi, A. & Tabbane, S. (2006). Other-cell-interference factor distribution model in downlink WCDMA systems. *Wireless Personal Communications*, Vol. 36, no. 3, Feb. 2006, 245-475

Moraitis, N. & Constantinou, P. (2004). Indoor channel measurements and characterization at 60 GHz for wireless local area network applications. *IEEE Transactions on Antennas and Propagation*, Vol. 52, no. 12, Dec. 2004, 3180-3189

Muammar, R. & Gupta, S. (1982). Cochannel interference in high-capacity mobile radio systems. *IEEE Transactions on Communications*, Vol. 30, no. 8, Aug. 1982, 1973-1978

Nawaz, S. J.; Qureshi, B. H.; Khan, N. M. & Abdel-Maguid, M. (2010). Effect of directional antenna on the spatial characteristics of 3-D macrocell environment, *Proceedings of 2nd International Conference on Future Computer and Communication (ICFCC 2004)*, Vol. 1, pp. 552-556, Wuhan, May 2010

Nie, C.; Wong, T. C. & Chew, Y. H. (2004). Outage analysis for multi-connection multiclass services in the uplink of wideband CDMA cellular mobile networks, *Proceedings of 3rd International IFIP-TC6 Networking Conference*, pp. 1426-1432, Athens, May 2004

Niemelä, J.; Isotalo, T. & Lempiäinen, J. (2005). Optimum antenna downtilt angles for macrocellular WCDMA network. *EURASIP Journal on Wireless Communications and Networking*, Vol. 2005, no. 5, Oct. 2005, 816-827

Oh, S. W. & Li, K. H. (2001). Evaluation of forward-link performance in cellular DS-CDMA with Rayleigh fading and power control. *International Journal of Communication Systems*, Vol. 14, no. 3, Apr. 2001, 243-250

Omiyi, P.; Haas, H. & Auer, G. (2007). Analysis of TDD cellular interference mitigation using busy-bursts. *IEEE Transactions on Wireless Communications*, Vol. 6, no. 7, Jul. 2007, 2721-2731

Östlin, E.; Zepernick, H.-J. & Suzuki, H. (2010). Macrocell path-loss prediction using artificial neural networks. *IEEE Transactions on Vehicular Technology*, Vol. 59, no. 6, July 2010, 2735-2747

Panagopoulos, A. D.; Kritikos, T. D. & Kanellopoulos, J. D. (2007). Adaptive uplink power control in adjacent DVB-RCS networks: Interference statistical distribution, *Proceedings of IEEE 18th International Symposium on Personal, Indoor and Mobile Radio Communications (PIMRC'07)*, pp. 1-4, Athens, Sep. 2007, doi:10.1109/PIMRC.2007.4394408

Parsons, J. D. (2000). *The Mobile Radio Propagation Channel*, 2nd ed., John Wiley & Sons Ltd, Chichester

Petrus, P.; Ertel, R. B. & Reed, J. H. (1998). Capacity enhancement using adaptive arrays in an AMPS system. *IEEE Transactions on Vehicular Technology*, Vol. 47, no. 3, Aug. 1998, 717-727

Pirinen, P. (2006). Cellular topology and outage evaluation for DS-UWB system with correlated lognormal multipath fading, *Proceedings of IEEE 17th International Symposium on Personal, Indoor and Mobile Radio Communications (PIMRC'06)*, pp. 1-5, Helsinki, Sep. 2006, doi:10.1109/PIMRC.2006.254343

Popescu, I.; Nikitopoulos, D.; Constantinou, P. & Nafornita, I. (2006). Comparison of ANN based models for path loss prediction in indoor environments, *Proceedings of 64th IEEE Vehicular Technology Conference (VTC-2006 Fall)*, pp. 1-5, Montreal, Sep. 2006, doi:10.1109/VTCF.2006.43

Raeside, D. E. (1976). Monte Carlo principles and applications. *Physics in Medicine and Biology*, Vol. 21, no. 2, Mar. 1976, 181-197.

Rubio, L.; Reig, J. & Cardona, N. (2007). Evaluation of Nakagami fading behaviour based on measurements in urban scenarios. *AEU - International Journal of Electronics and Communications*, Vol. 61, no. 2, Feb. 2007, 135-138

Shankar, P. M. (2004). Error rates in generalized shadowed fading channels. *Wireless Personal Communications*, Vol. 28, no. 3, Feb. 2004, 233-238

Simon, M. K. & Alouini, M.-S. (2005). *Digital Communication over Fading Channels*, 2nd ed., John Wiley & Sons, Inc., Hoboken

Stavroulakis, P. (2003). *Interference Analysis and Reduction for Wireless Systems*, Artech House, Inc., Norwood

Thiele, L. & Jungnickel, V. (2006). Out-of-cell channel statistics at 5.2 GHz, *Proceedings of First European Conference on Antennas and Propagation (EuCAP 2006)*, pp. 1-6, Nice, Nov. 2006, doi:10.1109/EUCAP.2006.4584756

Thiele, L.; Peter, M. & Jungnickel, V. (2006). Statistics of the Ricean k-factor at 5.2 Ghz in an urban macro-cell scenario, *Proceedings of 17th IEEE International Symposium on Personal, Indoor and Mobile Radio Communications (PIMRC'06)*, pp. 1-5, Helsinki, Sep. 2006, doi:10.1109/PIMRC.2006.254301

Tripathi, N. D.; Reed, J. H. & VanLandingham, H. F. (1998). Handoff in cellular systems. *IEEE Personal Communications*, Vol. 5, no. 6, Dec. 1998, 26-37

Webb, W. (2007). *Wireless Communications: The Future*, John Wiley & Sons, Ltd, Chichester

Withers, C. S. & Nadarajah, S. (2010). A generalized Suzuki distribution. *Wireless Personal Communications*, Jul. 2010, 24 pages, doi:10.1007/s11277-010-0095-4.

Xiao, L.; Greenstein, L.; Mandayam, N. & Periyalwar, S. (2008). Distributed measurements for estimating and updating cellular system performance. *IEEE Transactions on Communications*, Vol. 56, no. 6, June 2008, 991-998.

Xylomenos, G.; Vogkas, V. and Thanos, G. (2008). The multimedia broadcast/multicast service. *Wireless Communications and Mobile Computing*, Vol. 8, no. 2, Feb. 2008, 255-265

Zhang, Z.; Lei, F. & Du, H. (2006). More realistic analysis of co-channel interference in sectorization cellular communication systems with Rayleigh fading environments, *Proceedings of 2006 International Conference on Wireless Communications, Networking and Mobile Computing (WiCom'06)*, pp. 1-5, Wuhan, Sep. 2004, doi:10.1109/WiCOM.2006.184

Zhang, Z.; Wei, S.; Yang, J. & Zhu, L. (2004). CIR performance analysis and optimization of a new mobile communication cellular configuration, *Proceedings of 4th International Conference on Microwave and Millimeter Wave Technology (ICMMT 2004)*, pp. 826-829, Beijing, Aug. 2004

Zhou, Y.; Chin, F.; Liang, Y.-C. & Ko, C.-C. (2001). Capacity of multirate multicell CDMA wireless local loop system with narrowbeam antenna and SINR based power control. *International Journal of Wireless Information Networks*, Vol. 8, no. 2, Apr. 2001, 99-108

Zhou, Y.; Chin, F.; Liang, Y.-C. & Ko, C.-C. (2003). Performance comparison of transmit diversity and beamforming for the downlink of DS-CDMA system. *IEEE Transactions on Wireless Communications*, Vol. 2, no. 2, Mar. 2003, 320-334

5

An Insight into the Use of Smart Antennas in Mobile Cellular Networks

Carmen B. Rodríguez-Estrello and Felipe A. Cruz Pérez
Electric Engineering Department, CINVESTAV-IPN
Mexico

1. Introduction

3G and 4G cellular networks are designed to provide mobile broadband access offering high quality of service as well as high spectral efficiency[1]. The main two candidates for 4G systems are WiMAX and LTE. While in details WiMAX and LTE are different, there are many concepts, features, and capabilities commonly used in both systems to meet the requirements and expectations for 4G cellular networks. For instance, at the physical layer both technologies use Orthogonal Frequency Division Multiple Access (OFDMA) as the multiple access scheme together with space time processing (STP) and link adaptation techniques (LA)

In particular, Space Time Processing has become one of the most studied technologies because it provides solutions to ever increasing interference or limited bandwidth (Van Rooyen, 2002), (Paulraj & Papadias, 1997). STP implies the signal processing performed on a system consisting of several antenna elements in order to exploit both the spatial (space) and temporal (time) dimensions of the radio channel. STP techniques can be applied at the transmitter, the receiver or both. When STP is applied at only one end of the link, Smart Antenna (SA) techniques are used. If STP is applied at both the transmitter and the receiver, multiple-input, multiple-output (MIMO) techniques are used. Both technologies have emerged as a wide area of research and development in wireless communications, promising to solve the traffic capacity bottlenecks in 4G broadband wireless access networks (Paulraj & Papadias, 1997).

MIMO techniques and their application in wireless communication systems have been extensively studied (Ball et. Al, 2009), (Kusume et. Al, 2010), (Phasouliotis & So, 2009), (Nishimori et. al, 2006), (Chiani et. al, 2010), (Seki & Tsutsui, 2007), (Hemrungrote et. al, 2010), (Gowrishankar et. al, 2005), (Jingming-Wang & Daneshrad, 2010); however, critical aspects of using SA techniques in cellular networks remain fragmental (Alexiou et. al, 2007). In particular those aspects related with the influence of users' mobility and radio environment at system level in SA systems which use Spatial Division Multiple Access (SDMA) as a medium access technique.

---

[1] A measure often used to assess the efficiency of spectrum utilization is the number of voice channels per Mhz of available bandwidth per square kilometer (Hammuda, 1997). This defines the amount of traffic that can be carried and is directly related to the ultimate capacity of the network.

SDMA cellular systems have gained special attention to provide the services demanded by mobile network users in 3G and 4G cellular networks, because it is considered as the most sophisticated application of smart antenna technology (Balanis, 2005) allowing the simultaneous use of any conventional channel (frequency, time slot or code) by many users within a cell by exploiting their position. However, SDMA technology has not been widely integrated into cellular systems as many as it had been predicted because SDMA introduces new challenges at the system level modeling. In particular, the design of radio resource management algorithms is an open research topic (Alexiou, A. et. al, 2007), (Toşa, 2010) Furthermore, in order to measure the performance of different radio resource management algorithms it is necessary to develop an adequate system level model because in SDMA cellular system level performance is closely affected by the constantly changing radio environment due to the users' mobility.
Thus, the objective of this chapter is to give an overview of the smart antenna technology in mobile cellular systems emphasizing those features which are related with SDMA. This chapter highlights the critical aspects of system level modeling regarding to radio resource management algorithms; in particular, users' mobility and radio environment issues are considered.
The chapter is organized as follows: In the first part an overview of smart antenna technology is given. Then, the techniques and applications of smart antennas in cellular systems are explained. After that, some commercial systems that use smart antennas are described. Afterward, the proposed model to include channel characteristics at system level is presented. Finally, the impact of users' mobility and radio environment on the system performance is evaluated.

2. Overview of smart antenna technology

The term smart antennas generally refers to any antenna array joint with signal processing, which can adjust or adapt its own beam pattern in order to emphasize signals of interest and to minimize interfering signals (Gross, 2005), (B. Allen and M. Ghavami, 2005). Smart Antennas can modify their radiation pattern by means of an internal feedback control while the antenna system is operating.
Smart antennas have alternatively been labeled through the years as adaptive arrays or digital beam forming arrays. The development of adaptive arrays began in the late 1950s. The term "adaptive arrays" was first coined by Van Atta (Van Atta, 1959) in 1959 to describe a self phased array. Self phased arrays reflect all incident signals back in the direction of arrival by using phase conjugation. Self phased arrays are instantaneously adaptive arrays since they essentially reflect the incident signal in a similar fashion to the classic corner reflector (Balanis, 2005).
Lately, in 1965, an adaptive sidelobe canceller was developed by Howell and Applebaum (Applebaum, 1976). This technique allows for mitigating interference, raising the signal to interference ratio (SIR). Another type of adaptive beamformer was developed by (Widrow et. al, 1967); this adaptive beamformer uses a pilot signal as a reference. It operates by forming a beam towards the wanted source(s) whilst simultaneously directing nulls towards interference sources. The beam was steered via phase shifters, which were often implemented at RF stage. This general approach to phase shifting has been referred as electronic beamsteering because the phase change is made directly at each antenna element (Gross, 2005).

Modern smart antennas systems still continue using the previously described techniques, but taking advantage of digital signal processing. The digital processing is performed at base band frequency, instead of doing it at RF stage. More over new beamforming techniques have emerged based on digital signal processing.
Due to the characteristics of smart antennas, they were originally focused on military uses like radar. Then smart antennas have been also employed in satellite applications to reuse frequency channels in different geographic locations. Recently, smart antennas were used in fixed wireless communication systems as wireless local loop (WLL). Nowadays, smart antennas are used in mobile wireless communication systems to improve coverage, capacity and spectral efficiency (Alexiou, A. et. al, 2007), (Toşa, 2010), iBurst, (2004), (3GPP TR 25.913, 2009). In particular, in cellular systems the use of smart antennas allows lower cost deployments with cells of moderate large size.

3. Architecture of smart antenna systems

As it was established, modern smart antennas are antenna arrays aided by digital signal processor. Thus, a generic smart antenna consists of two major components: the antenna array and the digital signal processor as it is shown in Figure 1.

3.1 Antenna array

The antenna array is one of the constitutive parts of a smart antenna: an antenna array consists of *N* identical independent antenna receivers separated in the space allocated in a geometric form. Thus, the electrical size of the complete array is greater than the electrical size of an individual element. By increasing the electrical size, highly directive radiation patterns are formed. Moreover, the multiplicity of elements allows more precise control of the radiation pattern resulting in lower sidelobes or fine pattern shaping.

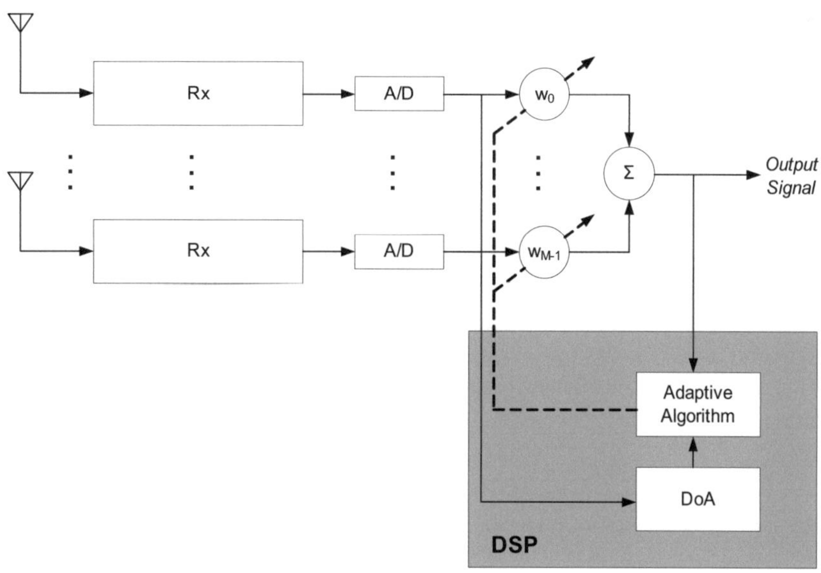

Fig. 1. A generic smart antenna system (Balanis, 20005)

The total radiated (received) field of the array at any point in the space is the vectorial sum of radiated (received) fields by each individual antenna (Balanis, 2005). Thus, the total radiated (received) field is determined by the product of individual element pattern and the array factor. The array factor is the spatial response to the received signals and it is affected by the geometrical relation of individual elements. Consequently, in order to provide directive patterns it is necessary that the fields from the elements of the array interfere constructively in the desired directions and interfere destructively in the remaining space. In an array of identical elements, there are five parameters of control that can be used to shape the overall pattern of antenna (Balanis, 2005):

The number of the elements of the array. The number of antenna elements in the array determines the degrees of freedom (to create nulls or maxima in the beampattern). That is, an antenna array with N elements allows forming N nulls or maxima.

The relative displacement between elements. The correlation among radiated (received) fields of the individual elements is influenced by the relative displacement between them. The displacement between elements has a direct relation with the physical and electrical array size. Displacement is measured in terms of wavelength because the response of the array is closely related with the operation frequency.

The separation between elements is also associated with a particular application. For instance, systems which use diversity need antenna arrays with relative displacements that ensure uncorrelated fading of the signals (more than one wave length). On the contrary, beamforming applications usually require relative displacements of less than a half wavelength. Relative displacement between elements is usually restricted by the available space for the antenna array.

The geometrical configuration of the overall array. The physical shape obtained with the allocation of the elements of the array is known as the geometry of the array. Location of the individual elements can vary widely, but the most common configurations are along a straight line, around a circle or in a planar way. The geometry of the antenna determines the relationship between the radiated fields of the individual elements. Thus, many important characteristics of the beampattern hang on the geometry. For example, linear and planar arrays generate grating lobes[2] if the relative displacement between elements is more than a half wavelength. Thus, linear arrays are typically used if sectored coverage is required. While circular or hexagonal arrays are used to provide 360 degrees continuous coverage.

The excitation (amplitude and phase) of the individual elements. Received (radiated) signals are weighted at each element and they are sum to form the beam pattern. Weights are in general complex in order to determine amplitude and time delay (phase) of the signal that feeds each antenna element. The weights can be approached in a fixed or adaptive way.

The relative pattern of the individual elements. Relative pattern is the radiated pattern of each element when it is not arranged in an array. Even though, it is possible to use any antenna as element in the array, individual elements used in arrays are generally half wavelength dipoles[3], to ensure that the radiation pattern of the overall array is completely determined by the geometry.

---

[2] Grating lobes are defined as: lobes, other than the main lobe, produced by an antenna array when the element spacing in the same plane is sufficiently large to permit the in-phase of radiated fields in more than one direction. (Balanis, 2005)

[3] Half wavelength dipoles present omnidirectional patterns. (Balanis, 2005)

3.2 Signal processing

The purpose of the signal processing stage is to adapt the beampattern to the radio environment conditions. This one is because signal processing has been used together with antenna arrays to act as spatial filters. Such filters can self-adjust to the characteristics of an incoming signal without external intervention, just as a closed loop control system. Thus, signal processing combined with antenna arrays produce a directive beam that can be repositioned (scanned) electronically by varying the excitation of the individual elements. Then, the objectives of the signal processing stage are (Balanis, 2005):

Estimate the direction of arrival (DoA) of all impinging signals. DoA estimation techniques can be categorized on the basis of the data analysis and implementation into four different areas (Liberti & Rappaport, 1999) (Krim & Viberg, 1996):

- *Conventional methods,* Conventional methods are based on beamforming and null steering. In this technique, DoA is determined tracking peaks by means of an exhaustive search on all possible directions. Examples of conventional methods are delay and sum, and Capon's minimum variance method.
- *Subspace-based methods.* Different from conventional methods, subspace methods exploit the structure of the received data; in particular the covariance matrix. This one results in a great improvement in resolution. Examples of these methods are MUltiple SIgnal Clasification (MUSIC) and the Estimation of signal parameters via rotational invariance technique (ESPIRIT) (Schmidt, 1979).
- *Parametric methods.* These methods are used in scenarios where the involved signals are highly correlated, or even coherent. Even though, parametric methods increase the efficiency and robustness of DoA, parametric estimation methods are computationally more complex than conventional and subspace methods. However, for Uniform Linear Arrays (ULA) there are a number of less demanding algorithms (Krim & Viberg, 1996). Maximum likelihood (deterministic and stochastic) algorithms are the most frequently used. These algorithms are based on finding the optimum data processing solution for the case of J unknown signals and K sensors ($K > J$) with only additive white noise errors (Schweppe,1968) (Ziskind & Wax, 1988)
- *Integrated methods,* which combine two or more methods. There are many combined methods which try to get the advantages of both methods while minimizing disadvantages. However, most of these methods are computationally very complex. (Parra et. al, 1995).

Calculate the appropriate weights to steer the maximum radiation of the antenna pattern toward the Signals of Interest (SoI) and to place nulls toward the Signals of No Interest (SNoI). The algorithms used to form a beampattern could be categorized in two kinds: those which are DoA-based and those which uses a reference signal or a training sequence (Balanis, 2005).

DoA-based beamforming algorithms. The information supplied by the DoA algorithm is processed by means of an adaptive algorithm to ideally steer the maximum radiation of the antenna pattern toward the SoI and place nulls in the pattern toward the SNoI.

Reference training beamforming algorithm. In these techniques, an excitation vector that minimizes a cost function is determined. The cost function is related to a performance measure and it is inversely associated to the quality of the signal at the array output. The most commonly used performance measures are Minimum Mean Square Error (MMSE), Maximum Signal to Noise Radio (MSNR) and Minimum Variance (MV) (Litva, 1996). If the cost function is minimized, then, the quality of the signal is maximized (Liberti & Rappaport, 1999). In order to minimize the cost function, reference training algorithms are used. These algorithms require solving a linear system of equations based on the information of training sequences. If the radio environment is stationary, the arrival angles of desired and undesired signals do not change. Consequently, weights are easily computed. However, if the radio environment is continuously changing, weights are needed to be computed with adaptive methods and they become computationally extremely demanding (Balanis, 2005). The most common used adaptive algorithms are Least Mean Squares (LMS), Sample Matrix Inversion (SMI), and Recursive Least Squares (RLS) (Gross, 2005).

Blind beamforming algorithms. These techniques exploit the characteristics of the signal, such as autocorrelation, when no reference signal is available. Blind beamforming techniques can be used as adaptive methods to constantly calculate the appropriate weights to steer the beam. An example of blind beamforming algorithm is the family of Constant Modulus (CM) algorithms, which takes advantage of the constant amplitude of phase-modulated signals (Gross, 2005). Decision-Directed algorithm and cyclostationary algorithms are other examples of blind beamforming (Litva, 1996).

4. Smart antenna systems categorization

Smart antennas systems can be categorized by considering the adaptation technique and their application to cellular systems. The adaptation technique refers to the capacity of the algorithm to track the user and to eliminate undesired signals and the application refers to the objective of using smart antennas in cellular systems.

4.1 Smart antenna adaptation techniques

Smart antennas generally encompass both switched and beamformed adaptive systems. Switched or fixed beam is the simplest one and it refers to the arrangement in which a finite number of predefined radiation patterns are formed in fixed directions. While the adaptive array approach denotes the capability of smart antennas to dynamically adjust the radiation pattern to improve the performance of the system according to a certain performance metric.

Switched beam. As an extension of sectorised concepts, the objective of switched beam systems is to form several available fixed beam patterns. In switched beam systems, the maxima of the fixed beampatterns are selected to ensure a uniform coverage of a region in the space. Normally, directions of the maximums are in equal angular increment.

Switched beam is technologically the simplest technique and can be implemented by using a number of fixed, independent, directional antennas or virtually with an antenna array and an analogue beamformer such as Butler matrix or roman lens (Butler, 1961). The operation of Butler matrix can be likened to a Fast Fourier Transform (FFT) and yields M mutually orthogonal beams. The orthogonality of the beams is defined by the angle minima of one beam pattern corresponding with the main beam angle of all of the other beams (Butler, 1961). A similar technique called grid of beams (GoB) can be used with digital beamforming systems which selects the best weights from a stored set (Tsoulos, 1999). This technique leads to a more complex implementation due to the drawbacks associated with digital beamforming.

In switched beam systems, the decision of which beam serves a specific user is made based upon the requirements of the system. For instance, one criterion which can be applied to select a beam is to maximize Signal to Interference Ratio (SIR); another criterion is to maximize the Received Signal Strength Indicator (RSSI), which consist in select the beam which provides the strongest signal.

Several works have studied switched beam systems at link and at system level (Mailloux, 1994), (Hansen, 1998), (Pattan, 2000). Many of these works have studied the switched beam systems at link level analyzing SNR and Bit Error Rate performance (Hu & Zhu, 2002), (Nasri et.al, 2008) (Ngamjanyaporn, 2005) (Lei et. al, 2005). In (Ho et. al, 1998) (Peng & Wang, 2005) the performance and feasibility at system level of switched beam systems has been investigated in terms of blocking probability.

Even though, switched beam systems provides an increment in the capacity of cellular systems, as the issue of trunking efficiency has become more pronounced, focus has recently shifted to more advance fully adaptive techniques.

Beamformed adaptive systems allow the antenna to steer the beam to any direction of interest while simultaneously nulling interfering signals. Adaptive array antenna systems continually monitor their coverage areas attempting to adapt to their changing radio environment which consists of mobile user and mobile interferers. In the simplest scenario (one users and no interferers) the systems adapts to the user's motion producing and effective antenna pattern that follows the user always providing maximum gain in the user's direction.

In adaptive systems, pattern optimization is done by real time active weighting of the received signal and can adapt to changes in the radio environment. Although in principle it is possible to adapt transmit patterns to optimize the transmission subject to some received signal or noise distribution, this is seldom done except for the formation of retrodirective beams which automatically transmit in the direction of the received signal or pilot tone (Balanis, 2005)

The switched beam technique is more attractive when compared to the adaptive null steering because with switched beam, no complicated multi-beam beamforming is needed and no significant changes to the existing cellular systems are required (Peng & Wang, 2005) (Shim & Choi, 1998).

4.2 Smart antenna applications in cellular systems (HSR, SFIR, SDMA)

The usages of smart antennas in cellular systems are focused on three different objectives: increasing coverage as a High Sensitivity Reception (HSR), reducing interference as spatial filters -Spatial Filter Interference Reduction- (SFIR), and spatially reusing basic radio resources (frequency bands, time slots, orthogonal codes, chunks) providing another way of multiple access -Spatial Division Multiple Access- (SDMA). These applications have been also considered as the stages of introduction of smart antenna technology in the evolution of cellular systems (Boukalov & Haggman, 2000).

High Sensitivity Reception (HSR). Smart antennas were originally used to provide range extension through the inherent directional gain obtained from using an array. The additional directional gain, D provided by an M element array is approximated by:

$$D \simeq 10\log_{10} M \tag{1}$$

This application is useful for rural cells which are required to cover larger areas than those in urban environments.

Spatial Filtering Interference Rejection (SFIR). Smart Antennas are also applied to achieve spatial filtering, enhancing the signal strength (beamforming) and weakening the interference power (nulling) in Wireless Communication Systems. Network capacity is increased by controlling the interference level received from other users and base stations. This operating mode is referred to as spatial filtering interference rejection (SFIR).

SFIR reduces susceptibility to multipath effects since spatial filtering reduces the channel delay spread. This is because signals arriving at angles outside of the main beam are attenuated. These signals will have longer path lengths and would normally contribute to the longer delays of the channel impulse response. The consequence of reducing delay spread is that equalization techniques are no longer required therefore simplifying the receiver design. Moreover, channel reuse patterns in cellular systems can be significantly tighter because the average interference resulting from co-channel signals in other cells is markedly reduced.

Spatial Division Multiple Access (SDMA). The most promising mode of smart antennas in cellular systems is referred to as space division multiple access (SDMA) where smart antennas are used to separate signals, allowing different subscribers to share the same resource provided their signals are spatially separable at the base station. This mode of operation allows multiple users to operate on the same slot, frequency and in the case of CDMA systems, the same code within a cell. The net result of the adaptive process is that SDMA systems can create a number of two way spatial channels on a single conventional channel (frequency, time, code). SDMA technology is not restricted to any particular modulation or air interface protocol and is compatible with all currently air interfaces deployed at this time.

Modern wireless communications systems deploy antenna arrays in SDMA configuration. Here a base station communicates with several active users by directing the beam towards them and it nulls users which cause interference. This has two beneficial effects: first the target users receive more power compared to the omnidirectional case and second the interference to the adjacent cells is decreased because only very selected directions are targeted.

5. Commercial smart antennas cellular systems

Although smart antennas have been a hot research topic in the last two decades, and smart antenna techniques have been proposed in 3G and 4G cellular systems (ITU-R M.1801, 2007), (3GPP TR 25.913, 2009), (Hoymann, 2006) as one of the leading technologies for achieving high spectral efficiency[4] by reusing basic radio resources, smart antenna systems are slowly becoming commercially available. The main reasons which explain why smart antenna systems have not been deployed completely are stated in (Alexiou et.al., 2007), (Kaiser, 2005), (Rajal, Dec 2005), (G. Okamoto, 2003). The conclusions are that the speed of DSP is not enough to real - time process the needed algorithms and the algorithms are computationally very demanding. Moreover the additional cost of using smart antenna at current systems is not sustainable.

[4] Spectral efficiency measures the ability of a wireless system to deliver information, within a given amount of radio spectrum. In cellular radio systems, spectral efficiency is measured in bits/second/Hertz/sector (bps/Hz/sector). Factors that contribute to the spectral efficiency include the modulation formats, "overhead" due to the signaling, multiple access method, etc. The reason to reference of the spatial dimension (per sector) is the self interference generated in the network, requiring the operator to allocate frequencies in blocks that are separated in space by one or more cells. This separation is represented by a reuse factor, where a lower number is representative of a more efficient system.

Some efforts have been made in order to standardize the use of smart antennas in wireless communication networks. Several standardization organizations such the Alliance of Telecommunications Industry Solutions (ATIS), the International Telecommunication Union (ITU), and the Institute of Electric and Electronic Engineers have standardized the use of smart antennas in wireless systems. Furthermore, some field tests for MIMO and SA technologies have been developed (TSUNAMI) (Tsoulos & Mark Beach, 1997). However, currently, only a few companies have successfully commercialized smart antenna systems for cellular base stations.

The first commercial system that uses smart antennas was iBurst. iBurst is a mobile broadband wireless access system that was first developed by ArrayComm, and subsequently adopted as the High Capacity - Spatial Division Multiple Access (HC-SDMA) radio interface standard (ATIS-0700004-2005). ITU also includes this system in ITU-R M.1678 and ITU-R M.1801.

ITU-R M.1678 recommendation addresses the use of adaptive antenna technology in the mobile service with the objective to improve spectrum efficiency significantly, improve the ability of mobile systems to coexist and facilitate cross-border and adjacent band sharing, and facilitate the deployment of new wireless networks, including broadband wireless access and radio local area network systems.

ITU-R M.1801 recommendation defines specific standards for broadband wireless access in the mobile service at radio interface. These specific standards are composed of common specifications developed by standards development organizations (SDOs). Using this Recommendation, manufacturers and operators should be able to determine the most suitable standards for their needs. These standards support a wide range of applications in urban, suburban and rural areas for both generic broadband internet data and real-time data, including applications such as voice and videoconferencing. The commercial name of this recommendation is High Capacity - Spatial Division Multiple Access (HC-SDMA).

In January 2006, the IEEE 802.20 Mobile Broadband Wireless Access Working Group adopted a technology proposal that includes the use of the HC-SDMA standard for the 625kHz Multi-Carrier Time Division Duplex (TDD) mode of the future IEEE 802.20 standard. Moreover the recommendation IEEE 802.16 proposes the use of smart antennas in WiMAX system and 3GPP group suggests also the use of smart antennas in LTE.

iBurst (HC-SDMA) is a wireless broadband technology developed by ArrayComm. The main objective of the iBurst system is to optimize the use of the available bandwidth with the help of smart antennas. Arraycomm and Kyocera are the main providers of the equipment for this technology. HC-SDMA offers up to 20 Mbps of aggregate usable IP-traffic capacity per sector in each 5 MHz TDD allocation and supports both fixed and fully mobile broadband users. As the standard's name implies, the key to the system's capacity is the spatial processing and interference management software, supporting up to 3 SDMA channels on each physical carrier in each cell. HC-SDMA works with TDD/TDMA/SDMA, 625 kHz channel spacing (iBurst, 2004).

Although, iBurst is an in-use commercial system, it actually experiences some drawbacks when it deals with users' mobility in the commercial implementation. Actually, ArrayComm reports in a white paper (iBurst, 2004), that "SDMA is most useful to operators with high capacity requirements, more limited spectrum, and tight constraints on client device costs and complexity. However, the effectiveness of SDMA declines gradually with increasing subscriber mobility".

ArrayComm has also implemented HSR and SFIR applications. Their solution for increased gain and interference management for GSM infrastructure includes products, designed to improve frequency re-use and overall network capacity. Moreover, ArrayComm's ranging extension solution for WiMAX provides a ~6 dB improvement in ranging channel link budget. This enables successful and practical use of all the traffic-channel range gains identified above.

Regarding to the Smart Antenna techniques, ArrayComm reports in (iBurst WP, 2004) that Beam switching technology has seen virtually no commercial use. Drawbacks include high sensitivity to the subscriber's location within the beam and interference from users outside the beam's primary target. Moreover, they consider that beam steering algorithms have also not seen successful use outside the laboratory. The radio environment in the world of commercial services is filled with multipath and scattering effects and non-line-of-site conditions that prevent techniques based on degree-of-arrival calculations from delivering useful results in most circumstances. Thus, more sophisticated algorithms are required as well as faster digital signal processors.

Other company that is working on smart antenna applications is Alvarion (SentieM WP, 2008). Alvarion's SentieM Mobile WiMAX is not currently available; however, it is on development. Alvarion's future SentieM Mobile WiMAX technology uses Spatial Division Multiple Access (SDMA) technology which provides the ability to use the same frequency (beam) at the same time for different users. This solution is unique in its ability to select the right user at the right time for the frequency sharing to work. Alvarion reports that "SentieM's SDMA solution will not require any information from the end-user terminal, avoiding the need for a complicated integration process with the subscriber device".

Ericsson has also has conducted extensive research and development of advanced base-station antennas for mobile communication (Derneryd & Johannisson, 1999). Their work comprises both adaptive and active antenna systems. With the introduction of active antenna products, such as Maxite products, small-sized base station units with high levels of equivalent radiated power (ERP) and low power consumption can be used.

Ericsson has also developed an HSR system which consists of two-dimensional antenna arrays for adaptive base-station systems. These arrays, which are developed for systems based on GSM and TDMA (IS 136) standards, work in the 900, 1800 or 1900 MHz frequency bands. Together with Mannesmann Mobilfunk GmbH (GSM) and AT&T Wireless Services (TDMA), Ericsson has conducted field trials in live networks to evaluate the performance of the adaptive systems. The results show that adaptive antenna systems increase capacity in 20%. The adaptive array antenna transmits and receives radio-frequency signals in directed narrow beams in the base of Butler matrices to produce horizontal beamforming networks. As a consequence of the use of fixed beamforming networks, besides increased capacity, the increase in antenna gain may also be exploited to offer greater coverage.

6. System-level model for mobile SDMA cellular systems

Although the use of smart antennas in cellular systems lead to a capacity increase, the deployment of smart antennas implies a more complicated design at radio stage and new radio resource management algorithms designed specifically for this kind of systems (Boukalov, 2000). In particular, in SDMA cellular systems, intra- and inter-cell co-channel interference is increased because of the intra-cell cochannel reuse. As a result, is not always possible to replicate basic channels[5] in the admission process. In addition, the effect of intra-

[5] Basic channels are considered time slots, frequency carriers and codes

cell interference becomes worst as the users' mobility increases. Consequently, users' SIR could be severely degraded due to users' mobility.

Previous published works have mathematically analyzed SDMA at system level by means of a multidimensional Markov model (Galvan-Tejada & Gardiner, 1999), (Galvan-Tejada & Gardiner, 2001), (Shuangmei et. al 2004); however, users' mobility together with co-channel interference due to the replicated channels within cells has not been considered. Thus, only new call blocking probability is calculated and call forced termination probability is disregarded. On the other hand, most of the studies addressing the impact of users' mobility at the system level performance of SDMA cellular networks have been done through discrete event computer simulations (Pabst et. al, 2007), (Czylwik et. al, 2001), (Cardieri & Rappaport, 2001). However, they are based only on geometrical considerations (i.e., hexagonal/circular shaped cells, linear users' movement, ideal beam patterns) while current co-channel interference conditions experienced by users are ignored at all. Just a few works have dealt with mobility in an analytical way (Tangeman, 1994), (Liu, 2004), (Liu, 2005) and none of the previous works have treated mobility and co-channel interference together.

In order to include users' mobility and the effect of interference at system level, it is necessary to develop and adequate teletraffic model. In (Rodríguez-Estrello & Cruz-Pérez, 2009) a system level analytical model which includes not only mobility but also co-channel interference for SDMA systems, Thus, in this section, the model proposed in (Rodríguez-Estrello & Cruz-Pérez, 2009) is taken as a basis to analyze SDMA system´s performance.

On the other hand, in order to evaluate the effects of mobility, the generalized mobility model proposed in (Zonoozi & Dassanayake, 1997) is used for random user mobility characterization due to its simplicity and versatility to represent several scenarios. The model in (Zonoozi & Dassanayake, 1997) is characterized by the parameter α that limits the range of maximum variation of the future moving direction relative to the current one.

6.1 Network topology

A real time (i.e., conversational) service homogeneous mobile multi-cellular system with smart antennas located at the center of cells is assumed. Figure 2 shows the network topology. SDMA is used as a multiple access scheme in conjunction with a basic multiple access scheme (TDMA, CDMA, OFDMA). Thus, two or more users could share a basic channel within a cell (intra-cell reuse); however, resources could be 'replicated' only if the SIR is above a threshold in both channels.

6.2 Proposed Model to Include Co-channel Interference

The effect of co-channel interference is captured through the system level model proposed in (Rodríguez-Estrello & Cruz-Pérez, 2009) by introducing two parameters that depends on the mobility and radio environment:

The acceptance probability. The acceptance probability is the probability that a basic resource could be replicated in the admission process. This probability reflects the probability that the SIR is above a given threshold. Notice that this probability depend on how many times the basic resource is replicated.

Poisson call interferential process. The proposed model in (Rodríguez-Estrello & Cruz-Pérez, 2009) is based on the physical process in which a call could be involved: after a new call or a handoff attempt is accepted by a base station (BS), if the call is served by a replicated channel, the link condition could become degraded mainly due to the intra-cell co-channel

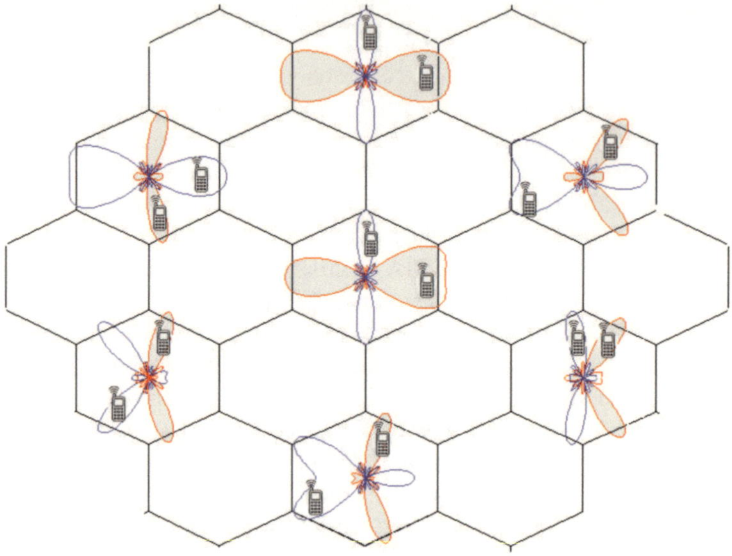

Fig. 2. Network Topology

interference causing a fall on the SIR of involved users and forced to terminate the call due to link unreliability. The period of time between the instant the call is originated and the moment the call is forced to terminate due to link unreliability is called "beam overlapping time". Thus, the proposed interferential process is a Poisson process which reflects the effect of link conditions and it has associated the beam overlapping time.

Beam overlapping time strongly depends on how many users share the same channel. Therefore, in general, the more users sharing a channel, the shorter beam overlapping time. Physically, beam overlapping time represents the period of time in which the users' SIR would be degraded due to the co-channel interference under the assumption that both cell dwell time and the unencumbered service time are of infinite duration.

6.3 Call Admission Control (CAC)

In SDMA cellular systems, call admission control refers the replicating order of basic radio resources. Two replication policies are possible:

Duplicate at last. (DL) Allocate basic channels one by one; once all N basic channels have been assigned, allocate the replicas of the basic channels (if the channel conditions are good enough to permit replicate).

Duplicate at first. (DF) Allocate basic channels and replicate them even if the all basic channels have not been previously allocated. If it is not possible to assign a replicated channel, allocate next primary channel.

6.4 Teletraffic model

6.4.1 General guidelines

For the mathematical analysis, the general guidelines of the model presented in (Lin et. Al, 1994) are adopted here to cast the system here considered in the framework of birth and death processes:

1. A homogeneous multi-cellular system with smart antennas located at the centre of each cell is assumed.
2. Each cell has a maximum number of basic radio resources C, each one of these basic channels could be replicated up to R times (intra-cell reuse factor) due to the SDMA capability.
3. The unencumbered call duration and cell dwell time are assumed to be negative exponentially distributed random variables with mean $1/\mu$ and $1/\eta$, respectively.
4. The probability that i channels can be replicated k times in the admission process is denoted by $p_{i,k}$. These probabilities have been extensively studied and characterized in the literature by means of mathematical analysis (Galvan-Tejada & Gardiner, 1999), (Galvan-Tejada & Gardiner, 2001), (Shuangmei et. al 2004), and it has been demonstrated that they can be expressed in terms of the probabilities that one channel could be replicated k times.

$$p_{i,k} = 1 - \left(1 - p_{1,k}\right)^{i} \tag{2}$$

5. It is considered that no reassignment nor prioritization scheme for handoff calls or new calls is used[6].
6. New call arrival process follows a Poisson process with mean arrival rate λ_n per cell.
7. Handoff call arrival process to every cell is also considered to be a Poisson process with mean arrival rate λ_h. The inter-cell handoff arrival rate is iteratively calculated using the method described in (Lin et. al, 1994) [7].
8. Call interferential process is considered to be a Poisson process with mean arrival rate γ_i, where i is the number of users sharing the same channel within a cell. Characterization of beam overlapping time has been previously reported in (Rodríguez-Estrello et. al, 2009)

6.4.2 Queuing formulation

As a homogenous case is assumed, the overall system performance can be analyzed by focusing on only one given cell and considering that its neighboring cells exhibit identical statistical behavior. Let us denote the state of a given cell as $\mathbf{K} = [k_1, k_2, \ldots k_R]$, where k_j (for i = 1, 2, ..., R) represents the number of active mobiles in the i-th level of repetition[8]; $i = 1$ represents the case in which only basic resources has been used, $i = 2$ represents the level in which two users are sharing one basic resource and so on. Thus, states in the system must accomplish with $k_1 \geq k_2 \geq \ldots \geq k_R$.

Multidimensional birth and death process is analyzed by "rate out equals rate in" equations (Cooper, 1990). The vector $\mathbf{e}_i$ is a unity vector of the same dimension of $\mathbf{K}$ and a "one" in the

---

[6] Considering a channel reservation scheme for handoff calls would bias the performance of the analyzed system and would not allow for the assessment of its "pure" performance. However, the system level performance evaluation of mobile cellular networks simultaneously considering co-channel interference, resource insufficiency and resource management schemes is desirable and can be easily achieved.

[7] Although handoff call arrival process generated by a single cell is clearly not Poissonian, the combined process from the six different neighboring cells can be adequately approximated by a Poisson process (Cheblus and W. Ludwin, 1995)

[8] Repetition level means the number of times that a basic resource is used within a cell.

i-th position. System steady state probabilities are denoted by vector π. The equilibrium state equations are (Cooper, 1990):

$$\pi_{\mathbf{K}} = \frac{\sum_{i=1}^{R} a_{i\mathbf{K}-\mathbf{e}_i} \pi_{\mathbf{K}-\mathbf{e}_i} + \sum_{i=1}^{R} b_{i\mathbf{K}+\mathbf{e}_i} \pi_{\mathbf{K}+\mathbf{e}_i}}{\sum_{i=1}^{R} a_{i\mathbf{K}} + \sum_{i=1}^{R} b_{i\mathbf{K}}} \tag{3}$$

for $\mathbf{K} = [k_1, k_2, \ldots k_R]$ such that $\mathbf{K} \in \Omega$. Where Ω is the valid state space, defined by:

$$\Omega = \{0 \le k_i \le C,\ k_1 \ge k_2 \ge \ldots \ge k_R \quad \text{for } i = 1 \ldots R\} \tag{4}$$

When duplicate at first policy (DF) is used, new call arrival for mobiles that will generate a transition from $\mathbf{K}$ to $\mathbf{K}+\mathbf{e}_i$ (where $i = \{1,2,3,\ldots,R\}$) given by:

$$a_{i\mathbf{K}} = \left\{ \prod_{j=0}^{R-i-1} \left(1 - p_{(k_{R-j-1} - k_{R-j}),(R-j)}\right) p_{(k_{i-1} - k_i),(i)} (\lambda + \lambda_h) \right. \quad \text{for } 0 \le k_i \le C,\ k_1 \ge k_2 \ge \ldots k_R \tag{5}$$

When duplicate at last policy (DL) is used, new call arrival for mobiles that will generate a transition from $\mathbf{K}$ to $\mathbf{K}+\mathbf{e}_i$ (where $i = \{1,2,3,\ldots,R\}$) given by:

$$a_{i\mathbf{K}} = \left\{ \prod_{j=1}^{i-1} \left(1 - p_{(k_{j-1} - k_j),j}\right) p_{(k_{i-1} - k_i),(i)} (\lambda + \lambda_h) \right. \quad \text{for } 0 \le k_i \le C,\ k_1 \ge k_2 \ge \ldots k_R \tag{6}$$

Notice that if $i = 1$, $k_{-1} = C$.

Call departure rate for users that will generate a transition from $\mathbf{K}$ to $\mathbf{K}-\mathbf{e}_i$ (where $i = \{1,2,3\ldots,R\}$) is given by:

$$b_{i\mathbf{K}} = \{ i(k_i - k_{i+1})(\mu + \eta + \gamma_i) \ \text{for } 0 \le k_i \le C,\ k_1 \ge k_2 \ge \ldots k_R \tag{7}$$

Notice that if $i = R$, $k_{R+1} = 0$. The departure rate includes the interferential rate γ_i depending on the number of mobiles using replicated channels within the analyzed cell[9].
Together with the normalization condition:

$$\sum_{\{\mathbf{K} \in \Omega\}} \pi_{\mathbf{K}} = 1 \tag{8}$$

The corresponding steady state probabilities are calculated by using the Gauss-Seidel method (Cooper, 1990).
As an example, Figure 3 shows a state transition diagram for an SDMA system that uses Duplicate at First policy with C basic resources that can be replicated twice.

---

[9] This rate is strongly affected by the parameters involved in the beam forming such as number of elements in the antenna, geometry of the array, beam forming algorithm, separation between elements in the array as well as users' mobility conditions.

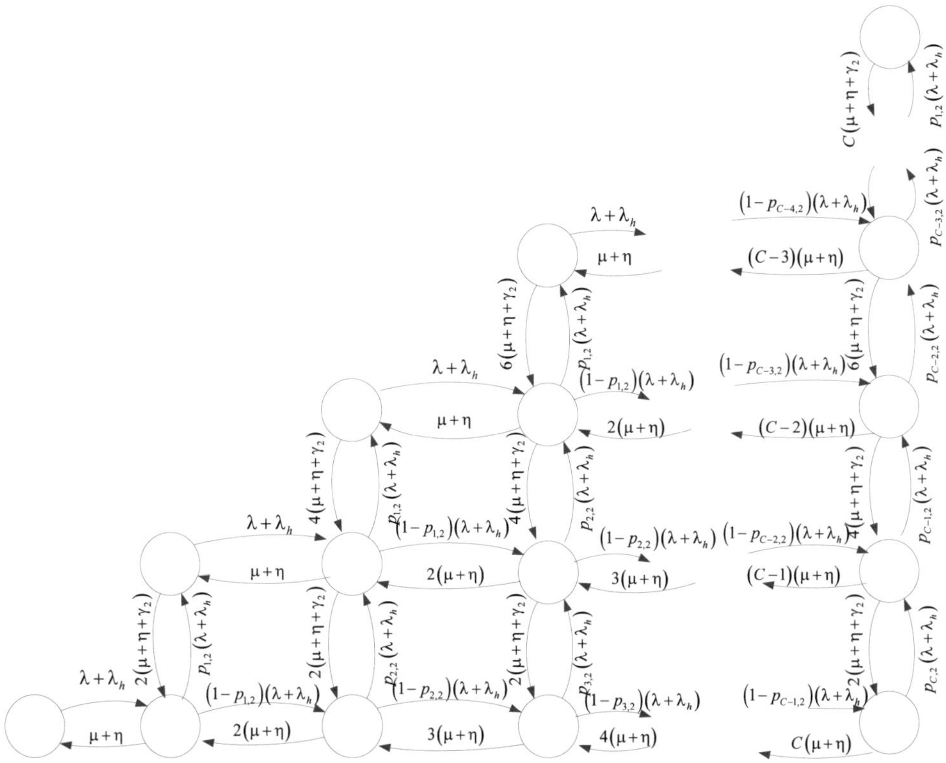

Fig. 3. SDMA system with Duplicate at First Policy.

6.4.3 QoS metrics

In this section, closed mathematical expressions for the most important QoS metrics are found.

Blocking probability / Handoff failure probability.

New calls are blocked with probability:

$$P_b = \sum_{\{\mathbf{K}\in\Omega;k_1=C\}} \prod_{j=0}^{R-1}\left(1-p_{(k_{R-j-1}-k_{R-j}),(R-j)}\right)\pi_{\mathbf{K}} \tag{9}$$

Handoff probability

Handoff probability is defined as the probability that a call requires a handoff. Mathematically it is expressed as the probability that the random variable associated with the residence time ($\mathbf{X_r}$) is less than the minimum of the random variables associated to the service time ($\mathbf{X_s}$) and the beam overlapping time ($\mathbf{X_{oi}}$); that is,

$$\mathbf{P}_{Hi} = \mathrm{P}\{\mathbf{X_r} < \min(\mathbf{X_s}, \mathbf{X_{oi}})\} \tag{10}$$

The handoff probability for a specific user depends on the number of users i that share the same channel. Thus, handoff probability is defined in terms of the system steady state probabilities as:

$$P_H = \sum_{\{\mathbf{k}\in\Omega\}} \pi_{\mathbf{K}} \sum_{i=1}^{R} \frac{i(k_i - k_{i+1})}{\sum_{i=1}^{R} k_i} \frac{\eta}{\eta + \mu + \gamma_i} \tag{11}$$

Call interruption probability

Call interruption probability is defined as the probability that a call is forced to terminate due to the interference and it depends on the number of users i sharing the same channel within the analyzed cell. It is mathematically expressed as

$$P_{Ii} = \mathrm{P}\{\mathbf{X_{oi}} < \min(\mathbf{X_s}, \mathbf{X_r})\} \tag{12}$$

Thus, call interruption probability can be calculated as:

$$P_I = \sum_{\{\mathbf{K}\in\Omega\}} \pi_{\mathbf{K}} \sum_{i=1}^{R} \frac{i(k_i - k_{i+1})}{\sum_{i=1}^{R} k_i} \frac{\gamma_i}{\eta + \mu + \gamma_i} \tag{13}$$

Probability of Call Forced Termination

Call forced termination probability (Pft) is considered as one of the most important QoS metrics for RP-based mobile cellular systems (Heredia-Ureta et.al., 2003), (Jiang & Rappaport, 2004). In particular, in SDMA cellular systems, call forced termination may result from either handoff failure or link unreliability due to the excessive co-channel interference. Since a combinatorial analysis in order to evaluate call forced termination probability is very complex, a signal flow diagram is used to characterize the call states transitions), (Jiang & Rappaport, 2004) , (Robichaud et.al, 1962).

Figure 4 shows the signal flow diagram for an SDMA cellular system with the capability of replicate a basic channel up to three times. Thus, the signal flow diagram is solved by considering that the call initiation state is the source node and the forced call termination is the sink node, and then by using the Mason's rule. Call state transitions include handoff attempts, successful call termination and interruption probability.

7. Performance evaluation of mobile SDMA cellular systems

In this section, the impact of mobility and radio environment in SDMA cellular networks is evaluated by means of teletraffic analysis considering the teletraffic model presented in the previous section. Blocking and Call Forced Termination probabilities are evaluated and compared when Duplicate at First and Duplicate at Last admission policies are used. Numerical evaluations are conducted considering that the system has $C = 10$ basic channels that can replicated up to three times. Offered traffic is varied from $a = 10$ to 50 Erlangs/cell. In addition, Blocking Calls Cleared (BCC) policy is considered. Mean service time $\mathbf{E}\{\mathbf{X_s}\} = 180$ s. In order to evaluate mobility mean cell dwell time is varied ($\mathbf{E}\{\mathbf{X_r}\}$) and to evaluate radio environment mean beam overlapping time ($\mathbf{E}\{\mathbf{X_{oi}}\}$) is varied.

7.1 The impact of mobility in SDMA cellular systems

Figures 5-8 show the impact of mobility in blocking and call forced termination probabilities for different scenarios. Mean cell dwell time is varied ($\mathbf{E}\{\mathbf{X_r}\} = 1000, 5000$, No mobility) Evaluations presented in this section do not consider link unreliability due to the excessive co-channel interference.

Node	Meaning
I	Call initialization
iC	Call initializes in a channel type *i*
FT	Forced termination
ST	Successful termination

Weight	Meaning
P_{ic}	Probability of using channel of type *i*
$P_{ic\text{->}jc}$	Probability of changing from channel type *i* to a channel type *j*
P_{Hi}	Probability of handoff failure for an active call in a channel type *i*
P_{Ii}	Probability of interruption due to link unreliability for an active call in a channel type *i*

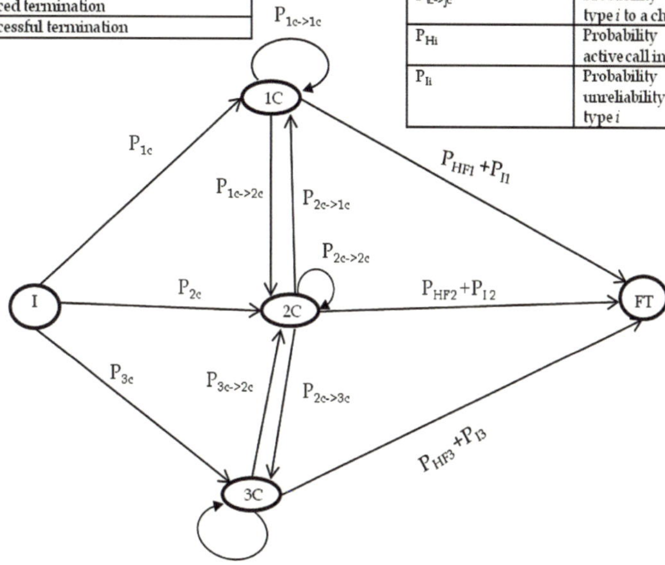

Fig. 4. SDMA system with Duplicate at First Policy .

From figures 5-6 it is possible to observe that the blocking probability is almost insensible to the residence time and to the call admission control policy. However, from figures 7-8 it is possible to observe that call forced termination probability is very sensible to the mobility. As the mean residence time decreases, call forced termination probability increases exponentially. This is because of the handoff probability also increases. Notice that when no

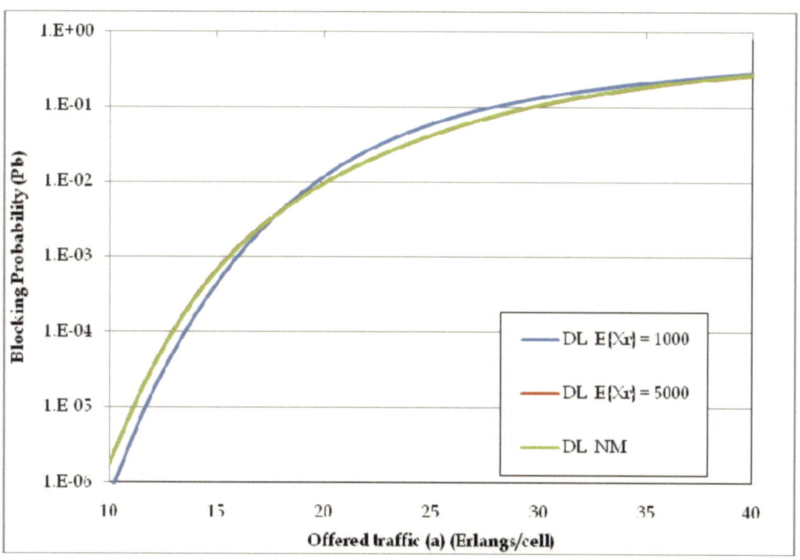

Fig. 5. Blocking Probability for a system with Duplicate at Last Policy. No link unreliability is considered.

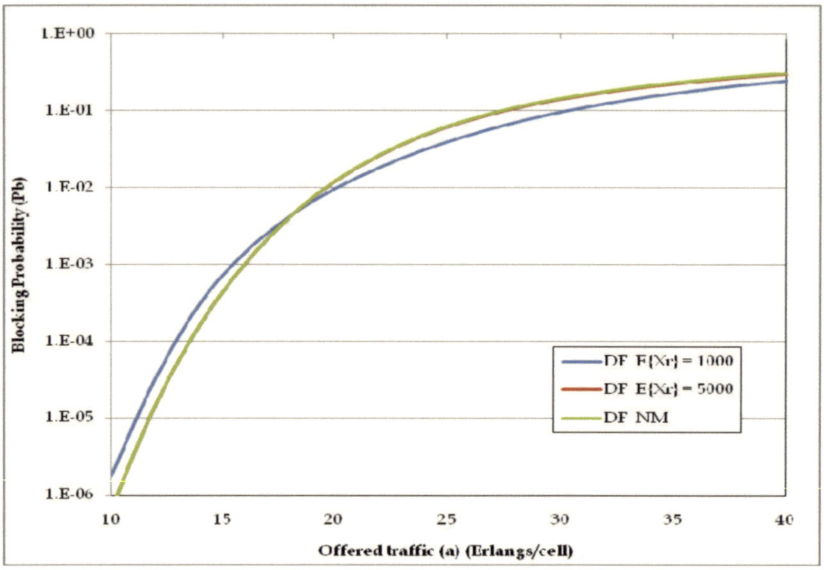

Fig. 6. Blocking Probability for a system with Duplicate at First Policy . No link unreliability is considered.

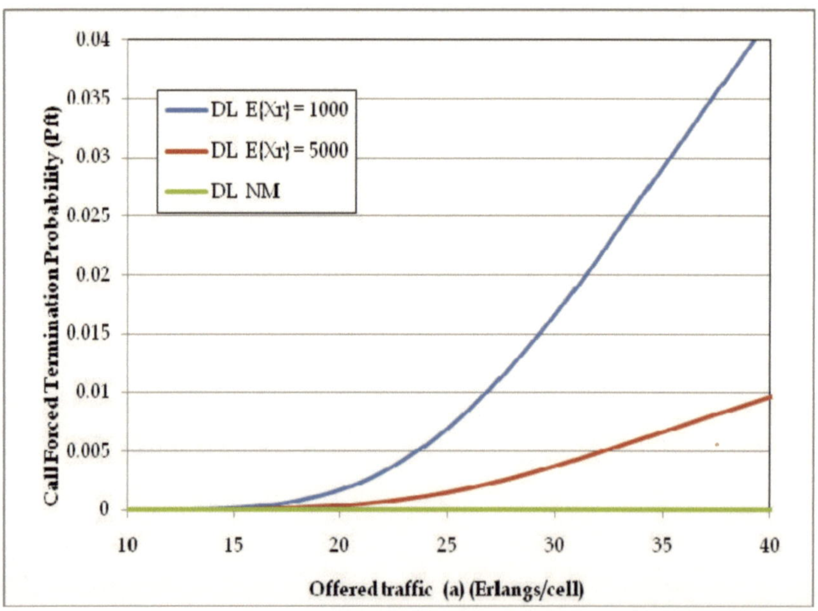

Fig. 7. Call Forced Termination Probability for a system with Duplicate at Last Policy. No link unreliability is considered.

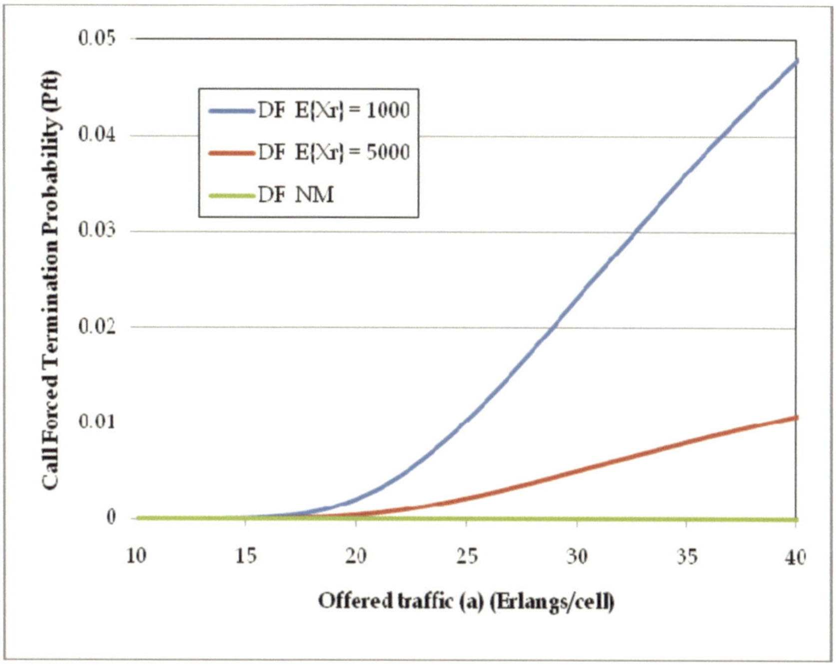

Fig. 8. Call Forced Termination Probability for a system with Duplicate at First Policy. No link unreliability is considered.

mobility is considered call forced termination is zero for all cases. This is because there are no causes of forced termination. Figures 7-8 show that the "Duplicate at First" policy is more sensible to the mobility. This is because in the scenario where there is more mobility, there are also more handoff requests.

7.2 The impact of radio environment in SDMA cellular systems

Figures 9-12 show the impact of radio environment in blocking and call forced termination probabilities for different scenarios. Mean beam overlapping time ($\mathbf{E}\{\mathbf{X}_{oi}\}$ = 4000, 8000, No link unreliability). Evaluations presented in this section do not consider link unreliability due to the excessive co-channel interference.

Figures 9-12 show how the link unreliability due to the co-channel interference brought within the cell because of the intra-cell reuse affects the system´s performance. Notice that the larger beam overlapping time represents the scenario where the channel conditions are better, that is where Signal to Interference Ratio is not very affected due to the intra-cell reuse

From figures 9-12 it is possible to observe that "Duplicate at Last" policy provides the best performance in terms of call forced termination probability. This behaviour is because the more mobility the more interference is carried within the cell.

8. Conclusions

In this chapter an outline of the smart antenna technology in mobile cellular systems was given. An historical overview of the development of smart antenna technology was

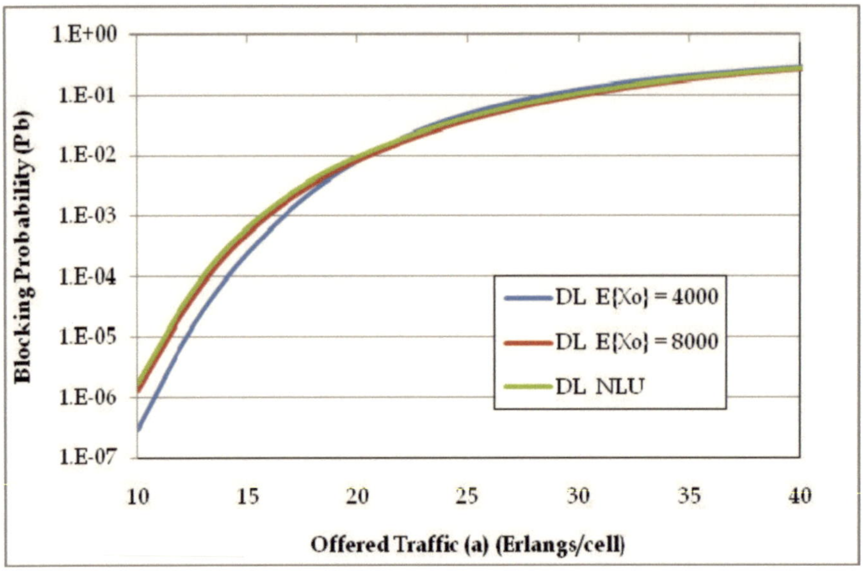

Fig. 9. Blocking Probability for a system with Duplicate at Last Policy. No mobility is considered

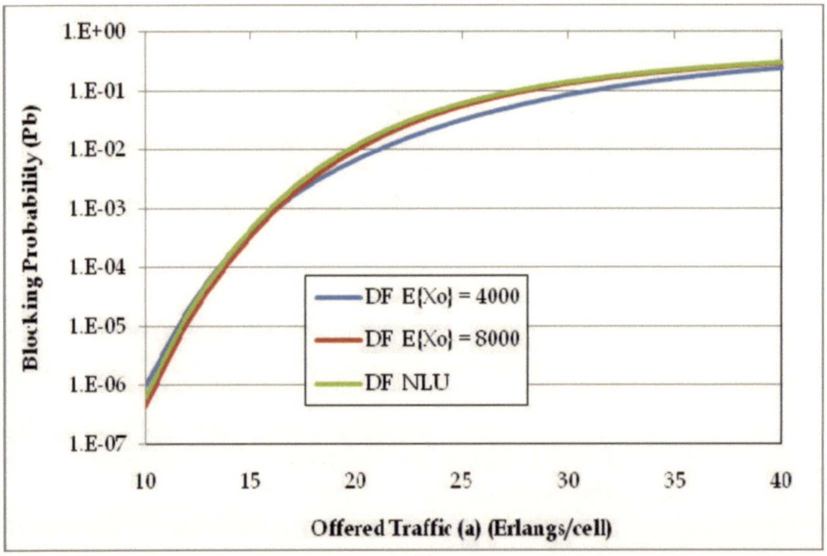

Fig. 10. Blocking Probability for a system with Duplicate at First Policy. No mobility is considered

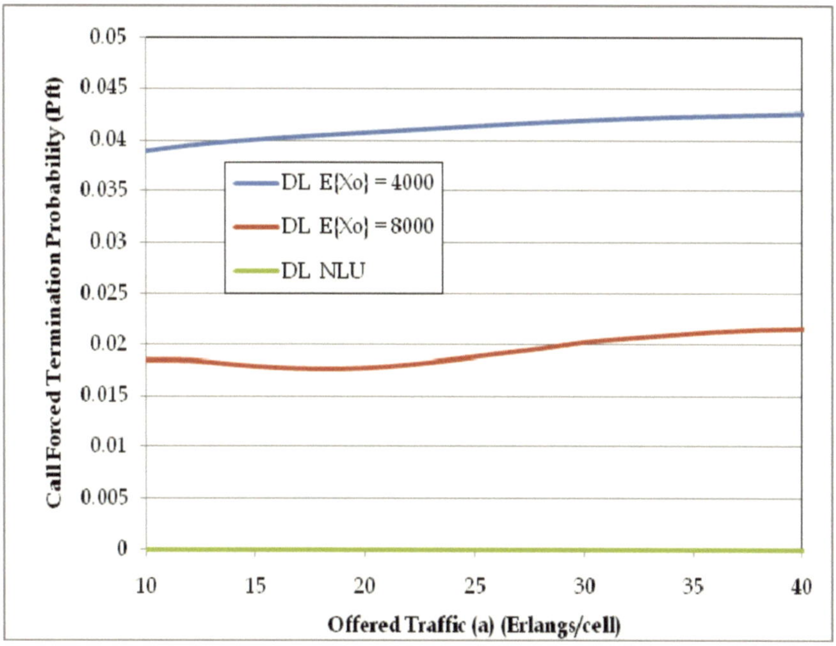

Fig. 11. Call Forced Termination Probability for a system with Duplicate at Last Policy. No mobility is considered

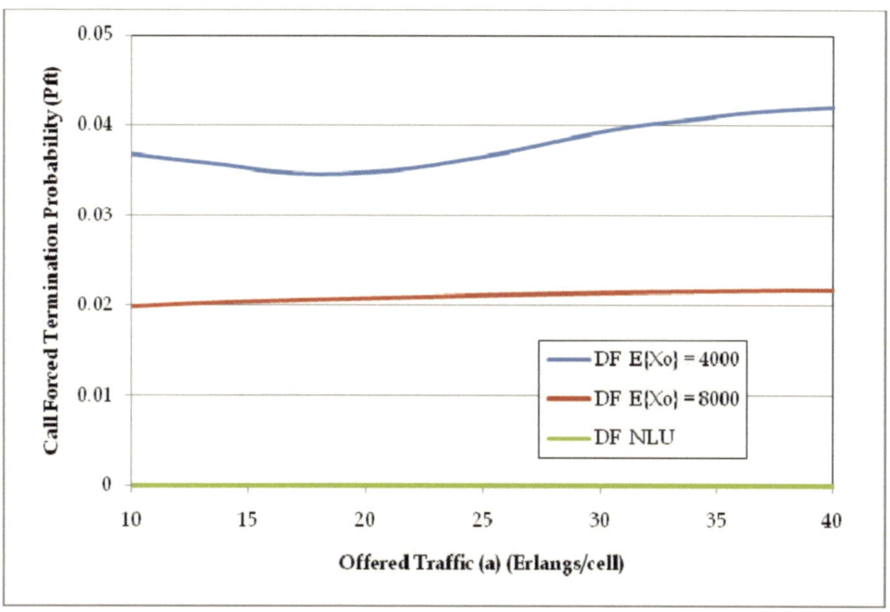

Fig. 12. Call Forced Termination Probability for a system with Duplicate at First Policy. No mobility is considered

presented. Main aspects of the smart antenna components (array antenna and signal processing) were described. Main configurations and applications in cellular systems were summarized and some commercial products were addressed.

Spatial Division Multiple Access was emphasized because it is the technology that is considered the last frontier in spatial processing to achieve an important capacity improvement. Critical aspects of SDMA system level modeling were studied. In particular, users' mobility and radio environment issues are considered. Moreover, the impact of these aspects in system´s performance were evaluated through the use of a new proposed system level model which includes mobility as well as channel conditions. Blocking and call forced termination probability were used as QoS metrics.

9. References

3GPP TR 25.913, (2009), Requirements for Evolved UTRA (EUTRA) and Evolved UTRAN (E-UTRAN).

Alexiou, A.; Navarro, M. & Heath, R.W., (2007), Smart Antennas for Next GenerationWireless Systems, *EURASIP Journal on Wireless Communications and Networking*, Hindawi Publishing Corporation, vol. 2007, Article ID 20427, 2 pages.

Allen, B. & Ghavami, M., (2005), *Adaptive Array Systems Fundamentals and Applications*, John Wiley & Sons Ltd, England.

Applebaum, S.P., (1976), Adaptive arrays. *IEEE Transactions on Antennas and Propagation*, vol. 24, pp. 585-598, September 1976.

Balanis, C.A. (2005), *Antenna Theory Analysis and Design*, Third Edition, John Wiley & Sons, Inc., Hoboken, New Jersey, 2005.

Ball, C.F.; Mullner, R.; Lienhart, J. & Winkler, H. (2009), Performance analysis of Closed and Open loop MIMO in LTE, Proceedings *European Wireless Conference 2009 (EW 2009)*, pp. 260 – 265, Aalborg, Denmark, May 2009.

Bettstetter, C., (2003), Mobility modelling in wireless networks categorization, smooth movement, and border effects, *Mobile Computing and Communications Review*, vol. 5, no. 3, pp. 55-67, 2003.

Boggia, G.; Camarda, P.; D'Alconzo, A.; De Biasi, A.; & Siviero, M. (2005). Drop call probability in established cellular networks: from data analysis to modelling, *in Proceedings of IEEE Vehicular Technology Conference 2005 Spring (VTC Spring'05)*, pp. 2775-2779, Stockholm, Sweden, May-Jun. 2005.

Boukalov, A.O. & Haggman, S.-G., (2000), System aspects of smart-antenna technology in cellular wireless communications-an overview, *IEEE Transactions on Microwave Theory and Techniques*, vol. 48, no. 6, pp. 919-1929, June 2000.

Butler J. & Lowe R., (1961), Beam-forming matrix simplifies design of electronically scanned antennas, *Electronic Devices*, vol. 9, no. 8, pp. 1730–1733, Apr. 1961.

Cardieri, P. & Rappaport, T., (2001), Channel allocation in SDMA cellular systems, *in Proceedings of IEEE Vehicular Technology Conference Fall 2001 (VTC Fall'01)*, vol. 1, pp. 399-403, Atlantic City, New Jersey, USA October 2001.

Cheblus, E. & Ludwin, W., (1995), Is handoff traffic really Poissionian?, *in Proceedings of IEEE International Conference on Universal Personal Communications (ICUPC'95)*, pp. 348-353, Tokyo, Japan, November 1995.

Chiani, M.; Win, M.Z. & Hyundong Shin, (2010), MIMO Networks: The Effects of Interference, *IEEE Transactions on Information Theory*, vol. 56, no.1, pp. 336-349, January 2010.

Cooper, R.B, (1990), *Introduction to Queueing Theory*, Washington D.C. CEE Press Book, 1990.

Czylwik A. & Dekorsky, A., (2001) System level simulations for downlink beamforming with different array topologies, *in Proceedings of IEEE Global Communications Conference 2001 (GLOBECOM'01)*, vol. 5, pp. 3222-3226, San Antonio, Texas, USA, November 2001.

Derneryd, A. & Johannisson, B., (1999), Adaptive base-station antenna arrays, *Ericsson Review No. 3*, vol.76, pp. 136-137, 1999, Online Available: http://www.ericsson.com/ericsson/corpinfo/publications/review/1999_03/files/1999033.pdf

Galvan-Tejada G.M. & Gardiner, J.G., (2001) Theoretical model to determine the blocking probability for SDMA systems, *IEEE Transactions on Vehicular Technology*, Vol. 50, No. 4, pp. 1279-1288, September 2001.

Galvan-Tejada, G.M. & Gardiner, J.G., (1999), Theoretical blocking probability for SDMA, Communications, *Institute of Electrical Engineers Proceedings*, vol. 146, no. 5, pp. 303-306, October 1999.

Galvan-Tejada, G.M. & Gardiner, J.G., (2001), Performance of a Wireless Local Loop System Based on SDMA for Different Propagation Conditions, *in Proceedings of IEEE Global Communications Conference 2001 (GLOBECOM '01)*, vol. 6, pp. 3594-3598, San Antonio, Texas USA December 2001.

Gowrishankar, R.; Demirkol, M.F. & Zhengquing, Y., (2005), Adaptive modulation for MIMO systems and throughput evaluation with realistic channel model, *in Proceedings of International Conference on Wireless Networks, Communications and Mobile Computing 2005*, vol. 2 pp 851 - 856, Cologne, Germany, August 2005.

Gross, F., (2005), *Smart Antennas for Wireless Communications, with Matlab*, Mc Graw Hill, USA, 2005.

Hammuda, H. (1997), *Cellular Mobile Radio Systems*, John Wiley & Sons. England, 1997.

Hansen, R.C., (1998) Phased Array Antennas, Wiley, New York, 1998.

Hemrungrote, S.; Hori, T.; Fujimoto, M. & Nishimori, K., (2010), Channel capacity characteristics of multi-user MIMO systems in urban area, *in Proceedings of IEEE Antennas and Propagation Society International Symposium 2010 (APSURSI 2010)*, Toronto, Ontario, Canada, July 2010.

Heredia-Ureta, H; Cruz-Pérez, F.A. & Ortigoza-Guerrero, L., (2003), Capacity optimization in multiservice mobile wireless networks with multiple fractional cannel reservation, IEEE Transactions on Vehicular Technology, vol. 52, no. 6, pp. 1519-1539, November 2003

Ho M.; Stuber, G.L. & Austin, M.D., (1998), Performance of switched-beam smart antennas for cellular radio systems, *IEEE Transactions on Vehicular Technology*, vol. 47, no.1, pp. 10-19, February 1998.

Hoymann, C., (2006), MAC Layer Concepts to Support Space Division Multiple Access in OFDM based IEEE 802.16s, *Wireless Personal Communications*, p. 23, May 2006

Hu, H. & Zhu, J, (2002), A Combined Beam Hopping and Single Beam Switched-Beam Smart Antennas Scheme and Its Performance Analysis, *in Proceedings of IEEE*

Region 10 Conference on Computers, Communications, Control and Power Engineering 2002 (TENCON 2002), Beijing, China, October 2002.

iBurst, (2004), iBurst Broadband Wireless System Overview, ArrayComm, *White paper*, October 2004, Online Available: http://www.arraycomm.com/docs/iBurstOverview.pdf

ITU-R M.1801, (2007), Radio interface standards for broadband wireless access systems, including mobile and nomadic applications, in the mobile service operating below 6 GHz

Jiang, H. & Rappaport, S.S. (1994), Hand-off analysis for CBWL schemes in cellular communications, in Proceedings of International Conference of Universal Personal Communications 1994 (UPC 1994), pp. 496-500, San Diego, CA, USA October 1994

Jingming-Wang & Daneshrad, B. (2005), A comparative study of MIMO detection algorithms for wideband spatial multiplexing systems, *in Proceedings of IEEE Wireless Communications and Networking Conference 2005 (WCNC 2005)*, vol. 1, pp. 408 – 413, New Orleans, L.A., U.S.A. 2005.

Kaiser, T., (2005), When will Smart antennas be ready for the market?, *IEEE Signal Processing Magazine*, vol. 22, no. 2, pp. 87-92, March 2005.

Krim, H. & Viberg, M., (1996), Two decades of array signal processing research: the parametric approach, *IEEE Signal Processing Magazine*, vol. 14, no. 4, pp. 67-94.

Kusume, K.; Dietl, G.; Abe, T.; Taoka, H. & Nagata, S., (2010), System Level Performance of Downlink MU-MIMO Transmission for 3GPP LTE-Advanced, *in Proceedings of Vehicular Technology Conference Spring 2010 (VTC Spring'10)*, Taipei, May 2010.

Liberti J.C. & Rappaport T.S., (1999), *Smart Antennas for Wireless Communications: IS-95 and Third Generation CDMA Applications*, Prentice Hall PTR, Upper Saddle River, New Jersey, 1999.

Lin, Y.B.; Mohan, S.; & Noerpel, A., (2004), Queuing priority channel assignment strategies for PCS and handoff initial access, *IEEE Transactions on Vehicular Technologies*, vol. 43. no. 3, pp. 704-712, August 1994.

Litva, (1996), *Digital Beamforming in Wireless Communications*, Artech House Mobile Communications Series,U.S.A 1996.

Liu, H. & Zeng, Q.-A., (2004), Performance Analysis of Handoff Scheme in Mobile Radio Systems with Smart Antennas, *in Proceedings of World Wireless Congress 2004 (WWC'04)*, San Francisco, California, USA, May 2004.

Liu, H.; Xu, Y. & Zeng, Q.-A., (2005), Modeling and performance analysis of future generation multimedia wireless and mobile networks using smart antennas, *in Proceedings of Wireless Communications and Networks Conference 2005 (WCNC'05)*, pp. 1286-1291, New Orleans, LA, USA, March 2005.

Mailloux, R.J., (1994), *Phased Array Antenna Handbook*, Artech House, Norwood, MA, 1994.

Nagai, Y. & Kobayashi, T., (2005), Statistical characteristics of pedrestrians' motion and effects on teletraffic of mobile communication networks, *in Proceedings of IEEE Wireless and Optical Communications Networks 2005 (WOCN'05)*, pp. 377-382, Dubai, United Arab Emirates, March 2005.

Nasri, R.; Kammoun, A.; Stephenne, A. & Affes, S., (2008), System – Level Evaluation of a Downlink OFDM Kalman – Based Switched-Beam System with Subcarrier Allocation Strategies, *in Proceedings of Vehicular Technology Conference Fall 2008 (VTC Fall'08)*, Calgary, Alberta, Canada, September, 2008.

Ngamjanyaporn, P.; Phongcharoenpanich, C.; Akkaraekthalin, P. & Krairiksh, M., (2005), Signal-to-interference ratio improvement by using a phased array antenna of switched-beam elements, *IEEE Transactions on Antennas and Propagation*, vol. 53, no. 5, pp. 1819-1828, May 2005.

Nishimori, K.; Kudo, R.; Takatoti, Y.; Ohta, A. & Tsunekawa, K., (2006), Performance Evaluation of Multi-User MIMO-OFDM Testbed in an Actual Indoor Environment, *in Proceedings of International Symposium on Personal, Indoor and Mobile Radio Communications 2006 (PIMRC 2006)*, Helsinki, Finland, September 2006.

Okamoto, G., (2003), Developments and advances in smart antennas for wireless communications, *Santa Clara University, Tech. Rep.*, 2003, Online Available: www.wmrc.com/bussinessbriefing/pdf/wireless_2003/Publication/okamoto.pdf

Pabst, R.; Ellenbeck, J.; Schinnenburg, M. & Hoymann, C., (2007) System level performance of cellular WIMAX IEEE 802.16 with SDMA-enhanced medium access, *in Proceedings of IEEE Wireless Communications and Networking Conference 2007 (WCNC'07)*, pp. 1820-1825, Hong Kong, March 2007.

Parra, I.; Xu, G. & Liu, H., (1995), A Least Squares Projective Constant Modulus Approach, *in Proceedings of IEEE International Symposium on Personal, Indoor and Mobile Radio Communications 1995 (PIMRC 1995)*, vol. 2, pp. 673-676, Toronto, Canada, September 1995.

Pattan, B., (2000), *Robust Modulation Methods and Smart Antennas in Wireless Communications*, Prentice Hall, New York, 2000.

Paulraj, A.J. & Papadias, C. B. (1997), Space-time processing for wireless communications, *IEEE Signal Processing Magazine.*, vol. 14, no. 6, pp. 49-83, November 1997.

Peng, M. & Wang, W., (2005), Comparison of capacity between adaptive tracking and switched beam smart antenna techniques in TDD-CDMA systems, *in Proceedings of IEEE International Symposium on Microwave, Antenna, Propagation and EMC Technologies for Wireless Communications, 2005 (MAPE 2005)*, vol. 1, pp.135-139, Beijing, China, August 2005.

Phasouliotis, A. & So, D.K.C., (2009), Performance Analysis and Comparison of Downlink MIMO MC-CDMA and MIMO OFDMA Systems, *in Proceedings of Vehicular Technology Conference Spring 2009 (VTC Spring'09)*, Barcelona, Spain, April 2009.

Rajal, F., (2005), Why have smart antennas not yet gained traction with wireless network operators?, *IEEE Antennas and Propagation Magazine*, vol. 47, no. 6, pp. 124-126, December 2005.

Robichaud, L.P.A.; Boisvert, M. & Robert, J., (1962), Signal Flow Graphs and Applications, Prentice Hall, 1962.

Rodríguez-Estrello, C.B. & Cruz-Pérez, F.A., (2009), System Level Analytical Model for SDMA Mobile Cellular Networks Considering Cochannel Interference, *Proceeding of IEEE International Symposium on Personal, Indoor Mobile Radio Communications 2009, (PIMRC 2009)*, September, Tokyo, Japan 2009

Rodríguez-Estrello, C.B.; Hernández-Valdéz, G. & Cruz-Pérez F.A., (2009), System Level Analysis of Mobile Cellular Networks Considering Link Unreliability, *IEEE Transactions on Vehicular Technology*, vol. 58, no.2, February 2009

Schmidt, R. O., (1979), Multiple Emitter Location and Signal Parameter Estimation, *in Proceedings of RADC Spectrum Estimation Workshop 1979*, Griffiss AFB, New York, pp. 243-258, 1979

Schweppe, F., (1968), Sensor-array data processing for multiple-signal sources, *IEEE Transactions on Information Theory*, vol. 14, no. 2, pp. 294 - 305, March 1968

Seki, H. & Tsutsui, M., (2007), Throughput Performance of Pre-coding MIMO Transmission with Multi-Beam Selection, *in Proceedings of IEEE International Conference on Communications 2007 (ICC '07)*, pp 2785-2790, Glasgow, Scotland, June 2007.

SentieM™ , (2009), SentieM™ Technologies Enhance Coverage, Capacity and Throughput, the Fundamentals for a Superior Mobile WiMAX™ Network, *Alvarion, Technology White Paper*, July 2009, Online Available: http://alvarion.net/images/stories/resourcecenter/whitepapers/Mobile_WiMAX_Tech_Edge_wp_LR_1_.216.pdf

Shim, D & Choi, S., (1998), Should the smart antenna be a tracking beam array or switching beam array?, *in Proceedings of IEEE Vehicular Technology Conference 1998 (VTC'98)*, vol. 1, pp. 494-498, Ottawa, Ontario, Canada, May 1998.

Shuangmei, C; Jianhua, L; Ying, Y. & Zehong, X., (2004), A novel method for performance analysis of SDMA, *in Proceedings of IEEE Vehicular Technology Conference Fall 2004 (VTC Fall'04)*, vol. 6, pp. 4180-4184, Los Angeles, California, USA, September 2004.

Tangeman, M., (2004), Influence of the user mobility on the spatial multiplex gain of an adaptive SDMA system, *in Proceedings of Personal Indoor Mobile Radio Communications 2004 (PIMRC'94)*, vol. 2, pp. 745-749, Hague, Holland, September 2004.

Toşa, F., (2010), Comparisons of beamforming techniques for 4G wireless communications systems, Proceeding of IEEE 8th International Conference on Communications (COMM 2010), Bucarest, Romania, June 2010.

Tsoulos, G.V., (1999), Smart antennas for mobile communication systems: Benefits and challenges, *IEEE Communications Journal*, vol. 11, no. 2, pp. 84–94, Apr. 1999.

Tsoulos, G.V.; Beach, M. & McGeehan, J., (1997), Wireless Personal Communications for the 2 1st Century, *European Technological Advances in Adaptive Antennas*, vol. 35 no. 9, pp 102-109, September 97

Van Atta, L. C., (1959), Electromagnetic Reflector, U.S. Patent 2 908 002, Serial no. 514 040, October 1959.

Van Rooyen, P. (2002), Advances in space–time processing techniques open up mobile apps, November, 2002, *EE|Times The global electronics engineering community News & Analysis*, Online Available: http://www.eetimes.com/electronics-news/4143480/Advances-in-space-time-processing-techniques-open-up-mobile-apps.

Widrow, B.; Mantey, P.E.; Griffiths, L.J. & Goode, B. B., (1967), Adaptive Antenna systems, *in Proceedings of the IEEE*, vol. 55, no. 12, December 1967.

Zhongding, L; Chin, F.P.S. & Ying-Chang, L, (2005), Orthogonal switched beams for downlink diversity transmission, *IEEE Transactions on Antennas and Propagation*, vol. 52, no.7, pp. 2169-2177, July 2005.

Ziskind, I. & Wax, M., (1988), Maximum Likelihood Localization of Multiple Sources by Alternating Projection, *IEEE Transactions on Acoustics*, Speech, and Signal Process., vol. 36, no. 10, pp.1553–1560, October 1988.

Zonoozi, M.M. & Dassanayake, P., (1997), User mobility modeling and characterization of mobility patterns, *IEEE Journal on Selected Areas in Communications*, vol. 15, no. 7, pp. 1239-1252, September 1997.

Part 2

Mathematical Models and Methods in Cellular Networks

6

Approximated Mathematical Analysis Methods of Guard-Channel-Based Call Admission Control in Cellular Networks

Felipe A. Cruz-Pérez[1], Ricardo Toledo-Marín[1]
and Genaro Hernández-Valdez[2]
[1]*Electrical Engineering Department, CINVESTAV-IPN*
[2]*Electronics Department, UAM-A*
Mexico

1. Introduction

Guard Channel-based call admission control strategies are a classical topic of exhaustive research in cellular networks (Lunayach et al., 1982; Posner & Guerin, 1985; Hong & Rappaport, 1986). Guard channel-based strategies reserve an amount of resources (bandwidth/number of channels/transmission power) for exclusive use of a call type (i.e., new, handoff, etc.), but they have mainly been utilized to reduce the handoff failure probability in mobile cellular networks. Guard Channel-based call admission control strategies include the Conventional Guard Channel (CGC) scheme[1] (Hong & Rappaport, 1986), Fractional Guard Channel (FGC) policies[2] (Ramjee et al., 1997; Fang & Zhang, 2002; Vázquez-Ávila et al., 2006; Cruz-Pérez & Ortigoza-Guerrero, 2006), Limited Fractional Guard Channel scheme (LFGC) (Ramjee et al., 1997; Cruz-Pérez et al., 1999), and Uniform Fractional Guard Channel (UFGC) scheme[3] (Beigy & Meybodi, 2002; Beigy & Meybodi, 2004). They have widely been considered as prioritization techniques in cellular networks for nearly 30 years because they are simple and effective resource management strategies (Lunayach et al., 1982; Posner & Guerin, 1985; Hong & Rappaport, 1986).
In this Chapter, both a comprehensive review and a comparison study of the different approximated mathematical analysis methods proposed in the literature for the performance evaluation of Guard-Channel-based call admission control for handoff prioritization in mobile cellular networks is presented.

---

[1] An integer number of channels is reserved.

[2] FGC policies are general call admission control policies in which an arriving new call will be admitted with probability β_i when the number of busy channels is i ($i = 0, ..., N-1$).

[3] LFGC finely controls communication service quality by effectively varying the average number of reserved channels by a fraction of one whereas UFGC accepts new calls with an admission probability independent of channel occupancy.

2. System model description

The general guidelines of the model presented in most of the listed references are adopted to cast the system considered here in the framework of birth and death processes. A homogeneous multi-cellular system with S channels per cell is considered. It is also assumed that both the unencumbered call duration and the cell dwell time for new and handed off calls have negative exponential probability density function (pdf). Hence, the channel holding time is also negative exponentially distributed. $1/\mu_n$ and $1/\mu_h$ denote the average channel holding time for new and handed off calls, respectively. Finally, it is also assumed that new and handoff call arrivals follow independent Poisson processes with mean arrival rates λ_n and λ_h, respectively.

In general, the mean and probability distribution of the cell dwell time for users with new and handed off calls are different (Posner & Guerin, 1985; Hong & Rappaport, 1986; Ramjee et al., 1997; Fang & Zhang, 2002). The channel occupancy distribution in a particular cell directly depends on the channel holding time (i.e.: the amount of time that a call occupies a channel in a particular cell). The channel holding time is given by the minimum of the unencumbered service time and the cell dwell time. On the other hand, the average time that a call (new or handed off) occupies a channel in a cell (here called effective average channel holding time) depends on the channel holding time of new and handed off calls and its respective admission rate. However, these quantities depend on each other and can only be approximately estimated. Thus, to achieve accurate results in the performance evaluation of mobile cellular systems with guard channel-based strategies, the precise estimation of the effective average channel holding time is crucial.

3. Approximated mathematical analysis methods proposed in the literature

In the first published related works, new call blocking and handoff failure probabilities were analyzed using one-dimensional Markov chain under the assumption that channel holding times for new and handoff calls have equal mean values. This assumption was to avoid large set of flow equations that makes exact analysis of these schemes using multidimensional Markov chain models infeasible. However, it has been widely shown that the new call channel holding time and handoff call channel holding time may have different distributions and, even more, they may have different average values (Hong & Rappaport, 1986; Fang & Zhang, 2002; Zhang et al., 2003; Cruz-Pérez & Ortigoza-Guerrero, 2006; Yavuz & Leung, 2006). As the probability distribution of the channel holding times for handed off and new calls directly depend on the cell dwell time, the mean and probability distribution of the channel holding times for handed off and new calls are also different. On the other hand, the channel occupancy distribution in a particular cell directly depends on the channel holding time (i.e. the amount of time that a call occupies a channel in a particular cell). To avoid the cumbersome exact multidimensional Markov chain model when the assumption that channel holding times for new and handoff calls have equal mean values is no longer valid, different approximated one-dimensional mathematical analysis methods have been proposed in the literature for the performance evaluation of guard-channel-based call admission control schemes in mobile cellular networks (Re et al., 1995; Fang & Zhang, 2002; Zhang et al., 2003; Yavuz & Leung, 2006; Melikov and Babayev, 2006; Toledo-Marín et al., 2007). In general, existing models in the literature for the performance analysis of GC-based strategies basically differ in the way the channel holding time or the offered load per cell used for the numerical evaluations is determined. Let us briefly describe and contrast these methods. Due to its better performance, the Yavuz and iterative methods are described more detailed.

3.1 Traditional approach

The "traditional" approach assumes that channel holding times for new and handoff calls have equal mean values (Hong & Rappaport, 1986) and it considers that the average channel holding time (denoted by $1/\gamma_{av\_trad}$) is given by

$$\frac{1}{\gamma_{av\_trad}} = \frac{\lambda_n}{\lambda_n + \lambda_h}\left(\frac{1}{\mu_n}\right) + \frac{\lambda_h}{\lambda_n + \lambda_h}\left(\frac{1}{\mu_h}\right) \tag{1}$$

However, this equation cannot accurately approximate the value of the average channel holding time in GC-based call admission strategies because new and handoff calls are not blocked equally.

3.2 Soong method

To improve the traditional approach, a different method using a simplified one-dimensional Markov chain model was proposed in (Zhang et al., 2003). Yan Zhang, B.-H- Soong, and M. Ma proposed mathematical expressions for the estimation of the conditional average numbers of new and handoff ongoing calls given a number of free channels and used them to calculate the call blocking probabilities. This method is referred here as the "Soong method".

3.3 Normalized approach

The issue of improving the accuracy of the traditional approximation was also addressed in (Fang & Zhang, 2002) by normalizing to one the channel holding time for new call arrival and handoff call arrival streams. By normalizing the channel holding time, this parameter is the same for both traffic streams. This is known as the "normalized approach".

3.4 Weighted mean exponential approximation

In (Re et al., 1995), the common channel holding time is approximated by weighting the summation of the new call mean channel holding time and the handoff call mean channel holding time and it is referred as the "weighted mean exponential approximation".

3.5 Melikov method

The authors in paper (Melikov & Babayev, 2006) also proposed an approximate result for the stationary occupancy probability. The bi-dimensional state space of the exact method is split into classes, assuming that transition probabilities within classes are higher than those between states of different classes. Then, phase merging algorithm (PMA) is applied to approximate the stationary occupancy probability distribution by the scalar product between the stationary distributions within a class and merged model. This method is referred here as the "Melikov method".

3.6 Yavuz method

On the other hand, in (Yavuz & Leung, 2006) the exact two-dimensional Markov chain model was reduced to a one-dimensional model by replacing the average channel holding

times for new and handoff calls by the so called average effective channel holding time (Yavuz & Leung, 2006). Based on the well-known Little's theorem, the average effective channel holding time was defined in (Yavuz & Leung, 2006) as the ratio of the expected number of arrivals of both call types to the expected number of occupied channels. However, the authors of (Yavuz & Leung, 2006) realized that this requires the knowledge of equilibrium occupancy probabilities and observed that the average channel holding time of each type of call is not directly considered in these equations when computing the approximate equilibrium occupancy probabilities since they are replaced by the average effective channel holding time. Hence, they proposed to initially set the approximate equilibrium occupancy probabilities with the values obtained by the normalized approach. This method is referred here as the "Yavuz method".
Inspired by the Litte's theorem, the inverse of the average effective channel holding time (denoted by $1/\mu_{eff}$) is defined as the ratio of expected number of both types of call arrivals to the expected number of occupied channels, that is,

$$\mu_{eff} = \frac{\sum_{j=0}^{S-1}\left(\lambda_n \beta_j q(j)\right) + \sum_{j=0}^{S-1}\left(\lambda_h q(j)\right)}{\sum_{j=0}^{S} j q(j)} \tag{2}$$

Let $q'(l)$, $l = 0, \ldots, S$ represent the occupancy probabilities. The probability that l channels are being used is approximated by the one-dimensional Kauffman recursive formula:

$$\left(\lambda_n \beta_{c-1} + \lambda_h\right) q'(l-1) = l \mu_{eff} q'(l) \quad ; l = 1, \ldots, S \tag{3}$$

where β_i represents the probability that an arriving new call is admitted when the number of busy channels is i ($i = 0, \ldots, S$-1). FGC policies use a vector $\mathbf{B} = [\beta_0, \ldots, \beta_{S-1}]$ to determine if new calls can be accepted and the components of this vector determine the strategy.

Using the normalization equation, $\sum_{j=0}^{S} q'(j) = 1$, equation (3) can be recursively solved for $q'(j)$,

$$q'(j) = \frac{\prod_{k=0}^{j-1}\left(\lambda_n \beta_k + \lambda_h\right)}{\mu_{eff}^{j} \cdot j!} q'(0) \quad ; 1 \le j \le S \tag{4}$$

where,

$$q'(0) = \left[1 + \sum_{j=1}^{S} \frac{\prod_{k=0}^{j-1}\left(\lambda_n \beta_k + \lambda_h\right)}{\mu_{eff}^{j} \cdot j!}\right]^{-1} \tag{5}$$

It is important to notice that to calculate the average effective channel holding time is necessary the knowledge of equilibrium occupancy probabilities. However, this probability distribution cannot be calculated if the average effective channel holding time is unknown. To solve this, the authors of (Yavuz & Leung, 2006) proposed to initially set the approximate equilibrium occupancy probabilities $q(j)$ with the values obtained by the normalized approach.

3.7 Iterative method

Contrary to the Yavuz and Leung approach, in (Toledo-Marín et al., 2007) it is proposed an iterative approximation analysis method that does not require consideration of an initial occupancy probability distribution because the approximate equilibrium occupancy probabilities are iteratively calculated by directly considering the average channel holding time of each type of call. In (Toledo-Marín et al., 2007), the average effective channel holding $1/\gamma$ is iteratively calculated by weighting, at each iteration, the mean channel holding time for the different types of calls by its corresponding effective admission probability (also referred to as effective channel occupancy probability). This method is referred here as the "Iterative method".

Let P_b and P_h represent, respectively, the new call blocking and handoff failure probabilities. Then,

$$\frac{1}{\gamma}=\frac{\lambda_n(1-P_b)\dfrac{1}{\mu_n}+\lambda_h(1-P_h)\dfrac{1}{\mu_h}}{\lambda_n(1-P_b)+\lambda_h(1-P_h)} \tag{6}$$

As a homogenous system is assumed, the overall system performance can be analyzed by focusing on one given cell. Let β_i (for i = 0,. . . , S-1) denote a non-negative number no greater than one (i.e., $0 \le \beta_i \le 1$) and β_S=0. FGC policies use a vector $\mathbf{B} = [\beta_0,\ldots,\beta_{S-1}]$ to determine if new calls can be accepted and the components of this vector determine the strategy (Cruz-Pérez et al., 1999; Vázquez-Ávila et al., 2006). Let us also denote the state of the given cell as j, where j represents the number of active users in the cell. Let P_j denote the steady state probability with j calls in progress in the cell of reference; then, for the FGC scheme, the equilibrium occupancy probabilities are given by:

$$P_j=\frac{\dfrac{\prod_{i=0}^{j-1}(\beta_i\lambda_n+\lambda_h)}{j!\gamma^j}}{\sum_{k=0}^{S}\dfrac{\prod_{i=0}^{k-1}(\beta_i\lambda_n+\lambda_h)}{k!\gamma^k}};\quad 0\le j\le S \tag{7}$$

The new call blocking and handoff failure probabilities are given, respectively, by:

$$P_b=\sum_{j=0}^{S}\left(1-\beta_j\right)P_j \tag{8}$$

$$P_h = P_S \tag{9}$$

The iteration algorithm works as follows:

Input: S, μ_n, μ_h, λ_n, λ_h, **B**.
Output: P_b, P_h.
Step 0: $P_b \leftarrow 0$, $P_h \leftarrow 0$, $\varepsilon \leftarrow 1$, $\gamma \leftarrow 0$.
Step 1: If $|\varepsilon| < 10^{-5}\gamma$ finish the algorithm, else go to Step 2.
Step 2: Calculate new γ using (6), calculate P_j using (7), and calculate P_b and P_h using (8) and (9), respectively.
Step 3: Calculate new ε as the difference between the new γ and the old γ, go to Step 1.

For all cases studied in this work, the above procedure converges. The algorithm initially assumes arbitrary values for the new call blocking and handoff failure probabilities. Finally, note that recursive formulas can be alternatively employed for the calculation of the new call blocking and handoff failure probabilities in Step 2 (Santucci, 1997; Haring et al., 2001; Vázquez-Ávila et al., 2006).

4. Numerical results

In this section, the performance of the different approximated mathematical analysis methods is compared in terms of the accuracy of numerical results for the new call blocking and handoff failure probabilities and their computational complexity. To the best authors' knowledge, the comprehensive review and performance comparison have not been performed before in the open literature. In particular, no performance comparison of the PMA-based (referred to as Melikov) method against any other approximated analytical method has been previously reported. In (Yavuz & Leung, 2006), the performance of the Yavuz method is compared against the Exact (Li & Fang, 2008), Traditional (Hong & Rappaport, 1986), and Normalized (Fang & Zhang, 2002) methods; and in (Toledo-Marín et al., 2007), the performance of the One-Dimensional Iterative (referred to as Iterative) method is additionally compared against the Yavuz and Soong (Zhang et al., 2003) methods.

In this Section, numerical results for the new call blocking and handoff failure probabilities of the normalized, Melikov, Yavuz, and Iterative analytical methods are compared. As shown in the listed references, the other approximation methods show very poor performance in terms of its accuracy relative to the exact method and, therefore, are not considered here. In addition, all of these methods are compared against the exact solution (Exact method) given by the computation of a two-dimensional Markov chain and numerically solved by using the Gauss-Seidel method. In the evaluations, it is assumed that each cell has $S = 30$ channels. For the sake of comparison two different ranges of values for the traffic load are considered: 0-15 Erlangs/cell (light traffic load scenario) and 110-160 Erlangs/cell (heavy traffic load scenario). For the sake of clarity and similar to (Yavuz & Leung, 2006), the values of the new call and handoff rates, and the channel holding time for handoff calls are fixed and have been arbitrarily chosen. These values are shown in Table 1. Similar numerical results have been obtained for other scenarios. The range of the offered traffic per cell a is determined by the arrival rate and channel holding time of new calls, given by:

$$a = \lambda_n / \mu_n \tag{10}$$

Figures in this section plot the new call blocking and handoff failure probabilities versus the offered load per cell with the number of reserved channels for handoff prioritization (N) as parameter. It is observed that the Iterative method gives the best approximation to the exact

Evaluation scenario	λ_n	λ_h	$1/\mu_n$(s)	$1/\mu_h$(s)
Low traffic load	1/30	1/20	1500 - 100	200
Heavy traffic load	1/5	1/20	800 - 450	200

Table 1. System parameters values for the considered scenarios.

solution followed by the Yavuz method; this is particularly true for a low and moderate number of reserved channels, which typically is a scenario of practical interest (Vázquez-Ávila et al., 2006). The Soong method offers the worst approximation. All the approximations, except the Soong method, give exact solutions in the case of no handoff prioritization (i.e., $N = 0$), as shown in (Toledo-Marín et al., 2007). It is important to note that differences between approximation approaches and the exact solution rise with the increment of the number of guard channels and/or the offered load. Finally, it is important to note that the iterative method is applicable to any GC-based strategy and recursive formulas (Vázquez-Ávila et al., 2006) can be alternatively used for the calculation of the new call blocking and handoff failure probabilities.

4.1 Light traffic load scenario

In this section, under light-traffic-load conditions, the performance of the different approximated mathematical analysis methods for the performance evaluation of Guard-Channel-based call admission control for handoff prioritization in mobile cellular networks is investigated. In this Chapter, light traffic load means that the used values of the offered traffic load result in new call blocking probabilities less than 5%, which are probabilities of practical interest.

Figs. 2 and 3 (4 and 5) show the new call blocking probability (handoff failure probability) as function of traffic load for the cases when 1 and 2 channels are, respectively, reserved for handoff prioritization. Fig. 1 shows the new call blocking and handoff failure probabilities as function of traffic load for the case when no channels are reserved for handoff prioritization (i.e., N=0). Due to the fact that handoff failure and new call blocking probabilities are equal for the case when N=0, then, Fig. 1, also correspond to the handoff failure probability. From Fig. 1, it is observed that all the approximated methods give exact solutions in the case of no handoff prioritization (i.e., $N = 0$).

On the other hand, from Figs. 2-5, it is observed that differences between approximated approaches and the exact solution increase with the increment of the number of guard channels and/or the offered load. Notice, also, that these differences are more noticeable when the handoff failure probability is considered. It is interesting to note from Figs. 2-5 that, contrary to the *iterative*, *Yavuz* and *Melikov* methods, the *normalized* method underestimate both new call blocking and handoff failure probabilities.

In order to directly quantify the relative percentage difference between the exact and the different approximated methods, Figs. 6 and 7 plot in 3D graphics these percentage differences for the blocking and handoff failure probabilities, respectively. These differences are plotted as function of both offered load and the average number of reserved channels. It is observed that, irrespective of the number of reserved channels, the *iterative* and *Yavuz* methods have similar performance and give the best approximation to the exact solution followed by the *normalized* method. The *Melikov* method offers, in general, the worst approximation followed by the *normalized* method. For instance, for the range of values presented in Fig. 6 (Fig. 7), it is observed that the maximum difference between the exact method and the *iterative, Yavuz, normalized* and *Melikov* methods is respectively 2.44%,

2.55%, 5.77%, and 24.4% (7.56%, 7.30%, 46%, and 167%) when the new call blocking probability (handoff failure probability) is considered.

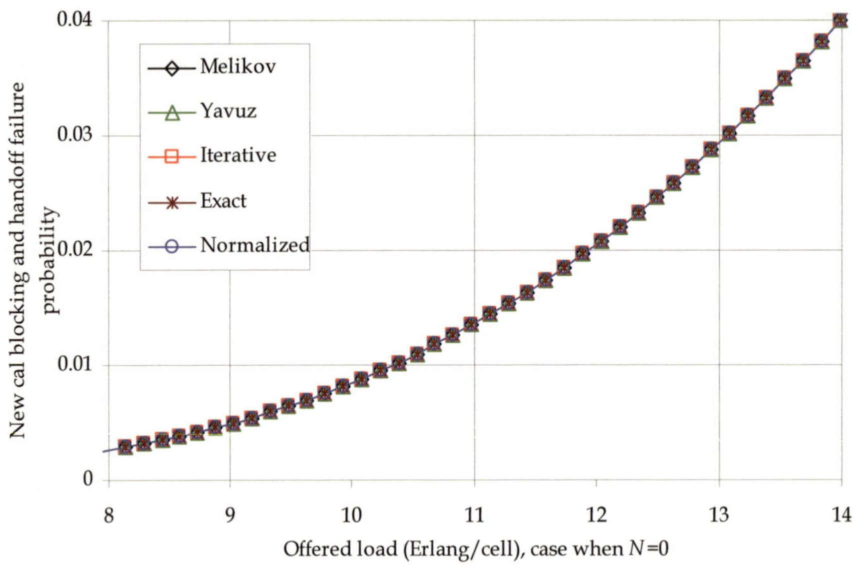

Fig. 1. New call blocking and handoff failure probability versus offered traffic per cell when $N = 0$, light traffic load scenario.

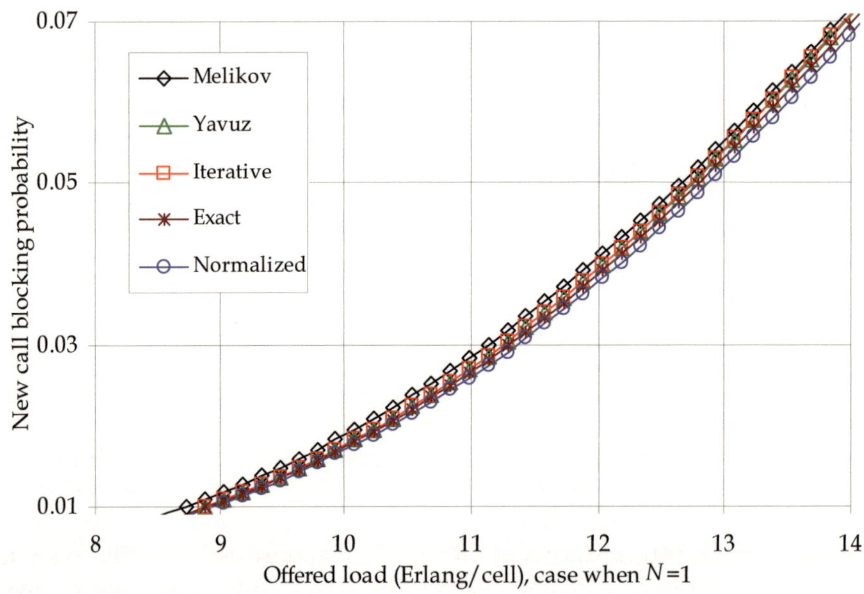

Fig. 2. New call blocking probability versus offered traffic per cell when $N = 1$, light traffic load scenario.

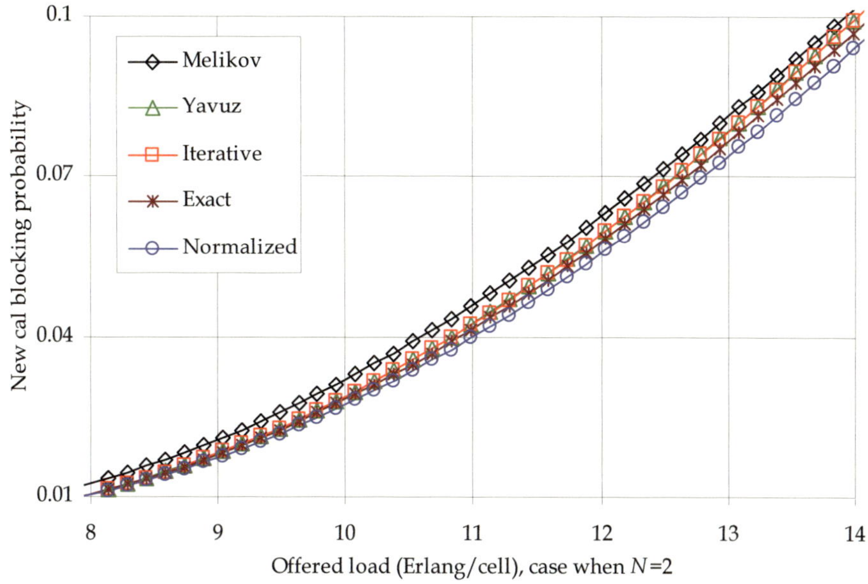

Fig. 3. New call blocking probability versus offered traffic per cell when $N = 2$, light traffic load scenario.

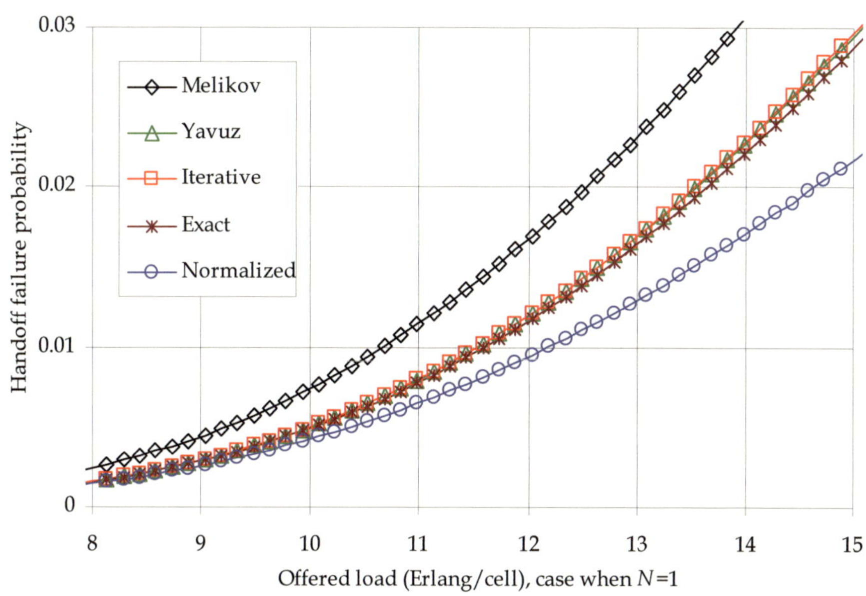

Fig. 4. Handoff failure probability versus offered traffic per cell when $N = 1$, light traffic load scenario.

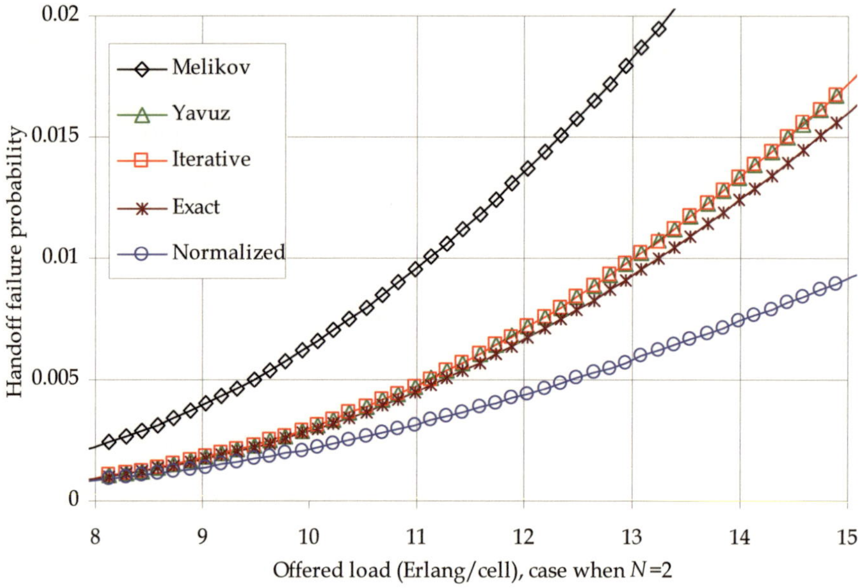

Fig. 5. Handoff failure probability versus offered traffic per cell when $N = 2$, light traffic load scenario.

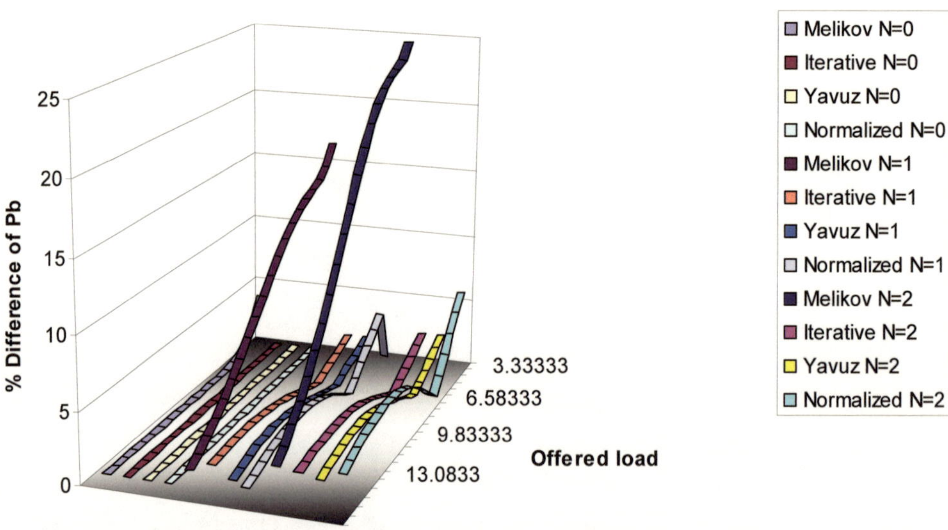

Fig. 6. Percentage difference between the new call blocking probabilities obtained with the exact and the different approximated methods, light traffic load scenario.

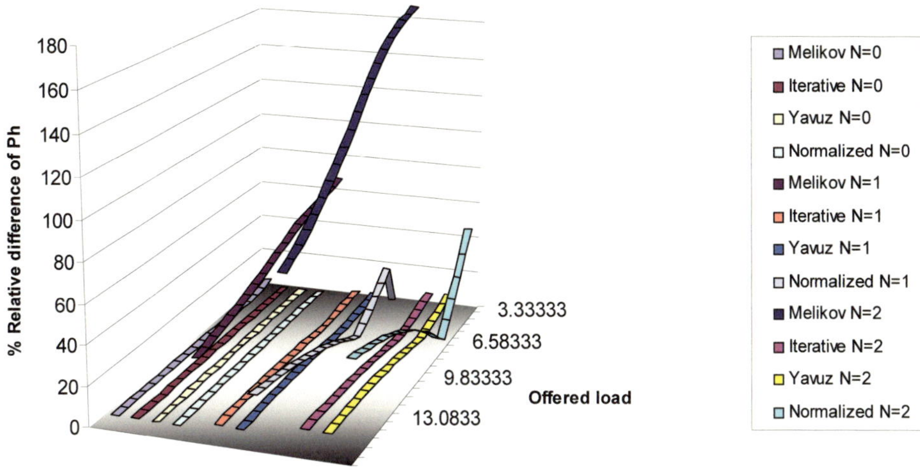

Fig. 7. Percentage difference between the handoff failure probabilities obtained with the exact and the different approximated methods, light traffic load scenario.

4.2 Heavy traffic load scenario

In this section, under heavy-traffic-load conditions, the performance of the different approximated mathematical analysis methods for the performance evaluation of Guard-Channel-based call admission control for handoff prioritization in mobile cellular networks is investigated. In this Chapter, heavy traffic load means that the used values of the offered traffic load result in new call blocking probabilities grater than 70%.

Figs. 9 and 10 (11 and 12) show the new call blocking probability (handoff failure probability) as function of traffic load for the cases when 1 and 2 channels are, respectively, reserved for handoff prioritization. Fig. 8 shows the new call blocking and handoff failure probabilities as function of traffic load for the case when no channels are reserved for handoff prioritization (i.e., N=0). From Fig. 8, it is observed that all the approximated methods give exact solutions in the case of no handoff prioritization (i.e., $N = 0$).

On the other hand, from Figs. 9-12, it is observed that differences between approximated approaches and the exact solution increase with the increment of the number of guard channels and/or the offered load. Notice, also, that these differences are more noticeable when the handoff failure probability is considered. It is interesting to note from Figs. 9 and 10 (11 and 12) that, contrary to the *iterative*, *Yavuz* and *Melikov* (*normalized*) methods, the *normalized* (*Melikov*) method overestimate new call blocking (handoff failure) probabilities.

On the other hand, Figs. 13 and 14 plot in 3D graphics the relative percentage difference between the exact and the different approximated methods for the blocking and handoff failure probabilities, respectively. These differences are plotted as function of both offered load and the average number of reserved channels. As expected, from these figures it is observed that all the approximated methods give exact solutions in the case of no handoff prioritization (i.e., $N = 0$). Figs. 8-11 show that the iterative method presents the best accurate results. Also, from Figs. 8 and 10, it is interesting to note that, referring to the blocking probability, the normalized approach performs slightly better than the Yavuz one; the opposite occurs when the handoff failure probability is considered (see Figs. 9 and 11). For instance, for the range of values presented in Fig. 10 (Fig. 11), it is observed that the maximum difference between the exact method and the iterative, Yavuz, normalized, and Melikov methods is respectively 0.074%, 2.77%, 1.33%, and 3.25% (4.41%, 7.59%, 64.8%, and 165%) when the new call blocking probability (handoff failure probability) is considered.

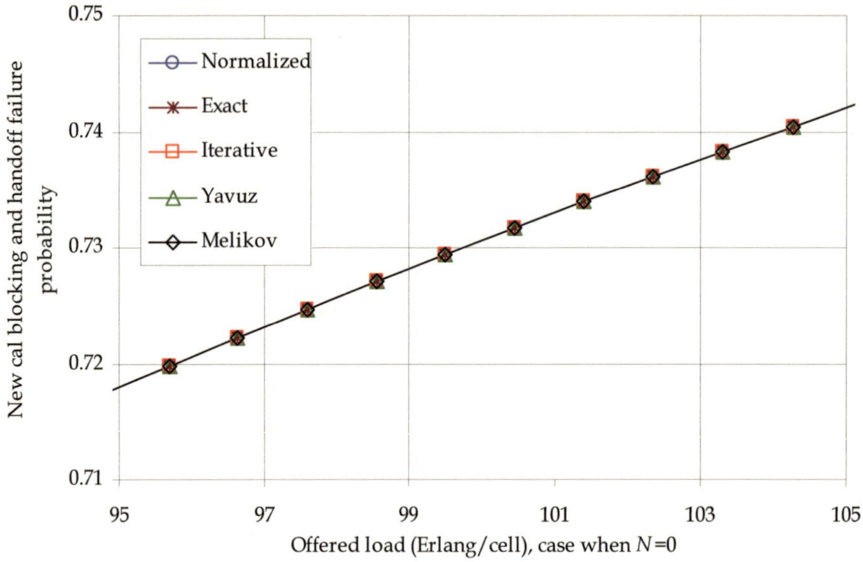

Fig. 8. New call blocking probability versus offered traffic per cell when $N = 0$, heavy traffic load scenario.

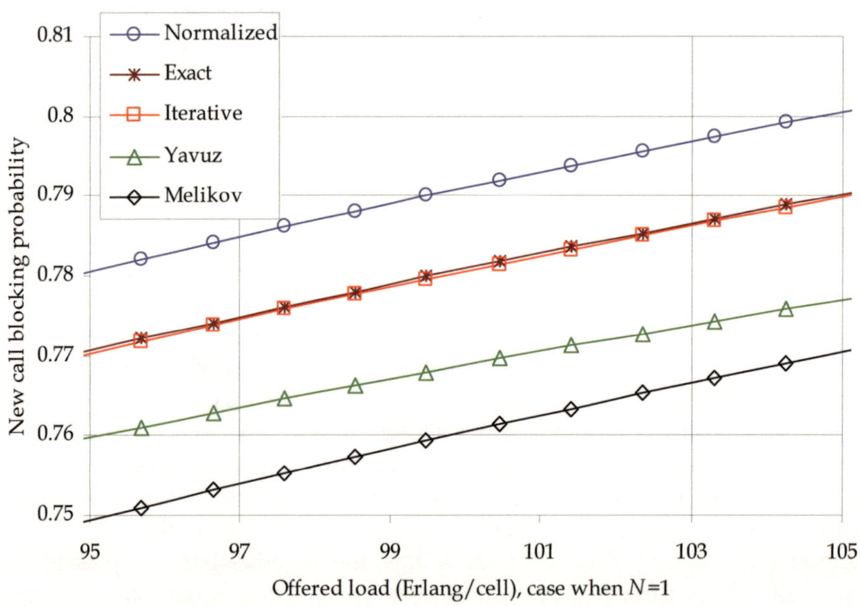

Fig. 9. New call blocking probability versus offered traffic per cell when $N = 1$, heavy traffic load scenario.

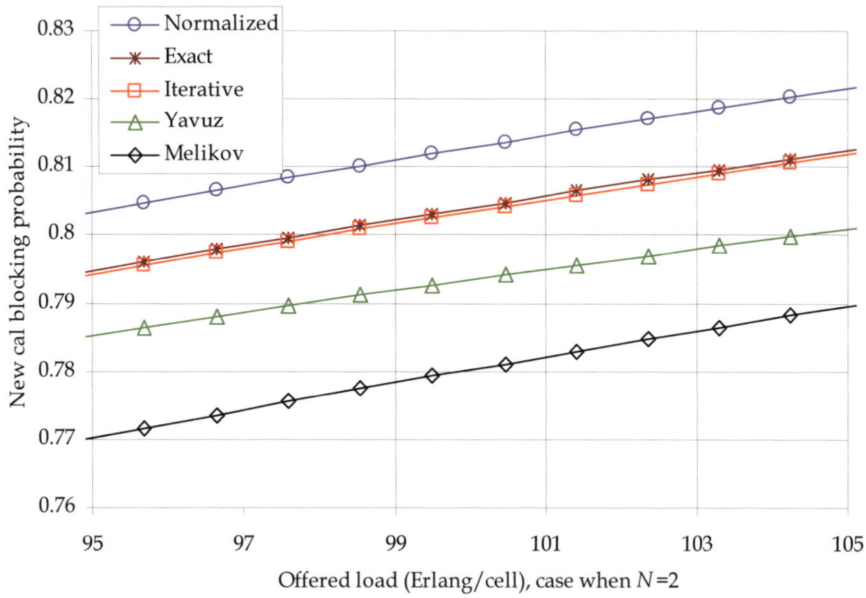

Fig. 10. New call blocking probability versus offered traffic per cell when $N = 2$, heavy traffic load scenario.

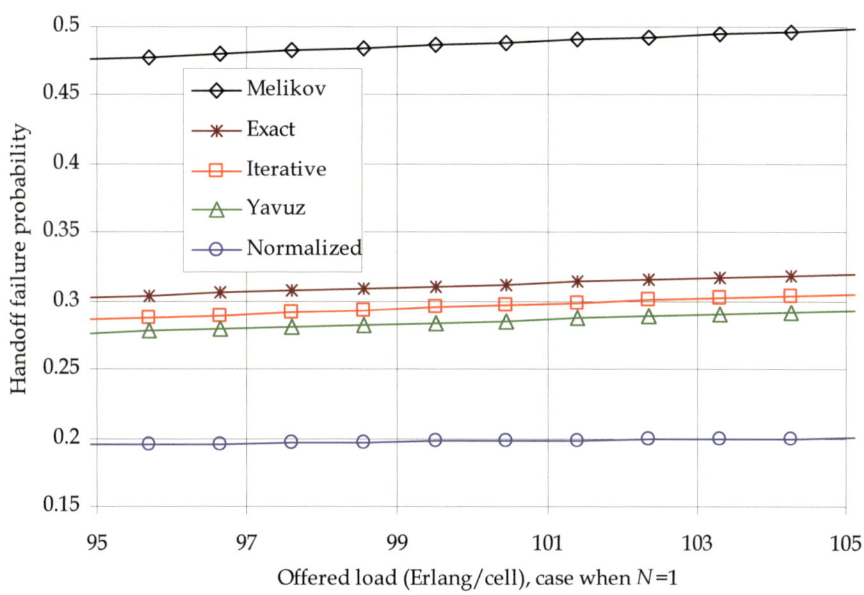

Fig. 11. Handoff failure probability versus offered traffic per cell when $N = 1$, heavy traffic load scenario.

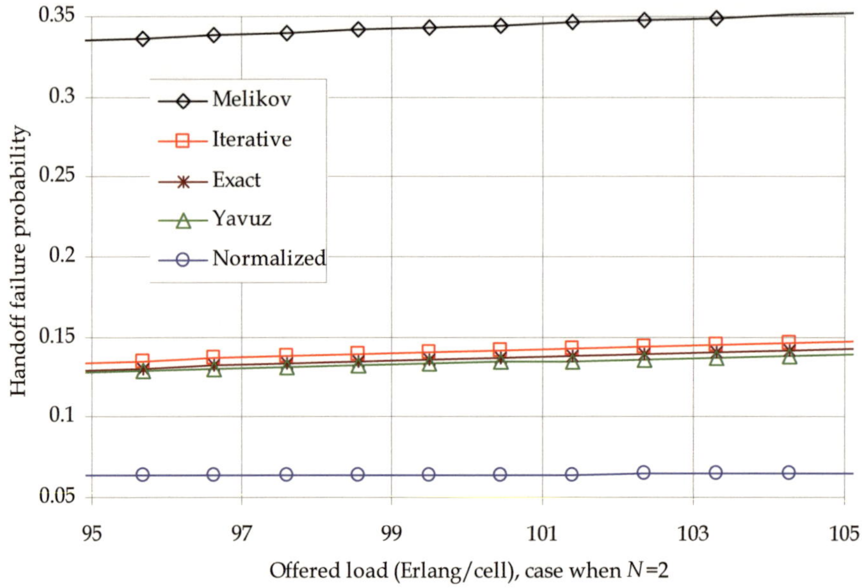

Fig. 12. Handoff failure probability versus offered traffic per cell when $N = 2$, heavy traffic load scenario.

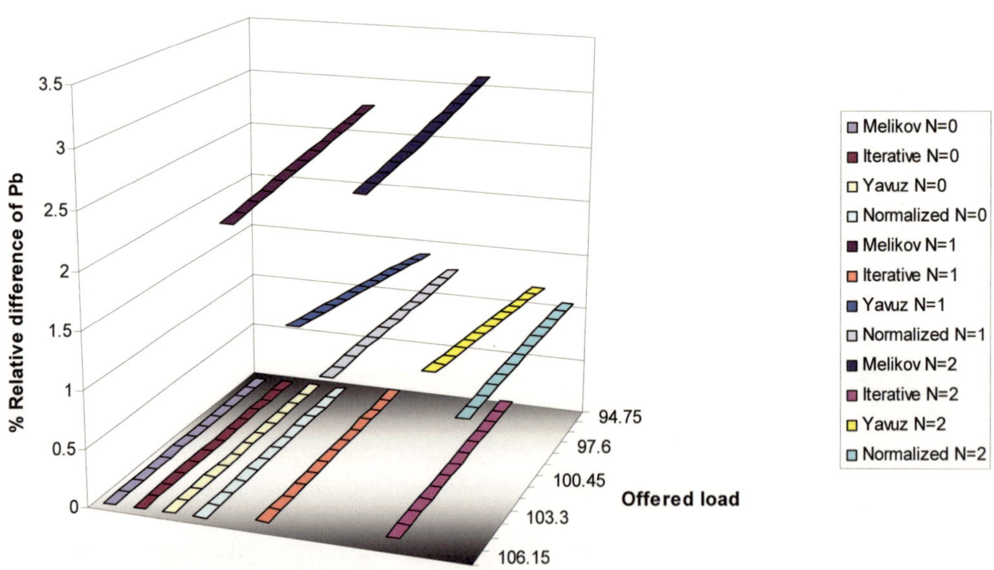

Fig. 13. Percentage difference between the new call blocking probabilities obtained with the exact and the different approximated methods, heavy traffic load scenario.

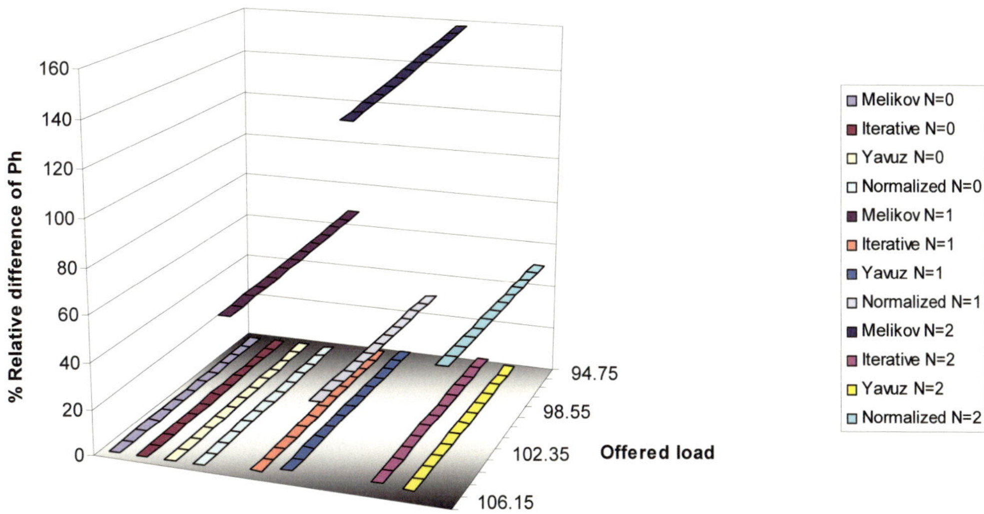

Fig. 14. Percentage difference between the handoff failure probabilities obtained with the exact and the different approximated methods, heavy traffic load scenario.

4.3 Comparison of computation complexity

In this section, the performance of the different approximated mathematical analysis methods is compared in terms of their computational complexity. As stated in (Yavuz & Leung, 2006), the reason why an acceptable approximation method is needed to evaluate the performance of a CAC scheme when an exact solution with a numerical method based on multidimensional Markov chain modeling exists is to avoid solving large sets of flow equations and, therefore, the curse of dimensionality. To give the reader a better idea regarding the "CPU time" and the amount of "memory" used for evaluating the performance of the approximated methods studied in this chapter, consider the Table V shown in (Yavuz & Leung, 2006). Yavuz and Leung implement one direct and two widely used iterative methods, which are the direct (LU decomposition), Jacobi (iterative), and Gauss-Seidel (iterative) methods, to compare their computational costs with that of the Yavuz method.

As shown in Table V of (Yavuz & Leung, 2006), as the number of channels increases, the values of CPU time for the numerical solution methods (both direct and iterative) become significantly greater than the corresponding values for the Yavuz method. The same observation can also be made for the used memory. This should not be surprising since the Yavuz method has much smaller number of states in its respective models and, also, those models have a closed-form formulation.

It is important to remark that the Yavuz method can be considered as a particular case of the iterative one. Both methods are based on the computation of an average effective channel holding time ($1/\gamma$). However, in the Yavuz method, in order to compute the average effective channel holding time, consideration of an initial estimation of occupancy probabilities is required. Moreover, the average channel holding time of each type of call (i.e., new and handed off calls) is not directly considered in these equations when computing the approximate equilibrium occupancy probabilities since they are replaced by the average effective channel holding time. On the other hand, the iterative method computes the equilibrium occupancy probabilities by directly considering the average channel of each type of call (Toledo-Marín et al., 2007). Because of these facts, it has been observed that the Iterative method has similar CPU time values to the corresponding ones for the Yavuz method.

5. Conclusions

Numerical results show that the differences between approximated approaches and the exact solution, in general, increase with the increment of the number of guard channels and/or the offered traffic load. Furthermore, the iterative approximated analytical method is identified as the most suitable for different evaluation conditions/scenarios. In general, at the cost of increasing the computational complexity (compared with the normalized method), the iterative and Yavuz methods provide the best approximation to the exact solution for both light to moderate traffic load and low to moderate average number of reserved channels (in this case, both methods provide similar results), which is a typical scenario of practical interest (Vázquez-Ávila et al., 2006). On the other hand, the iterative method provides the best accurate results at heavy offered traffic loads.

Even though guard-channel based call admission control schemes have been analyzed considering circuit-switched based network architectures, they will continue to be useful when applied with suitable scheduling techniques to guarantee quality of service at the packet level since most applications such as interactive multimedia are inherently connection oriented. Thus, the study of guard-channel based call admission control will continue to be a relevant topic in cellular networks for a long time. Additionally, it is important to note that the considered approximated analytical methods are applicable to any GC-based strategy and, recursive formulas[4], as those derived in (Santucci, 1997; Haring et al., 2001; Vázquez-Ávila et al., 2006), can be alternatively used for the calculation of the new call blocking and handoff failure probabilities.

6. References

Beigy H. and Meybodi M. R., "Uniform fractional guard channel policy," in Proc. 6th *SCI'*2002, vol. 15, Orlando, FL, July 2002.

Beigy H. and Meybodi M. R., "A new fractional channel policy," *J. High Speed Networks*, vol. 13, no. 1, pp. 25-36, Spring 2004.

---

[4] Recursive formulas allow simple and stable computing of (new call and/or handoff) blocking probabilities, especially when the number of channels is large.

Cruz-Pérez F.A., Lara-Rodríguez D., and Lara M., "Fractional channel reservation in mobile communication systems," *IEE Elect. Lett.*, vol. 35, no. 23, pp. 2000-2002, Nov. 1999.

Cruz-Pérez F.A. and Ortigoza-Guerrero L., Part II: Mobility Management, Chapter 11: "Fractional Resource Reservation in Mobile Cellular Systems," pp. 335-362, for the book "Resource, Mobility and Security Management in Wireless Networks and Mobile Communications," Auerbach Publications, CRC Press, USA. Editors: Yan Zhang, Honglin Hu, and Masayuki Fujise. First edition Oct. 25, 2006. ISBN: 0849380367, p. 632.

Fang Y. and Zhang Y., "Call admission control scheme and performance analysis in wireless mobile networks," *IEEE Trans. Veh. Technol.*, vol. 51, no. 2, pp. 371-382, 2002.

Haring G., Marie R., Puigjaner R., and Trivedi K., "Loss formulas and their application to optimization for cellular networks," *IEEE Trans. Veh. Technol.*, vol. 50, pp. 664-673, May 2001.

Hong D. and Rappaport S. S., "Traffic model and performance analysis for cellular mobile radio telephone systems with prioritized and nonprioritized handoff procedures," *IEEE Trans. Veh. Technol.*, vol. 35, no. 3, pp. 77–92, Aug. 1986.

Li W. and Fang Y., "Performance evaluation of wireless cellular networks with mixed channel holding times," *IEEE Trans. Wireless Commun.*, vol. 7, no. 6, pp. 2154-2160, June 2008.

Lunayach R. S., Rao S., and Gupta S. C., "Analysis of a mobile radio communication system with two types of customers and priority," *IEEE Trans. Commun.*, vol. 30, pp. 2470–2475, Nov. 1982.

Melikov A. Z. and Babayev A. T., "Refined approximations for performance analysis and optimization of queueing model with guard channels for handovers in cellular networks," *Computer Commun.*, vol. 29, no. 9, pp. 1386-1392, May 2006.

Posner E. and Guerin R., "Traffic policies in cellular radio that minimize blocking of handoff calls," in Proc. *ITC*, Kyoto, Japan, Sept. 1985, pp. 294–298.

Ramjee R., Nagarajan R., and Towsley D., "On optimal call admission control in cellular networks," *Wirel. Netw.*, vol. 3, no. 1, pp. 29–41, 1997.

Re E.D., Fantacci R., and Giambene G., "Efficient dynamic channel allocation techniques with handover queuing for mobile satellite networks," *IEEE J. Sel. Areas Commun.*, vol. 13, no. 2, pp. 397-405, 1995.

Santucci F., "Recursive algorithm for calculating performance of cellular networks with cutoff priority," *IEE Elect. Lett.*, vol. 33, no. 8, pp. 662-664, Apr. 1997.

Toledo-Marin R., Cruz-Pérez F.A., and Ortigoza-Guerrero L., "Iterative approximation analysis of guard-channel-based strategies in mobile cellular networks," *IET Electron. Lett.*, vol. 43, no. 7, p.399-401, March 2007.

Vázquez-Ávila J.L., Cruz-Pérez F.A., and Ortigoza-Guerrero L., "Performance analysis of fractional guard channel policies in mobile cellular networks," *IEEE Trans. Wirel. Commun.*, vol. 5, no. 2, pp. 301–305, 2006.

Yavuz E.A., and Leung V.C.M., "Computationally efficient method to evaluate the performance of guard-channel-based call admission control in cellular networks", *IEEE Trans. Veh. Technol.*, vol. 55, no. 4, pp. 1412-1424, July 2006.

Zhang Y., Soong B.-H., and Ma M., "Approximation approach on performance evaluation for guard channel scheme," *Electron. Lett.*, vol. 39, no. 5, pp. 465-467, 2003.

7

Numerical Approach to Performance Analysis of Multi-Parametric CAC in Multi-Service Wireless Networks

Agassi Melikov[1] and Mehriban Fattakhova[2]
[1]*Institute of Cybernetics, National Academy of Sciences of Azerbaijan*
[2]*National Aviation Academy of Azerbaijan,*
Azerbaijan

1. Introduction

Cellular wireless network (CWN) consists of radio access points, called base stations (BS), each covering certain geographic area. With the distance power of radio signals fade away (fading or attenuation of signal occurs) which makes possible to use same frequencies over several cells, but in order to avoid interference, this process must be carefully planned. For better use of frequency recourse, existing carrier frequencies are grouped, and number of cells, in which this group of frequencies is used, defines so called frequency reuse factor. Therefore, in densely populated areas with large number of mobile subscribers (MS) small dimensioned cells (micro-cells and pico-cells) are to be used, because of limitations of volumes and frequency reuse factor.

In connection with limitation of transmission spectrum in CWN, problems of allocation of common spectrum among cells are very important. Unit of wireless spectrum, necessary for serving single user is called channel (for instance, time slots in TDMA are considered as channels). There are three solutions for channels allocation problem: Fixed Channel Allocation (FCA), Dynamic Channel Allocation (DCA) and Hybrid Channel Allocation (HCA). Advantages and disadvantage of each of these are well known. At the same time, owing to realization simplicity, FCA scheme is widely used in existing cellular networks. In this paper models with FCA schemes are considered.

Quality of service (QoS) in the certain cell with FCA scheme could be improved by using the effective call admission control (CAC) strategies for the heterogeneous traffics, e.g. see [1]-[3]. Use of such access strategy doesn't require much resource, therefore this method could be considered operative and more defensible for solution of resource shortage problem.

Apart from original (or new) calls (o-calls) flows additional classes of calls that require special approach also exist in wireless cellular networks. These are so-called handover calls (h-calls). This is specific only for wireless cellular networks. The essence of this phenomenon is that moving MS, that already established connection with network, passes boundaries between cells and gets served by new cell. From a new cell's point of view this is h-call, and since the connection with MS has already established, MS handling transfer to new cell must be transparent for user. In other words, in wireless networks the call may occupy channels from different cells several times during call duration, which means that channel occupation period is not the same as call duration.

Mathematical models of call handling processes in multi-service CWN can be developed adequately enough based on theory of networks of queue with different type of calls and random topology. Such models are researched poorly in literature, e.g. see [4]-[6]. This is explained by the fact, that despite elegance of those models, in practice they are useful only for small dimensional networks and with some limiting simplifying assumptions that are contrary to fact in real functioning wireless networks. In connection with that, in majority of research works models of an isolated cell are analyzed.

In the overwhelming majority of available works one-dimensional (1-D) queuing models of call handling processes in an isolated cell of mono-service CWN are proposed. However these models can not describe studying processes in multi-service CWN since in such networks calls of heterogeneous traffics are quite differ with respect to their bandwidth requirement and arrival rate and channel occupancy time. In connection with that in the given paper two-dimensional (2-D) queuing models of multi-service networks are developed.

In order to be specific we consider integrated voice/data CWN. In such networks real time voice calls (v-calls) are more susceptible to possible losses and delays than non-real time data (original or handover) call (d-calls). That is why a number of different CAC strategies for prioritization of v-calls are suggested in various works, mostly implying use of guard channels (or cutoff strategy) for high priority calls [7], [8] and/or threshold strategies [9] which restrict the number of low priority calls in channels.

In this paper we introduce a unified approach to approximate performance analysis of two multi-parametric CAC in a single cell of un-buffered integrated voice/data CWN which differs from known works in this area. Our approach is based on the principles of theory of phase merging of stochastic systems [10].

The proposed approach allows overcoming an assumption made in almost all of the known papers about equality of handling intensities of heterogeneous calls. Due to this assumption the functioning of the CWN is described with one-dimensional Markov chain (1-D MC) and authors managed simple formulas for calculating the QoS metrics of the system. However as it was mentioned in [11] (pages 267-268) and [12] the assumption of the same mean channel occupancy time even for both original and handover calls of the same class traffic is unrealistic. The presented models are more general in terms of handling intensities and the equality is no longer required.

This paper is organized as follows. In Section 2, we provide a simple algorithm to calculate approximate values of desired QoS metrics of the model of integrated voice/data networks under CAC based on guard channels strategy. Similar algorithm is suggested in Section 3 for the same model under CAC based on threshold strategy. In Section 4, we give results of numerical experiments which indicate high accuracy of proposed approximate algorithms as well as comparison of QoS metrics in different CAC strategies. In Section 5 we provide some conclusion remarks.

2. The CAC based on guard channels strategy

It is undisguised that in an integrating voice/data CWN voice calls of any type (original or handover) have high priority over data calls and within of each flow handover calls have high priority over original calls.

As a means of assigning priorities to handover v-calls (hv-call) in such networks a back-up scheme that involves reserving a particular number of guard channels of a cell expressly for calls of this type are often utilized. According to this scheme any hv-call is accepted if there exists at least one free channel, while calls of remain kind are accepted only when the number of busy channels does not exceed some class-dependent threshold value.

We consider a model of an isolated cell in an integrated voice/data CWN without queues. This cell contains N channels, $1<N<\infty$. These channels are used by Poisson flows of hv-calls, original v-calls (ov-calls), handover d-calls (hd-calls) and original d-calls (od-calls). Intensity of x-calls is λ_x, $x \in \{ov,hv,od,hd\}$. As in almost all cited works the values of handover intensities are considered known hereinafter. Although it is apparent that definition of their values depending on intensity of original calls, shape of a cell, mobility of an MS and etc. is rather challenging and complex. However, if we consider the case of a uniform traffic distribution and at most one handover per call, the average handover intensity can be given by the ratio of the average call holding time to the average cell sojourn time [13].

For handle of any narrow-band v-call (either original or handover) one free channel is required only while one wide-band d-call (either original or handover) require simultaneously $b \geq 1$ channels. Here it is assumed that wide-band d-calls are inelastic, i.e. all b channels are occupied and released simultaneously (though can be investigated and models with elastic d-calls).

Note that the channels occupancy time considers the both components of occupancy time: the time of calls duration and their mobility. Distribution functions of channel occupancy time of heterogeneous calls are assumed be independent and exponential, but their parameters are different, namely intensity of handling of voice (data) calls equals μ_v (μ_d), and generally speaking $\mu_v \neq \mu_d$. If during call handling handover procedure is initiated, the remaining handling time of this call in a new cell (yet as an h-call) is also exponentially distributed with the same mean due to memoryless property of exponential distribution.

In a given CAC the procedure by which the channels are engaged by calls of different types is realized in the following way. As it was mentioned before, if upon arrival of an hv-call, there is at least one free channel, this call seizes one of free channels; otherwise this call is blocked. With the purpose of definition of proposed CAC for calls of other types three parameters N_1, N_2 and N_3 where $1 \leq N_1 \leq N_2 \leq N_3 \leq N$ are introduced. It is assumed that N_1 and N_2 are multiples of b.

Arrived ov-call is accepted if the number of busy channels is less than N_3, otherwise it is blocked. Arrived od-call (respectively, hd-call) is accepted only in the case at most N_1-b (respectively, N_2-b) busy channels, otherwise it is blocked.

Consider the problem of finding the major QoS metrics of the given multi-parametric CAC strategy - blocking (loss) probabilities of calls of each type and overall channels utilization. For simplicity of intermediate mathematical transformations first we shall assume that $b=1$. The case $b>1$ is straightforward (see below).

By adopting an assumption for the type of distribution laws governing the incoming traffics and their holding times it becomes possible to describe the operation of an isolated cell by means of a two-dimensional Markov chain (2-D MC), i.e. in a stationary regime the state of the cell at an arbitrary moment of time is described by a 2-D vector $\mathbf{n}=(n_d, n_v)$, where n_d (respectively, n_v) is the number of data (respectively, voice) calls in the channels. Then the state space of the corresponding Markov chain describing this call handling scheme is defined thus:

$$S := \left\{ \boldsymbol{n} : n_d = \overline{0, N_2}, n_v = \overline{0, N}, n_d + n_v \le N \right\}. \tag{2.1}$$

Elements of generating matrix of this MC $q(\boldsymbol{n},\boldsymbol{n}'), \boldsymbol{n},\boldsymbol{n}' \in S$ are determined from the following relations:

$$q(\boldsymbol{n},\boldsymbol{n}') = \begin{cases} \lambda_d & \text{if } n_d + n_v \le N_1 - 1, \boldsymbol{n}' = \boldsymbol{n} + \boldsymbol{e}_1, \\ \lambda_{hd} & \text{if } N_1 \le n_d + n_v \le N_2 - 1, \boldsymbol{n}' = \boldsymbol{n} + \boldsymbol{e}_1, \\ \lambda_v & \text{if } n_d + n_v \le N_3 - 1, \boldsymbol{n}' = \boldsymbol{n} + \boldsymbol{e}_2, \\ \lambda_{hv} & \text{if } N_3 \le n_d + n_v \le N - 1, \boldsymbol{n}' = \boldsymbol{n} + \boldsymbol{e}_2, \\ n_d\mu_d & \text{if } \boldsymbol{n}' = \boldsymbol{n} - \boldsymbol{e}_1, \\ n_v\mu_v & \text{if } \boldsymbol{n}' = \boldsymbol{n} - \boldsymbol{e}_2, \\ 0 & \text{in other cases,} \end{cases} \tag{2.2}$$

where $\lambda_d := \lambda_{od} + \lambda_{hd}$, $\lambda_v := \lambda_{ov} + \lambda_{hv}$, $e_1 = (1,0)$, $e_2 = (0,1)$.
State diagram of the model and the system of global balance equations (SGBE) for the steady state probabilities $p(\boldsymbol{n})$, $\boldsymbol{n} \in S$ are shown in [14]. Existence of stationary regime is proved by the fact that all states of finite-dimensional state space S are communicating.
Desired QoS metrics are determined via stationary distribution of initial model. Let P_x denote the blocking probability of the x-calls, $x \in \{\text{hv,ov,hd,od}\}$. Then by using PASTA theorem [15] we obtain:

$$P_{hv} := \sum_{\boldsymbol{n} \in S} p(\boldsymbol{n}) \delta(n_d + n_v, N) \ , \tag{2.3}$$

$$P_{ov} := \sum_{\boldsymbol{n} \in S} p(\boldsymbol{n}) I(n_d + n_v \ge N_3) \ , \tag{2.4}$$

$$P_{hd} := \sum_{\boldsymbol{n} \in S} p(\boldsymbol{n}) I(n_d + n_v \ge N_2) \ , \tag{2.5}$$

$$P_{od} := \sum_{\boldsymbol{n} \in S} p(\boldsymbol{n}) I(n_d + n_v \ge N_1) \ , \tag{2.6}$$

where $I(A)$ denoted the indicator function of event A and $\delta(i,j)$ represents Kronecker's symbols.
The mean number of busy channels $\tilde{N}$ is also calculated via stationary distribution as follows:

$$\tilde{N} := \sum_{k=1}^{N} k p(k) \ , \tag{2.7}$$

where $p(k) = \sum_{\boldsymbol{n} \in S} p(\boldsymbol{n}) \delta(n_d + n_v, k), k = \overline{1, N}$, are marginal probability mass functions.
Stationary distribution is determined as a result of solution of an appropriate SGBE of the given 2-D MC. However, to solve the last problem one requires laborious computation efforts for large values of N since the corresponding SGBE has no explicit solution. Very

often the solution of such problems is evident if the corresponding 2-D MC has reversibility property [16] and hence for it there exists stationary distribution of multiplicative form. Given SGBE has a multiplicative solution only in a special case when $N_1=N_2=N_3=N$ (even in this case there are known computational difficulties). However, by applying Kolmogorov criteria [16] it is easily verified that the given 2-D MC is not reversible. Indeed, according to mentioned criteria the necessary reversibility condition of 2-D MC consists in the fact that if there exists the transition from state (i,j) into state (i',j'), then there must also be the reverse transition from the state (i',j') to the state (i,j). However, for MC considered this condition is not fulfilled. So by the relations (2.2) in the given MC there exists transition $(n_d,n_v)\to(n_d-1,n_v)$ with intensity $n_d\mu_d$ where $n_d+n_v\geq N_2$, but the inverse transition not existing.
In [14] a recursive technique has been proposed for solution of mentioned above SGBE. It requires multiple inversion calculation of certain matrices of sufficiently large dimensions that in itself is complex calculating procedure. To overcome the mentioned difficulties, new efficient and refined approximate method for calculation of stationary distribution of the given model is suggested below. The proposed method, due to right selection of state space splitting of corresponding 2-D MC allows one to reduce the solution of the problem considered to calculation by explicit formulae which contain the known (even tabulated) stationary distributions of classical queuing models.
For correct application of phase merging algorithms (PMA) it is assumed below that $\lambda_v>>\lambda_d$ and $\mu_v>>\mu_d$. This assumption is not extraordinary for an integrating voice/data CWN, since this is a regime that commonly occurs in multimedia networks, in which wideband d-calls have both longer holding times and significantly smaller arrival rates than narrowband v-calls, e.g. see [17, 18]. Moreover, it is more important to note, that shown below final results are independent of traffic parameters, and are determined from their ratio, i.e. the developed approach can provide a refined approximation even when parameters of heterogeneous traffics are only moderately distinctive.
The following splitting of state space (2.1) is examined:

$$S=\bigcup_{k=0}^{N_2} S_k,\ \ S_k\bigcap S_{k'}=\varnothing,\ \ k\neq k', \tag{2.8}$$

where $S_k:=\{\boldsymbol{n}\in S: n_d=k\}$.
Note 1. The assumption above meets the major requirement for correct use of PMA [10]: state space of the initial model must split into classes, such that transition probabilities within classes are essentially higher than those between states of different classes. Indeed, it is seen from (2.2) that the above mentioned requirement is fulfilled when using splitting (2.8).
Further state classes S_k combine into separate merged states $<k>$ and the following merging function in state space S is introduced:

$$U(\boldsymbol{n})=<k> \text{ if } \boldsymbol{n}\in S_k, k=\overline{0,N_2}\,. \tag{2.9}$$

Function (2.9) determines merged model which is one-dimensional Markov chain (1-D MC) with the state space $\widetilde{S}:=\{<k>:\ k=\overline{0,N_2}\}$. Then, according to PMA, stationary distribution of the initial model approximately equals:

$$p(k,i)\approx\rho_k(i)\pi(<k>),\ (k,i)\in S_k,\ k=\overline{0,N_2}\,, \tag{2.10}$$

where $\{\rho_k(i):(k,i)\in S_k\}$ is stationary distribution of a split model with state space S_k and $\{\pi(<k>):<k>\in\widetilde{S}\}$ is stationary distribution of a merged model, respectively.
State diagram of split model with state space S_k is shown in fig.1, a. By using (2.2) we conclude that the elements of generating matrix of this 1-D birth-death processes (BDP) $q_k(i,j)$ are obtained as follows:

$$q_k(i,j)=\begin{cases}\lambda_v & \text{if } i\le N_3-k-1, j=i+1,\\ \lambda_{hv} & \text{if } N_3-k\le i<N, j=i+1,\\ i\mu_v & \text{if } j=i-1,\\ 0 & \text{in other cases.}\end{cases}$$

So, stationary distribution within class S_k is same as that $M|M|N\text{-}k|N\text{-}k$ queuing system where service rate of each channel is constant, μ_v and arrival rates are variable quantities

$$\begin{cases}\lambda_v \text{ if } i<N_3-k,\\ \lambda_{hv} \text{ if } j\ge N_3-k.\end{cases}$$

Hence desired stationary distribution is

$$\rho_k(i)=\begin{cases}\dfrac{\nu_v^i}{i!}\rho_k(0) & \text{if } 1\le i\le N_3-k,\\ \left(\dfrac{\nu_v}{\nu_{hv}}\right)^{N_3-k}\dfrac{\nu_{hv}^i}{i!}\rho_k(0) & \text{if } N_3-k+1\le i\le N-k,\end{cases} \tag{2.11}$$

where

$$\rho_k(0)=\left(\sum_{i=0}^{N_3-k}\frac{\nu_v^i}{i!}+\left(\frac{\nu_v}{\nu_{hv}}\right)^{N_3-k}\sum_{i=N_3-k+1}^{N-k}\frac{\nu_{hv}^i}{i!}\right)^{-1}, \nu_v:=\lambda_v/\mu_v, \nu_{hv}:=\lambda_{hv}/\mu_v .$$

Then, from (2.2) and (2.11) by means of PMA elements of generating matrix of a merged model $q(<k>,<k'>)$, $<k>,<k'>\in\widetilde{S}$ are found:

$$q(<k>,<k'>)=\begin{cases}\lambda_d\sum_{i=0}^{N_1-k-1}\rho_k(i)+\lambda_{hd}\sum_{i=N_1-1}^{N_2-k-1}\rho_k(i) & \text{if } 0\le k\le N_1-1, k'=k+1,\\ \lambda_{hd}\sum_{i=0}^{N_2-k-1}\rho_k(i) & \text{if } N_1\le k\le N_2-1, k'=k+1,\\ k\mu_d & \text{if } k'=k-1,\\ 0 & \text{in other cases.}\end{cases} \tag{2.12}$$

The latter formula allows determining stationary distribution of a merged model. It coincides with an appropriate distribution of state probabilities of a 1-D BDP, for which transition intensities are determined in accordance with (2.12). Consequently, stationary distribution of a merged model is determined as (see fig.1, b):

$$\pi(<k>) = \frac{\pi(<0>)}{k!\mu_d^k}\prod_{i=1}^{k} q(<k-1>,<k>),\ k=\overline{1,N_2}, \tag{2.13}$$

where $\pi(<0>) = \left(1+\sum_{k=1}^{N_2}\frac{1}{k!\mu_d^k}\prod_{i=1}^{k} q(<k-1>,<k>)\right)^{-1}$.

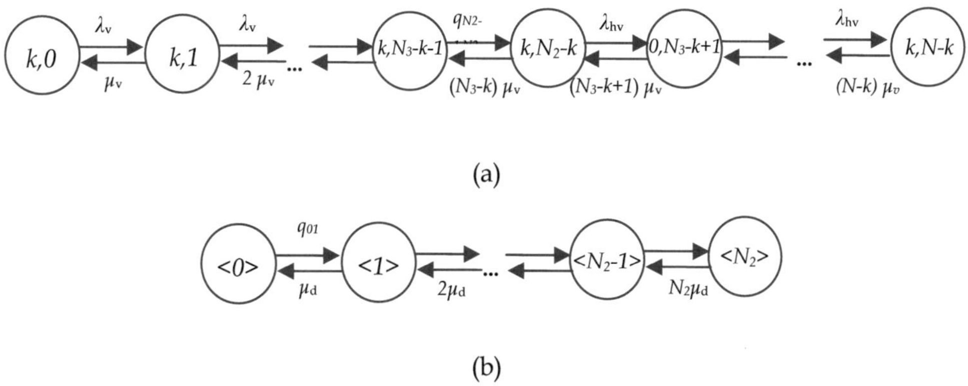

Fig. 1. State diagram of split model with state space S_k, $k=0,1,\ldots,N_2$ (a) and merged model (b).

Then by using (2.11) and (2.13) from (2.10) stationary distribution of the initial 2-D MC can be found. So, summarizing above given and omitting the complex algebraic transformations the following approximate formulae for calculation of QoS metrics (2.3)-(2.7) can be suggested:

$$P_{hv} \approx \sum_{k=0}^{N_2}\pi(<k>)\rho_k(N-k); \tag{2.14}$$

$$P_{ov} \approx \sum_{k=0}^{N_2}\pi(<k>)\sum_{i=N_3-k}^{N-k}\rho_k(i); \tag{2.15}$$

$$P_{hd} \approx \sum_{k=0}^{N_2}\pi(<k>)\sum_{i=N_2-k}^{N-k}\rho_k(i); \tag{2.16}$$

$$P_{od} \approx \sum_{k=0}^{N_1-1}\pi(<k>)\sum_{i=N_1-k}^{N-k}\rho_k(i) + \sum_{k=N_1}^{N_2}\pi(<k>); \tag{2.17}$$

$$\tilde{N} \approx \sum_{i=1}^{N} i \sum_{k=0}^{f_{N_2}(i)}\pi(<k>)\rho_k(i-k). \tag{2.18}$$

Hereinafter $f_k(x)=\begin{cases} x & \text{if } 1\le x\le k, \\ k & \text{if } k\le i\le N. \end{cases}$

Now we can develop the following algorithm to calculate the QoS metrics of investigated multi-parametric CAC for the similar model with wide-band d-calls, i.e. when $b>1$.

Step 1. For $k = 0,1,\ldots,[N_2/b]$ calculate the following quantities

$$\rho_k(i)=\begin{cases} \dfrac{\nu_v^i}{i!}\rho_k(0) & \text{if } 1\le i\le N_3-kb, \\ \left(\dfrac{\nu_v}{\nu_{hv}}\right)^{N_3-kb}\dfrac{\nu_{hv}^i}{i!}\rho_k(0) & \text{if } N_3-kb+1\le i\le N-kb, \end{cases}$$

where $\rho_k(0)=\left(\sum_{i=0}^{N_3-kb}\dfrac{\nu_v^i}{i!}+\left(\dfrac{\nu_v}{\nu_{hv}}\right)^{N_3-kb}\sum_{i=N_3-kb+1}^{N-kb}\dfrac{\nu_{hv}^i}{i!}\right)^{-1}$;

$$\pi(<k>)=\frac{\pi(<0>)}{k!\mu_d^k}\prod_{i=1}^{k}q(<k-1>,<k>),$$

where $\pi(<0>)=\left(1+\sum_{k=1}^{[N_2/b]}\dfrac{1}{k!\mu_d^k}\prod_{i=1}^{k}q(<k-1>,<k>)\right)^{-1}$,

$$q(<k>,<k'>)=\begin{cases} \lambda_d\sum_{i=0}^{N_1-kb-1}\rho_k(i)+\lambda_{hd}\sum_{i=N_1-kb}^{N_2-kb-1}\rho_k(i) & \text{if } 0\le k\le[N_1/b]-1, k'=k+1, \\ \lambda_{hd}\sum_{i=0}^{N_2-kb-1}\rho_k(i) & \text{if } [N_1/b]\le k\le[N_2/b]-1, k'=k+1, \\ k\mu_d & \text{if } k'=k-1, \\ 0 & \text{in other cases.} \end{cases}$$

Step 2. Calculate the approximate values of QoS metrics:

$$P_{hv}\approx\sum_{k=0}^{[N_2/b]}\pi(<k>)\rho_k(N-kb);$$

$$P_{ov}\approx\sum_{k=0}^{[N_2/b]}\pi(<k>)\sum_{i=N_3-kb}^{N-kb}\rho_k(i);$$

$$P_{hd}\approx\sum_{k=0}^{[N_2/b]}\pi(<k>)\sum_{i=N_2-kb}^{N-kb}\rho_k(i);$$

$$P_{od}\approx\sum_{k=0}^{[N_1/b]-1}\pi(<k>)\sum_{i=N_1-kb}^{N-kb}\rho_k(i)+\sum_{k=[N_1/b]}^{[N_2/b]}\pi(<k>);$$

$$\tilde{N} \approx \sum_{i=1}^{N} i \sum_{k=0}^{f_{[N_2/b]}(i)} \pi(<k>)\rho_k(i-k).$$

Henceforth $[x]$ denote the integer part of x.

Now consider some important special cases of the investigated multi-parametric CAC (for the sake of simplicity consider case $b=1$).

1. CAC based on Complete Sharing (CS). Under given CAC strategy, no distinction is made between v-calls and d-calls for channel access, i.e. it is assumed that $N_1=N_2=N_3=N$. In other words, we have 2-D Erlang's loss model. It is obvious, that in this case blocking probabilities of calls from heterogeneous traffics are equal each other, i.e. this probability according to PASTA theorem coincides with probability of that the arrived call of any type finds all channels of a cell occupied. Then from (2.11)-(2.18) particularly we get the following convolution algorithms for calculation of QoS metrics in the given model:

$$P_{hv} = P_{ov} = P_{hd} = P_{od} \approx \sum_{k=0}^{N} E_B(\nu_v, N-k)\pi(<k>), \tag{2.19}$$

$$\tilde{N} \approx \sum_{i=1}^{N} i \sum_{k=0}^{i} \theta_{i-k}(\nu_v, N-k)\pi(<k>) \tag{2.20}$$

Here

$$\pi(<k>) = \frac{\nu_d^k}{k!} \prod_{i=0}^{k-1} (1 - E_B(\nu_v, N-i))\pi(<0>),\ k = \overline{1,N}, \tag{2.21}$$

where $\pi(<0>) = \left(1 + \sum_{k=1}^{N} \frac{\nu_d^k}{k!} \prod_{i=0}^{k-1} (1 - E_B(\nu_v, N-i))\right)^{-1}$.

Henceforth $E_B(\nu, m)$ denote the Erlang's B-formula for the model $M/M/m/m$ with load ν erl, and $\theta_i(\nu, m)$, $i=0,1,\dots,m$, denote the steady state probabilities in the same model, i.e.

$$\theta_i(\nu, m) = \left(\frac{\nu_v^i}{i!}\right)\left(\sum_{j=0}^{m} \frac{\nu_v^j}{j!}\right)^{-1},\ i = \overline{0,m};\ E_B(v,m) := \theta_m(\nu, m). \tag{2.22}$$

Note that developed above analytic results for the CS-strategy is similar in spirit to proposed in [18] algorithm for nearly decomposable 2-D MC.

2. CAC with Single Parameter. Given strategy tell the difference between v-calls and d-calls but do not take into account distinctions between original and handover calls within each traffic, i.e. it is assumed that $N_1=N_2$ and $N_3=N$ where $N_2<N_3$. For this case from (2.11)-(2.18) we get the following approximate formulae for calculating the blocking probabilities of v-calls (P_v) and d-calls (P_d) and mean number of busy channels:

$$P_v = P_{hv} = P_{ov} \approx \sum_{k=0}^{N_2} E_B(\nu_v, N-k)\pi(<k>), \tag{2.23}$$

$$P_d = P_{hd} = P_{od} \approx \sum_{k=0}^{N_2} \pi(<k>) \sum_{i=N_2-k}^{N-k} \theta_i(\nu_v, N-k), \tag{2.24}$$

$$\tilde{N} \approx \sum_{i=1}^{N} i \sum_{k=0}^{f_{N_2}(i)} \theta_{i-k}(\nu_v, N-k)\pi(<k>) . \tag{2.25}$$

Here

$$\pi(<k>) = \frac{\nu_d^k}{k!} \prod_{i=1}^{k} \Lambda(i)\pi(<0>), \; k = \overline{1, N_2}, \tag{2.26}$$

where $\pi(<0>) = \left(1 + \sum_{j=1}^{N_2} \frac{\nu_d^j}{j!} \prod_{i=1}^{j} \Lambda(i)\right)^{-1}$, $\Lambda(i) := \theta_0(\nu_v, N-i+1) \sum_{j=0}^{N_2-i} \frac{\nu_v^j}{j!}$.

3. Mono-service CWN with guard channels. Last results can be interpreted for the model of isolated cell in mono-service CWN with guard channels for h-calls, i.e. for the model in which distinctions between original and handover calls of single traffic is taken into account. Brief description of the model is following. The network supports only the original and handover calls of single traffic that arrive according Poisson processes with rates λ_o and λ_h, respectively. Assume that the o-call (h-call) holding times have an exponential distribution with mean μ_o (μ_h) but their parameters are different, i.e. generally speaking $\mu_o \neq \mu_h$, see [11] and [12].

In a cell mentioned one-parametric CAC strategy based on guard channels scheme is realized in the following way [19]. If upon arrival of an h-call, there is at least one free channel, this call seizes one of free channels; otherwise h-call is dropped. Arrived o-call is accepted only in the case at least $g+1$ free channels (i.e. at most $N-g-1$ busy channels), otherwise o-call is blocked. Here $g \geq 0$ denotes the number of guard channels that are reserved only for h-calls.

By using the described above approach and omitting the known intermediate transformations we conclude that QoS metrics of the given model are calculated as follows:

$$P_o \approx \sum_{k=0}^{N-g} \pi(<k>) \sum_{i=N-g-k}^{N-k} \theta_i(\nu_h, N-k), \tag{2.27}$$

$$P_h \approx \sum_{k=0}^{N-g} E_B(\nu_h, N-k)\pi(<k>), \tag{2.28}$$

$$\tilde{N} \approx \sum_{i=1}^{N} i \sum_{k=0}^{f_{N-g}(i)} \theta_{i-k}(\nu_h, N-k)\pi(<k>). \tag{2.29}$$

Here

$$\pi(<k>) = \frac{\nu_o^k}{k!} \prod_{i=1}^{k} \Lambda(i)\pi(<0>), \; k = \overline{1, N-g}, \tag{2.30}$$

where $\nu_o = \lambda_o / \mu_o$, $\nu_h = \lambda_h / \mu_h$;

$$\pi(<0>) = \left(1 + \sum_{j=1}^{N-g} \frac{\nu_o^j}{j!} \prod_{i=1}^{j} \Lambda(i)\right)^{-1}, \ \Lambda(i) = \theta_0(\nu_h, N-i+1) \sum_{j=0}^{N-g-i} \frac{\nu_h^j}{j!}.$$

Formulas (2.27)-(2.30) coincided with ones for CAC with single parameter in integrated voice/data networks if we set $g:=N-N_2$, $\nu_o:=\nu_d$, $\nu_h:=\nu_v$. And in case $g=0$ we get the results for CAC based on CS-strategy, see (2.19)-(2.21). Also from (2.28) we get the following un-improvable limits for P_h

$$E_B(\nu_h, N) \le P_h \le E_B(\nu_h, g).$$

In the proposed algorithms the computational procedures contains the well-known Erlang's B-formula as well as expressions within that formula which has even been tabulated [20]. Thus, complexity of the proposed algorithms to calculate QoS metrics of investigated multi-parametric CAC based on guard channels are almost congruous to that of Erlang's B-formula. Direct calculations by Erlang's B-formula bring known difficulties at large values of N because of large factorials and exponents. To overcome these difficulties the known effective recurrent formulae can be used, e.g. see [8].

3. The CAC based on threshold strategy

Now consider an alternative CAC in integrated voice/data networks which based on threshold strategy. More detailed description of the given CAC is follows. As in CAC based on guard channels, we assume that arrived an hv-call is accepted as long as at least one free channel is available; otherwise it is blocked. With the purpose of definition of CAC based on threshold strategy for calls of other types three parameters R_1, R_2 and R_3 where $1 \le R_1 \le R_2 \le R_3 \le N$ are introduced. Then proposed CAC defines the following rules for admission of heterogeneous calls: an od-call (respectively, hd-call and ov-call) is accepted only if the number of calls of the given type in progress is less than R_1 (respectively, R_2 and R_3) and a free channel is available; otherwise it is blocked.

For the sake of simplicity we shall assume that $b=1$. The case $b>1$ is straightforward (see section 2). The state of the system under given CAC at any time also is described by 2-D vector $\boldsymbol{n}=(n_d, n_v)$, where n_d (respectively, n_v) is the number of data (respectively, voice) calls in the channels. Then state space of appropriate 2-D MC is given by:

$$S := \left\{\boldsymbol{n} : n_d = \overline{0, R_2},\ n_v = \overline{0, N};\ n_d + n_v \le N\right\} \tag{3.1}$$

Note 2. Hereinafter, for simplicity, we use same notations for state spaces, stationary distribution and etc. in different CAC strategy. This should not cause misunderstanding, as it will be clear what model is considered from the context.

The elements of generating matrix of the appropriate 2-D MC in this case is determined as follows:

$$q(\boldsymbol{n},\boldsymbol{n}')=\begin{cases}\lambda_d & \text{if } n_d \le R_1-1,\ \boldsymbol{n}'=\boldsymbol{n}+\boldsymbol{e}_1,\\ \lambda_{hd} & \text{if } R_1 \le n_d \le R_2-1,\ \boldsymbol{n}'=\boldsymbol{n}+\boldsymbol{e}_1,\\ \lambda_v & \text{if } n_v \le R_3-1,\ \boldsymbol{n}'=\boldsymbol{n}+\boldsymbol{e}_2,\\ \lambda_{hv} & \text{if } R_3 \le n_v \le N-1,\ \boldsymbol{n}'=\boldsymbol{n}+\boldsymbol{e}_2,\\ n_d\mu_d & \text{if } \boldsymbol{n}'=\boldsymbol{n}-\boldsymbol{e}_1,\\ n_v\mu_v & \text{if } \boldsymbol{n}'=\boldsymbol{n}-\boldsymbol{e}_2,\\ 0 & \text{in other cases.}\end{cases} \tag{3.2}$$

Blocking probability of hv-calls and mean number of busy channels are defined similarly to (2.3) and (2.7), respectively. The other QoS metrics are defined as following marginal distributions of initial chain:

$$P_{ov} := \sum_{\boldsymbol{n}\in S} p(\boldsymbol{n}) I(n_v \ge R_3)\ , \tag{3.3}$$

$$P_{hd} := \sum_{\boldsymbol{n}\in S} p(\boldsymbol{n})\delta(n_d, R_2) + \sum_{\boldsymbol{n}\in S} p(\boldsymbol{n})\delta(n_d+n_v, N) I(n_d < R_2)\ , \tag{3.4}$$

$$P_{od} := \sum_{\boldsymbol{n}\in S} p(\boldsymbol{n}) I(n_d \ge R_1) + \sum_{\boldsymbol{n}\in S} p(\boldsymbol{n})\delta(n_d+n_v, N) I(n_d < R_1)\ . \tag{3.5}$$

Unlike CAC based on guard channel strategy, it is easily to show that under this one there is no circulation flow in the state diagram of the underlying 2-D MC, i.e. it is reversible [16]. In other words, there is general solution of the system of local balance equations (SLBE) in this chain. Therefore, we can express any state probability $p(n_d,n_v)$ by state probability $p(0,0)$ by choosing any path between these states in the state diagram. So, in case $R_2+R_3 \le N$ we get following multiplicative solution for stationary distribution of the underlying 2-D MC:

$$p(n_d,n_v)=\begin{cases}\dfrac{\nu_d^{n_d}}{n_d!}\cdot\dfrac{\nu_v^{n_v}}{n_v!}\cdot p(0,0), & \text{if } n_d \le R_1,\ n_v \le R_3,\\[2mm] \dfrac{\nu_d^{n_d}}{n_d!}\cdot\dfrac{\nu_{hv}^{n_v}}{n_v!}\left(\dfrac{\nu_v}{\nu_{hv}}\right)^{R_3}\cdot p(0,0), & \text{if } n_d \le R_1,\ R_3 < n_v \le N,\\[2mm] \dfrac{\nu_{hd}^{n_d}}{n_d!}\cdot\dfrac{\nu_v^{n_v}}{n_v!}\cdot\left(\dfrac{\nu_d}{\nu_{hd}}\right)^{R_1}\cdot p(0,0), & \text{if } R_1 < n_d \le R_2,\ n_v \le R_3,\\[2mm] \dfrac{\nu_{hd}^{n_d}}{n_d!}\cdot\dfrac{\nu_{hv}^{n_v}}{n_v!}\cdot\left(\dfrac{\nu_d}{\nu_{hd}}\right)^{R_1}\cdot\left(\dfrac{\nu_v}{\nu_{hv}}\right)^{R_3}\cdot p(0,0), & \text{if } R_1 < n_d \le R_2,\ R_3 < n_v \le N,\end{cases} \tag{3.6}$$

where $p(0,0)$ is determined from normalizing condition:

$$p(0,0)=\left(\sum_{n\in S_1}\frac{\nu_d^{n_d}}{n_d!}\cdot\frac{\nu_v^{n_v}}{n_v!}+\left(\frac{\nu_v}{\nu_{hv}}\right)^{R_3}\sum_{n\in S_2}\frac{\nu_d^{n_d}}{n_d!}\cdot\frac{\nu_{hv}^{n_v}}{n_v!}+\left(\frac{\nu_d}{\nu_{hd}}\right)^{R_1}\sum_{n\in S_3}\frac{\nu_{hd}^{n_d}}{n_d!}\cdot\frac{\nu_v^{n_v}}{n_v!}+\left(\frac{\nu_d}{\nu_{hd}}\right)^{R_1}\left(\frac{\nu_v}{\nu_{hv}}\right)^{R_3}\sum_{n\in S_4}\frac{\nu_{hd}^{n_d}}{n_d!}\cdot\frac{\nu_{hv}^{n_v}}{n_v!}\right)^{-1}$$

Here we use the following notations: $\nu_d := \lambda_d/\mu_d$, $\nu_{hd} := \lambda_{hd}/\mu_d$;

$$S_1 := \{\boldsymbol{n} \in S : n_d \le R_1, n_v \le R_3\},\ S_2 := \{\boldsymbol{n} \in S : n_d \le R_1, R_3 + 1 \le n_v \le N\},$$
$$S_3 := \{\boldsymbol{n} \in S : R_1 + 1 \le n_d \le R_2, n_v \le R_3\},\ S_4 := \{\boldsymbol{n} \in S : R_1 + 1 \le n_d \le R_2, R_3 + 1 \le n_v \le N\}.$$

In the case $R_2+R_3>N$ stationary distribution has the following form:

$$p(n_d, n_v) = \begin{cases} \dfrac{\nu_d^{n_d}}{n_d!} \cdot \dfrac{\nu_v^{n_v}}{n_v!} \cdot p(0,0), & \text{if } 0 \le n_d \le R_1, 0 \le n_v \le R_3, \\ \dfrac{\nu_{hd}^{n_d}}{n_d!} \cdot \dfrac{\nu_v^{n_v}}{n_v!} \cdot \left(\dfrac{\nu_d}{\nu_{hd}}\right)^{R_2} \cdot p(0,0), & \text{if } R_1 + 1 \le n_d \le R_2, 0 \le n_v \le N - n_d, \\ \dfrac{\nu_d^{n_d}}{n_d!} \cdot \dfrac{\nu_{hv}^{n_v}}{n_v!} \cdot \left(\dfrac{\nu_v}{\nu_{hv}}\right)^{R_3} \cdot p(0,0), & \text{if } 0 \le n_d \le N - R_3 - 1, R_3 + 1 \le n_v \le N, \end{cases} \tag{3.7}$$

where $p(0,0) = \left(\sum_{n \in T_1} \dfrac{\nu_d^{n_d}}{n_d!} \cdot \dfrac{\nu_v^{n_v}}{n_v!} + \left(\dfrac{\nu_d}{\nu_{hd}}\right)^{R_1} \sum_{n \in T_2} \dfrac{\nu_{hd}^{n_d}}{n_d!} \cdot \dfrac{\nu_v^{n_v}}{n_v!} + \left(\dfrac{\nu_v}{\nu_{hv}}\right)^{R_3} \sum_{n \in T_3} \dfrac{\nu_d^{n_d}}{n_d!} \cdot \dfrac{\nu_{hv}^{n_v}}{n_v} \right)^{-1}$;

$$T_1 := \{\boldsymbol{n} \in S : 0 \le n_d \le R_1, 0 \le n_v \le R_3\},\ T_2 := \{\boldsymbol{n} \in S : R_1 + 1 \le n_d \le R_2, 0 \le n_v \le N - n_d\},$$

$$T_3 := \{\boldsymbol{n} \in S : 0 \le n_d \le N - R_3 - 1, R_3 + 1 \le n_v \le N\}.$$

The exact method to determine the steady state probabilities in terms of a multiplicative representation (3.6) (or (3.7)) for large values of N encounters numerical problems such as imprecision and overflow. These are related to the fact that with such a method the entire state space has to be generated, and large factorials and powers close to the zero of the quantities (for low loads) or large values (for high loads) have to be calculated, i.e. there arises the problem of exponent overflow or underflow. Hence we can use developed approximate method to determine the QoS metrics of the model under using the proposed CAC based on threshold strategy even when state space (3.1) is large.

As in section 2, we assume that $\lambda_v >> \lambda_d$ and $\mu_v >> \mu_d$ and examine the following splitting of the state space (3.1):

$$S = \bigcup_{k=0}^{R_2} S_k,\ S_k \bigcap S_{k'} = \varnothing,\ k \ne k',$$

where $S_k := \{\boldsymbol{n} \in S : n_d = k\}$.

Next classes of states S_k are combined into individual merged states $<k>$ and in (3.1) the merged function with range $\tilde{S} := \{<k>:\ k = 0,1,\ldots,R_2\}$ which is similar to (2.9) is introduced. As in exact algorithm in order to find stationary distribution within splitting classes S_k we will distinguish two cases: 1) $R_2+R_3 \le N$ and 2) $R_2+R_3>N$. In first case the elements of generating matrix of appropriate 1-D BDP are same for all splitting models, i.e.

$$q_k(i,j) = \begin{cases} \lambda_v & \text{if } i \le R_3 - 1,\ j = i + 1, \\ \lambda_{hv} & \text{if } R_3 \le i \le N - 1,\ j = i + 1, \\ i\mu_v & \text{if } j = i - 1, \\ 0 & \text{in other cases.} \end{cases}$$

From last formula we conclude that stationary distribution within class S_k is same as that $M|M|N\text{-}k|N\text{-}k$ queuing system with state-dependent arrival rates and constant service rate of each channel, i.e.

$$\rho_k(i)=\begin{cases}\dfrac{\nu_v^i}{i!}\rho_k(0) & \text{if } 1\le i\le R_3,\\ \left(\dfrac{\nu_v}{\nu_{hv}}\right)^{R_3}\dfrac{\nu_{hv}^i}{i!}\rho_k(0) & \text{if } R_3+1\le i\le N-k,\end{cases} \tag{3.8}$$

where $\rho_k(0)=\left(\sum_{i=0}^{R_3}\dfrac{\nu_v^i}{i!}+\left(\dfrac{\nu_v}{\nu_{hv}}\right)^{R_3}\sum_{i=R_3+1}^{N-k}\dfrac{\nu_{hv}^i}{i!}\right)^{-1}$.

So, from (3.2) and (3.8) we conclude that elements of generating matrix of the merged model are

$$q(<k>,<k'>)=\begin{cases}\lambda_d(1-\rho_k(N-k)) & \text{if } 0\le k\le R_1-1, k'=k+1,\\ \lambda_{hd}(1-\rho_k(N-k)) & \text{if } R_1\le k\le R_2-1, k'=k+1,\\ k\mu_d & \text{if } k'=k-1,\\ 0 & \text{in other cases.}\end{cases} \tag{3.9}$$

Distribution of merged model is calculated by using (3.9) and has the following form:

$$\pi(<k>)=\frac{\pi(<0>)}{k!\,\mu_d^k}\prod_{i=1}^{k}q(<k-1>,<k>),\ k=\overline{1,R_2}, \tag{3.10}$$

where $\pi(<0>)=\left(1+\sum_{k=1}^{R_2}\dfrac{1}{k!\,\mu_d^k}\prod_{i=1}^{k}q(<k-1>,<k>)\right)^{-1}$.

Finally the following approximate formulae to calculate the desired QoS metrics under using the proposed CAC based on threshold strategy are obtained:

$$P_{hv}\approx\sum_{k=0}^{R_2}\pi(<k>)\rho_k(N-k); \tag{3.11}$$

$$P_{ov}\approx\sum_{k=0}^{R_2}\pi(<k>)\sum_{i=R_3}^{N-k}\rho_k(i); \tag{3.12}$$

$$P_{hd}\approx\pi(<R_2>)+\sum_{k=0}^{R_2-1}\pi(<k>)\rho_k(N-k); \tag{3.13}$$

$$P_{od}\approx\sum_{k=R_1}^{R_2}\pi(<k>)+\sum_{k=0}^{R_1-1}\pi(<k>)\rho_k(N-k); \tag{3.14}$$

$$N_{av} \approx \sum_{k=1}^{N} k \sum_{i=0}^{f_{R_2}(k)} \pi(<i>)\rho_i(k-i). \qquad (3.15)$$

In second case (i.e. when $R_2+R_3>N$) distributions for splitting models with state space S_k for $k=0,1,\ldots,N-R_3-1$ are calculated by using relations (3.8) while distributions for splitting models with state space S_k for $k=N-R_3,\ldots,R_2$ coincides with distributions of model $M/M/N-k/N-k$ with load ν_v erl, see (2.22). And all stage of developed procedure to calculate the QoS metrics are same with first case except the calculating of P_{ov}. Last QoS metric in this case is calculated as follows:

$$P_{ov} \approx \sum_{k=0}^{N-R_3} \pi(<k>) \sum_{i=R_3}^{N-k} \rho_k(i) + \sum_{k=N=R_3+1}^{R_2} \pi(<k>)\rho_k(N-k). \qquad (3.16)$$

Now consider some special cases. First of all note that CAC based on CS-strategy is a special case of proposed one when $R_1=R_2=R_3=N$. It is important to note that if we set in developed approximate algorithm the indicated value of parameters we obtain exactly the results which were established in section 2, see (2.19)-(2.21).

1. CAC with Single Parameter. As in section 2, let us examine subclass of investigated CAC in which distinction is made only between voice and data traffics, i.e. it is assumed that $R_1=R_2$ and $R_3=N$ where $R_2<R_3$. For this case from (3.8)-(3.15) we get the following simple approximate formulae for calculating the blocking probabilities of v-calls (P_v) and d-calls (P_d):

$$P_v = P_{hv} = P_{ov} \approx \sum_{k=0}^{R_2} \pi(<k>)E_B(\nu_v, N-k), \qquad (3.17)$$

$$P_d = P_{hd} = P_{od} \approx \sum_{k=0}^{R_2-1} \pi(<k>)E_B(\nu_v, N-k) + \pi(<R_2>). \qquad (3.18)$$

Here

$$\pi(<k>) = \frac{\nu_d^k}{k!} \prod_{i=0}^{k-1} (1 - E_B(\nu_d, N-i)), k = \overline{1, R_2}, \qquad (3.19)$$

where $\pi(<0>) = \left(1 + \sum_{k=1}^{R_2} \frac{\nu_d^k}{k!} \prod_{i=0}^{k-1} (1 - E_B(\nu_d, N-i))\right)^{-1}$.

Mean number of busy channels is calculated as follows:

$$\tilde{N} \approx \sum_{i=1}^{N} i \sum_{k=0}^{f_{R_2}(i)} \theta_{i-k}(\nu_v, N-k)\pi(<k>) . \qquad (3.20)$$

Note that if in formulas (3.17)-(3.20) set $R_2=N$ then obtains the results for CAC based on CS-strategy, see (2.19)-(2.21).

2. Mono-service CWN with individual pools for heterogeneous calls. In a given CAC the entire pool of N channels is divided into three pools, an individual pool consisting of r_o

channels (for the o-calls alone), r_h channels (for the o-calls alone) and a common pool consisting of N-r_o-r_h channels (for the o- and the h-calls). Assume that $N>r_o+r_h$, since in case $N=r_o+r_h$ there is trivial CAC based on Complete Partitioning (CP) strategy, i.e. initial system is divided into two separate subsystems where one of them contains r_h channels for handling only h-calls whereas second one with r_o channels handle only o-calls.

If there is at least one free channel (either in the appropriate individual or common pool) at the moment an o-call (h-call) arrives, it is accepted for servicing; otherwise, the call is lost. Note that the process by which the channels are engaged by heterogeneous calls is realized in the following way. If there is one free channel in own pool at the moment an o-call (h-call) arrives, it engages a channel from the own individual pool, while if there is no free channel in the own individual pool, the o-call (h-calls) utilize channels from the common pool. Upon completion of servicing of an o-call (h-call) in the individual pool, the relinquished channel is transferred to the common pool if there is an o-call (h-call) present there, while the channel in the common pool that has finished servicing the o-call (h-call) is switched to the appropriate individual pool. This procedure is called channel reallocation method [21].

From described above model we conclude that it correspond to general CAC based on threshold strategy in case $R_1=R_2=N$-r_h and $R_3=N$-r_o. Therefore, taking into account (3.8)-(3.15) we find the following approximate formulae to calculate the QoS metrics of the given model:

$$P_o \approx E_B(\nu_h, N-r_o)\sum_{k=0}^{r_o}\pi(<k>)+\sum_{k=r_o+1}^{N-r_h-1}E_B(\nu_h, N-k)\pi(<k>)+\pi(<N-r_h>), \tag{3.21}$$

$$P_h \approx E_B(\nu_h, N-r_o)\sum_{k=0}^{r_o}\pi(<k>)+\sum_{k=r_o+1}^{N-r_h}E_B(\nu_h, N-k)\pi(<k>), \tag{3.22}$$

$$\tilde{N} \approx \sum_{k=1}^{N-r_h}k\sum_{i=0}^{k}\pi(<i>)\rho_i(k-i)+\sum_{k=N-r_h+1}^{N}k\sum_{i=r_o-N+k}^{N-r_h}\pi(<i>)\rho_i(k-i)\,, \tag{3.23}$$

where

$$\rho_k(i)=\begin{cases}\theta_i(\nu_h, N-r_o), & \text{if } 0\le k\le r_o,\, 0\le i\le N-r_o,\\ \theta_i(\nu_h, N-k), & \text{if } r_o+1\le k\le N-r_h,\, 0\le i\le N-k;\end{cases} \tag{3.24}$$

$$\pi(<k>)=\begin{cases}\dfrac{\nu_o^k}{k!}\pi(<0>), & \text{if } 1\le k\le r_o,\\ \dfrac{\nu_o^k}{k!}\displaystyle\prod_{i=N-k+1}^{N-r_o}(1-E_B(\nu_h, i))\pi(<0>), & \text{if } r_o+1\le k\le N-r_h,\end{cases} \tag{3.25}$$

$$\pi(<0>)=\left(\sum_{i=0}^{r_o}\frac{\nu_o^k}{k!}+\sum_{k=r_o+1}^{N-r_h}\frac{\nu_o^k}{k!}\prod_{i=N-k+1}^{N-r_h}(1-E_B(\nu_h, i))\right)^{-1}.$$

Note that in special case $r_o=0$ the proposed CAC coincides with the one investigated in [9]. It is evident from derived formulas that in case approximate calculation of QoS metrics we don't have to generate the entire state space of the initial model and calculate its stationary distribution in order to calculate the QoS metrics of the CAC based on individual pools for heterogeneous calls. These parameters may be found by means of simple computational procedures which contain the Erlang's B-formula and terms within that formula. Note that for $r_o=r_h=0$ this scheme becomes fully accessible by both types of calls, i.e. CAC based on CS-strategy takes place.

4. Numerical results

For realization of the above derived algorithms a software package was developed to investigate the behavior of the QoS metrics as a function of the variation in the values of cell's load and structure parameters as well as CAC parameters. First briefly consider some results for the CAC based on guard channels strategy in integrated voice/data model with four classes of calls.

The developed approximate formulas allow without essential computing difficulties to carry out the authentic analysis of QoS metrics in any range of change of values of loading parameters of the heterogeneous traffic, satisfying to the assumption concerning their ratio (i.e. when $\lambda_v>>\lambda_d$ and $\mu_v>>\mu_d$) and also at any number of channels of cell. Some results are shown in figures 2-4 where $N=16$, $N_3=14$, $N_2=10$, $\lambda_{ov}=10$, $\lambda_{hv}=6$, $\lambda_{od}=4$, $\lambda_{hd}=3$, $\mu_v=10$, $\mu_d=2$. Behavior of the studied curves fully confirms all theoretical expectations.

In the given model at the fixed value of the total number of channels (N) it is possible to change values of three threshold parameters (N_1, N_2 and N_3). In other words, there is three degree of freedom. Let's note, that the increase in value of one of parameters (in admissible area) favorably influences on blocking probability of calls of corresponding type only (see fig.2 and 3). So, in these experiments, the increase in value of parameter N_1 leads to reduction of blocking probability of od-calls but other blocking probabilities (i.e. P_{hv}, P_{ov} and P_{hd}) increase. At the same time, the increase in value of any parameter leads to increase in overall channels utilization (see fig.4).

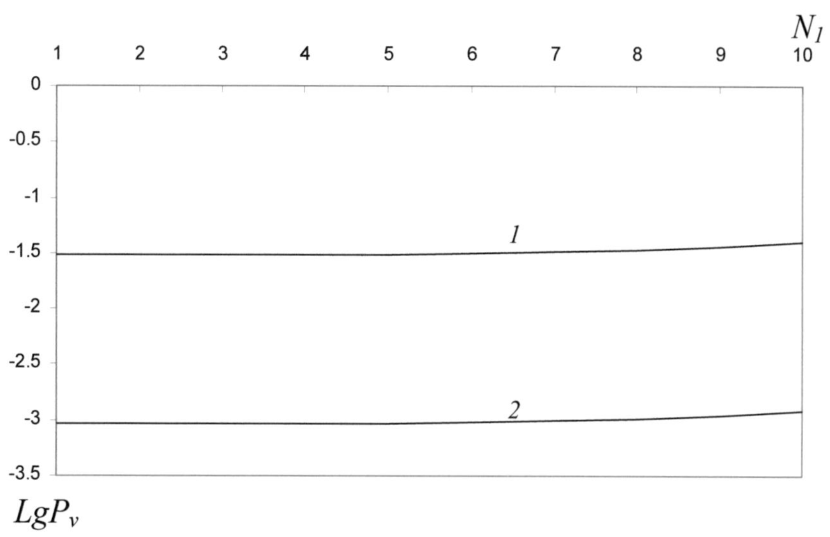

Fig. 2. Blocking probability of v-calls versus N_1: 1 - P_{ov}; 2 - P_{hv}.

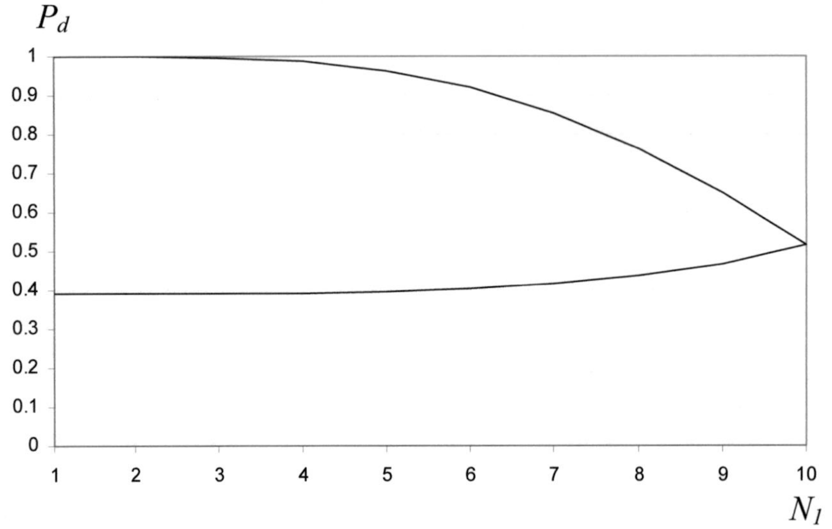

Fig. 3. Blocking probability of d-calls versus N_1: *1-*P_{od}*; 2-* P_{hd}.

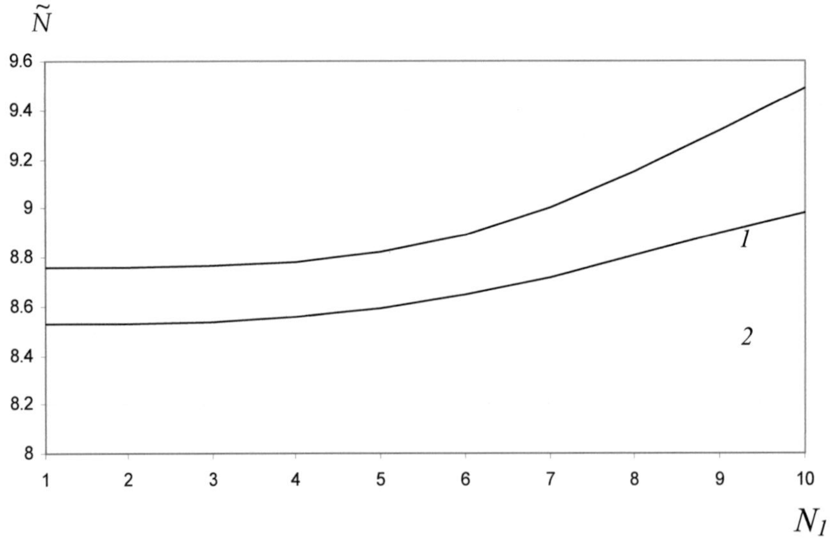

Fig. 4. Average number of busy channels versus N_1: *1 -* $N_3=15$*; 2 -* $N_3=11$.

Other direction of researches consists in an estimation of accuracy of the developed approximate formulas to calculate the QoS metrics. Exact values (EV) of QoS metrics are determined from SGBE. It is important to note, that under fulfilling of the mentioned assumptions related to ratio of loading parameters of heterogeneous traffic the exact and approximate values (AV) almost completely coincide for all QoS metrics. Therefore these comparisons here are not shown. At the same time, it is obvious that finding the exact values of QoS metrics on the basis of the solution of SGBE appears effective only for models with the moderate dimension.

It is important to note sufficiently high accuracy of suggested formulae even for the case when accepted assumption about ratio of traffic loads is not fulfilled. To facilitate the computation efforts, as exact values of QoS metrics we use their values that calculated from explicit formulas, see [22], pages 131-135. In mentioned work appropriate results are obtained for the special case $b=1$ and $\mu_v=\mu_d$. Let's note, that condition $\mu_v=\mu_d$ contradicts our assumption $\mu_v >> \mu_d$. The comparative analysis of results is easy for executing by means of tables 1-3 where initial data are $N=16$, $N_3=14$, $N_2=10$, $\lambda_{ov}=10$, $\lambda_{hv}=6$, $\lambda_{od}=4$, $\lambda_{hd}=3$, $\mu_v=\mu_d=2$. Apparently from these tables, the highest accuracy of the developed approximate formulas is observed at calculation of QoS metric for v-calls since for them the maximal difference between exact and approximate values does not exceed 0.001 (see tabl.1). Small deviations take place at calculation of QoS metrics for d-calls, but also thus in the worst cases the

N_1	P_{ov}		P_{hv}	
	EV	AV	EV	AV
1	0.03037298	0.03465907	0.00092039	0.00119181
2	0.03037774	0.03469036	0.00092054	0.00119309
3	0.03040249	0.03482703	0.00092129	0.00119878
4	0.03048919	0.03521813	0.00092392	0.00121521
5	0.03072036	0.03604108	0.00093092	0.00125021
6	0.03122494	0.03741132	0.00094621	0.00130942
7	0.03217389	0.03932751	0.00097497	0.00139396
8	0.03377398	0.04168754	0.00102345	0.00150073
9	0.03627108	0.04432985	0.00109912	0.00162373
10	0.03997025	0.04706484	0.00121112	0.00175503

Table 1. Comparison for v-calls in CAC based on guard channels.

N_1	P_{od}		P_{hd}	
	EV	AV	EV	AV
1	0.99992793	0.99985636	0.39177116	0.35866709
2	0.99925564	0.99855199	0.39183255	0.35886135
3	0.99612908	0.99271907	0.39215187	0.35969536
4	0.98645464	0.97565736	0.39327015	0.36203755
5	0.96398536	0.93891584	0.39625194	0.36685275
6	0.92198175	0.87621832	0.40276033	0.37462591
7	0.85564333	0.78660471	0.41500057	0.38506671
8	0.76370389	0.67487475	0.43563961	0.39731190
9	0.64880652	0.55004348	0.46784883	0.41028666
10	0.51556319	0.42295366	0.51556319	0.42295366

Table 2. Comparison for d-calls in CAC based on guard channels.

N_1	EV	AV
1	8.75786133	8.52991090
2	8.75908958	8.53136014
3	8.76473770	8.53753920
4	8.78196778	8.55476428
5	8.82125679	8.58985731
6	8.89293266	8.64583980
7	9.00241811	8.71992002
8	9.14705952	8.80533833
9	9.31596095	8.89429324
10	9.49204395	8.97976287

Table 3. Comparison for average number of busy channels in CAC based on guard channels.

absolute error of the proposed formulas does not exceed 0.09, that are quite comprehensible in engineering practice (see tabl.2). Similar results are observed for an average number of occupied channels of cell (see tabl.3). It is important to note, that numerous numerical experiments have shown, that at all admissible loads accuracy of the proposed approximate formulas grows with increase in the value of total number of channels.

It is clear that in terms of simplicity and efficiency, the proposed approach is emphatically superior to the approach based on the use of a balance equations for the calculate QoS metrics of the given CAC in the model with non-identical channel occupancy time.

Let's note, that high accuracy at calculation of QoS metrics for v-calls is observed even at those loadings which do not satisfy any of accepted above assumptions concerning ratio of intensities of heterogeneous traffic. So, for example, at the same values of number of channels and parameters of strategy, at $\lambda_{ov}=4$, $\lambda_{hv}=3$, $\lambda_{od}=10$, $\lambda_{hd}=6$, $\mu_v=\mu_d=2$ (i.e. when assumptions $\lambda_v>>\lambda_d$, $\mu_v>>\mu_d$ are not fulfilled) the absolute error for mentioned QoS metric does not exceed 0.002. Similar results are observed and for an average number of occupied channels of cell. However, the proposed approximate formulas show low accuracy for d-calls since for them the maximal absolute error exceeds 0.2.

Numerical experiments with the CAC based on threshold strategy are carried out also. Due to limitation of volume of work these results here are not resulted. As in CAC based on guard channels, the increase in value of one of parameters (in admissible area) favorably affect the blocking probability of calls of corresponding type only. So, the increase in value of parameter R_1 leads to reduction of blocking probability of od-calls but other blocking probabilities (i.e. P_{hv}, P_{ov} and P_{hd}) increase. At the same time, the increase in value of any parameter leads to increase in overall channels utilization.

The very high precision of the proposed approximate method should also be noted. Thus, in this case comparative analysis of approximate results and the results obtained using a multiplicative solution (for small values of channels) shows that their differences is negligible. Morever, in some cases these results completely coincide. But in terms of simplicity and efficiency, the proposed approximate approach is emphatically superior to the approach based on the use of a multiplicative solution. For the sake of brief these results are not shown here.

At the end of this section we conducted research on comparative analysis of QoS metrics of two schemes: CAC based on guard channels scheme and CAC based threshold strategy.

Comparison was done in the broad range of number of channels and load parameters. In each access strategy the total number of channels is fixed and controllable parameters are N_1, N_2, N_3 (for CAC based on guard channels scheme) and R_1, R_2, R_3 (for CAC based on threshold strategy). As it mentioned above, behavior of QoS metrics with respect to indicated controllable parameters in different CAC are same.

Some results of comparison are shown in fig.5-9 where label *1* and *2* denotes QoS metrics for CAC based on guard channels and CAC based on threshold strategies, respectively. The input data of model are chosen as follows: $N=16$, $R_3=14$, $R_2=12$, $\lambda_{ov}=10$, $\lambda_{hv}=6$, $\lambda_{od}=4$, $\lambda_{hd}=3$, $\mu_v=10$, $\mu_d=2$. In graphs the parameter of the CAC based on guard channels (i.e. N_1) is specified as X-line and as it has been specified above, it corresponds to parameter R_1 of the CAC based on threshold strategy.

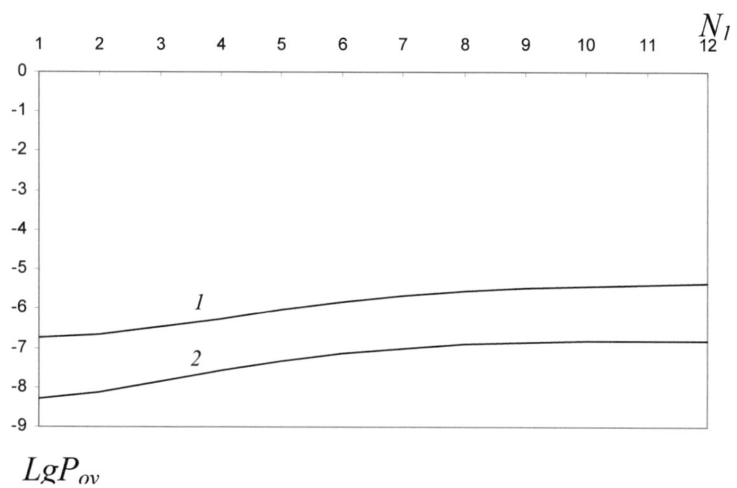

Fig. 5. Comparison for P_{ov} under different CAC.

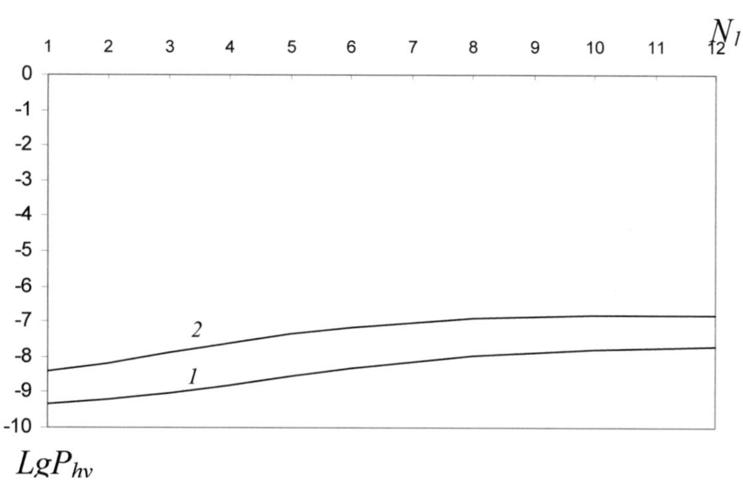

Fig. 6. Comparison for P_{hv} under different CAC.

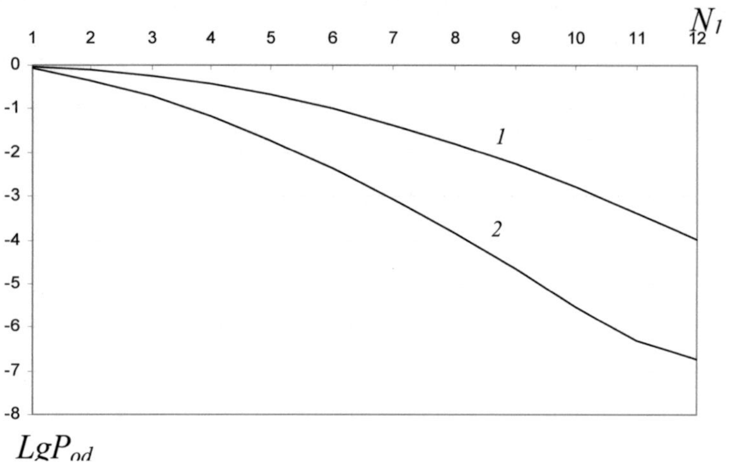

Fig. 7. Comparison for P_{od} under different CAC.

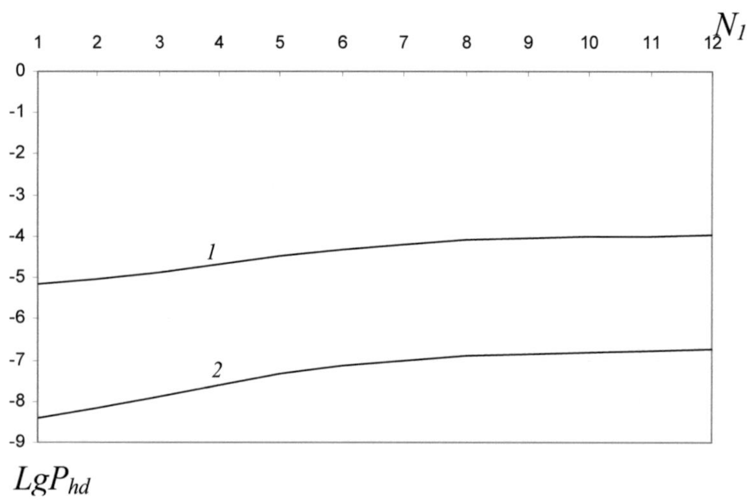

Fig. 8. Comparison for P_{hd} under different CAC

From these graphs we conclude that, for the chosen initial data three QoS metrics, except for blocking probability of hv-calls, essentially better under using CAC based on threshold strategy. The average number of occupied channels in both strategies is almost same. However, quite probably, that at other values of initial data QoS metrics (either all or some of them) in CAC based on guard channels will be better than the CAC based on threshold strategy.

It is important to note that with the given number of channels, loads and QoS requirements either of CAC strategy may or may not meet the requirements. For instance, in the model of mono-service CWN for the given values of $N=100$, $v_o=50$ erl, $v_h=35$ erl following requirements $P_o \le 0.1$, $P_h \le 0.007$ and $\tilde{N} \ge 80$ are not met with CAC based on guard channels irrespective of value of parameter g (number of guard channels), whereas CAC based on individual pool only for h-calls (i.e. $r_o=0$) meets the requirements at $r_h=40$. However, for the

same given initial data, requirements $P_o \leq 0.3$, $P_h \leq 0.0001$ and $\tilde{N} \geq 60$ are only met by CAC based on guard channel scheme at $g=20$, and never met by CAC based on individual pool strategy irrespective of value of its parameter r_h. Thus it is possible to find optimal (in given context) strategy at the given loads without changing number of channels.
Apparently, both strategies have the same implementation complexity. That is why the selection of either of them at each particular case must be based on the answer to the following question: does it meet the given QoS requirements? These issues are subjects to separate investigation.

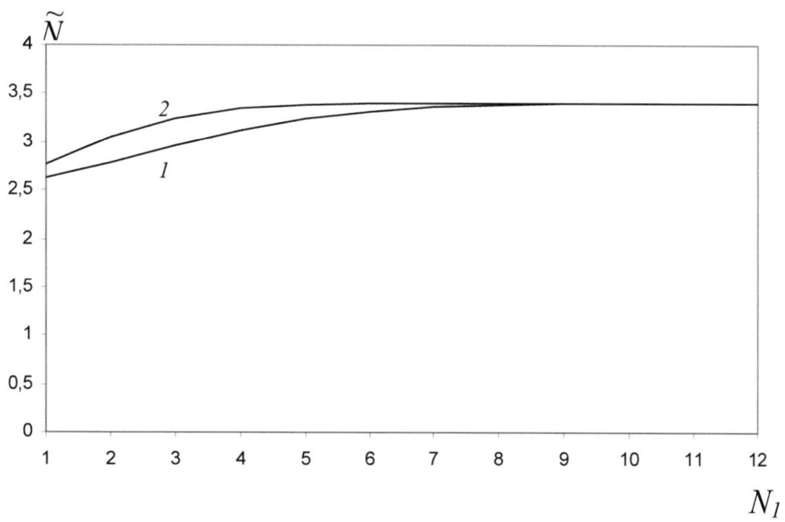

Fig. 9. Comparison for $\tilde{N}$ under different CAC.

5. Conclusion

In this paper effective and refined approximate approach to performance analysis of un-buffered integrated voice/data CWN under different multi-parametric CAC has been proposed. Note that many well-known results related to mono-service CWN are special cases of proposed ones. In the almost all available works devoted mono-service CWN the queuing model is investigated with assumption that both handover and original calls are identical in terms of channel occupancy time. This assumption is rather limiting and unreal. Here models of un-buffered integrated voice/data CWN are explored with more general parameter requirements. Performed numerical results demonstrate high accuracy of the developed approximate method.
It is important to note that the proposed approach may be facilitate the solution of problems related to selecting the optimal (in given sense) values of parameters of investigated multi-parametric CAC. These problems are subjects to separate investigation.

6. References

[1] Leoung C W, Zhuang W (2003) Call admission control for wireless personal communications. Computer Communications 26: 522-541

[3] DasBit S, Mitra S (2003) Challenges of computing in mobile cellular environment – a survey. Computer Communications 26: 2090-2105
[3] Ahmed M (2005) Call admission control in wireless networks. IEEE Communications Surveys 7(1): 50-69
[4] Boucherie R J, Mandjes M (1998) Estimation of performance measures for product form cellular mobile communications networks. Telecommunication Systems 10: 321-354
[5] Boucherie R J, Van Dijk N M (2000) On a queuing network model for cellular mobile telecommunications networks. Operation Research 48(1): 38-49
[6] Li W, Chao X (2004) Modeling and performance evaluation of a cellular mobile networks. IEEE/ACM Transactions on Networking 12(1): 131-145
[7] Hong D, Rapoport S S (1986) Traffic model and performance analysis of cellular mobile radio telephones systems with prioritized and non-prioritized handoff procedures. IEEE Transactions on Vehicular Technology 35(3): 77-92
[8] Haring G, Marie R, Puigjaner R, Trivedi K (2001) Loss formulas and their application to optimization for cellular networks. IEEE Transactions on Vehicular Technology 50(3): 664-673
[9] Gavish B, Sridhar S (1997) Threshold priority policy for channel assignment in cellular networks. IEEE Transactions on Computers 46(3): 367-370
[10] Korolyuk V S, Korolyuk V V (1999) Stochastic model of systems. Kluwer Academic Publishers, Boston
[11] Yue W, Matsumoto Y (2002) Performance analysis of multi-channel and multi-traffic on wireless communication networks. Kluwer Academic Publishers, Boston
[12] Fang Y, Zhang Y (2002) Call admission control schemes and performance analysis in wireless mobile networks. IEEE Transactions on Vehicular Technology 51(2): 371-382
[13] Nanda S (1993) Teletraffic models for urban and suburban microcells: Cell sizes and hand-off rates. IEEE Transactions on Vehicular Technology 42(4): 673-682
[14] Ogbonmwan S E, Wei L (2006) Multi-threshold bandwidth reservation scheme of an integrated voice/data wireless network. Computer Communications 29(9): 1504-1515
[15] Wolff R W (1992) Poisson arrivals see time averages. Operations Research 30(2):223-231
[16] Kelly F P (1979) Reversibility and stochastic networks. John Wiley & Sons, New York
[17] Casares-Giner V (2001) Integration of dispatch and interconnect traffic in a land mobile trunking system. Waiting time distributions. Telecommunication Systems 10: 539-554
[18] Greenberg A G, Srikant R, Whitt W (1999) Resource sharing for book-ahead and instantaneous-request calls. IEEE/ACM Transactions on Networking 7(1): 10-22
[19] Melikov A Z, Babayev A T (2006) Refined approximations for performance analysis and optimization of queuing model with guard channels for handovers in cellular networks. Computer Communications 29(9): 1386-1392
[20] Freeman R L (1994) Reference manual for telecommunications engineering. Wiley, New York
[21] Melikov A Z, Fattakhova M I, Babayev A T (2005) Investigation of cellular communication networks with private channels for service of handover calls. Automatic Control and Computer Sciences 39(3): 61-69
[22] Chen H, Huang L, Kumar S, Kuo C C (2004) Radio resource management for multimedia QoS supports in wireless networks. Kluwer Academic Publishers, Boston

8

Call-Level Performance Sensitivity in Cellular Networks

Felipe A. Cruz-Pérez[1], Genaro Hernández-Valdez[2] and Andrés Rico-Páez[1]
[1]*Electrical Engineering Department, CINVESTAV-IPN*
[2]*Electronics Department, UAM-A*
Mexico

1. Introduction

The development of analytically tractable teletraffic models for performance evaluation of mobile cellular networks under more realistic assumptions has been the concern of recent works (Corral-Ruíz et al. 2010; Fang a; 2005, Fang b; 2005; Kim & Choi 2009; Pattaramalai, 2009; Rico-Páez et al., 2007; Rodríguez-Estrello et al., 2009; Rodríguez-Estrello et al., 2010; Wang & Fan, 2007; Yeo & Yun, 2002; Zeng et al., 2002). The general conclusion of those works is that, in order to capture the overall effects of cellular shape, cellular size, users' mobility patterns, wireless channel unreliability, handoff schemes, and characteristics of new applications, most of the time interval variables (i.e., those used for modeling time duration of different events in telecommunications - for example, cell dwell time, residual cell dwell time, unencumbered interruption time, unencumbered service time) need to be modeled as random variables with general distributions. In this research direction, phase-type distributions have got a lot of attention because of the possibility of using the theory of Markov processes[1] (Fang, 1999, Christensen et al., 2004). Moreover, there have been major advances in fitting phase-type distributions to real data (Alfa & Li, 2002). Among the phase-type probability distributions, the use of hyper-Erlang distribution is of special interest due to its universality property (i.e., it can be used to accurately approximate the behavior of any non negative random variable) and also because of the fact that it provides accurate description of real distributions of different time variables in mobile cellular networks (Fang, 1999; Corral-Ruíz et al., 2010; Yeo & Yun, 2002).

When a probability distribution different to the negative exponential one is the best choice to fit the real distribution of a given time interval variable, not only its expected value but also its higher order moments are relevant. Nonetheless, the study of the effect of moments higher than the expected value has been largely ignored. The reason is twofold: 1) because of the relatively recent use of probability distributions different to the exponential one and 2) because the related works have been focused on developing mathematical models rather than numerically evaluating system performance.

In this chapter, the important task of identify and analyze the influence of moments higher than the expected value of both cell dwell time and unencumbered interruption time on

---

[1] Markovian properties are essential in generating tractable queuing models for mobile cellular networks.

network performance is addressed. Specifically, in this chapter, system performance sensitivity to the first three standardized moments (i. e., expected value, coefficient of variation, and skewness) of both cell dwell time and unencumbered interruption time in cellular networks is investigated. These time interval variables are assumed to be phase-type distributed random variables. System performance is evaluated in terms of new call blocking, handoff failure, and forced call termination probabilities, carried traffic, handoff call arrival rate, and the mean channel holding time for new and handed off calls.

2. System model description

In this section, the basic concepts and general guidelines for the mathematical analysis developed in subsequent sections are given.

2.1 Basic concepts

In a mobile cellular network, radio links for communication are provided by base stations whose radio coverage defines a cell. Every time a mobile user whishes to initiate a call, the mobile terminal attempts to obtain a radio channel for the connection. If no channel is available, the call is blocked and cleared from the network. This is called a new call blocking. Nonetheless, if a channel is available, it is used for the connection, and released under any of the following situations: the call is successfully completed, the call is forced to terminate due to the wireless link unreliability, or the mobile user moves out of the cell. The channel holding time is defined as the amount of time that a call occupies a channel in a particular cell.

Moreover, when a mobile user moves from one cell to another during an ongoing communication, the call requires a new channel to be reserved in the new cell. This procedure of changing channels is called a handoff. If no channel is available in the new cell during the handoff, the call is said to be forced to terminate due to resource insufficiency. This phenomenon is called a handoff call blocking. New call blocking and forced call termination probabilities are being considered as important design parameters for evaluating the level of quality of service (QoS) offered by a wireless network. It has been observed that priority to handoff calls over new call initiation enables to improve the QoS. In a well-established cellular network and from the call forced termination point of view, handoff call blocking can be usually a negligible event (Boggia et al., 2005). Thus, the main cause of call forced termination is due to the unreliable nature of the wireless communication channel[2] (Boggia et al., 2005).

2.2 Basic assumptions

A homogeneous multi-cellular system with omni-directional antennas located at the centre of each cell is assumed; that is, the underlying processes and parameters for all cells within the cellular network are the same, so that all cells are statistically identical. Each cell has a

---

[2] Physical link is said to be unreliable if the experienced signal to interference ratio (SIR) is lower in value than a minimum required value (SIR threshold) for more than a specified period of time (time threshold). During the course of a call, the physical link between base station and mobile station may suffer link unreliability due to propagation impairments such as multi-path fading, shadowing or path loss, and interference (Rodríguez-Estrello et al., 2010).

maximum number S of radio channels assigned to it and can therefore support at most S calls simultaneously. Since a sudden forced termination during a call session will be more upsetting than a failure to connect, a fractional cutoff priority scheme is used to give handoff calls priority over new calls. For this purpose, a real number N of channels in each cell is reserved for handoff prioritization (Vázquez-Ávila et al., 2006). As it has been widely accepted in the related literature (Orlik & Rappaport, 1998; Lin et al., 1994), both the new call arrivals and handoff attempts are assumed to follow independent Poisson processes with mean arrival rate λ_n and λ_h, respectively, per cell. Some other assumptions and definitions are presented in the Section 2.3.

2.3 Definition of time interval variables

In this section the different time interval variables involved in the teletraffic model of a mobile cellular network are defined.

First, the *unencumbered service time* per call x_s (also known as the *requested call holding time* (Alfa and Li, 2002) or *call holding duration* (Rahman & Alfa, 2009)) is the amount of time that the call would remain in progress if it experiences no forced termination. It has been widely accepted in the literature that the unencumbered service time can adequately be modeled by a negative exponentially distributed random variable (RV) (Lin et al., 1994; Hong & Rappaport, 1986, Del Re et al., 1995). The RV used to represent this time is $\mathbf{X}_s$ and its mean value is $E\{\mathbf{X}_s\} = 1/\mu$.

Now, *cell dwell time* or *cell residence time* $x_d^{(j)}$ is defined as the time interval that a mobile station (MS) spends in the j-th (for $j = 0, 1, \ldots$) handed off cell irrespective of whether it is engaged in a call (or session) or not. The random variables (RVs) used to represent this time are $\mathbf{X}_d^{(j)}$ (for $j = 0, 1, \ldots$) and are assumed to be independent and identically generally phase-type distributed. For homogeneous cellular systems, this assumption has been widely accepted in the literature (Lin et al., 1994; Hong & Rappaport, 1986, Del Re et al., 1995; Orlik & Rappaport, 1998; Fang & Chlamtac, 1999, Li & Fang, 2008; Alfa & Li, 2002; Rahman & Alfa, 2009). $1/\eta$ is the mean cell dwell time. In this Chapter, cell dwell time is modeled as a phase-type distributed RV.

The *residual cell dwell time* x_r is defined as the time between the instant that a new call is initiated and the instant that the user is handed off to another cell. Notice that residual cell dwell time is only defined for new calls. The RV used to represent this time is $\mathbf{X}_r$. Thus, the probability density function (pdf) of $\mathbf{X}_r$, $f_{\mathbf{X}_r}(t)$, can be calculated in terms of $\mathbf{X}_d$ using the excess life theorem (Lin et al., 1994)

$$f_{\mathbf{X}_r}(t) = \frac{1}{E[\mathbf{X}_d]}\left[1 - F_{\mathbf{X}_d}(t)\right] \tag{1}$$

where $E[\mathbf{X}_d]$ and $F_{\mathbf{X}_d}(t)$ are, respectively, the mean value and cumulative probability distribution function (CDF) of $\mathbf{X}_d$.

Finally, the mathematical model used to consider link unreliability is based on the proposed call interruption process proposed in (Rodríguez-Estrello et al., 2009). In (Rodríguez-Estrello et al., 2009), an interruption model and a potential associated time to this process, which is called "*unencumbered call interruption time,*" is proposed. Unencumbered call interruption time $x_i^{(j)}$ is defined as the period of time from the epoch the mobile terminal establish a link

with the j-th handed-off cell (for j = 0, 1, 2, …) until the instant the call would be interrupted due to the wireless link unreliability assuming that the mobile terminal has neither successfully completed his call nor has been handed off to another cell. The RVs used to represent this time are $\mathbf{X}_i^{(j)}$ (for j = 0, 1, 2, …). These RVs are assumed to be independent and phase-type distributed (Rodríguez-Estrello et al., 2009; Rodríguez-Estrello et al., 2010).
Relationships between the different time interval variables defined in this section are illustrated in Fig. 1. Specifically, Fig. 1 shows the time diagram for a forced terminated call due to link unreliability.

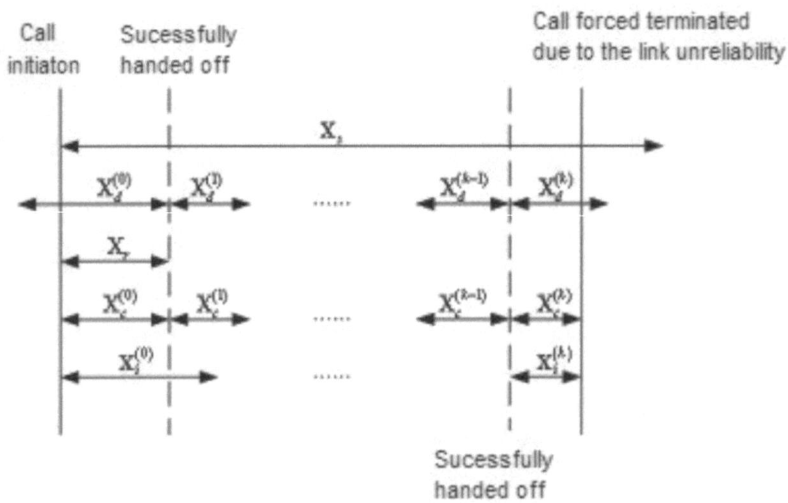

Fig. 1. Time diagram for a forced terminated call due to link unreliability.

3. Teletraffic analysis

In this section, the teletraffic analytical method for system-level evaluation of mobile cellular networks is presented. To avoid analytical complexity and for an easy interpretation of our numerical results, exponentially distributed unencumbered interruption time and hyper-Erlang distributed cell dwell time are considered. Nonetheless, numerical results for the following cases are also presented and discussed in Section 5 (Performance Evaluation): 1) Cell dwell time hyper-exponential distributed and unencumbered interruption time exponential distributed, 2) Cell dwell time exponential distributed and unencumbered interruption time hyper-Erlang distributed, 3) Cell dwell time exponential distributed and unencumbered interruption time hyper-exponential distributed.

3.1 Residual cell dwell time characterization

The methodology by which we can model the residual cell dwell time distribution is described in this section.
Suppose that cell dwell time follows an $m^{(h)}$-th order hyper-Erlang distribution function with parameters $\alpha_i^{(h)}$, $u_i^{(h)}$, and $\eta_i^{(h)}$ (for i = 1, 2,…, $m^{(h)}$). Let us represent by $\alpha_i^{(h)}$ the probability of

choosing the phase i (for i = 1, 2,..., $m^{(h)}$). Thus, the pdf of cell dwell time can be written as follows

$$f_{\mathbf{X}_d}(t)=\sum_{i=1}^{m^{(h)}}\alpha_i^{(h)}\frac{\left(\eta_i^{(h)}\right)^{u_i^{(h)}}t^{u_i^{(h)}-1}}{\left(u_i^{(h)}-1\right)!}e^{-\eta_i^{(h)}t};\ \eta_i^{(h)}>0,\ t\geq 0,\ 0\leq\alpha_i^{(h)}\leq 1,\ \sum_{i=1}^{m^{(h)}}\alpha_i^{(h)}=1 \tag{2}$$

where $u_i^{(h)}$ is a positive integer and $\eta_i^{(h)}$ is a positive constant. Note that the hyper-Erlang distribution is a mixture of $m^{(h)}$ different Erlang distributions, and each of them has a shape parameter $u_i^{(h)}$ and a rate parameter $\eta_i^{(h)}$. The rate parameter is related to the mean cell dwell time as follows: $\eta_i^{(h)} = \eta u_i^{(h)}$ (Fang, 1999). The value $\alpha_i^{(h)}$ represents the weight of each Erlang distribution. Using (1), the pdf of residual cell dwell time can be computed as follows

$$f_{\mathbf{X}_r}(t)=\frac{1}{\sum_{i=1}^{m^{(h)}}\frac{\alpha_i^{(h)}u_i^{(h)}}{\eta_i^{(h)}}}\sum_{i=1}^{m^{(h)}}\sum_{j=0}^{u_i^{(h)}-1}\alpha_i^{(h)}\frac{\left(\eta_i^{(h)}t\right)^j}{j!}e^{-\eta_i^{(h)}t} \tag{3}$$

It is straightforward to show that the probability distribution function (pdf) of residual cell dwell time can be rewritten in the following compact form

$$f_{\mathbf{X}_r}(t)=\sum_{i=1}^{m^{(n)}}\alpha_i^{(n)}\frac{\left(\eta_i^{(n)}\right)^{u_i^{(n)}}t^{u_i^{(n)}-1}}{\left(u_i^{(n)}-1\right)!}e^{-\eta_i^{(n)}t} \tag{4}$$

where

$$\alpha^{(n)}_{\sum_{x=1}^{i-1}u_x^{(h)}+j}=\frac{\alpha_i^{(h)}\prod_{l=1,\ l\neq i}^{m^{(h)}}\eta_l^{(h)}}{\sum_{k=1}^{m^{(h)}}\left(\alpha_k^{(h)}u_k^{(h)}\prod_{l=1,\ l\neq k}^{m^{(h)}}\eta_l^{(h)}\right)},\ u^{(n)}_{\sum_{x=1}^{i-1}u_x^{(h)}+j}=j,\ \eta^{(n)}_{\sum_{x=1}^{i-1}u_x^{(h)}+j}=\eta_i^{(h)},\ m^{(n)}=\sum_{i=1}^{m^{(h)}}u_i^{(h)} \tag{5}$$

for $i=1,2,\ldots,m^{(h)}$, $j=1,2,\ldots,u_i^{(h)}$.

From (4), it is not difficult to notice that residual cell dwell time is an $m^{(n)}$-th order hyper-Erlang random variable with shape and rate parameters $u_i^{(n)}$ and $\eta_i^{(n)}$ (for i = 1, 2,..., $m^{(n)}$). Also, $\alpha_i^{(n)}$ represents the probability of choosing the phase i (for i = 1, 2,..., $m^{(n)}$). The diagram of phases and stages of cell dwell time and residual cell dwell time is shown in Fig. 2. In Fig. 2, y={n} represents the case when residual cell dwell time is considered, while y={h} represents the case when cell dwell time is considered.

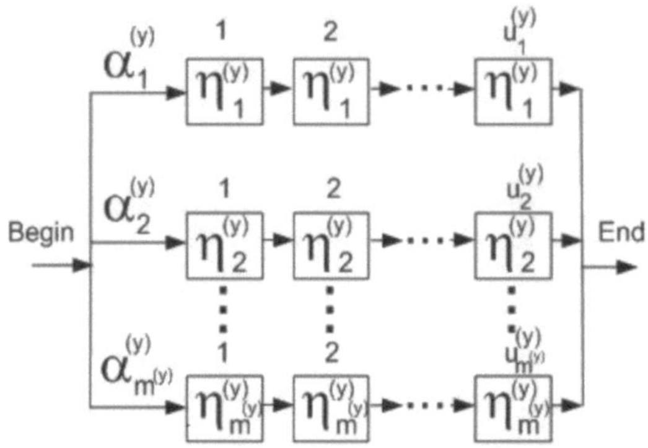

Fig. 2. Diagram of phases and stages of the probability distribution of residual cell dwell time ($y=\{n\}$) and cell dwell time ($y=\{h\}$).

3.2 Queuing formulation

In the context of a wireless network, each cell may be modelled as a queuing system where new and handoff arrivals correspond to connection requests or call origination, the departures correspond to disconnection due to call termination, force termination due to wireless link unreliability, or handoff throughout adjacent cells. The servers represent the available channels, whereas the clients represent the active mobile terminals. Since a homogenous case is assumed where all the cells in the service area are statistically identical, the overall system performance can be analyzed by focusing on only one given cell. In this Chapter, we analyze system performance by following the approach proposed in (Rico-Páez, et al., 2007; Rico-Páez et al., 2009; Rico-Páez et al., 2010) to capture the general distributions for both cell dwell time and unencumbered call interruption time. In this section, the special case when UIT is exponential distributed and CDT is hyper-Erlang distributed is considered. The mean value of UIT is $1/\gamma$. For modelling this system through a multidimensional birth and death process, a total number of $\sum_{x=1}^{m^{(n)}} u_x^{(n)} + \sum_{x=1}^{m^{(h)}} u_x^{(h)}$ state variables are needed. Let us define $k^{(y)}_{\sum_{x=1}^{i-1} u_x^{(y)}+j}$ as the number of users in stage i and phase j of residual cell dwell time ($y=\{n\}$) and cell dwell time ($y=\{h\}$).

To simplify mathematical notation, let us define the vector $\mathbf{K}^{(y)}$ as follows

$$\mathbf{K}^{(y)} = \left[k_1^{(y)}, k_2^{(y)}, \ldots, k^{(y)}_{\sum_{i=1}^{m^{(y)}} u_i^{(y)}} \right]$$

Also, let us define the vector $\mathbf{e}_i^{(y)}$ as a unit vector of dimension $m^{(y)}$ whose all entries are 0 except the i-th one which is 1 (for $i = 1, 2, \ldots, \sum_{i=1}^{m^{(y)}} u_i^{(y)}$). Let us define the current state of the analyzed cell as the vector $[\mathbf{K}^{(n)}, \mathbf{K}^{(h)}]$. Table I provides the rules that determine transition rates to the different successor states (shown in the second column). As stated before, we assume that all the cells are probabilistically equivalent. That is, the new call arrival rate in each cell is equal, and the rate at which mobiles enter a given cell is equal to the rate at which they interrupt its connection (due to either a handed off call event or link unreliability) to that cell. Thus, equating rate out to rate in for each state, the statistical-equilibrium state equations are given by (Cooper, 1990):

$$P\left(\left[\overrightarrow{\mathbf{K}^{(n)}},\overrightarrow{\mathbf{K}^{(h)}}\right]\right)=\frac{g_a^{(n)}+g_b^{(n)}+g_c^{(n)}+g_d^{(n)}+g_a^{(h)}+g_b^{(h)}+g_c^{(h)}+g_d^{(h)}}{\begin{array}{l}\sum_{i=1}^{m^{(n)}}a_i^{(n)}\left(\left[\overrightarrow{\mathbf{K}^{(n)}},\overrightarrow{\mathbf{K}^{(h)}}\right]\right)+\sum_{i=1}^{m^{(n)}}\sum_{j=1}^{u_i^{(n)}-1}b^{(n)}_{\sum_{x=1}^{i-1}\left(u_x^{(n)}-1\right)+j}\left(\left[\overrightarrow{\mathbf{K}^{(n)}},\overrightarrow{\mathbf{K}^{(h)}}\right]\right)+\sum_{i=1}^{m^{(n)}}c_i^{(n)}\left(\left[\overrightarrow{\mathbf{K}^{(n)}},\overrightarrow{\mathbf{K}^{(h)}}\right]\right)\\ +\sum_{i=1}^{m^{(n)}}\sum_{j=1}^{u_i^{(n)}-1}d^{(n)}_{\sum_{x=1}^{i-1}\left(u_x^{(n)}-1\right)+j}\left(\left[\overrightarrow{\mathbf{K}^{(n)}},\overrightarrow{\mathbf{K}^{(h)}}\right]\right)+\sum_{i=1}^{m^{(h)}}\sum_{j=1}^{u_i^{(h)}-1}b^{(h)}_{\sum_{x=1}^{i-1}\left(u_x^{(h)}-1\right)+j}\left(\left[\overrightarrow{\mathbf{K}^{(n)}},\overrightarrow{\mathbf{K}^{(h)}}\right]\right)\\ +\sum_{i=1}^{m^{(h)}}a_i^{(h)}\left(\left[\overrightarrow{\mathbf{K}^{(n)}},\overrightarrow{\mathbf{K}^{(h)}}\right]\right)+\sum_{i=1}^{m^{(h)}}c_i^{(h)}\left(\left[\overrightarrow{\mathbf{K}^{(n)}},\overrightarrow{\mathbf{K}^{(h)}}\right]\right)+\sum_{i=1}^{m^{(h)}}\sum_{j=1}^{u_i^{(h)}-1}d^{(h)}_{\sum_{x=1}^{i-1}\left(u_x^{(h)}-1\right)+j}\left(\left[\overrightarrow{\mathbf{K}^{(n)}},\overrightarrow{\mathbf{K}^{(h)}}\right]\right)\end{array}} \tag{6}$$

$$\left\{\left(\begin{array}{l}0\le k_i^{(n)}\le S-N \;\text{ for } i=1,2,\ldots,\sum_{x=1}^{m^{(n)}}u_x^{(n)}\\ 0\le k_i^{(h)}\le S \;\text{ for } i=1,2,\ldots,\sum_{x=1}^{m^{(h)}}u_x^{(h)}\end{array}\right)\cap\left(\sum_{i=1}^{\sum_{x=1}^{m^{(n)}}u_x^{(n)}}k_i^{(n)}+\sum_{i=1}^{\sum_{x=1}^{m^{(h)}}u_x^{(h)}}k_i^{(h)}\le S\right)\right\}$$

where

$$g_a^{(n)}=\sum_{i=1}^{m^{(n)}}\left[a_i^{(n)}\left(\left[\overrightarrow{\mathbf{K}^{(n)}}-\overrightarrow{\mathbf{e}^{(n)}_{\sum_{x=1}^{i-1}u_x^{(n)}+1}},\overrightarrow{\mathbf{K}^{(h)}}\right]\right)P\left(\left[\overrightarrow{\mathbf{K}^{(n)}}-\overrightarrow{\mathbf{e}^{(n)}_{\sum_{x=1}^{i-1}u_x^{(n)}+1}},\overrightarrow{\mathbf{K}^{(h)}}\right]\right)\right] \tag{7}$$

$$g_b^{(n)}=\sum_{i=1}^{m^{(n)}}\sum_{j=1}^{u_i^{(n)}-1}\left[b^{(n)}_{\sum_{x=1}^{i-1}\left(u_x^{(n)}-1\right)+j}\left(\left[\overrightarrow{\mathbf{K}^{(n)}}+\overrightarrow{\mathbf{e}^{(n)}_{\sum_{x=1}^{i-1}u_x^{(n)}+j}}-\overrightarrow{\mathbf{e}^{(n)}_{\sum_{x=1}^{i-1}u_x^{(n)}+j+1}},\overrightarrow{\mathbf{K}^{(h)}}\right]\right)P\left(\left[\overrightarrow{\mathbf{K}^{(n)}}+\overrightarrow{\mathbf{e}^{(n)}_{\sum_{x=1}^{i-1}u_x^{(n)}+j}}-\overrightarrow{\mathbf{e}^{(n)}_{\sum_{x=1}^{i-1}u_x^{(n)}+j+1}},\overrightarrow{\mathbf{K}^{(h)}}\right]\right)\right] \tag{8}$$

$$g_c^{(n)}=\sum_{i=1}^{m^{(n)}}\left[c_i^{(n)}\left(\left[\overrightarrow{\mathbf{K}^{(n)}}+\overrightarrow{\mathbf{e}^{(n)}_{\sum_{x=1}^{i}u_x^{(n)}}},\overrightarrow{\mathbf{K}^{(h)}}\right]\right)P\left(\left[\overrightarrow{\mathbf{K}^{(n)}}+\overrightarrow{\mathbf{e}^{(n)}_{\sum_{x=1}^{i}u_x^{(n)}}},\overrightarrow{\mathbf{K}^{(h)}}\right]\right)\right] \tag{9}$$

$$g_d^{(n)}=\sum_{i=1}^{m^{(n)}}\sum_{j=1}^{u_i^{(n)}-1}\left[d^{(n)}_{\sum_{x=1}^{i-1}\left(u_x^{(n)}-1\right)+j}\left(\left[\overrightarrow{\mathbf{K}^{(n)}}+\overrightarrow{\mathbf{e}^{(n)}_{\sum_{x=1}^{i-1}u_x^{(n)}+j}},\overrightarrow{\mathbf{K}^{(h)}}\right]\right)P\left(\left[\overrightarrow{\mathbf{K}^{(n)}}+\overrightarrow{\mathbf{e}^{(n)}_{\sum_{x=1}^{i-1}u_x^{(n)}+j}},\overrightarrow{\mathbf{K}^{(h)}}\right]\right)\right] \tag{10}$$

$$g_a^{(h)}=\sum_{i=1}^{m^{(h)}}\left[a_i^{(h)}\left(\left[\overrightarrow{\mathbf{K}^{(n)}},\overrightarrow{\mathbf{K}^{(h)}}-\overrightarrow{\mathbf{e}^{(h)}_{\sum_{x=1}^{i-1}u_x^{(h)}+1}}\right]\right)P\left(\left[\overrightarrow{\mathbf{K}^{(n)}},\overrightarrow{\mathbf{K}^{(h)}}-\overrightarrow{\mathbf{e}^{(h)}_{\sum_{x=1}^{i-1}u_x^{(h)}+1}}\right]\right)\right] \tag{11}$$

$$g_b^{(h)} = \sum_{i=1}^{m^{(h)}} \sum_{j=1}^{u_i^{(h)}-1} \left[b^{(h)}_{\sum_{x=1}^{i-1}\left(u_x^{(h)}-1\right)+j} \left(\left[\overrightarrow{\mathbf{K}^{(n)}}, \overrightarrow{\mathbf{K}^{(h)} + \mathbf{e}^{(h)}_{\sum_{x=1}^{i-1} u_x^{(h)}+j} - \mathbf{e}^{(h)}_{\sum_{x=1}^{i-1} u_x^{(h)}+j+1}} \right] \right) P\left(\left[\overrightarrow{\mathbf{K}^{(n)}}, \overrightarrow{\mathbf{K}^{(h)} + \mathbf{e}^{(h)}_{\sum_{x=1}^{i-1} u_x^{(h)}+j} - \mathbf{e}^{(h)}_{\sum_{x=1}^{i-1} u_x^{(h)}+j+1}} \right] \right) \right] \tag{12}$$

$$g_c^{(h)} = \sum_{i=1}^{m^{(h)}} \left[c_i^{(h)} \left(\left[\overrightarrow{\mathbf{K}^{(n)}}, \overrightarrow{\mathbf{K}^{(h)} + \mathbf{e}^{(h)}_{\sum_{x=1}^{i} u_x^{(h)}}} \right] \right) P\left(\left[\overrightarrow{\mathbf{K}^{(n)}}, \overrightarrow{\mathbf{K}^{(h)} + \mathbf{e}^{(h)}_{\sum_{x=1}^{i} u_x^{(h)}}} \right] \right) \right] \tag{13}$$

$$g_d^{(h)} = \sum_{i=1}^{m^{(h)}} \sum_{j=1}^{u_i^{(h)}-1} \left[d^{(h)}_{\sum_{x=1}^{i-1}\left(u_x^{(h)}-1\right)+j} \left(\left[\overrightarrow{\mathbf{K}^{(n)}}, \overrightarrow{\mathbf{K}^{(h)} + \mathbf{e}^{(h)}_{\sum_{x=1}^{i-1} u_x^{(h)}+j}} \right] \right) P\left(\left[\overrightarrow{\mathbf{K}^{(n)}}, \overrightarrow{\mathbf{K}^{(h)} + \mathbf{e}^{(h)}_{\sum_{x=1}^{i-1} u_x^{(h)}+j}} \right] \right) \right] \tag{14}$$

Of course, the probabilities must satisfy the normalization equation given by (15).

$$\sum_{k_1^{(n)}=0}^{S-N} \cdots \sum_{k^{(n)}_{\sum_{x=1}^{m^{(n)}} u_x^{(n)}}=0}^{S-N} \sum_{k_1^{(h)}=0}^{S} \cdots \sum_{k^{(h)}_{\sum_{x=1}^{m^{(h)}} u_x^{(h)}}=0}^{S} P\left(\left[\overrightarrow{\mathbf{K}^{(n)}}, \overrightarrow{\mathbf{K}^{(h)}} \right] \right) = 1 \tag{15}$$

$$\left\{ \left(\sum_{i=1}^{\sum_{x=1}^{m^{(n)}} u_x^{(n)}} k_i^{(n)} + \sum_{i=1}^{\sum_{x=1}^{m^{(h)}} u_x^{(h)}} k_i^{(h)} \le S \right) \cap \left(\sum_{i=1}^{\sum_{x=1}^{m^{(n)}} u_x^{(n)}} k_i^{(n)} \le S-N \right) \right\}$$

The corresponding steady state probabilities are calculated by means of the Gauss-Seidel Method (Cooper, 1990).

Let us assume that the channels reserved for handoff prioritization are given by $N^{(n)}=N$ and $N^{(h)}=0$, respectively, and the total number of channels is S. Transition rates shown in Table I are given by (for $y = \{n, h\}$)

$$a_i^{(y)}\left(\left[\mathbf{K}^{(n)}, \mathbf{K}^{(h)} \right] \right) = \begin{cases} \alpha_i^{(y)} \lambda^{(y)} & ; \sum_{l=1}^{\sum_{x=1}^{m^{(n)}} u_x^{(n)}} k_l^{(n)} + \sum_{l=1}^{\sum_{x=1}^{m^{(h)}} u_x^{(h)}} k_l^{(h)} < S - N^{(y)} \cap k^{(y)}_{\sum_{x=1}^{i-1} u_x^{(y)}+1} \ge 0 \\ 0 & ; \text{otherwise} \end{cases} \tag{16}$$

$$b^{(y)}_{\sum_{x=1}^{i-1}\left(u_x^{(y)}-1\right)+j}\left(\left[\mathbf{K}^{(n)}, \mathbf{K}^{(h)} \right] \right) = \begin{cases} k^{(y)}_{\sum_{x=1}^{i-1} u_x^{(y)}+j} \eta_i^{(y)} & ; \sum_{l=1}^{\sum_{x=1}^{m^{(n)}} u_x^{(n)}} k_l^{(n)} + \sum_{l=1}^{\sum_{x=1}^{m^{(h)}} u_x^{(h)}} k_l^{(h)} \le S \cap k^{(y)}_{\sum_{x=1}^{i-1} u_x^{(y)}+j} > 0 \cap k^{(y)}_{\sum_{x=1}^{i-1} u_x^{(y)}+j+1} \ge 0 \\ 0 & ; \text{otherwise} \end{cases} \tag{17}$$

$$c_i^{(y)}\left(\left[\mathbf{K}^{(n)},\mathbf{K}^{(h)}\right]\right)=\begin{cases} k^{(y)}_{i\sum_{x=1} u_x^{(y)}}\left(\mu+\eta_i^{(y)}+\gamma\right); & \sum\limits_{l=1}^{\sum_{x=1}^{m^{(n)}} u_x^{(n)}} k_l^{(n)}+\sum\limits_{l=1}^{\sum_{x=1}^{m^{(h)}} u_x^{(h)}} k_l^{(h)}\le S \cap k^{(y)}_{i\sum_{x=1} u_x^{(y)}}>0 \\ 0 & ; \quad \text{otherwise} \end{cases} \tag{18}$$

$$d^{(y)}_{i-1\atop \sum_{x=1}\left(u_x^{(y)}-1\right)+j}\left(\left[\mathbf{K}^{(n)},\mathbf{K}^{(h)}\right]\right)=\begin{cases} k^{(y)}_{i-1\atop \sum_{x=1} u_x^{(y)}+j}\left(\mu+\gamma\right); \sum\limits_{l=1}^{\sum_{x=1}^{m^{(n)}} u_x^{(n)}} k_l^{(n)}+\sum\limits_{l=1}^{\sum_{x=1}^{m^{(h)}} u_x^{(h)}} k_l^{(h)}\le S \cap k^{(y)}_{i-1\atop \sum_{x=1} u_x^{(y)}+j}>0 \cap k^{(y)}_{i-1\atop \sum_{x=1} u_x^{(y)}+j+1}\ge 0 \\ 0 \quad ; \text{ otherwise} \end{cases} \tag{19}$$

The new call blocking ($P^{(n)}$) and handoff failure ($P^{(h)}$) probabilities can be computed using (20).

$$P^{(y)}=\underbrace{\sum_{k_1^{(n)}=0}^{S-N^{(y)}}\cdots\sum_{k^{(n)}_{\sum_{x=1}^{m^{(n)}} u_x^{(n)}}=0}^{S-N^{(y)}}\sum_{k_1^{(h)}=0}^{S}\cdots\sum_{k^{(h)}_{\sum_{x=1}^{m^{(h)}} u_x^{(h)}}=0}^{S}}_{\left\{S-N^{(y)}\le\sum\limits_{i=1}^{\sum_{x=1}^{m^{(n)}} u_x^{(n)}} k_i^{(n)}+\sum\limits_{i=1}^{\sum_{x=1}^{m^{(h)}} u_x^{(h)}} k_i^{(h)}\le S\right\}} P\left(\left[\mathbf{K}^{(n)},\mathbf{K}^{(h)}\right]\right);\ \text{for } y=\{n\ ,h\} \tag{20}$$

Finally, the carried traffic (a_c) can be computed as follows

$$a_c=\underbrace{\sum_{k_1^{(n)}=0}^{s-N^{(y)}}\cdots\sum_{k^{(n)}_{\sum_{x=1}^{m^{(n)}} u_x^{(n)}}=0}^{s-N^{(y)}}\sum_{k_1^{(h)}=0}^{s}\cdots\sum_{k^{(h)}_{\sum_{x=1}^{m^{(h)}} u_x^{(h)}}=0}^{s}}_{\left\{\sum\limits_{i=1}^{\sum_{x=1}^{m^{(n)}} u_x^{(n)}} k_i^{(n)}+\sum\limits_{i=1}^{\sum_{x=1}^{m^{(h)}} u_x^{(h)}} k_i^{(h)}\le s\right\}}\left\{\left[\sum_{l=1}^{\sum_{x=1}^{m^{(n)}} u_x^{(n)}} k_l^{(n)}+\sum_{l=1}^{\sum_{x=1}^{m^{(h)}} u_x^{(h)}} k_l^{(h)}\right]P\left(\left[\mathbf{K}^{(n)},\mathbf{K}^{(h)}\right]\right)\right\} \tag{21}$$

Call forced termination probability can be calculated using the methodology developed in Section 4, while the handoff call attempt rate is calculated iteratively as explained in (Lin et al., 1994).

4. Forced termination probability

Forced termination may result from either link unreliability or due to handoff failure. In general, a dropped call suffers j (j=0, 1, 2, ...) successful handoffs and one forced interruption (due to either a handoff failure or link unreliability) before it is forced terminated. Thus, the forced termination probability in cell j can be expressed as follows

Event	Successor State	Rate
Call enters first phase of stage i of $\mathbf{X}_r$ (i=1,2,...,$m^{(n)}$)	$\left[\mathbf{K}^{(n)}+\mathbf{e}^{(n)}_{\sum_{x=1}^{i-1}u_x^{(n)}+1},\mathbf{K}^{(h)}\right]$	$a_i^{(n)}\left(\left[\mathbf{K}^{(n)},\mathbf{K}^{(h)}\right]\right)$
Call leaves phase j of stage i and enters phase $j+1$ of stage i of $\mathbf{X}_r$ (i=1, 2,..,$m^{(n)}$), (j=1, 2,..., $u_i^{(n)}$-1)	$\left[\mathbf{K}^{(n)}-\mathbf{e}^{(n)}_{\sum_{x=1}^{i-1}u_x^{(n)}+j}+\mathbf{e}^{(n)}_{\sum_{x=1}^{i-1}u_x^{(n)}+j+1},\mathbf{K}^{(h)}\right]$	$b^{(n)}_{\sum_{x=1}^{i-1}\left(u_x^{(n)}-1\right)+j}\left(\left[\mathbf{K}^{(n)},\mathbf{K}^{(h)}\right]\right)$
Call leaves last phase of stage i of $\mathbf{X}_r$ (i=1,2,...,$m^{(n)}$)	$\left[\mathbf{K}^{(n)}-\mathbf{e}^{(n)}_{\sum_{x=1}^{i}u_x^{(n)}},\mathbf{K}^{(h)}\right]$	$c_i^{(n)}\left(\left[\mathbf{K}^{(n)},\mathbf{K}^{(h)}\right]\right)$
Call leaves phase of stage i of $\mathbf{X}_r$ (i=1,2,...,$m^{(n)}$)	$\left[\mathbf{K}^{(n)}-\mathbf{e}^{(n)}_{\sum_{x=1}^{i}u_x^{(n)}+j},\mathbf{K}^{(h)}\right]$	$d^{(n)}_{\sum_{x=1}^{i-1}\left(u_x^{(n)}-1\right)+j}\left(\left[\mathbf{K}^{(n)},\mathbf{K}^{(h)}\right]\right)$
Call enters first phase of stage i of $\mathbf{X}_d$ (i=1,2,...,$m^{(h)}$)	$\left[\mathbf{K}^{(n)},\mathbf{K}^{(h)}+\mathbf{e}^{(h)}_{\sum_{x=1}^{i-1}u_x^{(h)}+1}\right]$	$a_i^{(h)}\left(\left[\mathbf{K}^{(n)},\mathbf{K}^{(h)}\right]\right)$
Call leaves phase j of stage i and enters phase $j+1$ of stage i of $\mathbf{X}_d$ (i=1, 2,...,$m^{(h)}$) (j=1, 2,..., $u_i^{(h)}$-1)	$\left[\mathbf{K}^{(n)},\mathbf{K}^{(h)}-\mathbf{e}^{(h)}_{\sum_{x=1}^{i-1}u_x^{(h)}+j}+\mathbf{e}^{(h)}_{\sum_{x=1}^{i-1}u_x^{(h)}+j+1}\right]$	$b^{(h)}_{\sum_{x=1}^{i-1}\left(u_x^{(h)}-1\right)+j}\left(\left[\mathbf{K}^{(n)},\mathbf{K}^{(h)}\right]\right)$
Call leaves last phase of stage i of $\mathbf{X}_d$ (i=1,2,..., $m^{(h)}$)	$\left[\mathbf{K}^{(n)},\mathbf{K}^{(h)}-\mathbf{e}^{(h)}_{\sum_{x=1}^{i}u_x^{(h)}}\right]$	$c_i^{(h)}\left(\left[\mathbf{K}^{(n)},\mathbf{K}^{(h)}\right]\right)$
Call leaves phase of stage i of $\mathbf{X}_d$ (i=1,2,...,$m^{(n)}$)	$\left[\mathbf{K}^{(n)},\mathbf{K}^{(h)}-\mathbf{e}^{(h)}_{\sum_{x=1}^{i}u_x^{(h)}+j}\right]$	$d^{(h)}_{\sum_{x=1}^{i-1}\left(u_x^{(h)}-1\right)+j}\left(\left[\mathbf{K}^{(n)},\mathbf{K}^{(h)}\right]\right)$

Table 1. Transition rules for the case when the cell residence time is hyper-Erlang distributed and unencumbered interruption time is negative exponential distributed.

$$P_{ft}^{(j)} = \begin{cases} P\left(\mathbf{X}_i \le \min\left(\mathbf{X}_s,\mathbf{X}_r\right)\right) + P\left(\mathbf{X}_r \le \min\left(\mathbf{X}_s,\mathbf{X}_i\right)\right)P_h & ; j=0 \\ P\left(\mathbf{X}_r \le \min\left(\mathbf{X}_s,\mathbf{X}_i\right)\right)\left(1-P_h\right)\left[P\left(\mathbf{X}_i \le \min\left(\mathbf{X}_s,\mathbf{X}_d\right)\right) + P\left(\mathbf{X}_d \le \min\left(\mathbf{X}_s,\mathbf{X}_i\right)\right)P_h\right] & ; j=1 \\ P\left(\mathbf{X}_r \le \min\left(\mathbf{X}_s,\mathbf{X}_i\right)\right)P\left(\mathbf{X}_d \le \min\left(\mathbf{X}_s,\mathbf{X}_i\right)\right)^{j-1}\left(1-P_h\right)^j \times & \\ \left[P\left(\mathbf{X}_i \le \min\left(\mathbf{X}_s,\mathbf{X}_d\right)\right) + P\left(\mathbf{X}_d \le \min\left(\mathbf{X}_s,\mathbf{X}_i\right)\right)P_h\right] & ; j>1 \end{cases} \tag{22}$$

where P_h represents the handoff failure probability. $P(\mathbf{X}_i \le \min(\mathbf{X}_s, \mathbf{X}_r)$ and $P(\mathbf{X}_i \le \min(\mathbf{X}_s, \mathbf{X}_d)$ represent interruption probabilities due to link unreliability for new and handoff calls, respectively. Function $\min(\cdot,\cdot)$ returns the minimum of two random variables.
The call forced termination can happen in any cell. Therefore, using the total probability theorem, the forced termination probability can be computed as follows

$$\begin{aligned} P_{ft} &= \sum_{j=0}^{\infty} P_{ft}^{(j)} \\ &= P\left(\mathbf{X}_i \le \min\left(\mathbf{X}_s,\mathbf{X}_r\right)\right) + P\left(\mathbf{X}_r \le \min\left(\mathbf{X}_s,\mathbf{X}_i\right)\right)P_h + P\left(\mathbf{X}_r \le \min\left(\mathbf{X}_s,\mathbf{X}_i\right)\right)\left(1-P_h\right)\times \\ &\quad \left[P\left(\mathbf{X}_i \le \min\left(\mathbf{X}_s,\mathbf{X}_d\right)\right) + P\left(\mathbf{X}_d \le \min\left(\mathbf{X}_s,\mathbf{X}_i\right)\right)P_h\right]\left[\frac{1}{1-P\left(\mathbf{X}_d \le \min\left(\mathbf{X}_s,\mathbf{X}_i\right)\right)\left(1-P_h\right)}\right] \end{aligned} \tag{23}$$

Let us define $\mathbf{Z}_s^{(w)}=\min(\mathbf{X}_s, \mathbf{X}_w)]$ with $w=\{r, i, d\}$. Notice that $\mathbf{Z}_s^{(w)}$ are non-negative RVs. Thus, the different probabilities in (23) can be calculated by using the following relationship (which uses the well known residual theorem) between two non-negative independent RVs (Fang a, 2005; Fang b, 2005)

$$\begin{aligned} P\left(\mathbf{X}_w \le \mathbf{Z}_s^{(w)}\right) &= \frac{1}{2\pi \mathrm{i}} \int_{\sigma-\mathrm{i}\infty}^{\sigma+\mathrm{i}\infty} \frac{f_{\mathbf{X}_w}^*(s)}{s} f_{\mathbf{Z}_s^{(w)}}^*(-s)\,ds \\ &= -\sum_{p\in\sigma_P} \operatorname*{Res}_{s=p}\left[\frac{f_{\mathbf{X}_w}^*(s)}{s} f_{\mathbf{Z}_s^{(w)}}^*(-s)\right] \end{aligned} \tag{24}$$

where $f_{\mathbf{X}_w}^*(s)$ and $f_{\mathbf{Z}_s^{(w)}}^*(s)$ represent, respectively, the Laplace transform of $\mathbf{X}_w$ and $\mathbf{Z}_s^{(w)}$ with $w=\{r, i, d\}$. σ_P is the set of poles of $f_{\mathbf{Z}_s^{(w)}}^*(-s)$. Equation (24) applies when the pdfs of $\mathbf{X}_w$ and $\mathbf{Z}_s^{(w)}$ are proper rational functions (Fang, 2005). This is the situation of the different cases studied in this paper.

5. Performance Evaluation

The goal of the numerical evaluations presented in this section is to understand and analyze the influence of standardized moments higher than the expected value of both cell dwell time (CDT) and unencumbered interruption time (UIT) on the performance of mobile cellular networks. At least otherwise stated, in numerical evaluations it is assumed that mean service time is $1/\mu$=180 s, total number of channels per cell S=8, offered traffic equal to 4.4 Erlangs per cell, and total number of channels reserved for handoff prioritization $N^{(n)}$=1. Figs. 3, 4, and 5 (6, 7, and 8) show, respectively, forced termination probability, blocking probability, and carried traffic as function of both skewness and coefficient of variation of the unencumbered call interruption time (cell dwell time). Figs. 3, 4, and 5 show numerical results for the particular case when UIT is either hyper-Erlang or hyper-exponential distributed and CDT is exponential distributed. On the other hand, Figs. 6, 7, and 8 show numerical results for the particular case when CDT is either hyper-Erlang or hyper-exponential distributed and UIT is exponential distributed. In Figs. 6-8, for the sake of comparison, two different values of the mean of the cell dwell time are considered: 100s (high mobility scenario) and 900 s (low mobility scenario). Also, two different values of the mean of the unencumbered call interruption time are considered in Figs. 3-5: 1500 s (low reliability scenario) and 5000 s (high reliability scenario)[3].

5.1 Influence of unencumbered interruption time statistics on system performance

In this Section, the influence of the expected value, coefficient of variation, and skewness of unencumbered interruption time on system performance is investigated.

From Fig. 3 it is observed that as the mean value of the UIT decreases the forced termination probability increases, indicating a detrimental effect of channel unreliability on system performance (remember that physically, the mean value of UIT represents a direct measure of link reliability). On the contrary, as Fig. 4 shows, the blocking probability decreases as link unreliability increases (i.e., when mean unencumbered call interruption time decreases), indicating a positive effect of channel unreliability on system performance. This behavior can be explained as follows. As link unreliability increases, more ongoing calls are forced to terminate, consequently, more resources are available for new calls decreasing, in this way, new call blocking probability.

Figs. 3 and 4 also show that, irrespective of the value of skewness, forced termination probability increases and new call blocking probability decreases as CoV of UIT increases. This behavior can be explained as follows. First, note that as the CoV increases, the variability of the UIT increases, thus the probability that UIT takes smaller values increases and, consequently, more calls are interrupted due to link unreliability. This fact contributes to both increase forced termination probability and decrease new call blocking probability. Furthermore, from Figs. 3 and 4 it is rather interesting to note that, for low values of skewness (say, less that 20), forced termination probability significantly increases and new call blocking probability decreases as CoV increases. For instance, Figs. 3 and 4 shows that, for the low reliability scenario, skewness equals 2, and UIT Hyper-Erlang distributed, the forced termination probability increases around 700% and new call blocking probability decreases 67% as the CoV of UIT changes from 1 to 20. Notice that the scenario where skewness equals 2 and CoV equals 1 corresponds to the case when UIT is negative

[3] Please note that both values of the mean of unencumbered call interruption time are significantly greater than the mean of cell dwell time. The reason of this is that communication systems are commonly designed to be reliable, thus mean unencumbered call interruption time should be typically greater than mean service time and mean cell dwell time.

exponential distributed. Finally, from Fig. 3 (Fig. 4) observe that forced termination (new call blocking) probability is a monotonically decreasing (increasing) function of skewness.
On the other hand, Fig. 5 shows that the carried traffic is an increasing function of both the skewness and mean value of UIT. Also, Fig. 5 shows that, for values of skewness smaller than around 30, carried traffic decreases as CoV of UIT increases. These observations indicate a detrimental effect of channel unreliability on carried traffic. Moreover, it is interesting to note that, for values of skewness grater than around 30 and for the same mean value and type of distribution of UIT, the carried traffic is almost insensitive to the CoV of UIT. The reason is as follows. Consider that the mean value, CoV and distribution type of UIT remain without change. Then, as the skewness of UIT increases, the tail on the right side of the UIT distribution function becomes longer (that is, the probability that UIT takes higher values increases and, consequently, less calls are interrupted due to link unreliability). In this manner, the influence of skewness on forced termination probability becomes negligible. At the same time, because of link unreliability is not considered to accept a call, the blocking probability is not sensitive to changes on neither skewness nor CoV of UIT statistics. As the carried traffic directly depends on both blocking and forced termination probabilities, the combined effect of these two facts lead us to the behavior observed in Fig. 5.
An interesting observation on the results illustrated in Figs. 3-5 is that, for the same scenario, skewness and CoV, there exists a non-negligible difference between the values taken by the different performance metrics when UIT is modeled as Hyper-Erlang and Hyper-exponential distributed random variable. Thus, it is evident that not only the expected value but also moments of higher order and the distribution model used to characterize UIT are relevant on system performance.

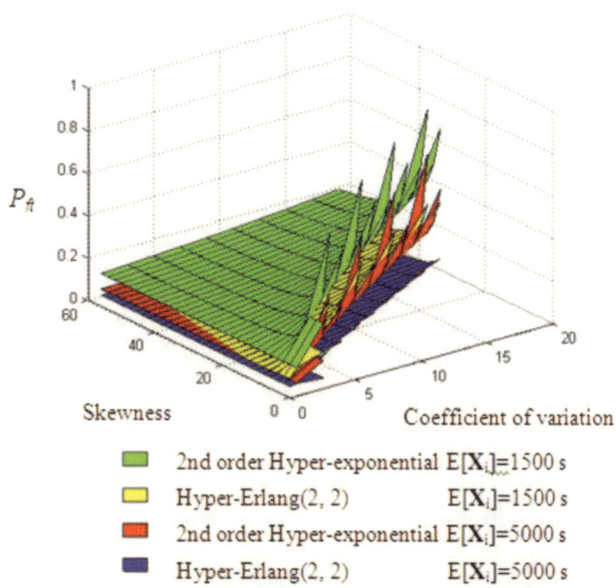

Fig. 3. Forced termination probability versus coefficient of variation and skewness of interruption time, with the pdf type and expected value of interruption time as parameters.

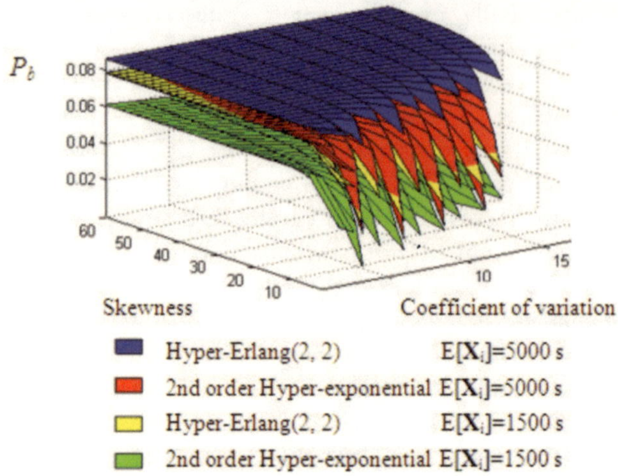

Fig. 4. New call blocking probability versus coefficient of variation and skewness of interruption time, with the pdf type and expected value of interruption time as parameters.

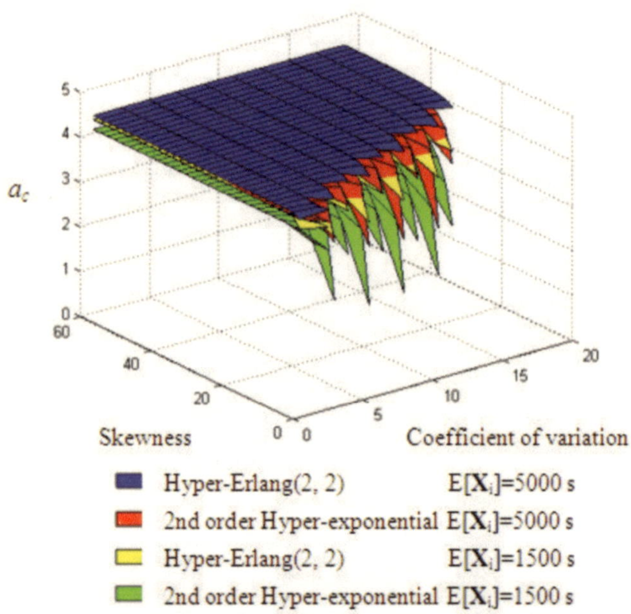

Fig. 5. Carried traffic versus coefficient of variation and skewness of interruption time, with the pdf type and expected value of interruption time as parameters.

5.2 Influence of cell dwell time statistics on system performance

In this Section, the influence of the expected value, coefficient of variation, and skewness of cell dwell time on system performance is investigated.

From Fig. 6 it is observed that as the mean value of the CDT decreases the forced termination probability increases, indicating a detrimental effect of mobility on system performance. This behavior can be explained by the fact that as the mean value of CDT decreases the average number of handoffs per call increases and, as consequence, the probability of a premature termination due to resource insufficiency increases. On the other hand, from Fig. 7, it is observed that the blocking probability increases as the mean value of CDT increases. This is because the larger the mean cell dwell time the slower users with ongoing calls move and, consequently, the rate at which radio resources are released decreases, causing a detrimental effect on blocking probability.

Fig. 6 also shows that, for the low mobility scenario, forced termination probability is practically insensitive to both skewness and CoV of CDT. This behavior is due to the fact that a low mobility scenario implies that most of the calls are completed (or blocked) in the cell where they were originated, reducing the average number of handoffs per call. Also, Fig. 6 shows that, for the high mobility scenario and irrespective of the value of skewness, forced termination probability decreases as CoV of CDT increases. This behavior can be explained as follows. First, note that as the CoV of CDT increases, the variability of the CDT increases, thus the probability that CDT takes higher values increases and, consequently, the average number of handoffs per call decreases, resulting in an improvement on the forced termination probability.

Furthermore, from Fig. 6 it is rather interesting to note that, for the high mobility scenario and low values of skewness (say, less that 20), forced termination probability is significantly improved as CoV of CDT increases. For instance, Fig. 6 shows that, for the high mobility scenario where CDT is Hyper-Erlang distributed, the forced termination probability decreases around 60% as the skewness and CoV of UIT change from 60 to 2 and from 1 to 20, respectively. Again, notice that the scenario where skewness equals 2 and CoV equals 1 corresponds to the case when CDT is negative exponential distributed.

On the other hand, Figs. 6 and 7 show that, for the high mobility scenario both forced termination and new call blocking probabilities are monotonically increasing functions of skewness of CDT. The reason is as follows. Consider that the mean value, CoV and distribution type of CDT remain without change. Then, as the skewness of CDT decreases, the tail on the right side of the CDT distribution function becomes longer (that is, the probability that CDT takes higher values increases and, consequently, less calls move to

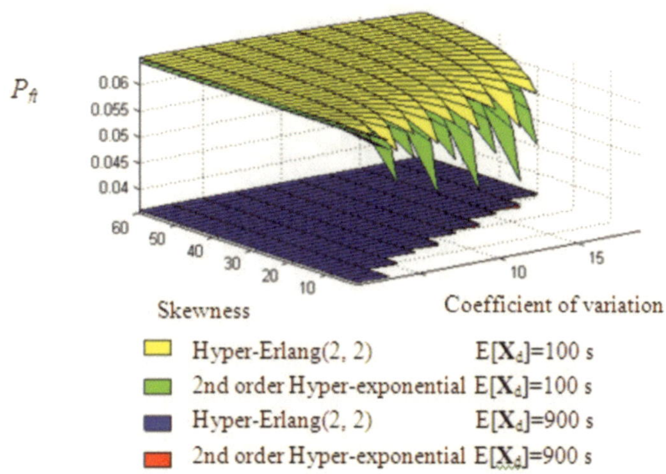

Fig. 6. Forced termination probability versus coefficient of variation and skewness of cell dwell time, with the pdf type and mean value of cell dwell time as parameters.

another cell). In this manner, the rate at which channels are used by handed off calls decreases in benefit of both new call blocking probability and handoff failure probability (and, thus, forced termination probability). This fact contributes to improve carried traffic in concordance with the results presented in Fig. 8. Fig. 8 shows that carried traffic is a decreasing function of skewness of CDT. Also, Fig. 8 shows that carried traffic increases as CoV of CDT increases. These observations indicate a beneficial effect of the variability of CDT on carried traffic.

Finally, as it was observed in the previous section, the results illustrated in Figs. 6-8 show that, for the same scenario, skewness and CoV of CDT, there exists a non-negligible difference between the values taken by the different performance metrics when CDT is modeled as hyper-Erlang and hyper-exponential distributed random variable. Thus, it is again evident that not only the expected value but also moments of higher order and the distribution model used to characterize cell dwell time are relevant on system performance.

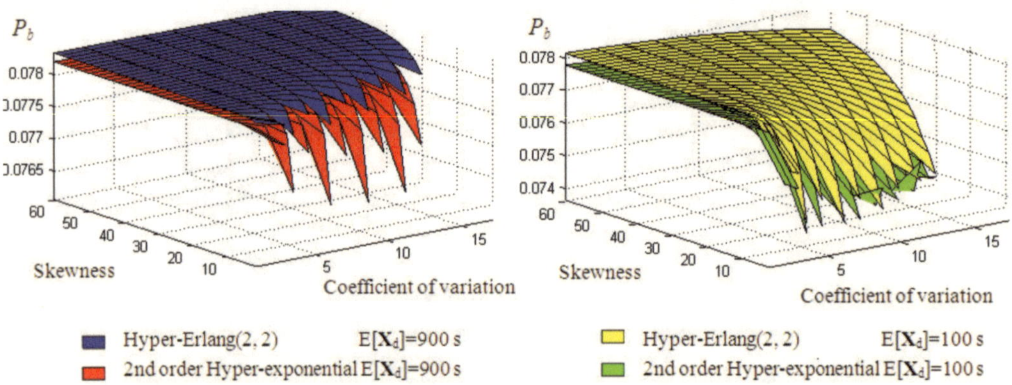

Fig. 7. New call blocking probability versus coefficient of variation and skewness of cell dwell time, with the pdf type and expected value of cell dwell time as parameters.

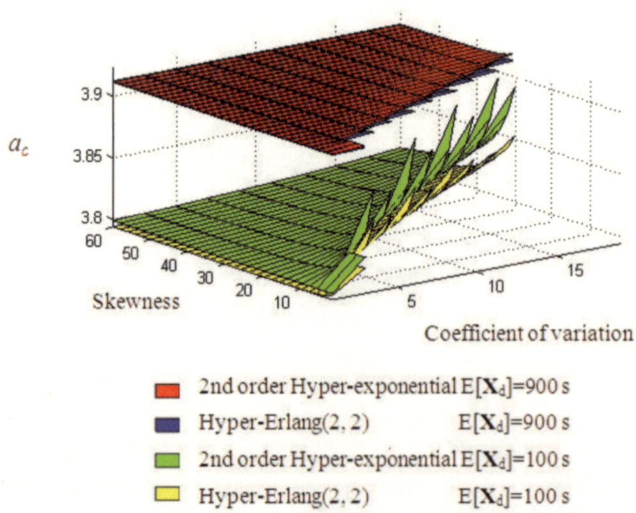

Fig. 8. Carried traffic versus coefficient of variation and skewness of cell dwell time, with the probability density function type and expected value of cell dwell time as parameters.

7. Conclusion

The study performed in this Chapter have allowed us to obtain new and important insights into the dependence of system performance on the first three standardized moments of both cell dwell time and unencumbered interruption time. Even though our numerical results are extracted from particular scenarios with certain set of parameter values, our contribution clearly shows that there exist relevant sensitive issues concerning higher order moments of both cell dwell time and unencumbered interruption time. We conclude that to accurately characterize the real distribution of the different random variables involved in the teletraffic model it is vital to consider not only the mean value but also higher order moments.

8. References

Alfa A. S., and Li W., "A homogeneous PCS network with Markov call arrival process and phase type cell dwell time," *Wireless Networks*, vol. 8, no. 6, pp. 597-605, 2002.

G. Boggia, P. Camarda, A. D'Alconzo, A. De Biasi and M. Siviero, "Drop call probability in established cellular networks: from data analysis to modelling," *in Proc. IEEE VTC'05-Spring*, Stockholm, Sweden, vol. 5, May-Jun. 2005, pp. 2775-2779.

Christensen T. K., Nielsen B.F., and Iversen V.B., "Phase-type models of channel-holding times in cellular communication systems," *IEEE Trans. Veh. Technol.*, vol. 53, no. 3, pp. 725-733, May 2004.

Cooper B., Introduction to Queuing Theory. CEE Press Books, Washington D.C. 1990.

Corral-Ruiz A.L.E., Cruz-Pérez F.A., and Hernández-Valdez G., "Teletraffic model for the performance evaluation of cellular networks with hyper-Erlang distributed cell dwell time," Proc. 71st *IEEE VTC'*10-Spring, Taipei, Taiwan, 16–19 May 2010.

Del Re E., Fantacci R., and Giambene G., "Handover and dynamic channel allocation techniques in mobile cellular networks," *IEEE Trans. Veh. Technol.*, vol. 44, no. 2, pp. 229-237, 1995.

Fang Y., "Hyper-Erlang distributions of traffic modeling in wireless and mobile networks," in *Proc. IEEE WCNC'99*, New Orleans, LA, Sept. 1999, pp.398-402.

Fang Y., a, "Modeling and performance analysis for wireless mobile networks: a new analytical approach," *IEEE Trans. on Networking*, vol. 13,no. 5, pp. 989-1002, Oct. 2005.

Fang Y., b, "Performance evaluation of wireless cellular networks under more realistic assumptions," *Wireless Commun. Mob. Comp.*, vol. 5, no. 8, pp. 867-885, Dec. 2005.

Fang Y. and Chlamtac I. "Teletraffic analysis and mobility modeling of PCS networks," *IEEE Trans. Commun.*, vol 47, no. 7, pp. 1062-1072, July 1999.

Fang Y., Chlamtac I., and Lin Y.-B., a, "Call performance for a PCS network," *IEEE J. Select. Areas Commun.*, vol. 15, no. 8, pp. 1568-1581, Oct. 1997.

Fang Y., Chlamtac I., and Lin Y.-B., b, "Modeling PCS networks under general call holding time and cell residence time distributions," *IEEE/ACM Trans. on Networking*, vol. 5, no. 6, pp. 893-906, Dec. 1997.

Guérin R., "Channel occupancy time distribution in a cellular radio system," *IEEE Trans. Veh. Technol.*, vol. VT-35, pp. 89–99, Aug. 1987.

Hong D. and Rappaport S. S., "Traffic model and performance analysis for cellular mobile radio telephone systems with prioritized and nonprioritized handoff procedures," *IEEE Trans. Veh. Technol.*, vol. 35, no. 3, pp. 77–92, Aug. 1986.

Khan F. and Zeghlache D., "Effect of cell residence time distribution on the performance of cellular mobile networks," in Proc. IEEE Veh. Tech. Conf.'97, Phoenix, AZ, 1997, pp. 949–953.

Kim K. and Choi H., "A mobility model and performance analysis in wireless cellular network with general distribution and multi-cell model," *Wireless Pers. Commun.*, published on line: 10 March 2009.

Li W. and Fang Y., "Performance evaluation of wireless cellular networks with mixed channel holding times," *IEEE Trans. Wireless Communs.*, vol. 7, no. 6, June 2008.

Lin Y.-B., Mohan S. and Noerpel A., "Queuing priority channel assignment strategies for PCS and handoff initial access," *IEEE Trans. Veh. Technol.*, vol. 43. no. 3, pp. 704-712, Aug. 1994.

Orlik P. V. and Rappaport S.S., "A model for teletraffic performance and channel holding time characterization in wireless cellular communication with general session and dwell time distributions," *IEEE J. Select. Areas Commun.*, vol. 16, no. 5, pp. 788-803, June 1998.

Pattaramalai S., Aalo V.A., and Efthymoglou G.P., "Evaluation of call performance in cellular networks with generalized cell dwell time and call-holding time distributions in the presence of channel fading," *IEEE Trans. Veh. Technol.*, vol. 58, no. 6, pp. 3002-3013, July 2009.

Rahman, M. M., Alfa A. S., "Computationally efficient method for analyzing guard channel schemes," *Telecommunication Systems*, vol. 41, pp. 1-11, published on line: 22 April 2009.

Rico-Páez A., Cruz-Pérez F.A., and Hernández-Valdez G., "Teletraffic Analysis Formulation based on Channel Holding Time Statistics," in Proc. IEEE International Conference on Wireless and Mobile Computing, Networking and Communications (WiMob'2009), Marrakech, Morocco, 12-14 October 2009.

Rico-Páez A., Cruz-Pérez F.A., and Hernández-Valdez G., "Performance Sensitivity to Higher Order Moments of Call Interruption and Cell Dwell Times in Cellular Networks," Proc. 72nd IEEE Vehicular Technology Conference (VTC'10-Fall), Ottawa, Canada, 6–9 September 2010.

Rico-Páez A., Rodríguez-Estrello C. B., Hernández-Valdez G., and Cruz-Pérez F. A., "Queueing Analysis of Mobile Cellular Networks Considering Wireless Channel Unreliability and Resource Insufficiency," *Lecture Notes in Computer Science* (LNCS 4516) -Managing Traffic Performance in Converged Networks-, L. Mason, T. Drwiega, and J. Yan (Editors): ITC20 2007, pp. 938-949, 2007.

Rodríguez-Estrello C. B., Hernández-Valdez G., and Cruz-Pérez F. A., "System level analysis of mobile cellular networks considering link unreliability," *IEEE Trans. Veh. Technol.*, vol. 58, no. 2, pp. 926-940, Feb. 2009.

Rodríguez-Estrello C. B., Hernández-Valdez G., and Cruz-Pérez F. A., Chapter: "Performance Modeling and Analysis of Mobile Wireless Networks," for the book "Mobile and Wireless Communications Physical Layer Development and Implementation," In-Tech Publisher, Salma Ait Fares and Fumiyuki Adachi (Ed.). Published: January 2010.

Vázquez-Ávila J.L., Cruz-Pérez F.A., and Ortigoza-Guerrero L., "Performance analysis of fractional guard channel policies in mobile cellular networks" *IEEE Trans. on Wireless Commun.*, vol. 5, no. 2, pp. 301-305, Feb. 2006.

Wang X. and Fan Pingzhi, "Channel holding time in wireless cellular communications with general distributed session time and dwell time," *IEEE Communications Letters*, vol. 11, no. 2, Feb. 2007.

Yeo K. and Jun C.-H., "Teletraffic analysis of cellular communication systems with general mobility based on hyper-Erlang characterization," *Computer & Industrial Engineering*, vol. 42, pp. 507-520, 2002.

Zeng H., Fang Y., and Chlamtac I., "Call blocking performance study for PCS Networks under more realistic mobility assumptions," *Telecommun. Systems*, vol. 19, no. 2, pp. 125-146, 2002.

9

Channel Assignment in Multihop Cellular Networks

Xue Jun Li and Peter Han Joo Chong
School of EEE, Nanyang Technological University
Singapore

1. Introduction

Recently, several work related to channel assignment for MCN-type systems were reported. Wu et al proposed diverting traffic from the congested cells to the non-congested cells (Wu et al., 2001; Wu et al., 2004), which is achieved by relaying traffic through unlicensed frequency band, such as the industrial, scientific and medical (ISM) band. For iCAR (Wu et al., 2001; Wu et al., 2004), ad hoc relay stations (ARSs), either being fixed (Wu et al., 2001) or mobile (Wu et al., 2004), are deployed for balancing traffic. The communication link between a mobile station (MS) and an ARS is established using the ISM band. Similarly, UCAN is used to increase the system throughput through relaying using the ISM band (Luo et al., 2003). None of the above papers describes how to select and allocate the relay channels for each hop in detail.

An ad hoc GSM (A-GSM) protocol was proposed in (Aggelou & Tafazolli, 2001) using the cellular frequency band for RSs to cover dead spots and increase the capacity. Aggelou and Tafazolli (Aggelou & Tafazolli, 2001) investigated the concept of A-GSM, which includes A-GSM network components, protocol architecture, and handover procedure to allow a MS to perform GSM-to-A-GSM and A-GSM-to-A-GSM handovers. The GSM-to-GSM connection uses the resource from the BS. The A-GSM-to-GSM connection routes through the RS and uses the resource from that RS. The coordinating of the allocation of resources of a relay node is controlled by a resource manager. However, it did not clearly address how the resources are allocated to the BS and RSs. It also did not address the channel assignment method for each type of connection. In addition, the analysis for a single GSM cell is done in (Aggelou & Tafazolli, 2001) so that the co-channel interference, which is one of the major issues in channel assignment in cellular networks, is not considered. Similarly, the MCN in (Hsu & Lin, 2002) assumes to use cellular frequency for relaying, whereas no clear description on how to allocate a channel to a MS for cellular or ad hoc mode is included.

2. Clustered multihop cellular networks

The key idea of CMCN (Li & Chong, 2006) is to achieve the characteristics of the macrocell/microcell hierarchically overlaid system (Rappaport & Hu, 1994; Yeung & Nanda, 1996) by applying the MANETs clustering (Yu & Chong, 2005) in *traditional* SCNs. In SCNs, the BS will cover the whole macrocell with a radius of r_M, as shown in Figure 1(a). The proposed CMCN divides the macrocell area into seven microcells with a radius of r_m in order to increase the spectrum efficiency as shown in Figure 1(b). Six *virtual* microcells with a coverage radius of r_m around the central microcell will be formed as six clusters.

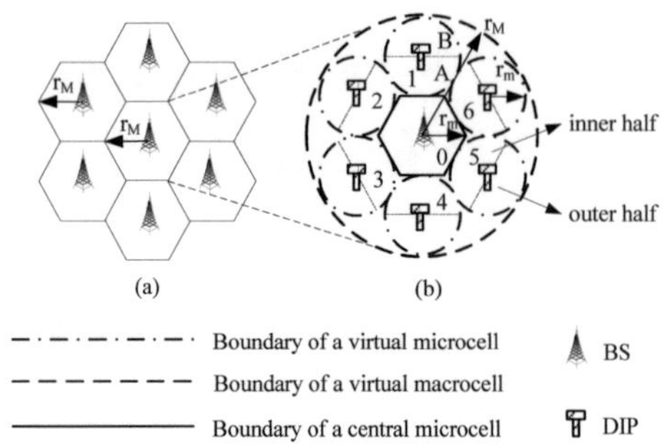

Fig. 1. (a) SCNs and (b) CMCN.

We proposed to use a DIP as a clusterhead in each *virtual* microcell for CMCN. A DIP is a wireless communication device, which has no wired interface. This is different from a BS, which may have a wired interface to a mobile switching center (MSC). Next, DIPs can be mobile and can be relocated anywhere to provide services while the locations of BSs are fixed due to the wired (or microwave) connection to MSC. The function of a DIP includes allocating channels to the MSs within its microcell, selecting a MS as a RS, and determining the routing path. Specially, DIPs can help the BS to perform the function of authentication, authorization and accounting (AAA). For example, a DIP is able to authorize a MS to relay the traffic for another MS. Different from ARSs in iCAR (Wu et al., 2001) or wireless ports in (Kudoh & Adachi, 2005) and (Liu et al., 2006), DIPs are not involved in data relaying. Hence, no worry about the capacity saturation problem, such as the load balancing considered in (Liu et al., 2006) for RSs, is concerned. Earlier researches have established that as long as there is a large number of MSs in the service area, it is not difficult for a DIP to find a RS. As a DIP only helps exchange the control/signaling information with a BS and MSs through the control channels, its complexity is much lower than a BS, so does the cost.

We assume that a DIP is installed in the center of a *virtual* microcell. The BS covers the central microcell and six DIPs cover the six *virtual* microcells. Each *virtual* microcell is divided into two regions: inner half and outer half. The inner half is near the central microcell. For example, as shown in Figure 1(b), for *virtual* microcell 1, area A is the inner half and area B is the outer half. This structure is named as *seven-cell* CMCN architecture.

In CMCN, BSs will have two levels of transmit power, P_{data} and $P_{control}$. As referred to Figure 1(b), P_{data} is used for a BS to transmit data packets including acknowledgement packets within its coverage area with a radius of r_m in the central microcell. P_{data} is also used for a MS to transmit data or control packets for a transmission range of radius, r_m. $P_{control}$ is used by the BS for transmitting the control/signaling packets between a BS and a DIP inside a *virtual* macrocell with a radius of r_M so that the BS is able to exchange the control/signaling information with every DIP.

3. Proposed fixed channel assignment scheme

In *traditional* SCNs, the channels assigned for uplink and downlink transmissions are balanced in every cell for symmetric traffic, such as voice calls. The number of channels assigned to the uplink and downlink for each call is the same. In practice, the FCA in SCNs normally assigns the same number of channels for uplink and downlink transmissions in each cell. However, in CMCN, taking both uplink and downlink transmissions into consideration, the channel assignment for uplink and downlink transmissions of a call is unbalanced. Thus we propose an AFCA to assign different number of uplink and downlink channels in each cell to provide the optimal capacity.

For AFCA in CMCN, each central/*virtual* microcell is assigned a set of channels for the uplink and downlink according to the AFCA rules. As shown in Figure 2, the channel assignment to calls in CMCN can be implemented as follows:

1) One-Hop Calls: If a MS, MS_1, originates a new call from the central microcell, it will require one uplink channel and one downlink channel. If all the uplink channels in the central microcell are occupied, or all the downlink channels in the central microcell are occupied, that new call will be blocked. Thus, a one-hop call takes one uplink channel and one downlink channel from the central microcell.

2) Two-Hop Calls: If a MS, MS_2, originates a new call from the inner half of the ith *virtual* microcell, for uplink transmission, it will request an uplink channel belonging to the ith *virtual* microcell from its DIP. Since the DIP is assigned a set of channels for use in its microcell, it will know the availability of channel within its *virtual* microcell to assign a free channel to that MS, MS_2. Then, the DIP will help find a MS, RS_0, in the central microcell as a RS, which will request another uplink channel belonging to the central microcell for relaying the call to BS. Thus, that new two-hop call will occupy one uplink channel from the ith *virtual* microcell and one uplink channel from the central microcell for uplink transmission. For downlink transmission, it requires two downlink channels from the central microcell, one for the BS to the RS, RS_0, and the other for RS_0 to the MS, MS_2. Therefore, as shown in Figure 2, a two-hop call, originated by MS_2, requires one uplink channel belonging to the central microcell and one uplink channel belonging to the ith *virtual* microcell, and two downlink channels belonging to the central microcell. That is why the channel assignment to the uplink and downlink is unbalanced in each *virtual* or central microcell. A new call will be blocked if either of the following conditions comes into existence: (i) there is no free uplink channel in the ith *virtual* microcell; (ii) there is no free uplink channel in the central microcell; (iii) there are less than two free downlink channels in the central microcell. When the call is completed, the MS, MS_2, in the *virtual* microcell will inform its DIP to release the channels for its call. Then, the DIP can update the channel status accordingly.

3) Three-Hop Calls: If a MS, MS_3, originates a new call from the outer half of the ith *virtual* microcell, for uplink transmission, it will require two uplink channels belonging to the ith *virtual* microcell from the DIP. This is because it takes two hops to reach the central microcell, as shown in Figure 2. Then, it will request one more uplink channel belonging to central microcell for a relay MS in the central microcell. For downlink transmission, it will require two downlink channels from the central microcell—one for the BS to the RS in the central microcell and the other for the RS in the central microcell to the RS in the ith *virtual* microcell. It will require one more downlink channel from i^{th} *virtual* microcell, for the downlink transmission from the RS in the ith *virtual* microcell to the MS. As shown in Figure 2, a three-hop call, originated by MS_3, requires two uplink channels in the ith *virtual* microcell and one uplink channel in the central microcell, and one downlink channel from the ith *virtual* microcell and two downlink channels from the central microcell. A call will be blocked if either of the following conditions fails: (i) there is at least one free uplink channel in the central microcell; (ii) there at least two free uplink channels in the ith *virtual* microcell; (iii) there are at least two free downlink channels in the central microcell; (4) there is at least one free downlink channel in the ith *virtual* microcell. Similarly, when the call is completed, the MS, MS_3, in the *virtual* microcell will inform its DIP to release the channels for its call. Then, the DIP can update the channel status accordingly.

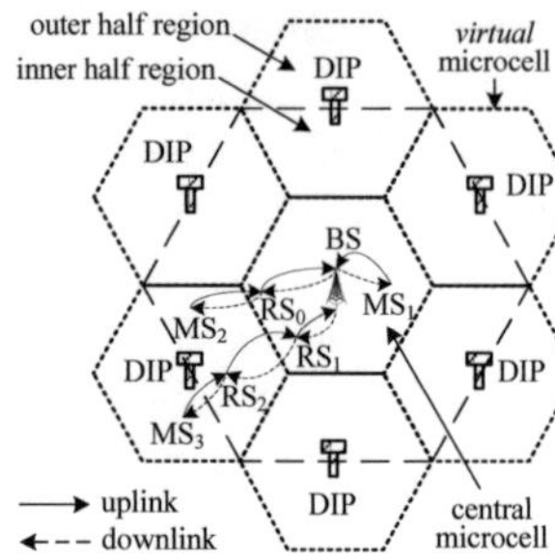

Fig. 2. Channel assignment for different types of calls.

In *traditional* SCNs as shown in Figure 1(a), if there are 70 uplink/downlink channels in the system, under the uniform FCA with a reuse factor of 7, each macrocell will be assigned 10 uplink/downlink channels. However, in our proposed CMCN architecture, as shown in Figure 1(b), each macrocell is split into one central microcell and six *virtual* microcells. Under the AFCA for CMCN, each (*virtual* or central) microcell is assigned N_U uplink and N_D downlink channels, whereas the FCA in SCNs allocates the same number of channels for the uplink and downlink. For inter-macrocell traffic of MCN (Hsu & Lin, 2002), as the BS in the central microcell is involved in all the calls including its own microcell and the surrounding six *virtual* microcells, more channels should be assigned to the central microcell in order to reduce the call blocking probability. If X_U uplink and X_D downlink channels are taken from each of the six surrounding *virtual* microcells, $6X_U$ uplink and $6X_D$ downlink channels can be added to the central microcell. The optimum uplink and downlink channel assignment to the system will be studied later.

3.1 Analytical models

To analyze the AFCA scheme in CMCN, we first study a hypothetical two-cell CMCN. Based on the methodology developed for the two-cell CMCN, we analyze the proposed AFCA scheme for the seven-cell CMCN. To make the problem tractable, we have assumed low mobility such that no handover or channel reassignment is required. Next, we have considered that it takes at most three hops for a MS to reach the BS. And the transmission range of a MS for each hop is the same.

In our analysis, we are considering a TDMA cellular system, in which the uplink and downlink channels are separated by half of the total allocated bandwidth. In traditional cellular systems, the uplink channel assignment and downlink channel assignment are symmetric. Thus, if the uplink is blocked, then the downlink is also blocked due to the symmetry between the uplink and the downlink channel assignment and symmetric uplink and downlink bandwidth allocation. Researchers usually study the channel assignment regardless of uplink and downlink in the literature. However, channel assignment in CMCN is different from that in *traditional* cellular systems that the uplink and downlink channel assignments are not symmetric. Specifically, for multihop calls, the downlink channel assignment requires more channels than the uplink channel assignment in the central microcells in CMCN. In addition, since uplink channel assignment and downlink channel assignment are independent in CMCN, the uplink blocking is independent of the downlink blocking. Hence, we are able to analyze the uplink and downlink performance separately. In fact, the uplink/downlink channel assignment does not imply the downlink/uplink channel available. On the contrary, a call is blocked if either uplink or downlink channel assignment is unsuccessful. Therefore, we have to consider the channel assignment for uplink and downlink transmission separately in order to study the distinct feature of CMCN. Otherwise, we are not able to find whether the call blocking is due to lack of uplink channels or it is due to lack of downlink channels.

3.2 Analytical models for uplink

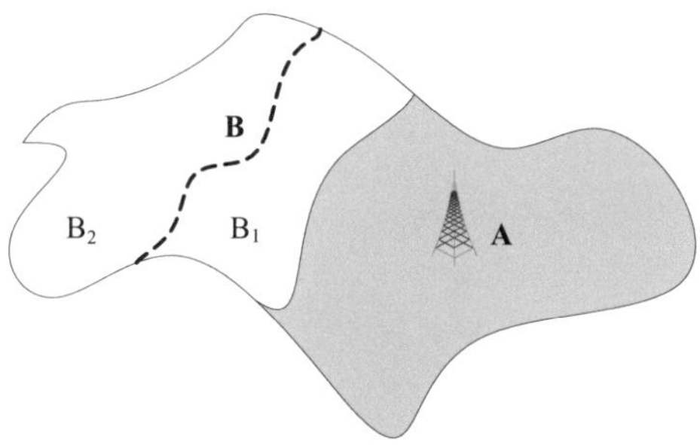

Fig. 3. A hypothetical two-cell CMCN.

We consider a hypothetical two-cell CMCN with the assumption that cell A and cell B have the same area. As shown in Figure 3, cell B is equally divided into two regions, B_1 and B_2. N_0 uplink channels and N_1 uplink channels are assigned to cell A and cell B, respectively. For uplink transmission, upon a call arrival in cell A, one channel will be assigned to the call, if any; otherwise, the call is blocked. This type of call is considered as one-hop call. Call arrivals in the sub-cell B_1 are considered as two-hop calls and call arrivals in sub-cell B_2 are considered as three-hop calls. For a two-hop call arrival, if there is at least one channel available in cell A and at least one channel available in cell B, then the call is accepted and two channels are assigned to the link, one from the cell A and one from the cell B; otherwise, the two-hop call is blocked. Similarly, when a three-hop call arrives, if there is at least one channel available in cell A and at least two channels available in cell B, the call is accepted; otherwise, the three-hop call is blocked.

The proposed AFCA scheme with a hypothetical two-cell CMCN for uplink transmission can be modeled by a three-dimensional Markov chain. For example, if N_0=4 and N_1=4, the Markov chain is shown in Figure 4. All the states form a set Π, which is given by

$$\Pi = \{(x,y,z) \mid 0 \le x \le N_0, 0 \le y \le N_1, 0 \le z \le \lfloor N_1/2 \rfloor, x+y+z \le N_0, y+2z \le N_1\} \quad (1)$$

where x, y and z stands for the number of one-hop, two-hop and three-hop calls, respectively. λ is the total call arrival rate. With uniform call arrivals, the call arrival rates are $\lambda_1 = \lambda/2$, $\lambda_2 = \lambda/4$ and $\lambda_3 = \lambda/4$ for one-hop, two-hop and three-hop calls, respectively. The corresponding service rates for three types of calls from state (x, y, z) are

$$\mu_1(x,y,z) = x\mu, \ \mu_2(x,y,z) = y\mu, \ \mu_3(x,y,z) = z\mu \quad (2)$$

where λ_i and $\mu_i(x, y, z)$ are the arrival rate and the service rate for i-hop calls, respectively. Thus, $\rho=\lambda/\mu$ gives the offered traffic per two-cell CMCN in Erlangs. Due to the limited space, we simply use μ_j to represent $\mu_j(x, y, z)$ in Figure 4, where j=1, 2 or 3.

By applying the global balance theory (Kleinrock, 1975), we can obtain a global-balance equation for each state in Π. For example, for state (1, 0, 1) in Figure 4, we have

$$\begin{aligned}&P(1,0,1)\times[\lambda_1+\lambda_2+\lambda_3+\mu_1(1,0,1)+0+\mu_3(1,0,1)]=\\&P(0,0,1)\times\lambda_1+0\times\lambda_2+P(1,0,0)\times\lambda_3+\\&P(2,0,1)\times\mu_1(2,0,1)+P(1,1,1)\times\mu_2(1,1,1)+P(1,0,2)\times\mu_3(1,0,2)\end{aligned} \tag{3}$$

where $P(x, y, z)$ is the probability of state (x, y, z). Then we have

$$\begin{aligned}&P(1,0,1)\times[\lambda_1+\lambda_2+\lambda_3+\mu_1(1,0,1)+0+\mu_3(1,0,1)]-\\&P(0,0,1)\times\lambda_1-0\times\lambda_2-P(1,0,0)\times\lambda_3-\\&P(2,0,1)\times\mu_1(2,0,1)-P(1,1,1)\times\mu_2(1,1,1)-P(1,0,2)\times\mu_3(1,0,2)=0\end{aligned} \tag{4}$$

The total state probability sums to unity, i.e.,

$$\sum_{(x,y,z)\in\Pi} P(x,y,z)=1 \tag{5}$$

For a set Π with m states, the array of global-balance equations can be transformed as

$$[A]_{m\times m}[P]_{m\times 1}=[B]_{m\times 1} \tag{6}$$

$$[A]=\begin{bmatrix}1 & 1 & \cdots & 1\\ t_{21} & t_{22} & \cdots & t_{2m}\\ \vdots & \vdots & \cdots & \vdots\\ \vdots & \vdots & t_{ii} & \vdots\\ \vdots & \vdots & \cdots & \vdots\\ t_{m1} & t_{m2} & \cdots & t_{mm}\end{bmatrix},\ [P]=\begin{bmatrix}P(x_1,y_1,z_1)\\ P(x_2,y_2,z_2)\\ \vdots\\ P(x_i,y_i,z_i)\\ \vdots\\ P(x_m,y_m,z_m)\end{bmatrix},\ [B]=\begin{bmatrix}1\\0\\ \vdots\\0\end{bmatrix} \tag{7}$$

Note that the states in Π are indexed, state i corresponds to (x_i, y_i, z_i). In matrix $[A]$, there are two types of transition rates:

- *Transition Rates into a State*: The element t_{ij} (when $i \neq j$) is the negative of the transition rate from state j to state i. For state j, we have

$$t_{ij}=\begin{cases}-\mu x_j,\ \text{if } x_j-x_i=1; & -\lambda_1,\ \text{if } x_j-x_i=-1;\\ -\mu y_j,\ \text{if } y_j-y_i=1; & -\lambda_2,\ \text{if } y_j-y_i=-1;\\ -\mu z_j,\ \text{if } z_j-z_i=1; & -\lambda_3,\ \text{if } z_j-z_i=-1;\\ 0,\ \text{otherwise.}\end{cases} \tag{8}$$

- *Transition Rates out of a State*: The element t_{ii} is equal to the sum of the transition rates out of the state i. For state i, we have

$$t_{ii}=-\sum_{j=1,j\neq i}^{m} t_{ji} \tag{9}$$

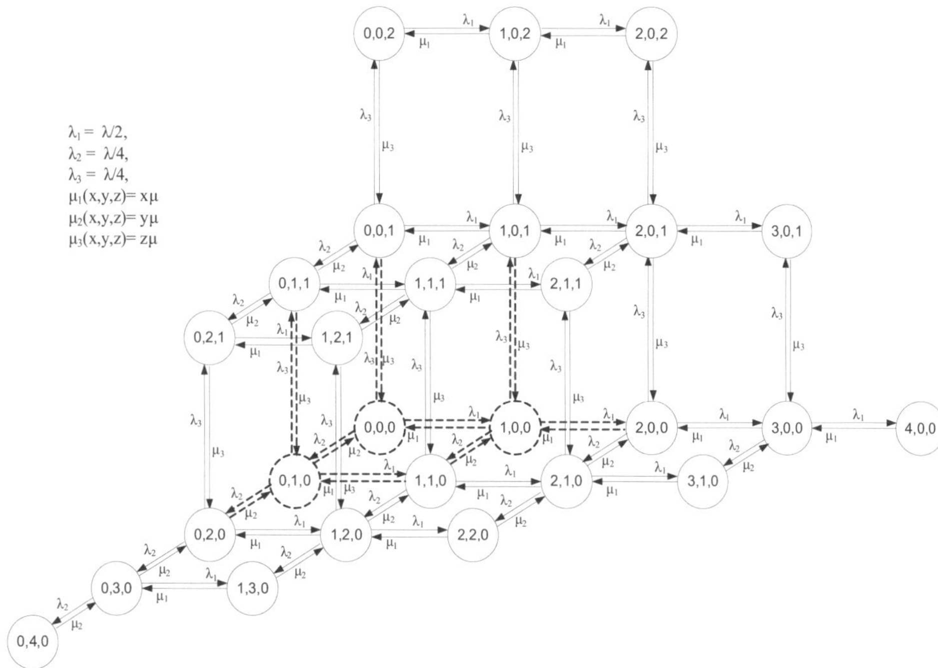

Fig. 4. Uplink exact model for the two-cell CMCN with N_0=4 and N_1=4.

By solving the matrix equation, we can get the matrix [P], which includes the probability for each state in Π. Denote P_{bi} as the average call blocking probability of ith-hop calls, which is defined as the ratio of the number of blocked i^{th}-hop calls and the total number of call arrivals for i^{th}-hop calls over the whole *virtual* macrocell area. Then, we have

$$P_{b1} = \sum_{x=0}^{N_0}\sum_{y=0}^{N_1}\sum_{z=0}^{\lfloor N_1/2 \rfloor} P(x,y,z)\big|_{(x,y,z)\in\Pi, x+y+z=N_0} \tag{10}$$

$$P_{b2} = P_{b1} + \sum_{x=0}^{N_0}\sum_{y=0}^{N_1}\sum_{z=0}^{\lfloor N_1/2 \rfloor} P(x,y,z)\big|_{(x,y,z)\in\Pi,\ y+2z=N_1} + \sum_{x=0}^{N_0}\sum_{y=0}^{N_1}\sum_{z=0}^{\lfloor N_1/2 \rfloor} P(x,y,z)\big|_{(x,y,z)\in\Pi,\ x+y+z=N_0,\ \text{and } y+2z=N_1} \tag{11}$$

$$P_{b3} = P_{b1} + \sum_{x=0}^{N_0}\sum_{y=0}^{N_1}\sum_{z=0}^{\lfloor N_1/2 \rfloor} P(x,y,z)\big|_{(x,y,z)\in\Pi,\ y+2z\geq N_1-1} - \sum_{x=0}^{N_0}\sum_{y=0}^{N_1}\sum_{z=0}^{\lfloor N_1/2 \rfloor} P(x,y,z)\big|_{(x,y,z)\in\Pi,\ x+y+z=N_0,\ \text{and } y+2z\geq N_1-1} \tag{12}$$

Finally, the average uplink blocking probability of the system is given by

$$P_{b,U} = P_{b1} \times \lambda_1/\lambda + P_{b2} \times \lambda_2/\lambda + P_{b3} \times \lambda_3/\lambda\ . \tag{13}$$

The seven-cell CMCN model shown in Figure 1(b) is used to analyze the performance of the proposed AFCA scheme. For simplicity, we assume a reuse factor of N_r=7 in the analysis. The methodology can be applied for other N_r values. Thus, each macrocell area is divided into N_r microcells. The channel assignment procedures will be different and the multi-dimensional Markov chain needs to be reconstructed.

Three types of new calls are originated from the system. They are calls originated from the central microcell, inner-half and outer-half *virtual* microcell as one-hop, two-hop and three-hop calls, respectively. With uniform call arrivals, we have $\lambda_1 = \lambda/7$, $\lambda_2 = 3\lambda/7$, $\lambda_3 = 3\lambda/7$; where λ_1, λ_2 and λ_3 are the call arrival rates for one-hop, two-hop and three-hop calls, respectively. And λ is the call arrival rate to a macrocell as shown in Figure 1(b). Hence the offered traffic load is $\rho_1 = \lambda_1/\mu$, $\rho_2 = \lambda_2/\mu$ and $\rho_3 = \lambda_3/\mu$ for one-hop, two-hop and three-hop calls, respectively. $1/\mu$ is average call duration. The offered traffic load per macrocell is given by $\rho = \lambda/\mu$.

For the seven-cell CMCN, the AFCA scheme can be modeled using a 13-dimensional Markov chain. The central microcell is assigned with N_0 channels and each of the six surrounding *virtual* microcells is assigned with N_1 channels. As shown in Figure 1(b), the seven microcell is numbered with 0, 1, 2, …, 6, starting from the central microcell. A state is defined as $(n_0^1, n_1^2, n_1^3, n_2^2, n_2^3, n_3^2, n_3^3, n_4^2, n_4^3, n_5^2, n_5^3, n_6^2, n_6^3)$. To make the analysis convenient, each state is numbered with an integer index s, and all the states form a set Π. For example, a state s corresponds to a distinct sequence of nonnegative integers $(n_0^1(s), n_1^2(s), n_1^3(s), n_2^2(s), n_2^3(s), n_3^2(s), n_3^3(s), n_4^2(s), n_4^3(s), n_5^2(s), n_5^3(s), n_6^2(s), n_6^3(s))$ where $n_i^j(s)$ is number of j-hop calls in microcell i at state s. The permissible states must satisfy the following constraints:

$$\begin{cases} n_0^1(s) + n_1^2(s) + n_1^3(s) + n_2^2(s) + n_2^3(s) + n_3^2(s) + n_3^3(s) \\ + n_4^2(s) + n_4^3(s) + n_5^2(s) + n_5^3(s) + n_6^2(s) + n_6^3(s) \le N_0 \\ 0 \le n_0^1(s) \le N_0; \quad 0 \le n_i^2(s) + 2n_i^3(s) \le N_1, \text{ for } i = 1, 2, \cdots, 6 \\ 0 \le n_i^2(s) \le N_1; \quad 0 \le n_i^3(s) \le \lfloor N_1/2 \rfloor, \text{ for } i = 1, 2, \cdots, 6 \end{cases} \quad (14)$$

A state transition can be caused by one of the following six events: an arrival of one-hop, two-hop or three-hop call; a departure of one-hop, two-hop or three-hop call. There are maximally 26 possible transitions for a state in the Markov chain model for seven-cell CMCNs. By comparing the state number of two states, we can obtain the transition rates. Applying the global-balance theory (Kleinrock, 1975), we can obtain one equation for each state. For example, as shown in Figure 5, for a state s_1 we may have,

$$\begin{aligned} &P(s_1)[\lambda_1 + \lambda_2/6 + \lambda_3/6 + \mu_1(s_1) + \mu_2(s_1) + \mu_3(s_1)] = P(s_2)\lambda_1 \\ &+ P(s_3)\lambda_2/6 + P(s_4)\lambda_3/6 + P(s_5)\mu_1(s_5) + P(s_6)\mu_2(s_6) + P(s_7)\mu_3(s_7) \end{aligned} \quad (15)$$

where $P(s_i)$ is the steady-state probability of state s_i, and $\mu_j(s_i)$ is the service rate for j-hop calls at state s_i. Similarly, the total state probability sums to unity,

$$\sum_{s \in \Pi} P(s) = 1 \quad (16)$$

Again for a state set Π with m states, the array of global-balance equations form

$$[A]_{m\times m}[P]_{m\times 1} = [B]_{m\times 1} \quad (17)$$

$$[A]=\begin{bmatrix} 1 & 1 & \cdots & 1 \\ t_{21} & t_{22} & \cdots & t_{2m} \\ \vdots & \vdots & \cdots & \vdots \\ \vdots & \vdots & t_{ii} & \vdots \\ \vdots & \vdots & \cdots & \vdots \\ t_{m1} & t_{m2} & \cdots & t_{mm} \end{bmatrix},\ [P]=\begin{bmatrix} P(1) \\ P(2) \\ \vdots \\ P(i) \\ \vdots \\ P(m) \end{bmatrix},\ [B]=\begin{bmatrix} 1 \\ 0 \\ \vdots \\ 0 \end{bmatrix} \tag{18}$$

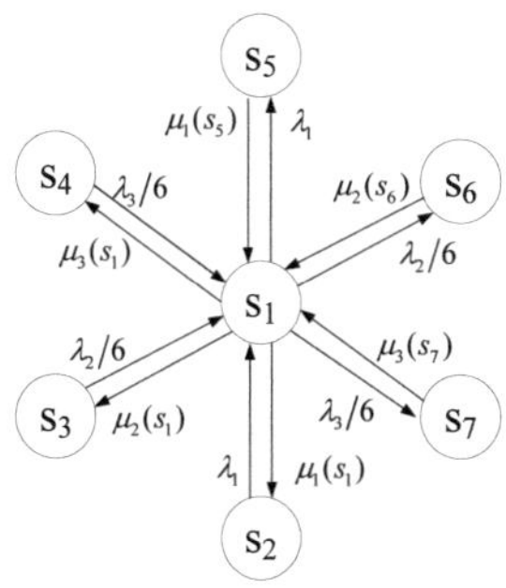

Fig. 5. Possible transitions for a state s_1.

In matrix $[A]$, there are two types of transition rates:

Transition Rates into a State: The element t_{ij} (when $i \neq j$) is the negative of the transition rate from state j to state i. The service rates for the three types of calls in microcell (central or *virtual*) k from state j are respectively $\mu_1(j)=n_0^1(j)\mu,\ \mu_2(j)=n_k^2(j)\mu,\ \mu_3(j)=n_k^3(j)\mu$ where k=1, 2, …, 6. Therefore,

$$t_{ij}=\begin{cases} -n_0^1(j)\mu, \text{if } n_0^1(j)-n_0^1(i)=1; & -\lambda_1, \text{if } n_0^1(j)-n_0^1(i)=-1; \\ -n_k^2(j)\mu, \text{if } n_k^2(j)-n_k^2(i)=1; & -\lambda_2/6, \text{if } n_k^2(j)-n_k^2(i)=-1; \\ -n_k^3(j)\mu, \text{if } n_k^3(j)-n_k^3(i)=1; & -\lambda_3/6, \text{if } n_k^3(j)-n_k^3(i)=-1; \\ 0, \text{otherwise.} \end{cases} \tag{19}$$

Transition Rates out of a State: The element t_{ii} is equal to the sum of the transition rates out of state i. For state i, we have,

$$t_{ii}=-\sum_{j=1, j\neq i}^{m} t_{ji} \tag{20}$$

By solving the matrix equation, we can get the matrix $[P]$. Then, the blocking probabilities, P_{b1}, P_{b2} and P_{b3} for one-hop call, two-hop call and three-hop call in each cell is given by, respectively,

$$P_{b1}=\sum_{s\in\Pi} P(s)\Big|_{n_0^1(s)+\sum_{i=1}^{6}[n_i^2(s)+n_i^3(s)]=N_0} \tag{21}$$

$$P_{b2}=P_{b1}+\frac{1}{6}\sum_{i=1}^{6}\sum_{s\in\Pi} P(s)\Big|_{n_i^2(s)+2n_i^3(s)=N_1} - \frac{1}{6}\sum_{i=1}^{6}\sum_{s\in\Pi} P(s)\Big|_{n_i^2(s)+2n_i^3(s)=N_1 \text{ and } n_0^1(s)+\sum_{k=1}^{6}\left[n_k^2(s)+n_k^3(s)\right]=N_0} \tag{22}$$

$$P_{b3} = P_{b1} + \frac{1}{6}\sum_{i=1}^{6}\sum_{s\in\Pi} P(s)\Big|_{n_i^2(s)+2n_i^3(s)\geq N_1-1} - \frac{1}{6}\sum_{i=1}^{6}\sum_{s\in\Pi} P(s)\Big|_{n_i^2(s)+2n_i^3(s)\geq N_1-1 \text{ and } n_0^1(s)+\sum_{k=1}^{6}\left[n_k^2(s)+n_k^3(s)\right]=N_0} \tag{23}$$

Then, the average call blocking probability, $P_{b,U}$, is given by (13).

The exact model may result in excessive long computational time, especially for large N_0 and N_1. To reduce the computational load, we find an approximated model with much fewer states to analyze the AFCA scheme and the detailed analysis is available in (Li & Chong, 2010).

3.2.1 Analytical models for downlink

To analyze the AFCA scheme in CMCN with downlink transmission, again we first study the hypothetical two-cell CMCN model. After that, we analyze the proposed AFCA scheme for seven-cell CMCN model based on the methodology developed for the hypothetical two-cell CMCN model.

We consider the two-cell CMCN as shown in Figure 3. The BS is located in cell *A*, which has the same area with cell *B*, the *virtual* microcell. Cell *B* is equally divided into two regions, B_1 and B_2, and B_1 is next to cell *A*. N_0 downlink channels and N_1 downlink channels are assigned to cell *A* and cell *B*, respectively.

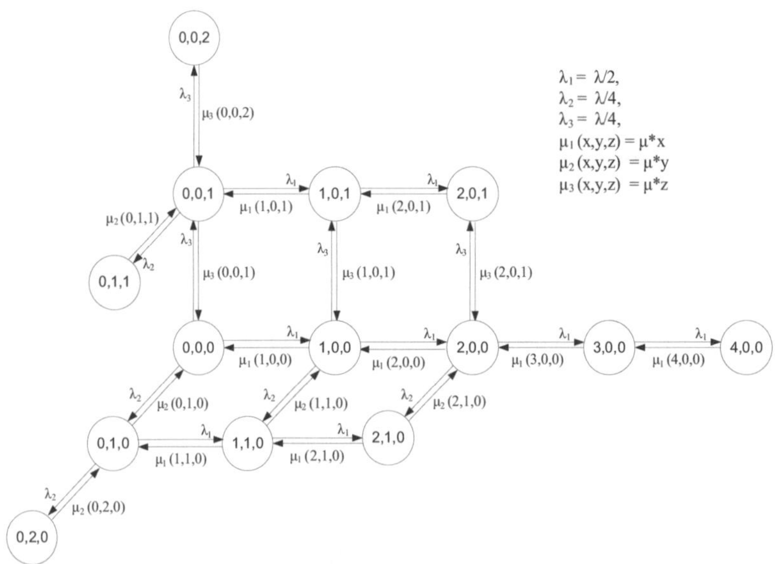

Fig. 6. Downlink exact model for the two-cell CMCN with N_0=4 and N_1=2.

For downlink transmission, a call arrival in cell *A*, considered as a one-hop call, will require one downlink channel, if any. Otherwise, the call is blocked. Call arrivals in the region B_1 are considered as two-hop calls, each of which takes two downlink channels from cell *A*. Call arrivals in region B_2 are considered as three-hop calls, each of which takes two downlink channels from cell *A* and one downlink channel from cell *B*. For a new two-hop call, if there are at least two downlink channels available in cell *A*, then the call is accepted; otherwise, the call is blocked. Similarly, when a three-hop call arrives, if there are at least two downlink channels available in cell *A* and at least one downlink channels available in cell *B*, the call is accepted. Otherwise, the call is blocked.

For downlink transmission, the AFCA scheme with the hypothetical two-cell CMCN can be modeled by a three-dimensional Markov chain. For example, if N_0=4 and N_1=2, the Markov chain can be drawn as Figure 6. All the states form a set Π,

$$\Pi = \{(x,y,z) \mid 0 \le x \le N_0, 0 \le y \le \lfloor N_0/2 \rfloor, 0 \le z \le N_1, x+2y+2z \le N_0\} \tag{24}$$

where x, y, z stands for the number of one-hop, two-hop and three-hop calls, respectively. With uniform call arrivals, the call arrival rates are $\lambda_1 = \lambda/2$, $\lambda_2 = \lambda/4$ and $\lambda_3 = \lambda/4$ for one-hop, two-hop and three-hop calls, respectively. λ is total call arrival rate to the two-cell CMCN. The corresponding service rates for three types of calls from state (x, y, z) are

$$\mu_1(x,y,z) = x\mu,\ \mu_2(x,y,z) = y\mu,\ \mu_3(x,y,z) = z\mu \tag{25}$$

where λ_i and $\mu_i(x, y, z)$ are the arrival rate and the service rate for i-hop calls, respectively. $1/\mu$ is the average call duration and $\rho = \lambda/\mu$ gives the offered traffic per two-cell CMCN in Erlangs.

Next, we apply the global balance theory (Kleinrock, 1975) and obtain a global-balance equation for each state in Π. For example, for state (1,0,0) in Figure 6, we have

$$\begin{aligned} &P(1,0,0)\times[\lambda_1+\lambda_2+\lambda_3+\mu_1(1,0,0)+0+0] = \\ &P(0,0,0)\times\lambda_1 + 0\times\lambda_2 + 0\times\lambda_3 + P(2,0,0)\times\mu_1(2,0,0) + P(1,1,0)\times\mu_2(1,1,0) + P(1,0,1)\times\mu_3(1,0,1) \end{aligned} \tag{26}$$

where $P(x, y, z)$ is the probability of state (x, y, z). Then we have

$$\begin{aligned} &P(1,0,0)\times[\lambda_1+\lambda_2+\lambda_3+\mu_1(1,0,0)+0+0] \\ &-P(0,0,0)\times\lambda_1 - 0\times\lambda_2 - 0\times\lambda_3 - P(2,0,0)\times\mu_1(2,0,0) - P(1,1,0)\times\mu_2(1,1,0) - P(1,0,1)\times\mu_3(1,0,1) = 0 \end{aligned} \tag{27}$$

The total state probability sums to unity, i.e.,

$$\sum_{(x,y,z)\in\Pi} P(x,y,z) = 1 \tag{28}$$

For a state set Π having m states, the array of global-balance equations form

$$[A]_{m\times m}[P]_{m\times 1} = [B]_{m\times 1} \tag{29}$$

$$[A] = \begin{bmatrix} 1 & 1 & \cdots & 1 \\ t_{21} & t_{22} & \cdots & t_{2m} \\ \vdots & \vdots & \cdots & \vdots \\ \vdots & \vdots & t_{ii} & \vdots \\ \vdots & \vdots & \cdots & \vdots \\ t_{m1} & t_{m2} & \cdots & t_{mm} \end{bmatrix},\ [P] = \begin{bmatrix} P(x_1,y_1,z_1) \\ P(x_2,y_2,z_2) \\ \vdots \\ P(x_i,y_i,z_i) \\ \vdots \\ P(x_m,y_m,z_m) \end{bmatrix},\ [B] = \begin{bmatrix} 1 \\ 0 \\ \vdots \\ 0 \end{bmatrix} \tag{30}$$

Note that the states in Π are indexed, state s corresponds to (x_s, y_s, z_s). In matrix $[A]$, there are two types of transition rates:

Transition Rates into a State: The element t_{ij} (when $i \ne j$) is the negative of the transition rate from state j to state i. For state j, we have

$$t_{ij} = \begin{cases} -\mu x_j, \text{ if } x_j - x_i = 1; & -\lambda_1, \text{ if } x_j - x_i = -1; \\ -\mu y_j, \text{ if } y_j - y_i = 1; & -\lambda_2, \text{ if } y_j - y_i = -1; \\ -\mu z_j, \text{ if } z_j - z_i = 1; & -\lambda_3, \text{ if } z_j - z_i = -1; \\ 0, \text{ otherwise.} \end{cases} \tag{31}$$

Transition Rates out of a State: The element t_{ii} is equal to the sum of the transition rates out of the state i. For state i, we may have

$$t_{ii} = -\sum_{j=1, j\neq i}^{m} t_{ji} \tag{32}$$

By solving the matrix equation, we can get the matrix $[P]$, which includes the probability for each state in Π. Then, the blocking probability for one-hop, two-hop and three-hop calls can be calculated as follows:

$$P_{b1} = \sum_{x=0}^{N_0}\sum_{y=0}^{\lfloor N_0/2\rfloor}\sum_{z=0}^{N_1} P(x,y,z)\big|_{(x,y,z)\in\Pi,\, x+2y+2z=N_0} \tag{33}$$

$$P_{b2} = \sum_{x=0}^{N_0}\sum_{y=0}^{\lfloor N_0/2\rfloor}\sum_{z=0}^{N_1} P(x,y,z)\big|_{(x,y,z)\in\Pi,\, x+2y+2z\geq N_0-1} \tag{34}$$

$$P_{b3} = P_{b2} + \sum_{x=0}^{N_0}\sum_{y=0}^{\lfloor N_0/2\rfloor}\sum_{z=0}^{N_1} P(x,y,z)\big|_{(x,y,z)\in\Pi,\, z=N_1} - \sum_{x=0}^{N_0}\sum_{y=0}^{\lfloor N_0/2\rfloor}\sum_{z=0}^{N_1} P(x,y,z)\big|_{(x,y,z)\in\Pi,\, x+2y+2z\geq N_0-1,\text{ and } z=N_1} \tag{35}$$

Finally, the average blocking probability, $P_{b,D}$, of the system for downlink is given by

$$P_{b,D} = P_{b1}\times\lambda_1/\lambda + P_{b2}\times\lambda_2/\lambda + P_{b3}\times\lambda_3/\lambda \tag{36}$$

The seven-cell CMCN model shown in Figure 1(b) is used to analyze the performance of the proposed AFCA scheme. For simplicity, again N_r=7 is used in the analysis. The methodology can be applied for other N_r. Thus, each macrocell area is divided into N_r microcells. The channel assignment procedures will be different and the multi-dimensional Markov chain needs to be reconstructed.

Three types of new calls are originated from the system. They are calls originated from the central, inner-half and outer-half microcell as one-hop, two-hop and three-hop calls, respectively. With uniform call arrivals, $\lambda_1 = \lambda/7$, $\lambda_2 = 3\lambda/7$, $\lambda_3 = 3\lambda/7$; where λ_1, λ_2 and λ_3 are the call arrival rates for one-hop, two-hop and three-hop calls, respectively. And λ is the call arrival rate to a macrocell area. Hence the offered traffic are $\rho_1 = \lambda_1/\mu$, $\rho_2 = \lambda_2/\mu$ and $\rho_3 = \lambda_3/\mu$ for one-hop, two-hop and three-hop calls, respectively. $1/\mu$ is the average call duration. The offered traffic load per macrocell is given by $\rho = \lambda/\mu$ Erlangs.

For the seven-cell CMCN, the proposed FCA scheme can be modeled using a 13-dimensional Markov chain. The central microcell is assigned with N_0 channels and each of the six *virtual* microcells is assigned with N_1 channels. As shown in Figure 1(b), the seven microcells are numbered with 0, 1, 2, …, 6, starting from the central microcell. A state is defined as $(n_0^1, n_1^2, n_1^3, n_2^2, n_2^3, n_3^2, n_3^3, n_4^2, n_4^3, n_5^2, n_5^3, n_6^2, n_6^3)$. To make the analysis convenient, each state is numbered with an integer index s, and all the states form a set Π. For example, state s corresponds to a distinct sequence of nonnegative integers $(n_0^1(s), n_1^2(s), n_1^3(s), n_2^2(s), n_2^3(s),$ $n_3^2(s), n_3^3(s), n_4^2(s), n_4^3(s), n_5^2(s), n_5^3(s), n_6^2(s), n_6^3(s))$ where $n_i^j(s)$ is number of j-hop calls in microcell i at state s. The permissible states must satisfy the following constraints:

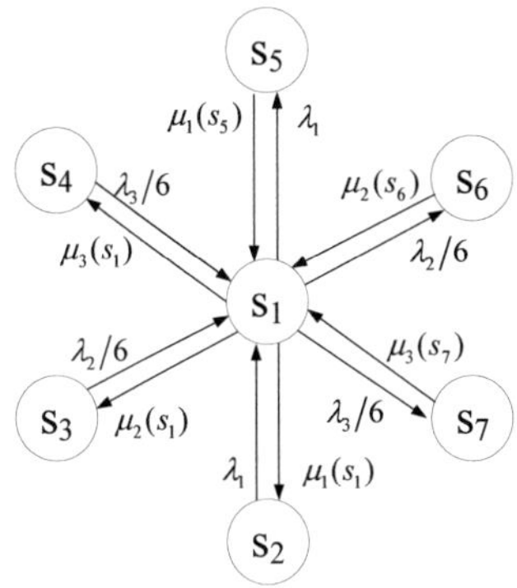

Fig. 7. Possible transitions for a state s_1.

$$\begin{cases} n_0^1(s)+2\times[n_1^2(s)+n_1^3(s)+n_2^2(s)+n_2^3(s)+n_3^2(s)+n_3^3(s) \\ +n_4^2(s)+n_4^3(s)+n_5^2(s)+n_5^3(s)+n_6^2(s)+n_6^3(s)]\leq N_0 \\ 0\leq n_0^1(s)\leq N_0;\ \ 0\leq 2n_i^2(s)\leq N_0,\ 0\leq n_i^3(s)\leq N_1,\ \text{for } i=1,2,\cdots,6 \end{cases} \tag{37}$$

A state transition can be caused by one of the following six events: an arrival/departure of a one-hop, two-hop or three-hop call. Applying the global-balance theory (Kleinrock, 1975), we can obtain one equation for each state. For example, for a state s_1 in Figure 7, we have,

$$\begin{aligned} &P(s_1)[\lambda_1+\lambda_2/6+\lambda_3/6+\mu_1(s_1)+\mu_2(s_1)+\mu_3(s_1)]=P(s_2)\lambda_1 \\ &+P(s_3)\lambda_2/6+P(s_4)\lambda_3/6+P(s_5)\mu_1(s_5)+P(s_6)\mu_2(s_6)+P(s_7)\mu_3(s_7) \end{aligned} \tag{38}$$

where $P(s)$ is the steady-state probability of state s, and $\mu_j(s)$ is the service rate for j-hop calls at state s. Similarly, the total state probability sums to unity,

$$\sum_{s\in\Pi}P(s)=1 \tag{39}$$

For a state set Π with m states, the array of global-balance equations form

$$[A]_{m\times m}[P]_{m\times 1}=[B]_{m\times 1} \tag{40}$$

$$[A]=\begin{bmatrix} 1 & 1 & \cdots & 1 \\ t_{21} & t_{22} & \cdots & t_{2m} \\ \vdots & \vdots & \cdots & \vdots \\ \vdots & \vdots & t_{ii} & \vdots \\ \vdots & \vdots & \cdots & \vdots \\ t_{m1} & t_{m2} & \cdots & t_{mm} \end{bmatrix},\ [P]=\begin{bmatrix} P(1) \\ P(2) \\ \vdots \\ P(i) \\ \vdots \\ P(m) \end{bmatrix},\ [B]=\begin{bmatrix} 1 \\ 0 \\ \vdots \\ 0 \end{bmatrix} \tag{41}$$

In matrix $[A]$, there are two types of transition rates:
Transition Rates into a State: The element t_{ij} (when $i \neq j$) is the negative of the transition rate from state j to state i. The service rates for the three types of calls in microcell k from state j are $\mu_1(j)=n_0^1(j)\mu,\ \mu_2(j)=n_k^2(j)\mu,\ \mu_3(j)=n_k^3(j)\mu$, where k=1, 2, …, 6. Then we have,

$$t_{ij}=\begin{cases}-n_0^1(j)\mu, \text{if } n_0^1(j)-n_0^1(i)=1; & -\lambda_1, \text{if } n_0^1(j)-n_0^1(i)=-1;\\ -n_k^2(j)\mu, \text{if } n_k^2(j)-n_k^2(i)=1; & -\lambda_2/6, \text{if } n_k^2(j)-n_k^2(i)=-1;\\ -n_k^3(j)\mu, \text{if } n_k^3(j)-n_k^3(i)=1; & -\lambda_3/6, \text{if } n_k^3(j)-n_k^3(i)=-1;\\ 0, \text{otherwise.} \end{cases} \tag{42}$$

Transition Rates out of a State: The element t_{ii} is equal to the sum of the transition rates out of state i. For state i, we have,

$$t_{ii}=-\sum_{j=1,j\neq i}^{m} t_{ji} \tag{43}$$

By solving the matrix equation, we can get the matrix $[P]$. Then, the blocking probabilities, P_{b1}, P_{b2} and P_{b3} for one-hop call, two-hop call and three-hop call in each cell are given by,

$$P_{b1}=\sum_{s\in\Pi} P(s)\Big|_{n_0^1(s)+2\times\sum_{i=1}^{6}[n_i^2(s)+n_i^3(s)]=N_0} \tag{44}$$

$$P_{b2}=\sum_{s\in\Pi} P(s)\Big|_{n_0^1(s)+2\times\sum_{i=1}^{6}[n_i^2(s)+n_i^3(s)]\geq N_0-1} \tag{45}$$

$$P_{b3}=P_{b2}+\frac{1}{6}\sum_{i=1}^{6}\sum_{s\in\Pi} P(s)\Big|_{n_i^3(s)=N_1}-\frac{1}{6}\sum_{i=1}^{6}\sum_{s\in\Pi} P(s)\Big|_{n_i^3(s)=N_1 \text{ and } n_0^1(s)+2\sum_{k=1}^{6}\left[n_k^2(s)+n_k^3(s)\right]\geq N_0-1} \tag{46}$$

The average call blocking probability, $P_{b,D}$, for downlink is given in (36).

The exact model may result in excessive long computational time, especially for large N_0 and N_1. To reduce the computational load, we find an approximated model with much fewer states to analyze the AFCA scheme and the detailed analysis is available in (Li & Chong, 2008).

3.2 Numerical results and discussions

First of all, we look at the validity of our analytical models for the proposed AFCA scheme. For the hypothetical two-cell CMCN, we do not assume any value of N. We have simply used any (N_0, N_1) channel combinations for a two-cell CMCN just for the purpose of validating the analytical model. These channel combinations (N_0, N_1) have no physical meaning for a two-cell CMCN. As shown in Figure 8, the simulation results match with the results obtained from the analytical model.

Then, we study the performance of the seven-cell CMCN. In this simulation study, we set the total number of system channels to be N = 70. For the seven-cell CMCN, we first look at the performance of uniform FCA, where each (central or *virtual*) microcell has equal number of channels with ($N_{U,c}$=10, $N_{U,v}$=10). The simulated system consists of one macrocell which is divided into seven microcells as shown in Figure 1(b). Calls arrive according to a Poisson process with a call arrival rate λ per macrocell. Call durations are exponentially distributed with a mean of $1/\mu$. Each simulation runs until 10 million calls are processed. The 95% confidence intervals are within ±10% of the average values shown.

Figure 9 shows the $P_{b,U}$ for seven CMCN with different (N_0, N_1) combinations. The performance of CMCN with (10, 10) has the same $P_{b,U}$ compared to SCNs, where a cell has

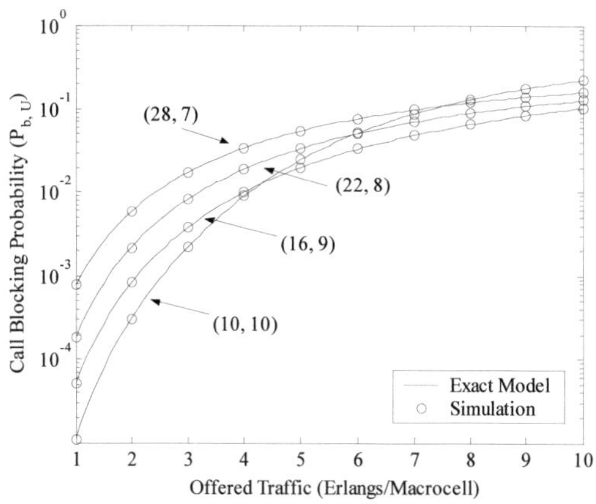

Fig. 8. Uplink performance of the hypothetical two-cell CMCN model.

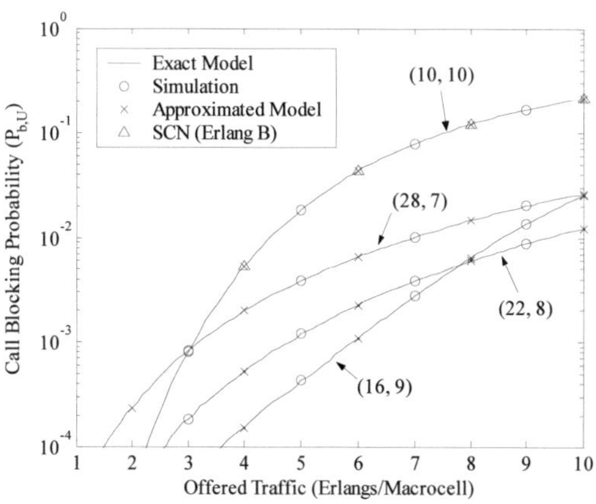

Fig. 9. Uplink performance of the AFCA for the seven-cell CMCN.

the same coverage area as the macrocell in CMCN, with 10 channels. This is expected because in CMCNs, every multihop call in the six surrounding *virtual* microcells will require one channel from the central microcell to access to the BS. In other words, the central microcell needs to support all the traffic from all seven microcells, which is equivalent to the traffic in a macrocell. Since N_1 is large enough in the *virtual* microcells, calls are rarely blocked due to insufficient channels in the *virtual* microcells. Thus, the bottleneck of the performance results with N_0 in the central microcell, which is used to support all the traffic in the seven microcells.

Next, we study the performance of CMCNs by increasing N_0 in the central microcell and reducing the N_1 in the six *virtual* microcells. Figure 9 shows the results of different channel combinations: (16, 9), (22, 8), (28, 7). From the results, it can be seen that our proposed CMCN with the AFCA scheme is able to reduce the $P_{b,U}$ significantly as compared to the SCNs. If $P_{b,U}$ is set at 1%, the AFCA scheme can support about 2 times more traffic. For example, the CMCN with (22, 8) supports 9.6 Erlangs; while a cell in SCNs only supports 4.5 Erlangs. Furthermore, the results obtained from the approximated model can provide a good estimation for the exact model. This again validates the correctness of both analytical models. Also, the simulation results match closely with the analytical results.

Table 1 shows the performance of the proposed CMCN with AFCA in different microcells under two channel combinations (10, 10) and (16, 9) at ρ=5 Erlangs. The number of call arrivals in different microcells is nearly the same because of uniform call arrivals. Performance in the six *virtual* microcells is statistically the same as they are homogenous. With (10, 10), the call blocking probability in microcell i, $P_{B,i}$, is the same for all (central or *virtual*) microcells because the call blocked is due to insufficient channel in the central microcell. The call blocking probability, $P_{B,0}$ in central microcell 0 is about 0.0182 under (10, 10), while it is reduced to 5.39×10^{-5} under (16, 9). This is because the increase in N_0 can reduce the call probability in the central microcell greatly. For each of the six *virtual* microcells, with N_0 increasing from 10 to 16 and N_1 reducing from 10 to 9, $P_{B,i}$, i=1, 2, ..., 6, also reduces sharply from 0.0182 to 5×10^{-4}. Since a multihop call not only occupies the channels in a *virtual* microcell, but also requires one channel from the central microcell, a multihop call can be blocked either by lack of channels in the central microcell or in a *virtual* microcell. More N_0 can provide connections to the calls originated from the *virtual* microcells so that $P_{B,i}$ of the *virtual* microcells is reduced.

scenario	Cell No.	No. of call arrival	No. of blocked calls	$P_{B,i}$	blocked calls due to insufficient channels in its own microcell	$P_{B\text{-}own,i}$
(10, 10)	0	1427909	25930	0.018159	25930	1.0
	1	1427448	25818	0.018087	77	0.002982
	2	1430117	26406	0.018464	61	0.00231
	3	1428187	26161	0.018318	44	0.001682
	4	1427711	25985	0.0182	60	0.002309
	5	1428577	26211	0.018348	51	0.001946
	6	1430051	26130	0.018272	58	0.00222
(16, 9)	0	1428030	77	5.39e-5	77	1.0
	1	1427287	721	5.05e-4	663	0.919556
	2	1430237	736	5.15e-4	658	0.894022
	3	1428246	647	4.53e-4	580	0.896445
	4	1427483	668	4.68e-4	614	0.919162
	5	1428799	693	4.85e-4	633	0.91342
	6	1429918	733	5.13e-4	664	0.905866

Table 1. Call blocking with different (N_0, N_1) combinations at ρ=5 Erlangs.

Furthermore, we define another parameter, $P_{B\text{-}own,i}$, which denotes the percentage that a blocked call is due to the lack of channels in microcell i. For $P_{B\text{-}own,0}$, all the calls blocked in the central microcell are due to lack of free channels in the central microcell. Thus, $P_{B\text{-}own,0}$ is 1. Table 1 shows that under (10, 10), multihop calls are rarely blocked due to lack of channels in their own *virtual* microcells because of sufficient channels in the *virtual* microcells; while under (16, 9) most of the multihop blocked call is due to lack of channels in their own *virtual* microcells.

The call blocking probabilities, P_{b1}, P_{b2} and P_{b3}, with different channel combinations for different ρ are shown in Figure 10. We notice that the P_{b1}, P_{b2} and P_{b3} are about the same under (10, 10), which is due to the fact that all types of calls are blocked for the same reason—inadequate channels are allocated to the central microcell. As we increase N_0 by 6 and slightly reduce N_1 by 1, the P_{b2} and P_{b3} are reduced because more channels in central microcell provide more channels for setting up connections for the two-hop and three-hop calls. P_{b3} is higher than P_{b2} because three-hop calls require one more channel than two-hop calls from its *virtual* microcell. For channel combination of (28, 7), as there are adequate channels available in the central microcell, P_{b1} is close to zero. P_{b2} and P_{b3} for (28, 7) are higher than those for (16, 9) because there are not enough channels in the *virtual* microcells to support multihop calls. Thus, N_1 will limit the QoS of the multihop calls. Beyond the optimum combination, if we further reduce N_1 and increase N_0, the performance will be degraded because more calls will be blocked in the *virtual* microcells.

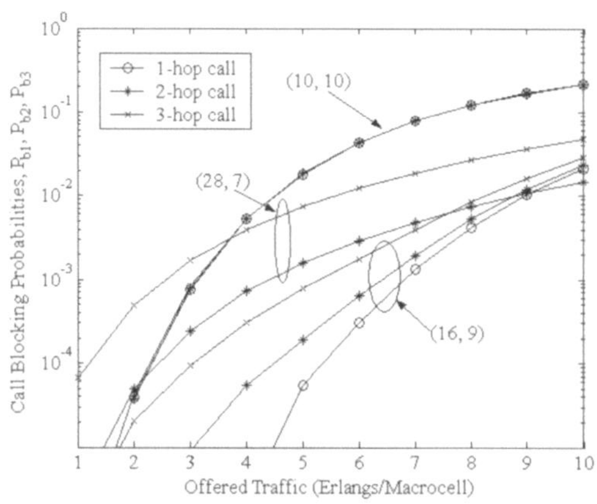

Fig. 10. Uplink call blocking probability for different types of calls.

For the hypothetical two-cell CMCN, analytical results match closely with the simulation results, which are not shown here because they are of less importance as compared to that of the seven-cell CMCN.

Figure 11 shows the downlink call blocking probability, $P_{b,D}$, for the seven-cell CMCN with different (N_0, N_1) for downlink transmission. We study the performance of CMCN by increasing N_0 in the central microcell and reducing the N_1 in the six *virtual* microcells. For the CMCNs, more channels are required from the central microcell for downlink transmission. Therefore, we can reduce the $P_{b,D}$ by increasing N_0. With N=70 and N_r=7 for uniform FCA in SCNs, each macrocell has 10 channels. From the Erlang B formula, the amount of traffic it can support at $P_{b,D}$ = 1% is about 4.5 Erlangs. For our CMCN, after dividing the macrocell into one central microcell and six *virtual* microcells, with channel combination of (34, 6), it can support about 10.5 Erlangs and the improvement factor is larger than 130%. We find that the optimum combination is (46, 4) and the capacity at $P_{b,D}$ = 1% is 14.2 Erlangs. If N_0 continues to increase by reducing N_1 after (46, 4), the capacity begins to decrease such as (52, 3) because there is not enough channels in surrounding *virtual* microcells to handle the traffic. Furthermore, the results obtained from the approximated model can provide a good estimation for the exact model. This again validates the correctness of both analytical models. Also, the simulation results match closely with the analytical results.

The capacity for CMCN, shown in Figure 11 is entirely based on the $P_{b,D}$. Since uplink and downlink channel assignment is different in CMCN, the system capacity will be different depending on (N_0, N_1) for uplink or downlink. We define the system call blocking probability, $P_{b,s}$, as the probability of lacking of either uplink or downlink channels for a new call. As uplink channel assignment and downlink channel assignment are independent, $P_{b,s}$ is mathematically equal to the larger value between $P_{b,U}$ and $P_{b,D}$. Figure 11 shows $P_{b,s}$ for uplink channel combination, UL($N_{U,c}$=10, $N_{U,v}$=10), with different downlink channel combination, DL($N_{D,c}$=N_0, $N_{D,v}$=N_1). For UL(10, 10) and DL(10, 10), it has the highest $P_{b,s}$. As we increase N_0 in the downlink channel combination, the system capacity at $P_{b,s}$ = 1% is increased. This is because $P_{b,D}$ is higher than uplink blocking probability, $P_{b,U}$, and thus, the system capacity is limited by downlink channel combination. However, further increase in N_0 for the central microcell in the downlink channel combination, such as increasing N_0 from 22 to 28, no more capacity is increased. This is because $P_{b,U}$ of UL(10, 10) is higher than $P_{b,D}$ for DL(28, 7), and the capacity is limited by UL(10, 10). Thus, further increase in N_0 in the downlink channel combination will not improve the capacity.

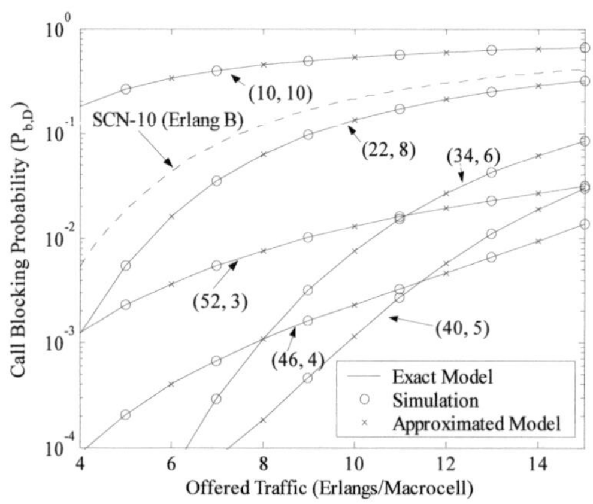

Fig. 11. Downlink performance of the AFCA for the seven-cell CMCN.

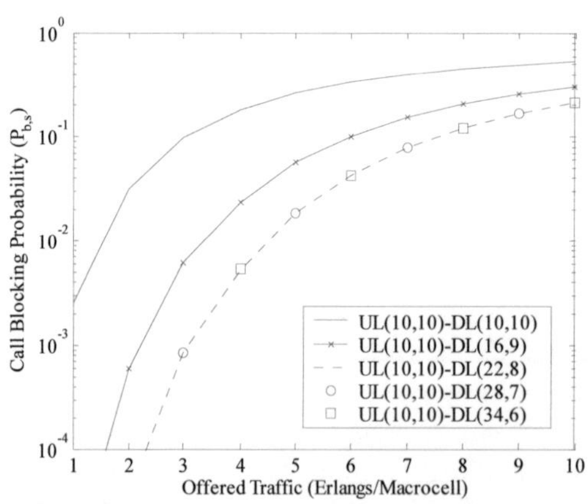

Fig. 12. Call blocking probability with both uplink and downlink transmissions.

Finally, we study the system capacity supported by the proposed AFCA scheme for uplink and downlink transmissions at $P_{b,s}$ = 1%. Table 2 shows the different values of system capacity supported by possible (N_0, N_1) for uplink or downlink. For symmetric FCA, if we use (10, 10) for both uplink and downlink channel combinations, the system capacity is 1.53 Erlangs, which is limited by the downlink capacity, as shown in Figure 11. Nevertheless, if we use (46, 4) instead of (10, 10) for both uplink and downlink channel combinations, the system capacity is 1.36 Erlangs, which is limited by the uplink capacity. From Table 2, it can be seen that the maximum capacity supported by symmetric FCA is about 6.92 Erlangs with (28, 7) for both uplink and downlink channel combinations. Therefore, we need to make use of the AFCA, in which the channel combinations (N_0, N_1) for uplink and downlink are different, in order to achieve larger system capacity. From Table 2, we suggest that with channel combination of UL(22, 8) and DL(34, 6) for downlink, the maximum system capacity can be obtained to be as large as 9.31 Erlangs. Beyond the optimum combination, if we further reduce N_1 and increase N_0, the performance will be degraded because more calls will be blocked in the *virtual* microcells.

Combinations (N_0, N_1)	Uplink Capacity (Erlangs)	Downlink Capacity (Erlangs)
(10, 10)	4.43	1.53
(16, 9)	8.51	3.33
(22, 8)	9.31	5.55
(28, 7)	6.92	7.89
(34, 6)	4.88	10.46
(40, 5)	2.92	12.92
(46, 4)	1.36	14.21
(52, 3)	0.33	9.08

Table 2. System capacity for uplink and downlink vs. channel combinations.

4. Proposed dynamic channel assignment scheme

Abovementioned results show that CMCN with AFCA can improve the system capacity. However, FCA is not able to cope with temporal changes in the traffic patterns and thus may result in deficiency. Moreover, it is not easy to obtain the optimum channel combination under the proposed AFCA, which is used to achieve the maximum system capacity. Therefore, dynamic channel assignment (DCA) is more desirable.

We proposed a multihop dynamic channel assignment (MDCA) scheme that works by assigning channels based on the interference information in the surrounding cells (Chong & Leung, 2001).

4.1 Multihop dynamic channel assignment

Figure 13 also shows the three most typical channel assignment scenarios:

1) *One-hop Calls*: One-hop calls refer to those calls originated from MSs in a central microcell, such as MS_1 in microcell *A* in Figure 13. It requires one uplink channel and one downlink channel from the microcell *A*. The call is accepted if microcell *A* has at least one free uplink channel and one free downlink channel. Otherwise, the call is blocked.

2) *Two-hop Calls*: Two-hop calls refer to those calls originated from MSs in the inner half region of a *virtual* microcell, such as MS_2 in region B_1 of microcell *B* in Figure 13. The BS is able to find another MS, RS_0, in the central microcell acting as a RS. For uplink transmission, a two-hop call requires one uplink channel from the microcell *B*, for the transmission from MS_2 to RS_0, and one uplink channel from the central microcell *A*, for the transmission from

RS_0 to the BS. For downlink transmission, a two-hop call requires two downlink channels from the central microcell *A*, for the transmission from the BS to RS_0, and from RS_0 to MS_2, respectively. A two-hop call is accepted if all the following conditions are met: (i) there is at least one free uplink channel in microcell *B*; (ii) there is at least one free uplink channel in the central microcell *A*; and (iii) there are at least two free downlink channels in the central microcell *A*. Otherwise, the call is blocked.

3) Three-hop Calls: Three-hop calls refer to those calls originated from MSs in the outer half region of a *virtual* microcell, such as MS_3 in region B_2 of microcell *B* in Figure 13. The BS is responsible for finding two other MSs, RS_1 and RS_2, to be the RSs for the call; RS_1 is in the central microcell *A* and RS_2 is in the region B_1. For uplink transmission, a three-hop call requires two uplink channels from microcell *B* and one uplink channel from the central microcell *A*. The three uplink channels are used for the transmission from MS_3 to RS_2, from RS_2 to RS_1 and RS_1 to the BS, respectively. For downlink transmission, a three-hop call requires two downlink channels from central microcell *A* and one downlink channel from microcell *B*. A three-hop call is accepted if all the following conditions are met: (i) there is at least one free uplink channel in the central microcell *A*; (ii) there at least two free uplink channels in the microcell *B*; (iii) there are at least two free downlink channels in the central microcell *A*; and (iv) there is at least one free downlink channel in microcell *B*. Otherwise, it is blocked.

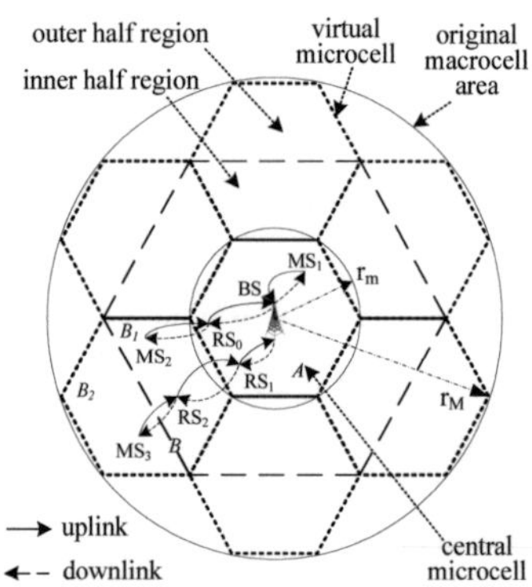

Fig. 13. Channel assignment in CMCN.

The channel assignment in CMCN to a call for the uplink and downlink is unbalanced. This is different from that in SCNs, where same number of channels is allocated to a call for uplink and downlink. Under the asymmetric FCA (AFCA) for CMCN (Li & Chong, 2006), each *virtual* or central microcell is allocated a fixed number of channels. The uplink and downlink channel combination are UL($N_{U,c}$, $N_{U,v}$) and DL($N_{D,c}$, $N_{D,v}$), respectively, where $N_{U,c}/N_{D,c}$ and $N_{U,v}/N_{D,v}$ are the number of uplink/downlink channels in the central and *virtual* microcells, respectively. The channel assignment procedure of AFCA is presented in Section 1.3, hence not revisited here.

4.2 Interference information table

The proposed MDCA scheme works on the information provided by the Interference Information Table (IIT) (Chong & Leung, 2001). Two global IITs are stored in mobile switching center (MSC) for the uplink and downlink channels. The channel assignment is conducted and controlled by the MSC, instead of a BS, because a MSC has more computational resource than a BS. This features a centralized fashion of MDCA, which results more efficient usage of the system channel pool. Consequently, the BS will only assign/release channels based on the instruction from the MSC.

Denote the set of interfering cells of any microcell A as $I(A)$. The information of $I(A)$ is stored in the Interference Constraint Table (ICT). ICT is built based on the cell configuration with a given reuse factor, N_r. For a given microcell A, different reuse factor N_r values will lead to different $I(A)$. Thus, we can implement MDCA with any N_r by changing $I(A)$ information in the ICT. For example, with $N_r = 7$ the number of interfering cells in $I(A)$ is 18, which includes those interfering cells in the first and second tiers. For example, Table 4 shows the ICT for the simulated network in Figure 14 with $N_r = 7$. Refer to Table 4, the cell number corresponds to the cell coverage of each cell in Figure 14.

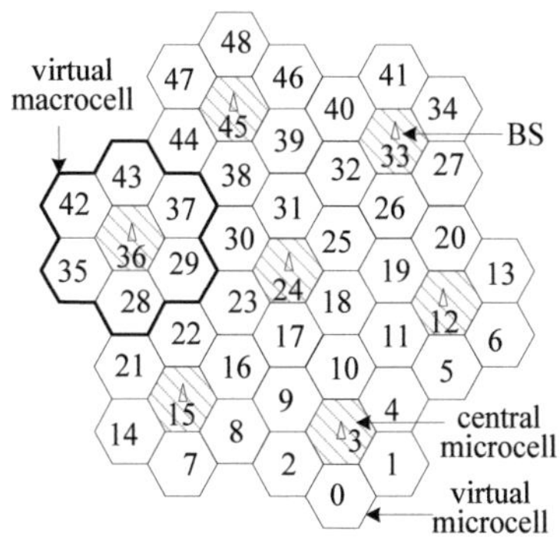

Fig. 14. The simulated 49-cell network.

	Channel				
Cell	1	2	3	…	N
0	L	L	2L	…	L
1		2L	U_{22}	…	U_{33}
2	L	L	2L	…	2L
3	L	U_{11}	2L	…	L
…	…	…	…	…	…
12		U_{11}	U_{11}	…	L
…	…	…	…	…	…
48	U_{22}	L	U_{33}	…	

Table 3. Interference Information Table for uplink.

Table 3 shows the uplink IIT for the CMCN shown in Figure 14, which includes the shared N system uplink channels in each cell. The downlink IIT is similar and hence not illustrated here. The content of an IIT is described as follows.

1) Used Channels: a letter '$U_{11/22/33}$' in the (microcell A, channel j) box signifies that channel j is a used channel in microcell A. The subscript indicates which hop the channel is used for; 'U_{11}', 'U_{22}', 'U_{33}' refer to the first-hop channel, the second-hop channel and the third-hop channel, respectively. The first-hop channel refers to the channel used between the BS and the destined MS inside the central microcell. The second-hop channel refers to the channel used between the MS (as a RS) in the central microcell and the destined MS in the inner half of the *virtual* microcell. The third-hop channel refers to the channel used between the MS (as a RS) in the inner half of the *virtual* microcell and the destined MS in the outer half of the *virtual* microcell.

2) Locked Channels: a letter 'L' in (microcell A, channel j) box signifies that microcell A is not allowed to use channel j due to one cell in $I(A)$ is using channel j. Similarly, 'nL' in (microcell A, channel j) box indicates n cells in $I(A)$ are using channel j.

3) Free Channels: an empty (microcell A, channel j) box signifies that channel j is a free channel for microcell A.

Cell	Central Microcell	Interfering Cells				
		1	2	3	…	18
0	3	40	46	2	…	34
1	3	41	0	3	…	28
2	3	46	48	8	…	41
…	…	…	…	…	…	…
48	45	45	47	7	…	40

Table 4. Interference Constraint Table for the simulated network.

4.3 Channel searching strategies

1) Sequential Channel Searching *(SCS)*: When a new call arrives, the SCS strategy is to always search for a channel from the lower to higher-numbered channel for the first-hop uplink transmission in the central microcell. Once a free channel is found, it is assigned to the first-hop link. Otherwise, the call is blocked. The SCS strategy works in the same way to find the uplink channels for second- or third-hop links for this call if it is a multihop call. The channel searching procedure is similar for downlink channel assignment as well.

2) Packing-based Channel Searching (PCS): The PCS strategy is to assign microcell A a free channel j which is locked in the largest number of cells in $I(A)$. The motivation behind PCS is to attempt to minimize the effect on the channel availability in those interfering cells. We use $F(A, j)$ to denote the number of cells in $I(A)$ which are locked for channel j by cells not in $I(A)$. Interestingly, $F(A, j)$ is equal to the number of cells in $I(A)$ with a label 'L' in channel j's column in the IIT. Then the cost for assigning a free channel j in microcell A is defined as

$$E(A,j) = I(A) - F(A,j) \tag{47}$$

This cost represents the number of cells in $I(A)$ which will not be able to use channel j as a direct result of channel j being assigned in microcell A. Mathematically, the PCS is to

$$\min_{j} E(A,j) = I(A) - F(A,j) \text{ subject to: } \; 1 \le j \le N. \tag{48}$$

Since $I(A)$ is a fixed value for a given N_r, the problem can be reformulated as

$$\max_{j} F(A,j) = \sum_{X \in I(A)} \delta(X,j) \text{ subject to: } \; 1 \le j \le N. \tag{49}$$

where $\delta(X, j)$ is an indicator function, which has a value of 1 if channel j is locked for microcell X and 0 otherwise. Specifically, to find a channel in microcell A, the MSC checks through the N channels and looks for a free channel in microcell A that has the largest $F(A, j)$ value. If there is more than one such channel, the lower-numbered channel is selected. For example, Table 5 shows a call in cell 15 requesting a first-hop channel. Channels 1, 2 and 3 are the three free channels in cell 15. Refer to , $I(15)$ = [2, 7, 8, 9, 13, 14, 16, 17, 20, 21, 22, 23, 27, 28, 29, 34, 47, 48] with $N_r = 7$. Since most of the cells in $I(15)$ are locked for channel 2, it is suitable to assign channel 2 as the first-hop channel in cell 15 because $F(15, 2) = 15$ is largest among the $F(15, j)$ values for $j = 1, 2$ and 3.

The best case solution is when $E(A, j) = 0$. However, it might not be always feasible to find such a solution. The proposed PCS strategy attempts to minimize the cost of assigning a channel to a cell that makes $E(A, j)$ as small as possible. Thus, it results in a sub-optimal solution.

	Channel				
Cell	1	2	3	...	N
...	...	...	...	...	...
2		L		...	L
...	...	...	...	...	...
7		L		...	L
8		L		...	L
9	L	L		...	2L
...	...	...	...	...	...
13		L		...	L
14		2L		...	L
15				...	U_{11}
16	L	L		...	2L
17	L	L		...	2L
....	...	...	...	...	...
20		2L		...	L
21		L		...	L
22	L			...	2L
23	L			...	2L
...	...	...	...	...	...
27		2L		...	L
28		L		...	L
29	L			...	2L
...	...	...	...	...	...
34		L		...	L
...	...	...	...	...	...
47		L		...	L
48		L	2L	...	L

Table 5. Packing-based Channel Searching for uplink.

Consider an uplink IIT and a downlink IIT with C cells and N uplink and N downlink channels. The cell of interest is cell m. The worst case scenario for channel assignment using the SCS strategy is for a three-hop call when there are only three free channels with the largest channel numbers left in cell m. The channel searching for the first-hop link requires N-2 operations. Similarly, the second-hop and third-hop links require N-1 and N operations, respectively. Next, for channel updating, the MSC needs to update 19 microcells (its own cell and 18 surrounding cells) with a total of 19 channel entries for each assigned channel. Then, a total of 19×3=57 steps are required for a three-hop call set-up. Finally, after the call is completed, another 57 steps are required for channel updates. Therefore, in the worst case scenario, a three-hop call requires a total of 3(N-1)+57×2, i.e. 3(N+37) steps. Therefore, the worst case algorithm complexity (Herber, 1986) for the SCS strategy is approximated to be $O(3N)$. The number of operations required for the uplink and downlink are the same.
The worst case algorithm complexity for the PCS strategy with N_r is estimated to be $O(12(N-1)[f(N_r)+1])$ (Herber, 1986), where $f(N_r)$ is number of cells in $I(A)$ for cell A with a given N_r (e.g. when N_r = 7, $f(N_r)$ = 18). This worst case algorithm complexity is calculated by estimating the number of steps required to assign channels to a three-hop call when all N channels are free. A three-hop call requires three uplink channels and three downlink channels. First, for a first-hop uplink, it takes N steps to check the channel status of all N channels in microcell A. Then, it takes $2f(N_r)$ steps to check the entry for each cell in $I(A)$ for a *free* channel j to calculate $F(A, j)$. Since all N channels are free, the total number of steps to obtain $F(A, *)$ for all N channels is $2f(N_r)N$. Finally, it takes N-1 steps to compare the N $F(A, *)$ values and find the largest $F(A, *)$. Similarly, the same approach can be applied for second- and third-hop uplink to obtain $F(B, *)$ and the complexity for uplink channel assignment is given by

$$O\left(\begin{Bmatrix} [N+2f(N_r)N+N-1] \\ +[N-1+2f(N_r)(N-1)+N-2] \\ +[N-2+2f(N_r)(N-2)+N-3] \end{Bmatrix}\right)=O\big(6(N-1)\left[f(N_r)+1\right]\big) \tag{50}$$

Since the computational complexity for downlink is the same as uplink, the total worst case algorithm complexity is simply equal to $O(12(N-1)[f(N_r)+1])$.

4.4 Channel updating

1) Channel Assignment: when the MSC assigns the channel j in the microcell A to a call, it will (i) insert a letter '$U_{11/22/33}$' with the corresponding subscript in the (microcell A, channel j) entry box of the IIT; and (ii) update the entry boxes for ($I(A)$, channel j) by increasing the number of 'L'.
2) Channel Release: when the MSC releases the channel j in the microcell A, it will (i) empty the entry box for (microcell A, channel j); and (ii) update the entry boxes for ($I(A)$, channel j) by reducing the number of 'L'.

4.5 Channel reassignment

When a call using channel i as a kth-hop channel in microcell A is completed, that channel i is released. The MSC will search for a channel j, which is currently used as the kth-hop channel of an ongoing call in microcell A. If $E(A, i)$ is less than $E(A, j)$, the MSC will reassign channel i to that ongoing call in microcell A and release channel j. CR is only executed for channels of the same type (uplink/downlink) in the same microcell. Thus, CR is expected to improve the channel availability to new calls. Mathematically, the motivation behind CR can be expressed as a reduction in the cost value:

$$\Delta E(A, i \rightarrow j) = E(A, i) - E(A, j) = F(A, j) - F(A, i) < 0 \tag{51}$$

4.6 Simulation results

The simulated network of an area consisting of 49 microcells is shown in Figure 15. The wrap-around technique is used to avoid the boundary effect (Lin & Mak, 1994), which results from cutting off the simulation at the edge of the simulated region. In reality, there are interactions between the cells outside the simulated region and the cells inside the simulated region. Ignorance of these interactions will cause inaccuracies in the simulation results. For example, in Figure 15, the shaped microcell 30 has 6 neighbor cells, while a boundary cell, e.g., the shaped microcell 42 has only 3 neighbor cells. Wrap-around technique "wraps" the simulation region such that the left side is "connected" to the right side and similarly for other symmetric sides. For example, for a hexagonal-shaped simulation region, there will be three pair of sides and they will be "connected" after applying the wrap-around technique. With wrap-around technique, in Figure 15, microcells 1, 4 and 5 will become "neighbor cells" (I & Chao, 1993) to microcell 42. Similar technique applies to other boundary cells. In this way, each of the microcells will have 6 "neighbor cells". Thus, the boundary effect is avoided.

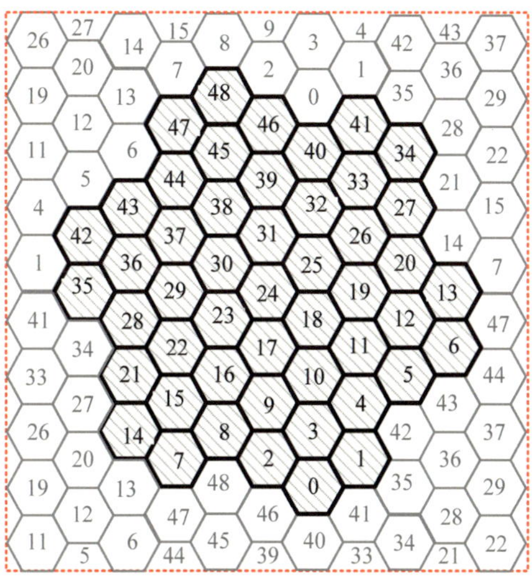

Fig. 15. The simulated network with wrap-around.

The number of system channels is N=70 (70 uplink channels and 70 downlink channels). We use N_r=7 as illustration, hence a channel used in cell A cannot be reused in the first and the second tier of interfering cells of A, i.e. two-cell buffering. Two traffic models are studied: the uniform traffic model generates calls which are uniformly distributed according to a

Poisson process with a call arrival rate λ per macrocell area, while the hot-spot traffic model only generates higher call arrival rate in particular microcells. Call durations are exponentially distributed with a mean of $1/\mu$. The offered traffic to a macrocell is given by $\rho=\lambda/\mu$. Each simulation runs until 100 million calls are processed. The 95% confidence intervals are within ±10% of the average values shown. For the FCA in SCNs, the results are obtained from Erlang B formula with $N/7$ channels per macrocell.

4.6.1 Simulation results with uniform traffic

Figure 16 shows both the uplink and downlink call blocking probability, i.e. $P_{b,U}$ and $P_{b,D}$. Notice that the $P_{b,U}$ is always higher than the $P_{b,D}$ due to the asymmetric nature of multihop transmission in CMCN that downlink transmission takes more channels from the central microcell than uplink transmission. The channels used in the central microcells can be reused in the other central microcells with minimum reuse distance without having to be concerned about the co-channel interference constraint, because two-cell buffering is already in place. The system capacity based on $P_{b,U}$ = 1% for MDCA with SCS and PCS are 15.3 and 16.3 Erlangs, respectively. With PCS-CR (channel reassignment), the capacity of MDCA is increased by 0.4 Erlangs.

Figure 17 shows the average call blocking probabilities for FCA and DCA-WI for SCNs (Chong & Leung, 2001), AFCA for CMCN (Li & Chong, 2006), MDCA with SCS, PCS and PCS-CR. DCA-WI, known as DCA with interference information, is a distributed network-based DCA scheme for SCNs. Under DCA-WI, each BS maintains an interference information table and assigns channels according to the information provided by the table. Only the $P_{b,U}$ for MDCA is shown because uplink transmission has lower capacity. At $P_{b,U}$ = 1%, the system capacity for the FCA and DCA-WI are 4.5 Erlangs and 7.56 Erlangs, respectively. AFCA with optimum channel combinations, UL($N_{U,c}$=22, $N_{U,v}$=8) and DL($N_{D,c}$=40, $N_{D,v}$=5), can support 9.3 Erlangs. The MDCA with SCS, PCS, and PCS-CR can support 15.3 Erlangs, 16.3 Erlangs and 16.7 Erlangs, respectively. As compared to DCA-WI and AFCA, the improvements of MDCA with PCS-CR are 120.9% and 79.6%, respectively.

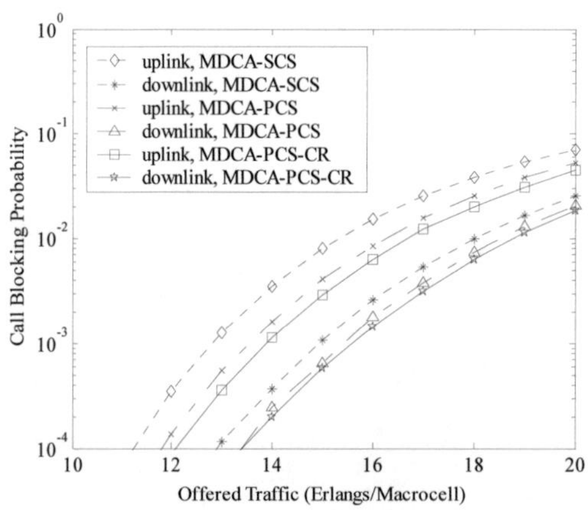

Fig. 16. Asymmetric capacity for uplink and downlink for CMCN using MDCA.

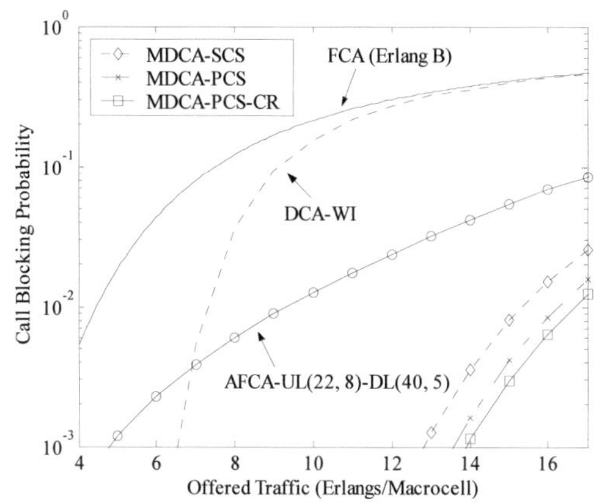

Fig. 17. Capacity comparison with N=70.

Figure 18 shows the uplink blocking probabilities, P_{b1}, P_{b2} and P_{b3}, for one-hop, two-hop and three-hop calls respectively. As expected, P_{b3} is generally higher than P_{b2}, and P_{b2} is higher than P_{b1}. The blocking probabilities for the three types of calls are lower for MDCA when using the PCS strategy as opposed to the SCS strategy. This is because the PCS strategy improves the channel availability and thus reduces the blocking probabilities of the three types of calls. The PCS-CR is not included in Figure 18 because the purpose CR will simply enhance the advantage of PCS by minimizing the effect of assigning a channel on the channel availability of the whole system.

Figure 19 illustrates the performance of MDCA with a larger number of system channels, when N=210. The Erlang B formula calculates that a SCN with N=210 can support only 20.3 Erlangs. The capacity for DCA-WI is 25.2 Erlangs. The capacity of CMCN with the optimum AFCA channel combination AFCA-UL(72, 23)-DL(144, 11) is 54.4 Erlangs at $P_{b,U}$ =1%. The MDCA using the SCS, PCS and PCS-CR strategies can support 61.5 Erlangs, 62.7 Erlangs and 63.7 Erlangs, respectively. Therefore, the MDCA sustains its advantage over conventional FCA, network-based DCA for SCNs and AFCA even for a large number of system channels.

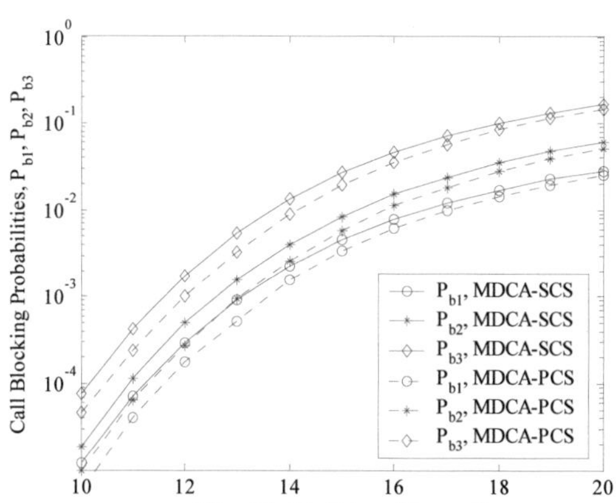

Fig. 18. Call blocking probability for different types of calls.

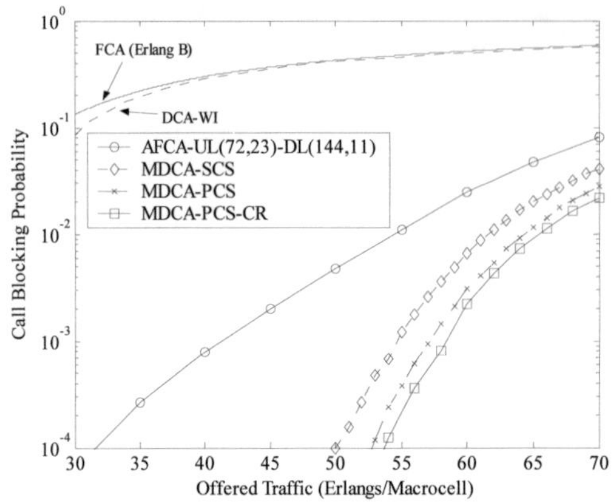

Fig. 19. Capacity comparison with N=210.

4.6.2 Simulation results with hot-spot traffic

First, as in (I & Chao, 1993), we adopted the same methodology to study the performance of MDCA with the static hot-spot traffic. Two scenarios are simulated. As shown in Figure 20, microcell 24 is chosen for the *isolated one hot-spot model* and microcells 2, 9, 17, 24, 31, 39, 46 are chosen to form the *expressway model*. First, each of the seven macrocells is initially loaded with a fixed nominal amount of traffic, which would cause 1% blocking if the conventional FCA were used. Next, we increase the traffic load in hot-spot microcells until the call blocking in any hot-spot microcell reaches 1%. Then we can obtain the capacity values for the hot-spot microcells areas.

With N = 70, each of the seven macrocells will be initially loaded at 4.46 Erlangs. In other words, each microcell is loaded with 0.637 Erlangs. We increase the traffic load for hot-spot cells, while keeping the traffic in non-hot-spot microcells at 0.637 Erlangs/Microcell. As shown in Figure 21, for the *isolated one hot-spot model*, FCA, AFCA and MDCA supports about 0.6 Erlangs, 9 Erlangs and 38 Erlangs per microcell, respectively. For the *expressway model*, FCA, AFCA and MDCA supports about 0.6 Erlangs, 1 Erlangs and 6 Erlangs per microcell, respectively. It can be seen that MDCA has a huge capacity to alleviate the blocking in hot-spot cells.

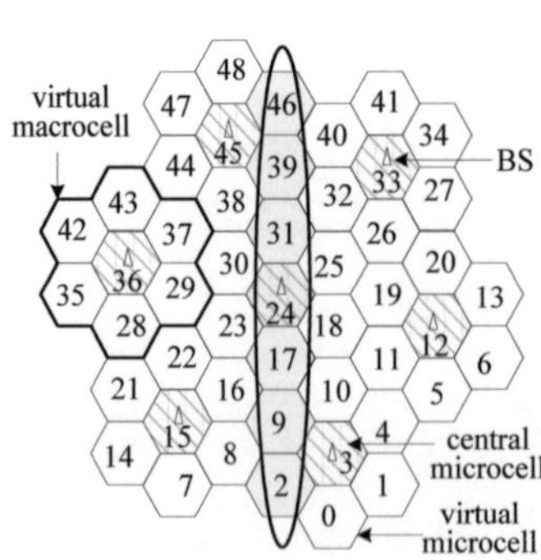

Fig. 20. The simulated hot-spot traffic cell model.

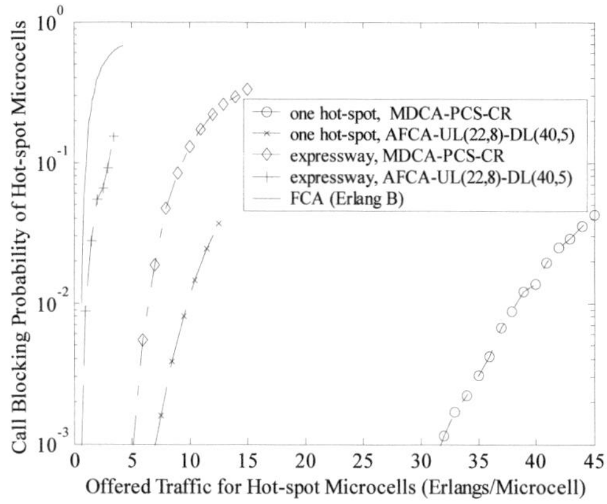

Fig. 21. Capacity comparison with hot-spot traffic for N=70.

Significant capacity improvements of MDCA have been observed with a larger N, e.g. N = 210, with uniform and hop-spot traffic. Same conclusion can be drawn that MDCA has a huge capacity to alleviate the blocking in hot-spot cells.

Finally, we investigate the performance of MDCA with a dynamic hot-spot traffic scenario and compare MDCA with AFCA. Under this traffic model, 7 hot-spot microcells are randomly selected from the 49 microcells shown in Figure 15. During the simulation, each data point is obtained by simulating the channel assignment for a period of with 1000 million calls. This period is divided into 10 equal intervals. For each interval, 7 hot-spot microcells are dynamically distributed over the 49-cell network by random selection. The average call blocking statistics are collected from the 7 hot-spot microcells from each interval. Notice that the selection of 7 hot-spot microcells is conducted for every interval and no two intervals will use the identical set of hot-spot microcells. At the end of the

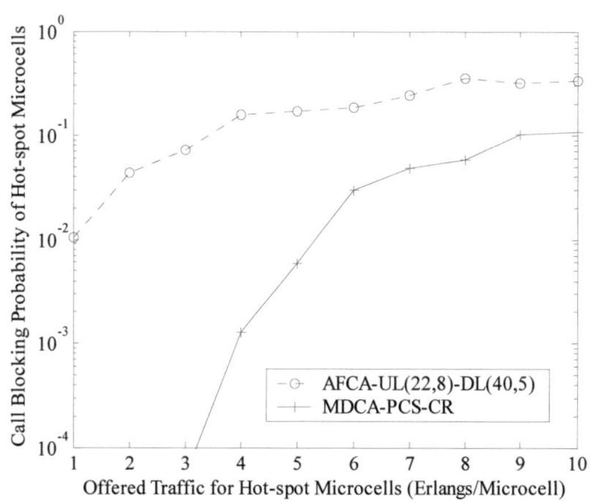

Fig. 22. Capacity comparison with dynamic hot-spot traffic for N=70.

simulation, we calculate the average call blocking probability over the 10 intervals. The traffic load in those non-hot-spot microcells is always 0.637 Erlangs/microcells according to the static hot-spot traffic model.
Figure 22 shows the capacity results for AFCA and MDCA with the dynamic hot-spot traffic scenario with N = 70 channels. MDCA and AFCA supports about 5.2 Erlangs and 1.0 Erlangs, respectively, at 1% call blocking. We can see that MDCA outperforms AFCA due to its flexibility of handling dynamic traffic distribution.

5. Conclusion

Clustered multihop cellular network (CMCN) is proposed as a compliment to *traditional* single-hop cellular networks (SCNs). A channel assignment, namely asymmetric fixed channel assignment (AFCA) is further proposed for the use in CMCNs. To analyze its performance, we have developed two multi-dimensional Markov chain models, including an exact model and an approximated model. The approximated model results in lower computational complexity and provides a good accuracy. Both models are validated through computer simulations and they matched with each other closely. Results show that the CMCN AFCA can increase the spectrum efficiency significantly. The system capacity can be improved greatly by increasing the number of channels assigned to the central microcell and decreasing the number of channels in the surrounding microcells. With optimum channel combination in the CMCN, the capacity can be doubled as compared to *traditional* SCNs.
We continued to investigate the feasibility of applying DCA scheme for MCN-type systems. A multihop DCA (MDCA) scheme with two channel searching strategies is proposed for clustered MCNs (CMCNs). Then, the computational complexity of the proposed MDCA with the two channel searching strategies is analyzed. A channel reassignment procedure is also investigated. Results show that MDCA can improve the system capacity greatly as compared to FCA and DCA-WI for SCNs and AFCA for CMCNs. Furthermore, MDCA can efficiently handle the hot-spot traffic.
In our analysis of fixed channel assignment scheme, we assumed that the MS population is infinite and RSs can be always found when a two-hop or three-hop call is concerned. Note that depending on the MS density, there would actually be an associated probability of finding a RS. It will cause serious difficulties with the analysis to incorporate the associated probability of finding a RS into the analytical models. Therefore, it has been left as part of our future work.

6. Reference

Adachi, Tomoko, & Nakagawa, Masao, (1998). A Study on Channel Usage in a Cellular-Ad-Hoc United Communication System for Operational Robots. *IEICE Transactions on Communications*. Vol. E81-B, No. 7 (July 1998), pp. 1500-07.

Aggelou, George Neonakis, & Tafazolli, Rahim, (2001). On the Relaying Capability of Next Generation Gsm Cellular Networks. *IEEE Personal Communications*. Vol. 8, No. 1 (February 2001), pp. 40-47.

Chong, P. H. J., & Leung, Cyril (2001). A Network-Based Dynamic Channel Assignment Scheme for Tdma Cellular Systems. *International Journal of Wireless Information Networks*. Vol. 8, No. 3 (July 2001), pp. 155-65.

Herber, S. Wilf. (1986). *Algorithms and Complexity*. 1st ed. Prentice-Hall, New Jersey, USA.

Hsu, Yu-Ching, & Lin, Ying-Dar, (2002). Multihop Cellular: A Novel Architecture for Wireless Data Communications. *Journal of Communications and Networks*. Vol. 4, No. 1 (March 2002), pp. 30-39.

I, Chih-Lin, & Chao, Pi-Hui. (1993). Local Packing - Distributed Dynamic Channel Allocation at Cellular Base Station. In Proceedings of IEEE GLOBECOM'93 (Houston, TX, USA, 29 November - 2 December 1993). 1, 293-301.

Kleinrock, Leonard. (1975). *Queueing System* 1st ed. John Wiley & Sons, New York.

Kudoh, E., & Adachi, F., (2005). Power and Frequency Efficient Wireless Multi-Hop Virtual Cellular Concept. *IEICE Transactions on Communications*. Vol. E88B, No. 4 (Apr 2005), pp. 1613-21.

Li, Xue Jun, & Chong, P. H. J., (2008). Asymmetric Fca for Downlink and Uplink Transmission in Clustered Multihop Cellular Network. *Wireless Personal Communications*. Vol. 44, No. 4 (March 2008), pp. 341-60.

---. (2006). A Fixed Channel Assignment Scheme for Multihop Cellular Network. In Proceedings of IEEE GLOBECOM'06 (San Francisco, CA, USA, 27 November - 1 December 2006). WLC 20-6, 1-5.

---, (2010). Performance Analysis of Multihop Cellular Network with Fixed Channel Assignment. *Wireless Networks*. Vol. 16, No. 2 (February 2010), pp. 511-26.

Lin, Yi-Bing, & Mak, Victor W., (1994). Eliminating the Boundary Effect of a Large-Scale Personal Communication Service Network Simulation. *ACM Transactions on Modeling and Computer Simulation*. Vol. 4, No. 2 (April 1994), pp. 165-90.

Lin, Ying-Dar, & Hsu, Yu-Ching. (2000). Multihop Cellular: A New Architecture for Wireless Communications. In Proceedings of IEEE INFOCOM'00 (Tel Aviv, Israel, 26-30 March 2000). 3, 1273-82.

Liu, Yajian, et al., (2006). Integrated Radio Resource Allocation for Multihop Cellular Networks with Fixed Relay Stations. *IEEE Journal on Selected Areas in Communications*. Vol. 24, No. 11 (November 2006), pp. 2137-46.

Luo, Haiyun, et al. (2003). Ucan: A Unified Cellular and Ad-Hoc Network Architecture. In Proceedings of ACM MOBICOM'03 (San Diego, CA, USA 14-19 September 2003). 1, 353-67.

Rappaport, Stephen S., & Hu, Lon-Rong, (1994). Microcellular Communication Systems with Hierarchical Macrocell Overlays: Traffic Performance Models and Analysis. *Proceedings of The IEEE*. Vol. 82, No. 9 (September 1994), pp. 1383-97.

Wu, Hongyi, et al. (2004). Managed Mobility: A Novel Concept in Integrated Wireless Systems. In Proceedings of IEEE MASS'04 (Fort Lauderdale, FL, 24-27 October 2004). 1, 537-39.

---, (2001). Integrated Cellular and Ad Hoc Relaying Systems: Icar. *IEEE Journal on Selected Areas in Communications*. Vol. 19, No. 10 (October 2001), pp. 2105-15.

Yeung, Kwan L., & Nanda, S., (1996). Channel Management in Microcell/Macrocell Cellular Radio Systems. *IEEE Transactions on Vehicular Technology*. Vol. 45, No. 4 (November 1996), pp. 601-12.

Yu, Jane Yang, & Chong, P. H. J., (2005). A Survey of Clustering Schemes for Mobile Ad Hoc Networks. *IEEE Communications Survey & Tutorials*. Vol. 7, No. 1 (First Quarter 2005), pp. 32-48.

10

Mobility and QoS-Aware Service Management for Cellular Networks

Omneya Issa
Communications Research Centre, Industry Canada
Canada

1. Introduction

As the technologies have evolved in cellular systems from 1G to 4G, the 4G system will contain all the standards that earlier generations have implemented. It is expected to provide a comprehensive packet-based solution where multimedia applications and services can be delivered to the subscriber on an anytime, anywhere basis with a satisfactory enough data rate and advanced features, such as, quality of service (QoS), low latency, high mobility, etc. Nevertheless, the 4G cellular system remains a wireless mobile environment, where resources are not given and their availability is prone to dynamic changes. Hence, the basis for QoS provisioning is to control the admission of new and handoff subscriber services in such a way to avoid future detriment perturbation of already connected ones. This task becomes a real challenge when service providers try to raise their profit, by maximizing the number of connected subscribers, while meeting their customer QoS requirements.

The problem can be summarized in that the cellular network should meet the service requirements of connected users using its underlying resources and features. These resources must be managed in order to fulfill the QoS requirements of service connections while maximizing the number of admitted subscribers. Furthermore, the solution(s) must account for the environmental and mobility issues that influence the quality of RF channels, such as, fading and interference. This is the role of service management in cellular networks.

In this chapter, we address service admission control and adaptation, which are the key techniques of service management in mobile cellular networks characterized by restricted resources and bandwidth fluctuation.

Several research efforts have been done for access control on wireless networks. The authors of (Kelif & Coupechoux, 2009) developped an analytical study of mobility in cellular networks and its impact on quality of service and outage probability. In (Kumar & Nanda, 1999), the authors have proposed a burst-mode packet access scheme in which high data rates are assigned to mobiles for short burst durations, based on load and interference measurements. It covers burst-mode only assuming that mobiles have only right to one service.

The authors of (Comaniciu et al., 2000) have proposed an admission control for an integrated voice/www sessions CDMA system based on average load measurements. It assumes that all data users have the same bit error rate (BER) requirements. A single cell environment is modeled and no interference is considered. In (Kwon et al., 2003), authors have presented a QoS provisioning framework where a distributed admission control algorithm guarantees the upper bound of a redefined QoS parameter called cell overload probability. Only a single class has been investigated; however, interference and fading are not taken into consideration. Also, the authors of (Kastro et al., 2010) proposed a model combining the information about the customer demographics and usage behavior together with call information, yielding to a customer-oriented resource management strategy for cellular networks to be applied during call initiation, handoff and allocation of mobile base stations. Although the model addressed well customer satisfaction within the studied cell, it did not consider interference to other cells.

The authors of (Aissa et al., 2004) proposed a way of predicting resource utilization increase, which is the total received/transmitted power, that would result when accepting an incoming call. Their admission control involves comparing the approximate predicted power with a threshold; this threshold is obtained by determining (offline) the permissible loading in a cell in a static scenario. However, the interference of other cells is not considered in the static scenario and no service adaptation is studied. In (Nasser & Hassanein, 2004; 2006), despite the fact that the authors have proposed a prioritized call admission control scheme and bandwidth adaptation algorithm for multimedia calls in cellular networks, their framework only supported a single class and only bandwidth is considered in adaptation, which is not tolerated by some multimedia services, such as, voice calls. They did not consider neighbor cell interference as well.

Other research efforts analyzed the soft handoff failure due to insufficient system capacity as done in this chapter. As an example, IS-95 and cdma2000 are compared with respect to the soft handoff performance in terms of outage, new call and handoff call blocking in (Homnan et al., 2000). In (Him & Koo, 2005), the call attempts of new and handoff voice/data calls are blocked if there is no channel available, and a soft handoff blocking probability is derived as well.

In what concerns the admission policies of handoff calls with respect to new calls, some schemes, such as the ones proposed in (Cheng & Zhuang, 2002; Kulavaratharasah & Aghvami, 1999), deploy a guard channel to reserve a fixed percentage of the BS's capacity for handoff users. Other schemes, called nonprioritized schemes in (Chang & Chen, 2006; Das et al., 2000), handle handoff calls exactly the same way they do with the new calls. Although these approaches are not specially designed, they can be adapted to 3G+ networks as it was briefly represented in (Issa & Gregoire, 2006) and will be discussed in this chapter.

The above survey has compared state-of-the art admission control proposals, highlighting the main factors of decision making, advantages and weaknesses of different approaches. This leads to pointing out that important challenges pertaining to the wireless environment are yet to be addressed. Therefore, this chapter proposes a strategy that accounts for most of these challenges, such as, cell loading, inter-cell and intra-cell interference, soft handoff as well as QoS requirements in making admission decisions. The strategy also considers the fact that, nowadays, mobile devices are not just restricted to cellular phones; instead, they became small workstations that allow for several simultaneous services per user connection. Factors such as service tolerance for degradation and QoS parameters allowed to be degraded are also exploited. The chapter is organized as follows: in sections 2 and 3 we describe the design details of our approach followed by the design evaluation in section 4, then we summarize the benefits of the proposal and present future work in section 5.

2. Admission control

Our scheme of service admission on either forward or reverse links is done by measuring the total received or transmitted power at the base station and calculating the available capacity according to QoS constraints, interference measurements and fading information gathered from Mobile stations (MSs) in the neighbouring cells. The advantage of building our admission control scheme on power constraints is that it incorporates both bandwidth and BER, represented in target Signal to interference ratio (SIR), since both bandwidth and SIR affect the required channel power and are important in guaranteeing QoS especially in a wireless mobile environment. In addition to power constraints, our admission control follows a policy-based criterion by giving the handoff services priority over the new service connections. In fact, soft handoff attempts should be considered differently from new call attempts because the rejection of handoff attempts from other cells could cause call dropping.

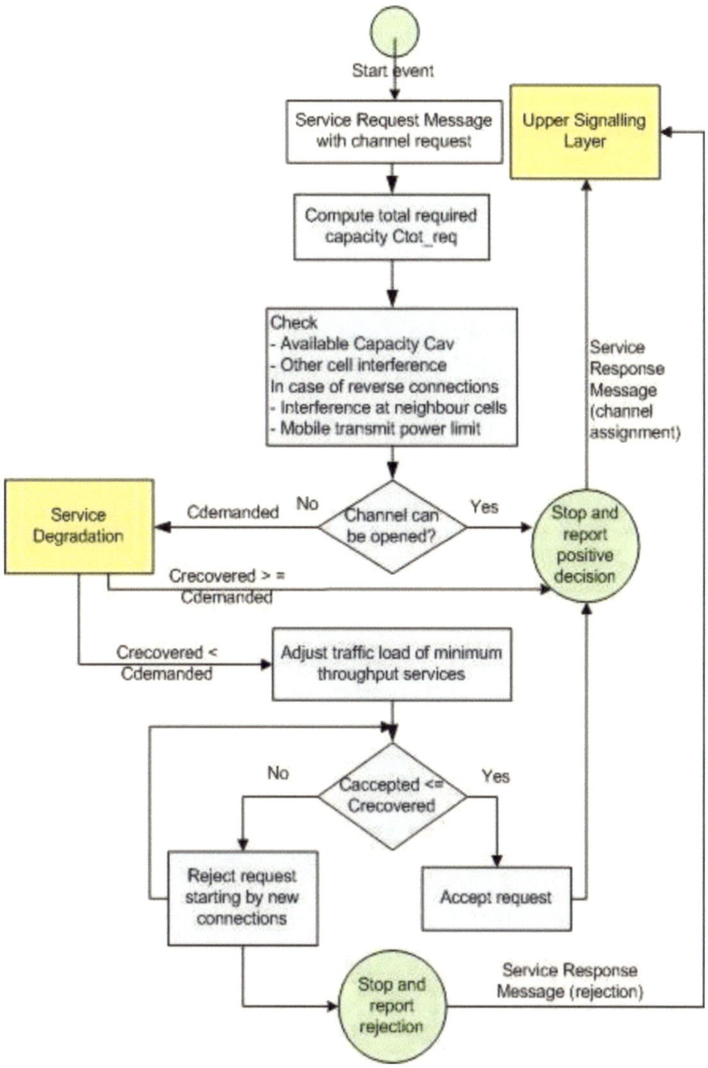

Fig. 1. Admission control scheme

Fig. 1 shows the admission control process. When a new service is required, the base station (BS) admission control (AC) module calculates the required capacity in terms of channel

power needed, then checks the available capacity taking into account the current service load, the mobile transmit power, the interference of other cells and the interference to neighbour cells. If the available capacity can not cover the initial requirements of the incoming services, the admission control scheme appeals to a degradation procedure for connected services. However, if the degradation process can not recover the needed capacity, the admission control module adjusts the requirements of the incoming services (only services requiring minimum throughput). However, when such adjustment is not enough, it starts rejecting new service requests. Service degradation is discussed in section 3.
We start by describing the verification of load and interference done at beginning of the admission control procedure before appealing to service degradation. The basic idea in resource estimation is actually the same for both uplink (reverse) and downlink (forward).

2.1 Uplink

Assuming one BS per cell, this capacity validation procedure is done as follows on the uplink:

$$Pt_K = \sum_j P_{j,K} + Other\_Cell\_Interference + No, \tag{1}$$

where Pt_K is the total received power by the BS in cell K, $P_{j,K}$ is the received power at cell K from MS_j and No is the background noise. As in (Kumar & Nanda, 1999), $P_{j,K}$ can be written as a function of SIR, i.e. the received ratio of signal bit energy to noise power spectral density $(Eb/No)_{j,K}$ for MS_j in cell K divided by its processing gain G_j,

$$P_{j,K} = \frac{1}{G_j} Pt_K \left(\frac{Eb}{No}\right)_{j,K}, \quad G_j = \frac{W}{R_j}, \tag{2}$$

W is the spreading bandwidth and R_j represents the transmission rate of MS_j. Including SIR in capacity measurements is very important since 3G+ network is interference limited.

2.1.1 Interference calculation

The other interference in cell K caused by neighbouring cells can be presented in an average sense as a fraction of the in-cell load (Gilhousen et al., 1991), on condition that the load is uniform across all cells. We relaxed this condition to the case where the load in different cells is different, but the average load over all cells is kept fixed to some value by the base station controller (BSC). So (1) can be rewritten as

$$\begin{aligned} Pt_K &= (1+\eta_K)\cdot \sum_j P_{j,K} + No \\ &= Pt_K \cdot \left(\sum_j \frac{1}{G_j} \left(\frac{Eb}{No}\right)_{j,K} \right) \cdot (1+\eta_K) + No, \end{aligned} \tag{3}$$

where the other cell interference factor η_K is defined as in (Kim et al., 2003)

$$\eta_K = \sum_{i \neq K} \left(\frac{1}{M_i} \sum_{x=1}^{M_i} \eta(x) \right), \tag{4}$$

M_i is the number of MSs in cell i and $\eta(x)$ is calculated as

$$\eta(x) = \frac{\rho_K(i,x)L_K(i,x)}{\rho_i(i,x)L_i(i,x)}, \tag{5}$$

where $\rho_K(i,x)$ is the fast (Rayleigh) fading and is given by

$$\rho_K(i,x) = \sum_{p=1}^{P} g^2_{K,x(i),p}, \tag{6}$$

with $g^2_{K,x(i),p}$ is the pth path gain between BS_K and MS_x in cell i, and $L_K(i,x)$ presents slow fading which is modeled as

$$L_K(i,x) = r^{-\delta}_{K,x(i)} \cdot 10^{\xi_{K,x(i)}/10}, \tag{7}$$

where the signal between BS_K and MS_x in cell i experiences an attenuation by the δth power of the distance $r_{K,x(i)}$ between BS_K and MS_x and log-normal shadowing (ξ is a zero-normal variant with standard variation σ).

The uplink capacity is directly affected by the noise rise generated by users in the uplink. The noise rise N_r is the increase in noise compared to the noise floor of the cell; thus:

$$N_r = \frac{Pt_K}{No}, \tag{8}$$

The concept of noise rise means that infinite noise rise must be considered when the load is 100% (e.g. the pole capacity). Hence, N_r can be written as a function the cell uplink load C_U; when C_U is close to unity, the noise rise approaches infinity as shown in (9):

$$N_r = \frac{1}{1 - C_U}, \tag{9}$$

From (8), the C_U can be written in function of the total received power as follows:

$$C_U = \frac{Pt_K - No}{Pt_K}, \tag{10}$$

Using (3),

$$C_U = (1 + \eta_K) \cdot \frac{\sum_j P_{j,K}}{Pt_K}, \tag{11}$$

Recall that from (2),

$$\frac{P_{j,K}}{Pt_K} = \frac{R_j}{W} \left(\frac{Eb}{No}\right)_{j,K}, \tag{12}$$

So (11) becomes:

$$C_U = (1 + \eta_K) \cdot \sum_j \frac{R_j}{W} \left(\frac{Eb}{No}\right)_{j,K}, \tag{13}$$

(13) assumes only one channel per MS. To extend it to the case, as for wideband CDMA, where an MS can have several channels with different target SIR's, data rates and activity factors, (13) can be written as

$$C_U = \left(\frac{1}{W} \sum_j \sum_{n=1}^{N_j} R_{n,j,K} \left(\frac{Eb}{No}\right)_{n,j,K} a_{n,j,K} \right) \cdot (1 + \eta_K), \tag{14}$$

where N_j is the number of channels of MS_j, and $(Eb/No)_{n,j,K}$ and $a_{n,j,K}$ are the received Eb/No and the activity factor for service of channel n of MS_j in cell K respectively.
So the admission condition for accepting the uplink connection(s) of MS_i in cell K is

$$\frac{(1+\eta_K)}{W}\left(\sum_{j\neq i}\sum_{n=1}^{N_j} R_{n,j,K}\left(\frac{Eb}{No}\right)_{n,j,K} a_{n,j,K}\right) + \frac{(1+\eta_K)}{W}\left(\sum_{n=1}^{N_i} R_{n,i,K}\left(\frac{Eb}{No}\right)_{n,i,K} a_{n,i,K}\right) \leq Th_U, \tag{15}$$

where N_i is the number of channels of MS_i, $(Eb/No)_{n,i,K}$ is the required Eb/No (SIR) for channel n of MS_i in cell K and η_K is computed by (4 7). Theoretically, Th_U is equal to 1, which is the pole capacity. However, an operator restricts the uplink load to a certain noise rise, hence, practically Th_U is kept below unity.

2.2 Downlink

The downlink cell capacity follows the same logic as the uplink. The total base station transmit power, Ptx_tot_K, is estimated. It is the sum of individual transmit powers required for downlink connections of MSs in a cell as shown below.

$$Ptx\_tot_K = \sum_j Ptx_{j,K}, \tag{16}$$

where $Ptx_{j,K}$ is the downlink power required for MSj in cell K, assuming only one service per mobile. $Ptx_{j,K}$ is given by (Sipila et al., 2000):

$$Ptx_{j,K} = \frac{R_j}{W}\left(\frac{Eb}{No}\right)_{j,K} \cdot \left((1-f)\,Ptx\_tot_K + \eta_j \cdot Ptx\_tot_K + No \cdot L_{K,j}\right), \tag{17}$$

where $L_{K,j}$ is the path loss from base station K to MS_j and f is the orthogonality factor modeling the intracell interference from non orthogonal codes of other MSs, using f=1 for fully orthogonal codes and 0 as not orthogonal. Note that the orthogonality factor f depends on the codes used for users inside a cell, and even if these codes are perfectly orthogonal, there is always some degree of interference between the signals of mobiles of the same cell due to multi path. Delayed copies received from a multipath fading are not orthogonal any more and cause multipath fading interference, which is modeled as a factor of the total base station transmit power. For simplicity, we do not consider the orthogonality factor of each code; we take f as the average orthogonality factor in the cell. η_j is the other cell to own cell received power ratio (inter cell interference) for MS_j modeled as a factor of the total downlink power and calculated as follows:

$$\eta_j = \sum_{i\neq K} \frac{\rho_{K,j} L_{K,j}}{\rho_{i,j} L_{i,j}}, \tag{18}$$

where $\rho_{K,j}$ and $L_{K,j}$ are the fast and slow (path loss) fading from the serving base station K to MS_j and $\rho_{i,j}$ and $L_{i,j}$ are the fast and slow (path loss) fading from another base station i to MS_j respectively.
The other to own cell interference on the downlink depends on mobile location and, therefore, is different for each mobile. However, the estimation of the downlink transmission power should be on average basis and not on the maximum transmission power at the cell

edge. The average transmission power per mobile is determined by considering the user at an average location in the cell. Thus, we may let η be the average other-to-own cell interference seen by the mobile as in (Sipila et al., 2000):

$$\eta = \frac{1}{J}\sum_{j}\eta_j, \tag{19}$$

where J is the number of MSs served by the base station.
By summing up $Ptx_{j,K}$ over the number of MSs, Ptx_tot_K can be derived as in (20). Note that in this estimation the soft handover must be included. thus, j must include the soft handover connections, which are modeled as additional connections in the cell, as well.

$$Ptx\_tot_K = \frac{No\sum_{j}\frac{R_j}{W}\left(\frac{Eb}{No}\right)_{j,K}L_{K,j}}{1-\sum_{j}\frac{R_j}{W}\left(\frac{Eb}{No}\right)_{j,K}\cdot((1-f)+\eta)}, \tag{20}$$

Using the same reasoning on the noise rise as for the uplink, we can define the cell downlink load C_D as follows (Sipila et al., 2000):

$$C_D = \sum_{j}\frac{R_j}{W}\left(\frac{Eb}{No}\right)_{j,K}\cdot((1-f)+\eta), \tag{21}$$

When C_D is close to 1, the base station transmit power approaches infinity. To extend (21) so that an MS can have several channels with different activity factors, we obtain the admission condition for the downlink connection of MSi by:

$$\begin{aligned}&\frac{(1-f+\eta)}{W}\left(\sum_{j\neq i}\sum_{n=1}^{N_j}R_{n,j,K}\left(\frac{Eb}{No}\right)_{n,j,K}a_{n,j,K}\right)\\&+\frac{(1-f+\eta)}{W}\left(\sum_{n=1}^{N_i}R_{n,i,K}\left(\frac{Eb}{No}\right)_{n,i,K}a_{n,i,K}\right)\leq Th_D,\end{aligned} \tag{22}$$

Theoretically, Th_D is equal to unity, which is called the pole capacity. However, lower Th_D values will be tested to limit the noise rise. Note that in case of no orthogonality (f = 0), (22) becomes similar to the uplink case.

2.3 Reverse connections and soft handoff

In reverse connections, a MS can have more than one leg in soft handoff. So, in addition to the procedure proposed above, to account for soft handoff, the following conditions must be satisfied:

- j in (15) is summed over the set of MSs that have cell K in their active set.
- (15) must be satisfied for each soft handoff leg of MS_i.
- Since adjacent-cell interference is critical in deciding for the admission of reverse connections, it is necessary to evaluate the interference at the non active cells caused by the admission of the reverse connection. So, interference constraints at neighbour cells that are not in the active or candidate set of MS_i should be satisfied.

Pilot strength information received at MS_i for cells in its neighbour list can indicate to BS the interference levels that will be seen at its neighbour BSs due to transmissions from MS_i. The MS reports pilot strength information for cells in its neighbour list in the Paging Channel. So, to avoid producing excessive interference at a neighbour cell NC, we constrain the path loss difference between the strongest active and strongest non-active pilots to a minimum Δ (Kumar & Nanda, 1999) such that

$$PS_0 - PS_{NC} \geq \Delta, \tag{23}$$

$$PS_0 = \max_K(PS_{iK}), \quad K \in AS,$$

$$PS_{NC} = \max_K(PS_{iK}), \quad K \notin AS \; and \; K \in NS,$$

where PS_{iK} is the pilot strength reported by MS_i from cell K, AS is the set of active and candidate pilots and NS is the set of neighbour pilots. PS_0 is the strength of the strongest active pilot and PS_{NC} is the strength of the strongest non-active pilot.

2.4 User equipment transmit power

Another factor that must be taken into account, is the limited transmission power available at the mobile stations. Here, we do not model battery lives, however, we check if the transmit power of the mobile station, required to meet the uplink target Eb/No, does not exceed the maximum mobile transmit power. As indicated in Fig.1, after the uplink cell load is checked, the required mobile transmit power is verified with respect to its maximum. The required mobile transmit power $Ptx_{MS_{i,K}}$ of MSi can be computed as follows (Sipila et al., 2000):

$$Ptx_{MS_{i,K}} = \frac{No\left(\sum_{n=1}^{N_i} R_{n,i,K}\left(\frac{Eb}{No}\right)_{n,i,K} a_{n,i,K}\right) L_{i,K}}{1 - C_U}, \tag{24}$$

where N_i is the number of uplink channels of MS_i. A mobile station, which is not able to transmit with the required amount of power to meet the required Eb/No due to maximum power limitations is not admitted (blocked). Note that incoming service requests can be blocked either because of noise rise limits (equations (15) and (22)) or in case of limited mobile transmit power. It is worth noting that the limitation of total base station power is already considered by applying lower values of the threshold Th_D in (22).

2.5 Recalculation of admitted load

A new channel can be assigned by sending a Channel Assignment Message with Service Response Message to MS if all the above conditions are met. However, if all the conditions are satisfied except the uplink condition (15) or the downlink one (22), the admission module asks the Degradation module to try to acquire the capacity left to satisfy the required one.
In order to compute the capacity required from the Degradation module, (15) and (22) are interpreted in terms of the available capacity. The available capacities C_{av_U} and C_{av_D} for accepting services on the uplink (U) and downlink (D) respectively can be given by:

$$C_{av_U} = \frac{W \cdot Th_U}{(1+\eta_K)} - \left(\sum_i \sum_{n=1}^{N_i} R_{U_{n,i,K}} \left(\frac{Eb}{No}\right)_{U_{n,i,K}} a_{U_{n,i,K}}\right), \tag{25}$$

$$C_{av_D} = \frac{W \cdot Th_D}{((1-f)+\eta)} - \left(\sum_i \sum_{n=1}^{N_i} R_{D_{n,i,K}} \left(\frac{Eb}{No}\right)_{D_{n,i,K}} a_{D_{n,i,K}} \right), \tag{26}$$

where i is summed over the set of existing (already connected) MSs.

Since the first term in (25) and (26) does not depend on the service requirements, for simplicity, we define a variable C that accounts for the second term that varies according to services; it will be referred to as 'service capacity'. Thus, the total required service capacity for new/handoff requests, for uplink and downlink, can be written as:

$$C_{tot\_req_{U/D}} = \left(\sum_j \sum_{n=1}^{N_j} R_{(U/D)_{n,j,K}} \left(\frac{Eb}{No}\right)_{(U/D)_{n,j,K}} a_{(U/D)_{n,j,K}} \right), \tag{27}$$

where j is summed over the set of the new/handoff MSs and N_j is the number of services on either the uplink or the downlink. If $C_{tot\_req_{U/D}}$ can not be satisfied by the available capacity, the admission control appeals service degradation with the uplink and downlink demanded service capacities, C_{d_U} and C_{d_D} respectively. The demanded capacity is the difference between the required capacity and the available one:

$$\begin{aligned} C_d &= C_{tot\_req} - C_{av}, \\ C_{d_U} &= \left(\sum_j \sum_{n=1}^{N_j} R_{U_{n,i,K}} \left(\frac{Eb}{No}\right)_{U_{n,i,K}} a_{U_{n,i,K}} \right) - \frac{W \cdot Th_U}{(1+\eta_K)}, \\ C_{d_D} &= \left(\sum_j \sum_{n=1}^{N_j} R_{D_{n,i,K}} \left(\frac{Eb}{No}\right)_{D_{n,i,K}} a_{D_{n,i,K}} \right) - \frac{W \cdot Th_D}{((1-f)+\eta)}, \end{aligned} \tag{28}$$

where j is summed over the set of existing MSs as well as the new/handoff ones.

If the capacity recovered by service degradation, $C_{recovered}$, can provide the demanded capacity, the service requests are accepted. However, to account for a worst case scenario where the Degradation module cannot deliver the demanded capacity, the AC module, in order to maximize the number of admitted services, recalculates the required capacity by trying to decrease the accepted load of new/handoff minimum throughput services only, since bounded delay services such as voice and video cannot tolerate such process. However, the accepted load of each minimum throughput service is reduced, if possible, by an equal share of unacquired capacity. In other words, each minimum throughput service will have an equal share of the total now available capacity ($C_{av} + C_{recovered} - C_{reqBD}$) with respect to the required capacity of all the new/handoff minimum throughput services. C_{reqBD} is the capacity required for bounded delay new/handoff services. Note that the service requirements are not decreased below the QoS limits, C_{min}, indicated in its QoS profile. Thus, the accepted capacity of service n is the greater of two values, its minimum required capacity and its share of the total available capacity:

$$C_{n\_accepted} = \max(C_{n\_min}, C_{n\_req} \cdot (C_{av} + C_{recovered} - C_{reqBD}) / \sum_{j=1}^{N} C_{j\_req}), \tag{29}$$

where $C_n = R_n (Eb/No)_n a_n$, where R_n and $(Eb/No)_n$ are the rate and the SIR of the channel needed for service n respectively, and a_n is the service activity factor. N is the number of

new/handoff minimum throughput services. It will be seen in the next section that the load of minimum throughput services is decreased by adjusting their rate R.
It can happen, even when reducing the capacity of minimum throughput services, that the accepted capacity exceeds available plus recovered capacities because we do not reduce the service capacity below the limits of its QoS profile. So the last possible course of action is to reject requests, as in Fig.1. We begin by rejecting new service requests since forced termination of handoff services has significant negative effect on the user's perception of network reliability, and, therefore, affect subscribers' expectation.
It is worth noting that the admission control scheme is simple to apply and has a low computation time. The time complexity for calculating conditions (15) and (22) is of order $O((Nc+1)M)$ and the one for reducing the accepted load of new/handoff minimum throughput services is of order $O(N_{minT})$ where Nc in the number of neighbour cells, M in the average number of MSs per cell and N_{minT} is the number of new/handoff minimum throughput services. So the time complexity of our admission module is of order $O((Nc+1)M+N_{minT})$, that is $O(M)$ since Nc is limited to a few cells. In fact, the complexity of the proposed approach is considered low with respect to the inclusion of a realistic interference computing. Other complex approaches adopt a global call admission control where the actual interference should be computed between all cells; their computational complexity is of order of $O(N^2)$ where N is the number of cells in the network. On the other hand, more simplistic admission control algorithms simply consider the calls currently active in the intended cell in order to accept or reject a new call. The inter-cell interference is either neglected or sometimes considered as a constant factor. These algorithms usually have the lowest computational complexity of $O(1)$, which is an optimal complexity, however, at the cost of not considering realistic conditions.

3. Degradation and improvement

The problem of high new call blocking and handoff dropping can be partially solved by QoS adaptation (degradation) which is much more bearable to users than a forced termination of their services. The improvement module is important for deciding the appropriate distribution of resources freed by terminating and ongoing services. Also, the admission control needs this module in order to make room for admitting new and handoff connections. In (Lee et al., 2000) authors have proposed an adaptive resource allocation mechanism that allocates connection resources for incoming calls utilizing bandwidth degradation and compensation. Also (Kwon et al., 2003) have proposed a bandwidth adaptation algorithm which seeks to minimize a redefined QoS parameter called cell overload probability. Both adaptation frameworks only take mobility into consideration; they do not account for fading and interference. Both approaches consider only the transmission rate as the only parameter to degrade. Moreover, all calls that exceed their average or minimum bandwidth are degraded to their average or minimum bandwidth respectively. This requires a large number of signalling messages.

3.1 Limiting signalling overhead

Service degradation/improvement implies changing resources assignment. This can be a time and bandwidth-consuming process, because the number of degraded/ improved services increases the amount of required signalling messages respectively. So we have decided to limit the number of degraded/improved services by selecting only MSs located in a certain zone of the cell.

In fact, other criteria than MS locations can be used to limit the number of adapted services such as the degraded time of services. But, since our objective is to maximize the number of admitted services while maintaining QoS, we should minimize interference as well. To achieve this, the cell is divided into two zones: zone 1 where the distance of MS from the BS, located in the center of the MS cell, is higher than a value Rsafe, and zone 2 where this distance is lower than Rsafe.

In our design, MSs located in zone 1 can be degraded. Since degradation means reducing resources, i.e. transmission power, so degrading services of MS located near the cell boundary, in zone 1, would decrease interference to adjacent cells. On the other hand, since improvement implies increasing transmission power, services of MS located in zone 2, far from cell boundary, can be improved without causing interference to neighbour cells. The prediction of MS location can be done based on the pilot signal strength from BS to MS on the Paging Channel or on the phase delay of the received signal. We call zone 2 the safe region because services are not degraded if their MS is in this zone.

The proposed policy for degradation and improvement mainly aims to limit interference and, hence, maximize the admission probability. Nevertheless, this also leads to unfairness in distributing radio resources among the mobiles of the same cell. MSs far from the BS are more likely subject to degradation while mobiles near the BS are favored for improvement.

Note that the nomenclature of signalling messages are based on cdma2000 signalling messages specified in (C.S0005-Ev2.0, 2010). This can be easily mapped to signalling messages of WCDMA and LTE.

3.2 QoS level adjustment

Service degradation/improvement (D/I) also implies the adjustment of the QoS level. QoS is adjusted by changing the data rate or the target BER according to service tolerance. Both kinds of changes lead to changing the transmission power because modifying the data rate modifies the channel processing gain, and modifying the BER leads to the alteration of the target Eb/No. We have decided to adjust the QoS of non real-time services (those requiring only a minimum throughput) by changing their data rate according to their profile since they cannot tolerate higher BER degradation. The QoS of real-time services is adjusted by modifying their Eb/No because they can tolerate such change, however, a variation in their data rate implies a delay change which is not tolerable by such services. We have decided not to alter the transmission power of voice services, whether by changing their bit rates or their target Eb/No, because imposing any degradation of voice traffic would harm its QoS requirements.

Moreover, our design respects the limited tolerance to variation of real-time services. So, in the Degradation module presented in Fig.2, we begin by degrading the minimum throughput services. The real-time services, such as near real-time video (NRTV) services, are not degraded unless there is no minimum throughput service left to be degraded. Also, when MSs finish their call or leave the cell, the Improvement module, shown in Fig.3, shares out the freed capacity C_{freed} beginning by improving the degraded NRTV services, then, the remaining capacity, C_{fr_left}, goes to the degraded minimum throughput services that are still in the safe region. Note that the degraded minimum throughput services get an equal share of the surplus capacity with respect to the required capacity of the degraded minimum throughput services. Thus, as in (30), the new allocated rate to a minimum throughput service n is the lower of two values, its required rate R_{n_req} and its share of the left freed capacity added to its

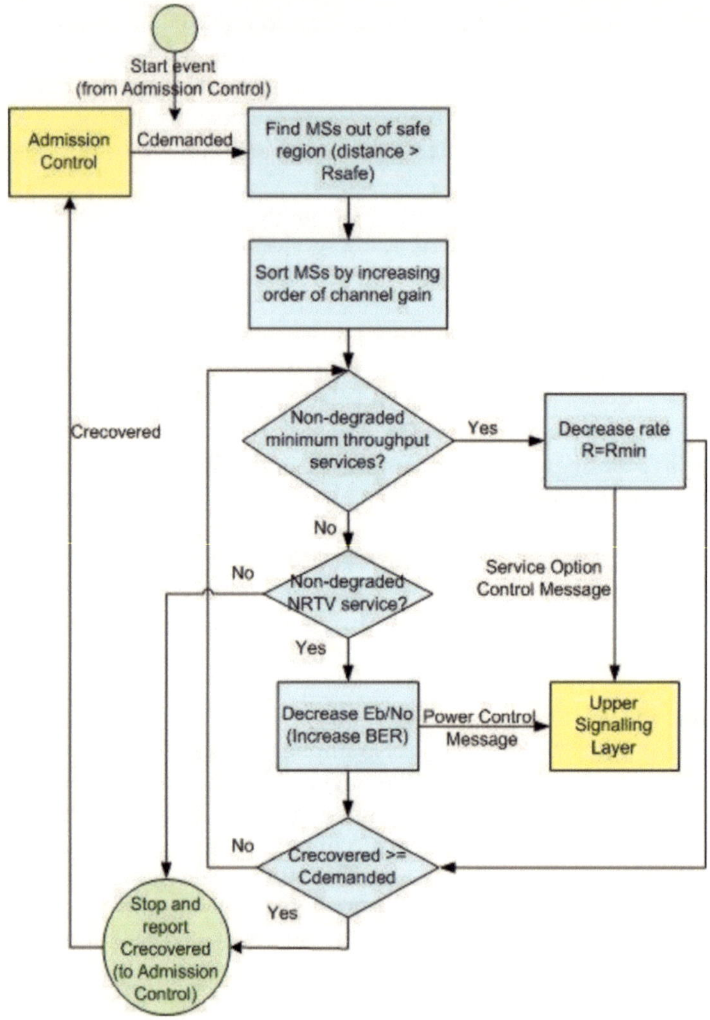

Fig. 2. The Degradation module

actual rate R_{n_a}:

$$R_n = \min(R_{n\_req}, (R_{n\_a} + (R_{n\_req} \cdot C_{fr\_left}) / \sum_{j=1}^{N} C_{j\_req})), \tag{30}$$

where N is the number of degraded minimum throughput services in zone 2.

3.3 Limiting interference

In addition to limiting signalling overhead, we have optimized our Degradation module to further minimize the interference to adjacent cells. Services to be degraded are sorted by ascending order of their channel gain, where the channel gain $G_{x,K}$ of channel x in cell K is given by

$$G_{x,K} = \rho_x(K, x) L_x(K, x), \tag{31}$$

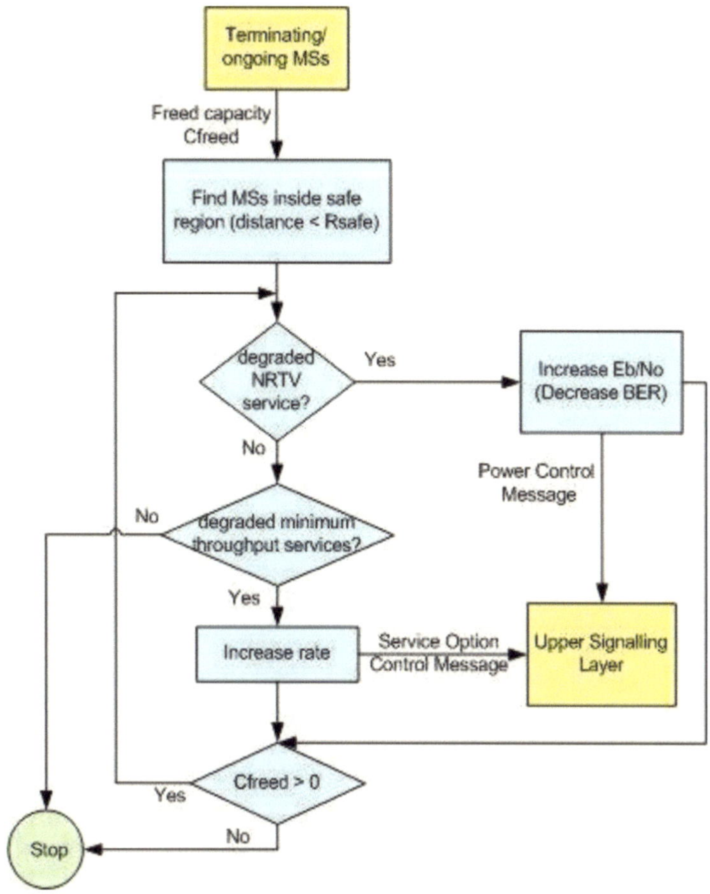

Fig. 3. The Improvement module

where ρ_x and L_x are calculated by (6) and (7). Since MSs located near cell boundary most likely have lower channel gain, allocating lower transmission power to these MSs would decrease interference to neighbour cells. So, we begin by degrading services having the lowest channel gain, hence, assigning to them lower transmission power and thus, limiting interference to adjacent cells.

When called by the admission control, the Degradation module presented in Fig.2 starts by finding the MSs in zone 1 then sorts them by increasing order of channel gain. It begins by adapting non-degraded minimum throughput services by decreasing their rate. The real-time services are degraded when there is no minimum throughput one left. Thereafter, the $C_{recovered}$ is reported to the admission control. Reciprocally, as shown in Fig.3, in case of terminating or ongoing MSs, the Improvement module equally distributes the freed capacity first among MSs in zone 2. It starts by the degraded real-time services then the minimum throughput ones according to (30).

The D/I modules have a low computation time. Locating MSs in safe region is of the order $O(N)$ where N in the number of MSs in the cell. Sorting selected MSs by their channel gain has a maximum computation order $O(N \log N)$. So the time complexity of the Improvement module is of order $O(N)$ and the upper bound of time complexity of the degradation module is of order $O(N \log N)$. This order of complexity is comparable to the order of any service adaptation approach that deploys some selection criteria among MSs in a single cell. It is lower than the complexity of the approaches that apply global adaptation among all the cells of the network. Their complexity is of the order of $O(M \times N)$ where M is the number of cells and N is the average number of MSs/cell.

4. Design evaluation

To test our proposed modules, we have built a simulation model that includes mobility, fading and interference.

Our cellular system consists of 19 hexagonal cells, with a BS in the center of each cell. The reported results are the ones of the central cell; the other 18 cells are source of interference around it. Typically, two rings are sufficient to generate significant interference to the MSs in the central cell.

Two environments are simulated: an indoor environment in an office with soft partitions and an outdoor environment in a shadowed urban area. In order to account for fading as well as mobility, slow and fast fading are simulated in the channel model, since they affect the signal strength of the MS channel measured by BS. Other parameters of fading and mobility simulation are shown in Table 1.

The simulation of soft handoff follows the IS 95 algorithm. At each MS, the active pilots are determined by computing the pilot signal strength received from all BSs. MS is assumed to be in handoff with all BSs having an average pilot signal strength within 6 dB of the strongest serving site, so Δ is set to 6 dB in (23). The parameters of active and candidate set evaluation are: $T_{Add} = -12dB$, $T_{Drop} = -16dB$ and $T_{T_{Drop}} = 3s$. In soft handoff, the transmit power is allocated on uplink and downlink according to the strongest base station pilot.

Parameter	Indoor (Office)	Outdoor (Vehicular)
Time between direction changes T	1 min	1 min
Cell radius Rc	50m	5km
Movement direction α	$\pi/4$	$\pi/4$
Velocity	5km/h	100km/h
Velocity standard deviation	3km/h	5km/h
Path loss exponent δ	1.6	3.0
Shadowing deviation σ	9.6	4.0
Number of paths P	2	2

Table 1. Simulation parameters

Requests are made for service connection establishment at a service arrival rate λ. Forward and reverse services are uniformly distributed on MSs; each MS can have up to 3 different services. Voice and NRTV are the simulated real time services with BER of 10^{-3} and 10^{-4} and target Eb/No of 7 and 8dB respectively (TR45.5, 1998). FTP and web sessions are the non real time services with BER of 10^{-6} and target Eb/No of 10dB. The activity factor of voice is 0.5 and 1.0 for the other services. Other traffic simulation parameters are provided in Table 2. Models for FTP, NRTV and web browsing traffic can be found in (TSG C.R1002, 2003); however, for simplicity, we modelled them as constant bit rate services. The minimum acceptable rate of FTP and web sessions is half their required average rate. The maximum Eb/No degradation corresponds to 0.5 dB.

Parameter	Voice	NRTV	FTP	web
Average bit rate (kb/s)	13.3	64	115.2	57.6
Call/Session duration (min)	2	5	5	2

Table 2. Call simulation parameters

We increased the average handoff rate to the central cell from 1 to 10 MS/10s by increasing the cell density from 25 to 400 MSs/cell in indoor cells and from 100 to 1200 MSs/cell in vehicular cells. The new call rate λ is set to be equal to the handoff rate in each simulation run. At a handoff rate of 10, the central and surrounding cells are overloaded with traffic according to the indoor and outdoor market requirements in busy hours mentioned in (TSG-C.R1002, 2003). Such high loads present a worst case interference scenario, so the results shown at those handoff rates can be considered as a low bound of wideband CDMA performance.
The overhead on the forward link due to the percentage of power dedicated to the pilot, sync and paging channels is assumed to be 17% and the maximum transmit power limit of a mobile is set to 24dBm (TR45.5, 1998). The background thermal noise, used in checking the limit of user equipment transmit power, is -100dB (Padovani, 1994). The maximum base station transmit power is assumed to be 20 W. The spreading bandwidth W is 3750 kb/s corresponding to a third-generation 3xRTT cdma2000 technology. The orthogonality factor is taken to be 0.8 (Yang & Lee, 1997). All results shown are the average of 20 simulation runs conforming to a confidence level of 95% with errors varying from 0.012 to 0.05, each having a simulation time of 1 hour.

4.1 Admission control performance

We begin by testing our admission control module without the use of the Degradation module. Two aspects are analyzed: the admission condition threshold Th and the new-call/ handoff admission policy. They are analyzed with respect to blocking, dropping and outage probabilities. The blocking probability is the probability to deny the access to a new call while the handoff drop probability concerns the possibility of not being able to satisfy the requirements of a new incoming handoff call; both are due to capacity limits. The rejected handoff call continues with its already connected legs; though causing more interference, it becomes soon in outage. As mentioned before, the blocking probability on the uplink covers new call blocking when either the required transmitted power exceeds the maximum transmit power limit or the uplink admission condition of noise rise is not satisfied. The range of uplink η_K and downlink η, obtained from simulations, is 0.16 to 0.58 and 0.21 to 0.73 respectively.
In order to analyze the effect of Th on both the uplink and downlink capacity, the two probabilities have been measured for three threshold values, 0.6, 0.75 and 0.9, as shown in Fig.4 and Fig.5. As the threshold increases, more services are admitted on both the uplink and downlink. This results in lower block and drop probabilities, however, with the cost of increasing mobile outage. Note that the mobile transmit power limit counts for 9-13% of the blocking probability of new calls and for 4-6.5% of the handoff drop probability on the uplink. It is worth noting that the outage rate increases drastically with a threshold of 0.9, especially, at high loads. It reaches 8% and 6% on uplink and downlink respectively. This is because the coverage area, over which the minimum acceptable Eb/No is obtained, reduces with increasing number of services. At low threshold, 0.6, the outage can be neglected, however, the block and drop probabilities increase significantly at medium and high loads. The 0.75

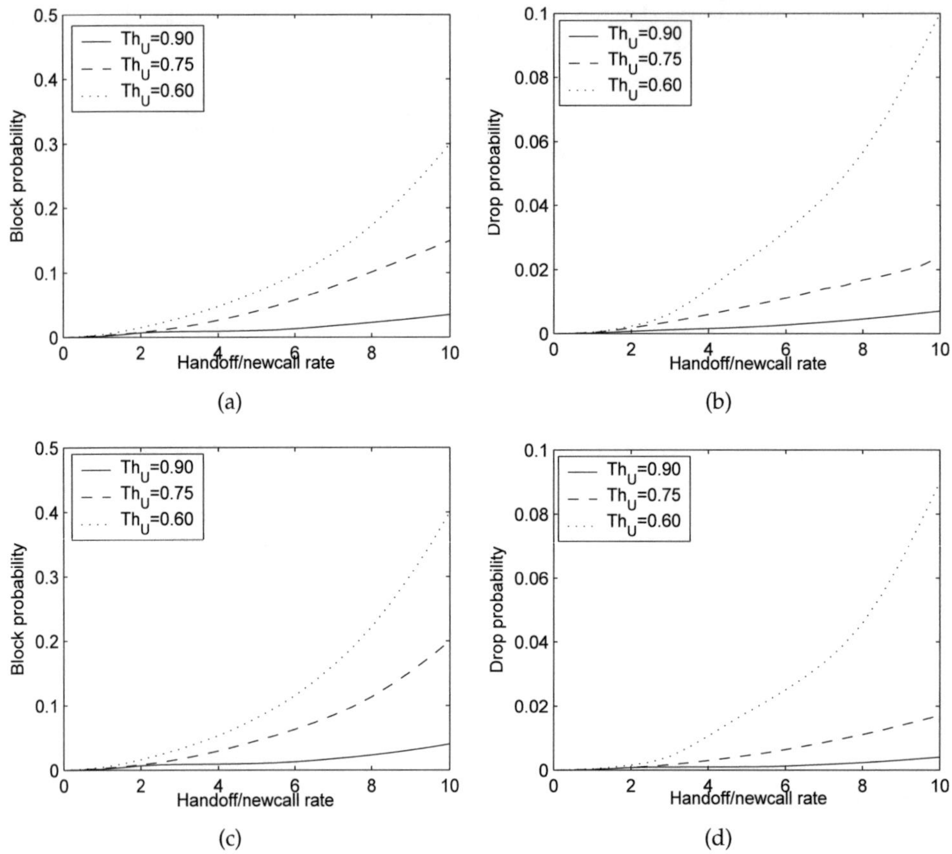

Fig. 4. Block and handoff drop probabilities on the uplink when varying the new/handoff rate (MS/10s) using different load thresholds in indoor (a,b) and outdoor (c,d) environments.

threshold gives a good coverage-capacity compromise. Moreover, with a threshold of 0.75, the outage rate at high loads is below 1% on both uplink and downlink, which is better than the 95% coverage required by ITU. Thus, the remaining results are obtained with a 0.75 threshold and the outage rate is not investigated further since it remains below 1% with this threshold value.

In what concerns the new-call/handoff admission policy, it can be seen in Fig.4 and Fig.5, that the proposed policy, which gives incoming handoff calls a priority over new calls, results in achieving handoff drop probabilities much lower than new-call blocking ones on both uplink and downlink. The handoff drop probability does not exceed 1% in medium loads and is around 2% in very high loads. Nevertheless, in order to assess it with respect to other possible policies, we compare the handoff drop and new-call block probabilities when deploying the same proposed admission conditions (for CDMA) on the same simulated environments, but with different policies. Two other policies have been tested: the guard channel (GC) approach and the equal priority (EP) scheme. With a GC policy, a certain cell capacity is reserved solely for incoming handoff calls and the left capacity is for common use for all calls. That is, the load threshold is further decreased by a guard factor for new calls. This strategy was suggested by (Cheng & Zhuang, 2002; Kulavaratharasah & Aghvami, 1999). In contrast, with the EP

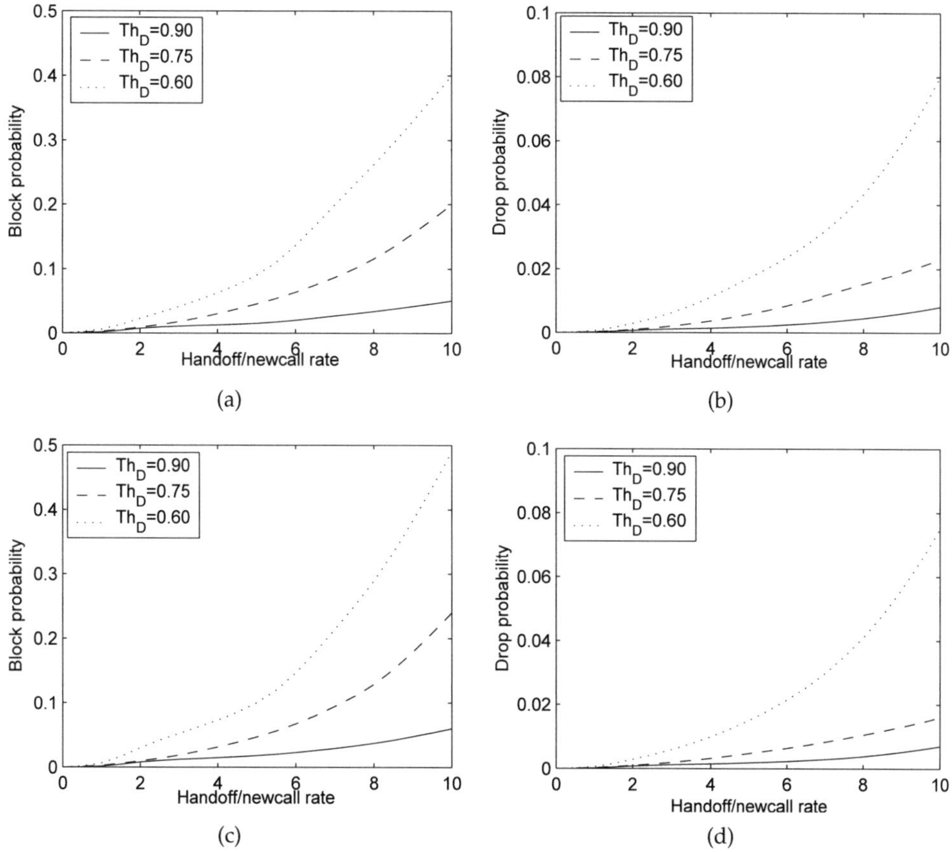

Fig. 5. Block and handoff drop probabilities on the downlink when varying the new/handoff rate (MS/10s) using different load thresholds in indoor (a,b) and outdoor (c,d) environments.

policy, both handoff and new calls are accepted if enough capacity exists to accomodate thier needs, no portion of the capacity is restricted for access of either type of call. This approach was selected by (Chang & Chen, 2006; Das et al., 2000). Our proposed policy gives incoming handoff calls a priority over new calls when call rejection becomes necessary, that is, when no capacity is available.

Note that for simplicity, from here after, the drop and block probabilities include both the uplink and downlink ones. Fig.6 shows that, in indoor environment, the handoff drop probability of our policy is below that of EP scheme by a difference that varies from 1% for a handoff rate of 1 to about 20% for a rate of 10. This is because our module gives the priority to handoff services compared to the EP scheme which does not differentiate handoff and new services. However, our block probability is higher than that of EP scheme by a difference that varies from 1% to 5%. It is clear that our gain in handoff admission surpasses the loss in new service admission.

Fig.6 demonstrates also the drop/block probability for 3 guard capacities of GC scheme. We observe that our handoff admission probability has a comparable performance with the GC scheme. It outperforms that of the 0.2 and 0.4 guard capacities by up to 7% and 2.5% respectively. However, the 0.6 guard capacity surpasses it by up to 2% for a handoff rate

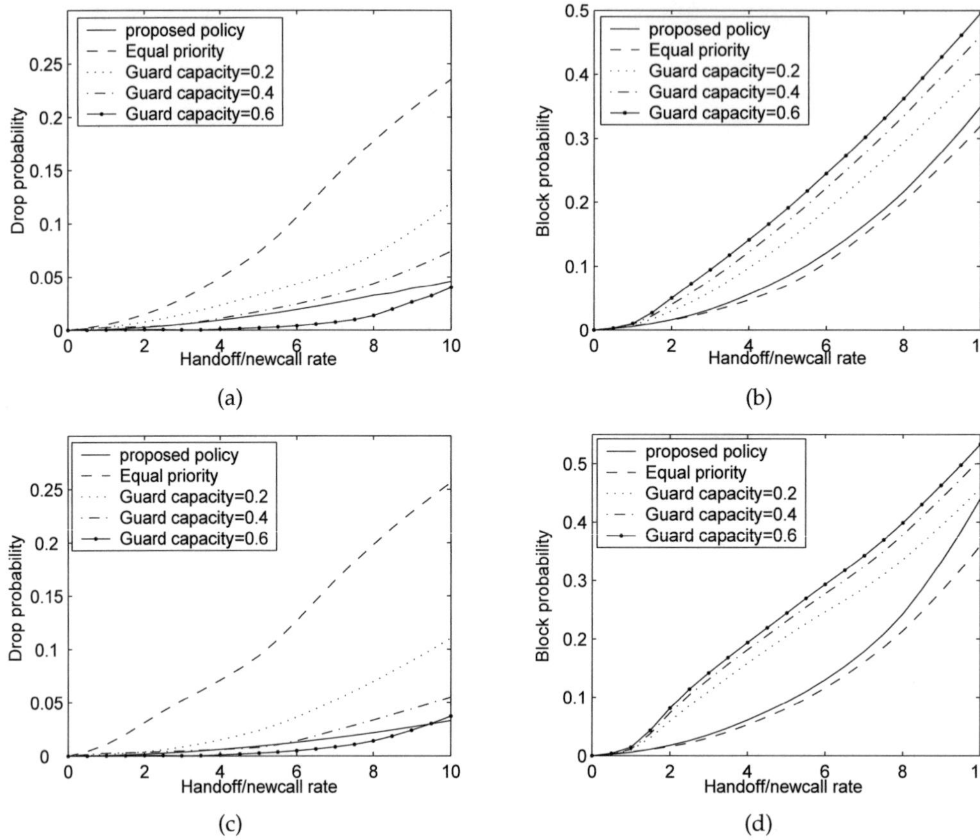

Fig. 6. Drop and block probability when varying the new/handoff rate (MS/10s) using different policies in indoor (a,b) and outdoor (c,d) environments.

that varies from 2 to 8. This difference drops to 0.5% in indoors and vanishes in outdoors at high handoff rates. As for the block probability of new services, it can be seen that our scheme outperforms all the guard capacities by up to 15% indoors and 18% outdoors for new call rates varying from 2 to 8. This is because, with a small handoff rate, the GC scheme results not only in high blocking of new services but also in low resource utilization because handoff services are allowed to use the reserved capacity exclusively. On the other hand, with a big number of handoff MSs that exceed the guard capacity, this scheme looses its advantage because it becomes difficult to guarantee the requirements of handoff users. The same observations can be noticed in outdoor environments. However, the drop probability of our approach is marginally better at high handoff rates with a difference of 1.4%. This is due to the fact that the outdoor cell is less dense than the indoor cell when using our motion model, which gives the AC module a little more capacity for admitting more handoff services.

We have combined both the block and drop probabilities in order to measure the total number of admitted services. Fig.7 shows that the proposed policy outperforms both the GC scheme and the EP approach in terms of total number of accepted services in the cell, either handoff or new ones, especially in high loads. It surpasses the EP approach by 14.2% and 13% in indoor and outdoor environments respectively. It outperforms the GC scheme by up to 12% and 15%

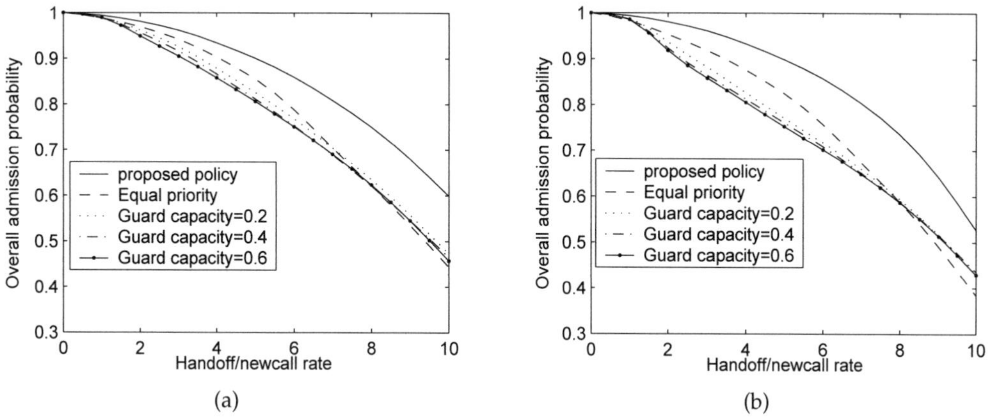

Fig. 7. Admission probability when varying the new/handoff rate (MS/10s) using different admission schemes in indoor (a) and outdoor (b) environments.

for 4-8 new/handoff rates. As also shown in Fig.7 it is clear that, at higher rates, this difference does not increase, where no capacity to be managed is left.
Recall that the time complexity of our AC module is $O(M)$ where M is the cell density. So, in the worst case where the cell density is 400 and 1200 MSs/cell in indoor and outdoor environments, the computation load is $O(1)$ in both environments. This is valid for forward services and reverse services that are not in soft handoff with other cells. However, for reverse connections that have, for instance, 2 soft handoff legs as in our simulations, this computing load would be multiplied by the number of handoff legs, which proves that soft handoff is computationally expensive as mentioned in (Kumar & Nanda, 1999).

4.2 Performance of D/I modules

Next, we evaluate the effect of deploying our D/I modules on the handoff/new admission probability resulting from our admission control scheme. First, we study the effect of varying Rsafe on the overall drop+block probability, then, for simplicity, two values are selected for Rsafe in order to study in details the benefits on admission probability as well as on cell throughput. Fig.8 shows the drop+block probability at a 7 new/handoff rate in indoor environment. Note that similar results were found in outdoors as well. When Rsafe=Rc, this corresponds to no degradation, while Rsafe=0 means that all MSs inside the cell are subject to degradation with no preference. It can be seen that as Rsafe decreases, the drop+block probability is reduced significantly. This is because as Rsafe decreases, zone 1 becomes larger and, hence, the probability of locating MSs that can be degraded rises, giving more possibility to acquire capacity for new and handoff calls. However, below 0.3Rc, the benefit of further decreasing of Rsafe on drop+block probability diminishes because the remaining safe area (zone 2) has become much smaller than zone 1. In what follows, we present results for Rsafe equal to 0.75Rc and 0.5Rc, which correspond to a safe zone of about half and quarter of the cell area respectively.
At low loads, the D/I scheme has a negligible effect on the admission performance. However, its contribution is manifest at high loads. Fig.9 shows that, when Rsafe is 0.5Rc, the drop probability is less than that shown in Fig.6 with 3.5% in indoor environment and 2.4% in outdoor environment. Moreover, it can be seen that, with the deployement of the Degradation module, the handoff admission probability surpasses the ones using guard capacities. This is

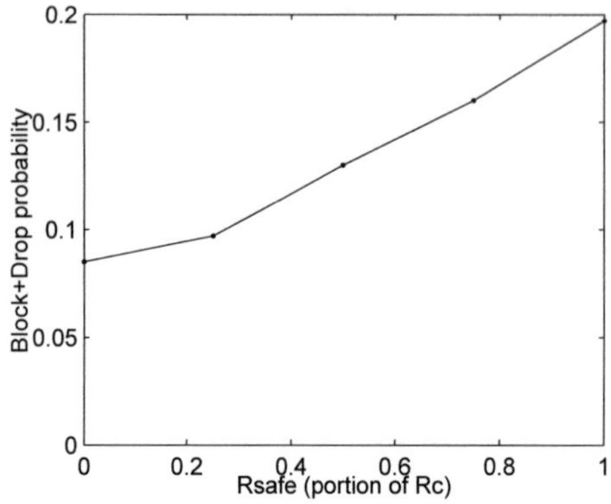

Fig. 8. Effect of varying Rsafe on the overall drop+block probability at a new/handoff rate of 7MSs/10s in indoor environments.

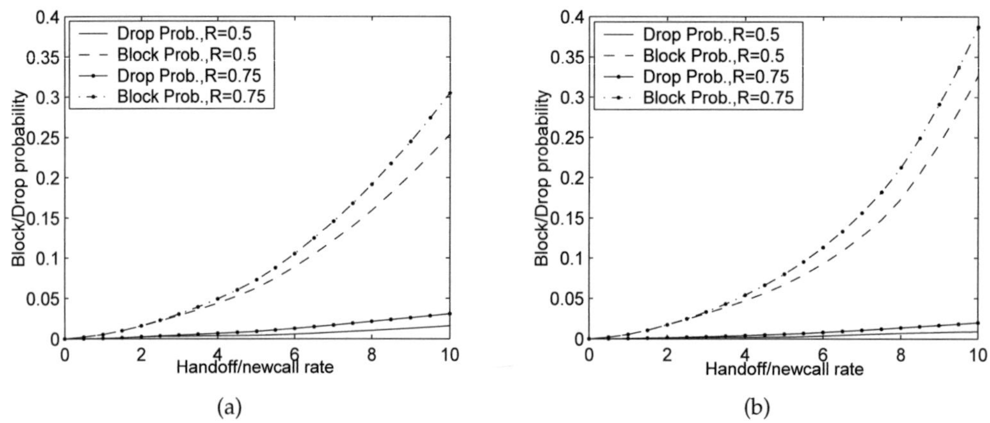

Fig. 9. Adaptation effect on drop and block probability when varying the new/handoff rate (MS/10s) in indoor (a) and outdoor (b) environments for Rsafe=0.5Rc and 0.75Rc.

because, with our design, there is no reservation of capacity for handoff services; instead, the call drop probability is decreased by degrading the QoS levels of services located near cell boundary, which reduces interference as well. As for new services, their block probability shows a significant improvement when compared to that shown in Fig.6; it has been reduced by a further 9.6% in indoor environments and 11.3% in outdoor environments. Furthermore, it can be seen that the new service admission probability is comparable to that of EP scheme shown in Fig.6 with the deployment of the Degradation module at Rsafe=0.75Rc and even better at 0.5Rc. Note that the observed outage when deploying D/I was always below 1%.
When Rsafe is set to 0.75Rc, the number of candidates for degradation decreases, which reduces the capacity that could be acquired for admitting new/handoff services. An improvement can still be observed in Fig.9. However, it is by far less than that of 0.5Rc. Fig.10 also shows the percentage of degraded MSs for both values of Rsafe. It can be seen that this percentage, in outdoor environments, goes up to 15.6% and 10% of total number of MSs for Rsafe of 0.5Rc and 0.75Rc respectively. This percentage drops to 7.5% and to 5.2%, in

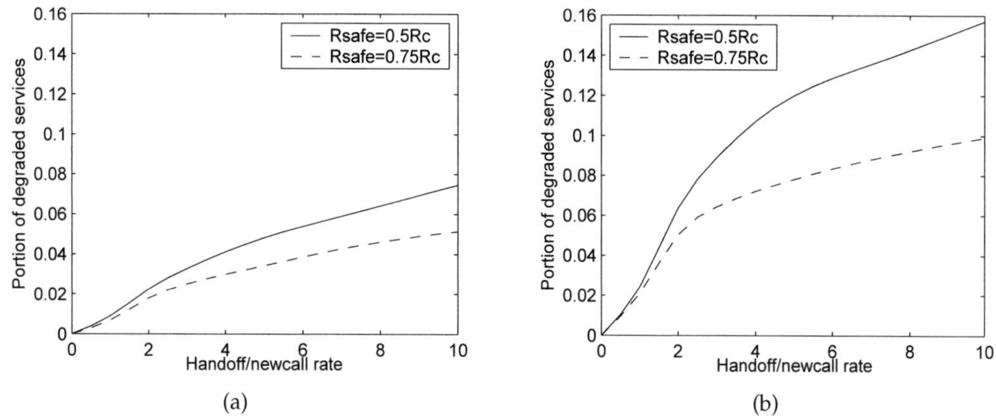

Fig. 10. Portion of degraded services in indoor (a) and outdoor (b) environments for Rsafe=0.5Rc and 0.75Rc when varying the new/handoff rate (MSs/10s).

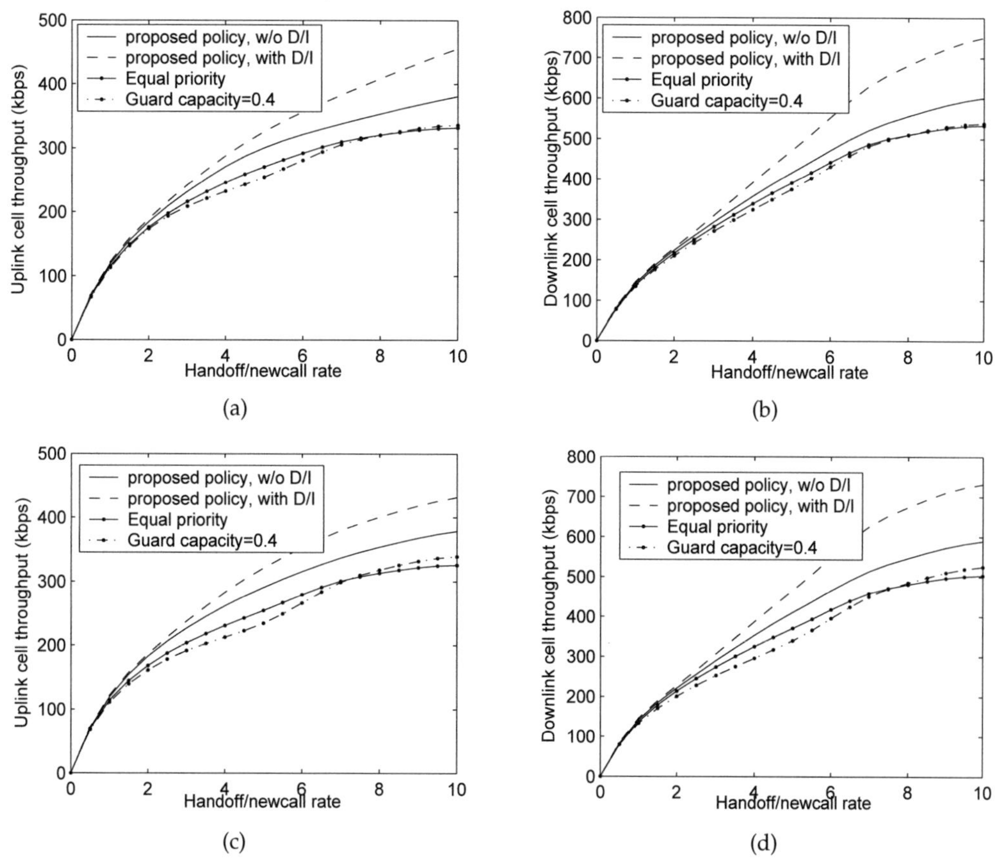

Fig. 11. Cell throughput on uplink and downlink when varying the new/handoff rate (MSs/10s) in indoor (a,b) and outdoor (c,d) environments.

indoor environments, for Rsafe of 0.5Rc and 0.75Rc respectively at high loads. This explains the difference in the observed improvement for both kinds of environments.

Note that, as the handoff rate increases, the proportion of degraded services increases till a point where the cell begins to be highly loaded. At this point, the AC module starts to decrease the rate of admitted minimum throughput services. Moreover, the Eb/No degradation of the near-real-time services is limited to 0.5 dB only, and degradation is only allowed if their measured signal to interference ratio is not already degraded. This limits the possibility of degradation for services since they are not degraded below their minimum acceptable requirements. Thus, in highly loaded situations, the proportion of degraded services increases as well but with a rate lower than that of lighter load situations. This also demonstrates that our design succeeds in limiting the number of degraded MSs and, hence, reducing the required signalling messages which saves time and capacity.

In order to verify the effect of D/I deployment on cell throughput, the throughput of the services inside the cell was measured, for Rsafe = 0.5Rc, and compared to the throughput of the admission control scheme without D/I. It was also compared to the throughput of EP policy and GC scheme having a guard capacity equal to 0.4. The cell throughput only includes the bit rate of the calls that stay in the cell till termination or ongoing to another cell without being in outage. It represents the average of the instantaneous aggregated bit rate of only the calls currently served by the base station. Fig.11 shows the throughput on uplink and downlink in indoor and outdoor environments when varying the new/handoff rate. At moderate loads, the D/I can only enhance by around 20-30kb/s each of the uplink and downlink throughputs. Nevertheless, in high loads, this enhancement is boosted up to 62kb/s and 151kb/s in indoors, and 51kb/s and 142kb/s in outdoors, on the uplink and downlink respectively. That is, an improvement of more than 210kb/s in the total cell throughput can be obtained in high loads. As also shown in Fig.11, the throughput of the proposed policy, with D/I, clearly outperforms those of EP and GC approaches. This demonstrates that the D/I deployment can rise the cell throughput as well as increasing the admission probability as seen above. However, this is achieved at the expense of unfairness between services, since degrading or improving the service level is not done uniformally across services, it depends on the MS location with respect to the safe area with aim of reducing interference.

The computation load of the Improvement module is the same as the one of the AC module without the soft handoff factor. However, the Degradation module has higher computation load of $O(N \log N)$ where N is the cell density. So, in the worst case where the cell density is 400 and 1200 MSs/cell in indoor and outdoor environments respectively, the computation load is $O(1)$ for forward and reverse services.

Another factor in evaluating the performance of the D/I modules is the response time for QoS adaptation. Since such QoS adjustment requires at most one signalling message per service, the time taken for a service to respond to such change is the time to send the control message to the MS of the service and processing it.

5. Conclusion and future work

In this chapter, we presented the design and evaluation of a service management scheme that is responsible for controlling the admission of new and handoff services and for service adaptation. The results show that our admission control proposal outperforms both the GC scheme and the EP approach in terms of total number of accepted services in the cell, either handoff or new ones, especially in high loads. It surpasses the EP approach by 14.2% and 13% and outperforms the GC scheme by 12% and 15% in indoor and outdoor environments respectively. Moreover, while limiting interference, signalling and computation overhead, the D/I modules succeeded in further improving the admission probability. The drop probability is lower than that when deploying the AC module only with 3.5% in indoor environment and 2.4% in outdoor environments. As for new services, their block probability shows a significant improvement, it is reduced by a further 9.6% in indoor environments and 11.3% in outdoor environments. The overall admission rate enhancement is achieved with low cost in terms of computition time and signalling messages, however, at the expense of unfairness among services.

In the research presented in this chapter, we did not consider automatic repeat request (ARQ) for retransmission on the radio link and forward error correction (FEC) techniques. These error correction mechanisms will be considered in a future work, since they can further enhance system capacity by decreasing target signal to noise ratios. Another research direction is to further examine new procedures for service admission on multiple cells level. This requires access coordination between BSs including sharing load information among neighbour cells, so that light loading in neighboring cells can be exploited to favor lower drop and block probabilities for handoff and new services respectively while still meeting interference constraints.

6. References

Aissa, S., Kori, J. & Mermelstein, P. (2004). Call admission on the uplink and downlink system based on total received and transmitted powers, *IEEE Transactions on Wireless Communications* Vol. 3(No. 6): 2407–2416.

Chang, J. & Chen, H. (2006). A borrowing-based call admission control policy for mobile multimedia wireless networks, *IEICE Transactions on Communications* Vol. E89-B(No. 10): 2722–2732.

Cheng, Y. & Zhuang, W. (2002). Diffserv resource allocation for fast handoff in wireless mobile internet, *IEEE Communications Magazine* Vol. 4(No. 5): 130–136.

Comaniciu, C., Mandayam, N. B., Famolari, D. & Agrawal, P. (2000). Qos guarantees for third generation (3g) cdma systems via admission and flow control, *Proceedings of VTC*, IEEE, Boston, MA, pp. 249–256.

C.S0005-Ev2.0 (2010). Upper layer (layer 3) signaling standard for cdma2000 spread spectrum systems, *3GPP2 Specifications* .

Das, S., Jayaram, R., Kakani, N. & Sen, S. (2000). A call admission and control scheme for qos provisioning in next generation wireless networks, *Journal of wireless Networks* Vol. 6(No. 1): 17–30.

Gilhousen, K., Jacobs, I., Padovani, R., Viterbi, A., Weaver, J. & Wheatley, C. (1991). On the capacity of a cellular cdma system, *IEEE Transactions on Vehicular Technology* Vol. 4: 303–312.

Him, K. & Koo, I. (2005). *CDMA Systems Capacity Engineering*, ARTECH HOUSE, INC.

Homnan, B., Kunsriruksakul, V. & Benjapolakul, W. (2000). The evaluation of soft handoff performance between is-95a and is-95b/cdma2000, *Proceedings of International Conference on Signal Processing and Communications*, IASTED, Marabella, Spain, pp. 38–42.

Issa, O. & Gregoire, J. (2006). A service admission scheme for cellular networks, *Proceedings of CCECE*, IEEE, Ottawa, Canada, pp. 750–753.

Kastro, Y., Isiklar, G. & Bener, A. (2010). Resource allocation in cellular networks based on marketing preferences, *Journal of wireless Networks* Vol. 16(No. 1): 27–38.

Kelif, J. & Coupechoux, M. (2009). On the impact of mobility on outage probability in cellular networks, *Proceedings of Wireless Communications and Networking Conference*, IEEE, Budapest, Hungary, pp. 1308–1313.

Kim, D., Hossain, E. & Bahragavan, V. (2003). Dynamic rate adaptation based on multidimentional multicode ds-cdma in cellular wireless networks, *IEEE Transactions on Communications* Vol. 51(No. 2): 247–260.

Kulavaratharasah, M. & Aghvami, A. (1999). Teletraffic performance evaluation of microcellular personal communication networks (pcn's) with prioritized handoff procedures, *IEEE Transactions on Vehicular Technology* Vol. 48(No. 1): 137–152.

Kumar, S. & Nanda, S. (1999). High data-rate packet communications for cellular networks using cdma: algorithms and performance, *IEEE Journal of Selelected Areas in Communnications* Vol. 17(No. 3): 472–492.

Kwon, T., Choi, Y., Bisdikian, C. & Naghshineh, M. (2003). Qos provisioning in wireless/mobile multimedia networks using an adaptive framework, *Journal of wireless Networks* Vol. 9(No. 1): 51–59.

Lee, J., Jung, T., Yoon, S., Youm, S. & Kang, C. (2000). An adaptive resource allocation mechanism including fast and reliable handoff in ip-based 3g wireless networks, *Proceedings of 3Gwireless*, IEEE, Silicon Valley, pp. 306–312.

Nasser, N. & Hassanein, H. (2004). Combined admission control algorithm and bandwidth adaptation algorithm in multimedia cellular networks for qos provisioning, *Proceedings of CCECE*, IEEE, Niagara Falls, Canada, pp. 1183–1186.

Nasser, N. & Hassanein, H. (2006). Robust dynamic call admission control framework for prioritized multimedia traffic in wireless cellular networks, *International Journal of High Performance Computing and Networking (IJHPCN)* Vol. 4(No. 1/2): 3–12.

Padovani, R. (1994). Reverse link performance of is-95 based cellular systems, *IEEE Personal Communications Magazine* Vol. 1: 28–34.

Sipila, K., Honkasalo, Z., j. Laiho-Steffens & Wacker, A. (2000). Estimation of capacity and required transmission power of wcdma downlink based on a downlink pole equation, *Proceedings of VTC*, IEEE, Tokyo, Japan, pp. 1002–1005.

TR45.5 (1998). The cdma2000 itu-r rtt candidate submission, *TIA* .

TSG-C.R1002 (2003). 1xev-dv evaluation methodology (v14), *3GPP2 Specifications* .

Yang, J. & Lee, W. (1997). Design aspects and system evaluations of is-95 based cdma systems, *IEEE Universal Personal Communications Record* Vol. 6: 381–385.

11

Radio Resource Management in Heterogeneous Cellular Networks

Olabisi E. Falowo and H. Anthony Chan
Department of Electrical Engineering, University of Cape Town
South Africa

1. Introduction

The evolution of cellular networks from one generation to another has led to the deployment of multiple radio access technologies (such as 2G/2.5G/3G/4G) in the same geographical area. This scenario is termed heterogeneous cellular networks. In heterogeneous cellular networks, radio resources can be jointly or independently managed. When radio resources are jointly managed, joint call admission control algorithms are needed for making radio access technology selection decisions. This chapter gives an overview of joint call admission control in heterogeneous cellular networks. It then presents a model of load-based joint call admission control algorithm. Four different scenarios of call admission control in heterogeneous cellular networks are analyzed and compared. Simulations results are given to show the effectiveness of call admission control in the different scenarios.

The coexistence of different cellular networks in the same geographical area necessitates joint radio resource management (JRRM) for enhanced QoS provisioning and efficient radio resource utilization. The concept of JRRM arises in order to efficiently manage the common pool of radio resources that are available in each of the existing radio access technologies (RATs) (Pérez-Romero *et al*, 2005). In heterogeneous cellular networks, the radio resource pool consists of resources that are available in a set of cells, typically under the control of a radio network controller or a base station controller.

There are a number of motivations for heterogeneous wireless networks. These motivations are (1) limitation of a single radio access technology (RAT), (2) users' demand for advanced services and complementary features of different RATs, and (3) evolution of wireless technology. Every RAT is limited in one or more of the following: data rate, coverage, security-level, type of services, and quality of service it can provide, etc. (Vidales *et al*, 2005). A motivation for heterogeneous cellular networks arises from the fact that no single RAT can provide ubiquitous coverage and continuous high QoS levels across multiple smart spaces, e.g. home, office, public smart spaces, etc. Moreover, increasing users' demand for advanced services that consume a lot of network resources has made network researchers developed more and more spectrally efficient multiple access and modulation schemes to support these services. Consequently, wireless networks have evolved from one generation to another. However, due to huge investment in existing RATs, operators do not readily discard their existing RATs when they acquire new ones. This situation has led to coexistence of multiple RATs in the same geographical area.

In wireless networks, radio resource management algorithms are responsible for efficient utilization of the air interface resources in order to guarantee quality of service, maintain the planned coverage area, and offer high capacity. In heterogeneous cellular networks, radio resource can be independently managed as shown in Figure 1 or jointly managed as shown in Figure 2. However, joint management of radio resources enhances quality of service and improves overall radio resource utilization in heterogeneous cellular networks.

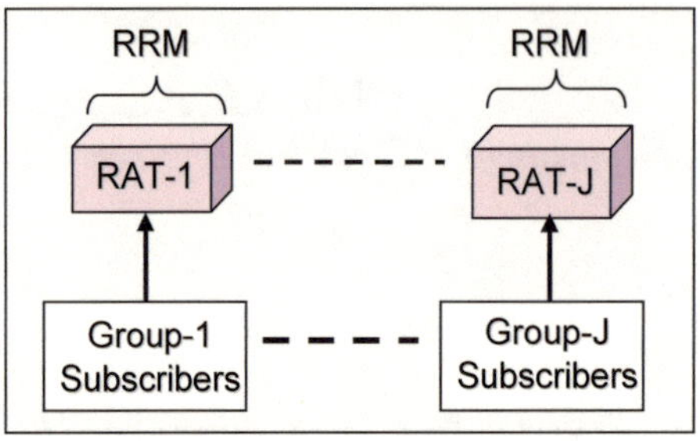

Fig. 1. Independent RRM in heterogeneous wireless networks.

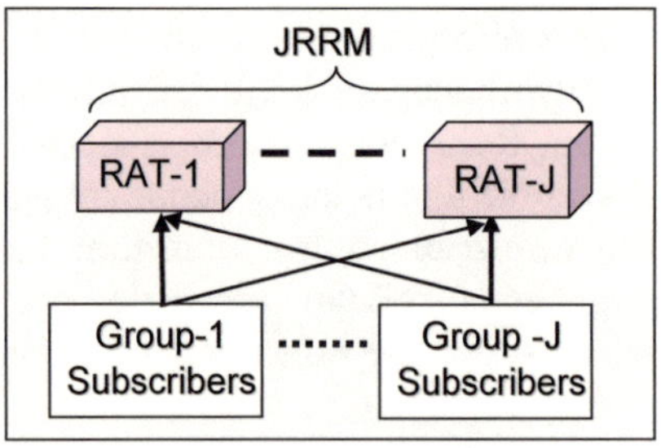

Fig. 2. Joint RRM in heterogeneous wireless networks

With joint radio resource management in heterogeneous cellular networks, mobile users will be able to communicate through any of the available radio access technologies (RATs) and roam from one RAT to another, using multi-mode terminals (MTs) (Gelabert et al, 2008), (Falowo & Chan, 2007), (Falowo & Chan, 2010), (Lee et al, 2009), (Niyato & Hossain, 2008). Figure 3, adapted from (Fettweis, 2009), shows a two-RAT heterogeneous cellular network with collocated cells.

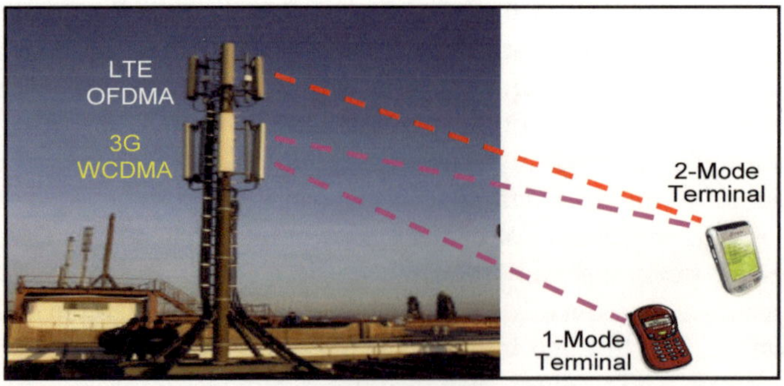

Fig. 3. A typical two-RAT heterogeneous cellular network with co-located cells.

Availability of multi-mode terminals is very crucial for efficient radio resource management in heterogeneous wireless networks. A mobile terminal can be single-mode or multi-mode. A single-mode terminal has just a single RAT interface, and therefore can be connected to only one RAT in the heterogeneous wireless network. A multi-mode terminal has more than one RAT interface, and therefore can be connected to any of two or more RATs in the heterogeneous wireless network.

As show in Figure 3, a subscriber using a two-mode terminal will be able to access network services through either of the two RATs. However, a subscriber using a single-mode terminal will be confined to a single RAT, and cannot benefit from joint radio resource management in the heterogeneous wireless network.

In heterogeneous cellular networks, radio resources are managed by using algorithms such as joint call admission control algorithms, joint scheduling algorithms, joint power control algorithms, load balancing algorithms, etc. This chapter focuses on joint call admission control (JCAC) algorithms in heterogeneous cellular networks.

The rest of this chapter is organized as follows. In Section 2, JCAC in heterogeneous cellular network is described. In Section 3, we present a JCAC model and assumptions. In Section 4, we investigate the performance of the JCAC algorithm through numerical simulations.

2. Joint Call Admission Control in heterogeneous cellular networks

JCAC algorithm is one of the JRRM algorithms, which decides whether an incoming call can be accepted or not. It also decides *which of the available radio access networks is most suitable to accommodate the incoming call*. Figure 4 shows call admission control procedure in heterogeneous cellular networks.

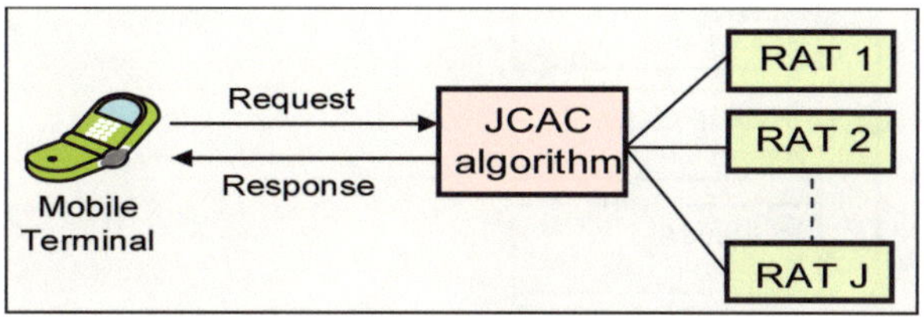

Fig. 4. Call admission control procedure in heterogeneous cellular networks.

A multi-mode mobile terminal wanting to make a call will send a service request to the JCAC algorithm. The JCAC scheme, which executes the JCAC algorithm, will then select the most suitable RAT for the incoming call.

Generally, the objectives of call admission control algorithm in heterogeneous cellular networks are:

1. Guarantee the QoS requirements (data rate, delay, jitter, and packet loss) of accepted calls.
2. Minimize number vertical handoffs,
3. Uniformly distribute network load as much as possible,
4. Minimize call blocking/dropping probability,
5. Maximize operators' revenue,
6. Maximize radio resource utilization

All the above objectives cannot be simultaneously realized by a single JCAC algorithm. Thus, there are tradeoffs among the various objectives.

2.1 RAT selection approaches used in JCAC algorithms

A number of RAT selection approaches have been proposed for JCAC algorithms in heterogeneous cellular networks. These approaches can be broadly classified as single-criterion or multiple-criteria. Single-criterion JCAC algorithms make call admission decisions considering mainly just one criterion, such as network load, service cost, service class, random selection, path loss measurement, RAT layer, and terminal modality. On the other hand, multiple-criteria JCAC algorithms make RAT selection decisions based on a combination of multiple criteria. The multiple criteria are combined using computational intelligent technique (such as fuzzy logic, Fuzzy-neural, Fuzzy MADM (Multiple Attribute Decision Making, etc.) or non-computational intelligent technique (such as cost function). Figure 5 summarizes the different approaches for making RAT selection decisions by JCAC algorithms.

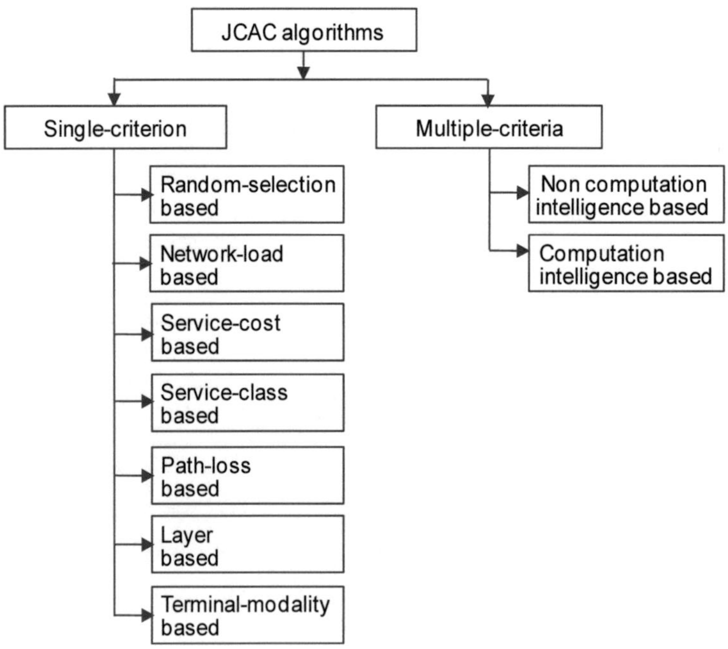

Fig. 5. RAT selection approaches for JCAC algorithm in heterogeneous cellular networks.

2.2 Bandwidth allocation techniques

In order to give different levels of priorities to different calls in wireless networks, it may be necessary to allocate certian block of basic bandwidth units (bbu) for new and handoff calls as well as for different classes of calls such as voice, video, etc,. In this section, bandwidth allocation strategies for wireless networks are reviewed. Bandwidth allocation strategies for wireless networks can be classified into four groups namely complete sharing, complete partitioning, handoff call prioritization, and service class prioritization. This classification is summarized in Table 1.

Bandwidth Allocation Strategy	Main Idea	Advantage	Disadvantage
Complete Sharing	An incoming call is accepted, regardless of the class/ type, as long as there is enough radio resource to accommodate it.	Implementation simplicity and high radio resource utilization	High handoff call dropping probability. No differential treatment for calls with stringent QoS requirements
Complete Partitioning	Available bandwidth is partitioned into pools and each pool is dedicated to a particular type of calls. An incoming call can only be admitted into a particular pool.	Implementation simplicity	Poor radio resource utilization
Handoff Call Prioritization	Handoff calls are given more access to radio resources than new calls. New calls may be blocked whereas handoff calls are still being admitted.	Low handoff call dropping probability	High new call blocking probability
Service-Class Prioritization	Certain classes of calls are given preferential treatment over some other classes of calls. For example, class-1 calls may be blocked whereas class-2 calls are still being admitted.	Differential treatments of calls based on QoS requirements	Implementation complexity

Table 1. Summary of Bandwidth Allocation Strategies for Wireless Networks.

2.2.1 Complete sharing

Complete sharing scheme is a first come first serve scheme and it is the simplest bandwidth allocation policy. It is a non-prioritization scheme in which new and handoff calls are treated the same way. An incoming call is accepted as long as there is enough radio resource to accommodate it. When the network gets to its maximum capacity, a new call will be blocked while a handoff call will be dropped. Two major advantages of complete sharing CAC scheme are implementation simplicity and good radio resource utilization. However, it has a high handoff call dropping probability because it does not give preference to any call. Consequently, complete sharing CAC scheme has a poor QoS performance (Ho, C. & Lea, C. 1999). Figure 6 is the state transition diagram for complete sharing scheme where $\lambda_n, \lambda_h, \mu_n$ *and* μ_h represent new call arrival rate, handoff call arrival rate, new call departure rate, and handoff call departure rate respectively.

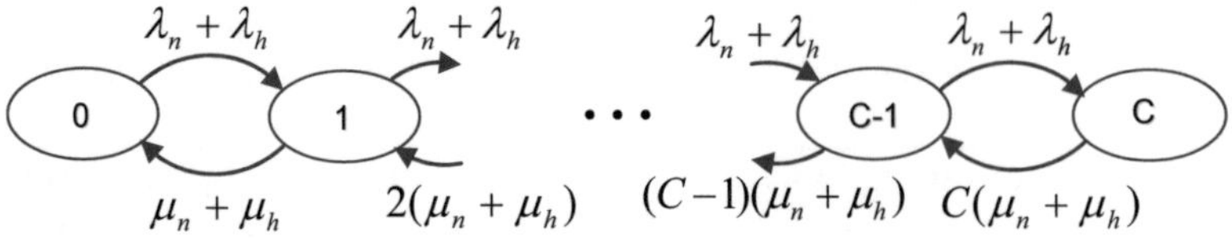

Fig. 6. State transition diagram for complete sharing policy.

2.2.2 Complete partitioning

In the complete partitioning CAC scheme, entire available bandwidth is partitioned into pools. Each pool is dedicated to a particular type of calls (new or handoff calls) and/or particular traffic class of calls. An incoming call is admitted if there is an available channel in the pool allocated for the type/class of the incoming call. This policy allocates a fixed bandwidth C_1 (C_2) to service s_1 (s_2) such that $C_1+C_2 <= C$. The acceptable states of this policy are a subset of the complete sharing case. This is a case of two independent queues, and the blocking probability is given by the well known Erlang-B formula.

Figure 7 and Figure 8 are the state transition diagrams of a system where the available resource (C) is partitioned into two (C_1 and C_2). C_1 is used for new calls (Figure 7) whereas C_2 is used for handoff calls (Figure 8).

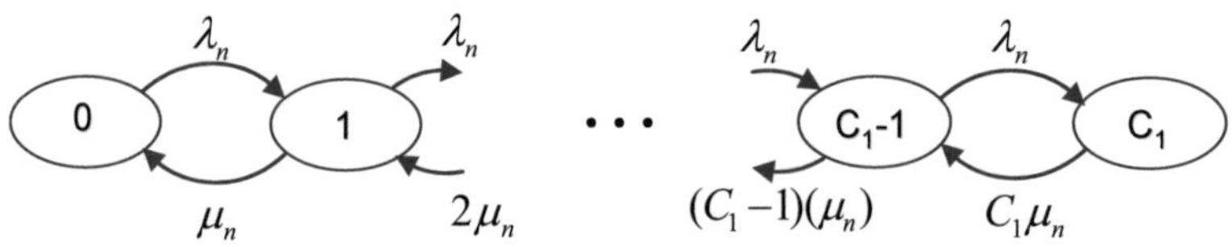

Fig. 7. State transition diagram for complete partitioning policy: first partition.

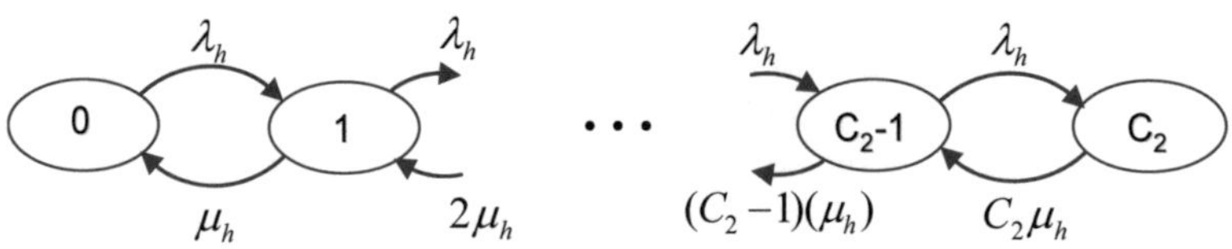

Fig. 8. State transition diagram for complete partitioning policy: second partition.

2.2.3 Handoff call prioritization

Due to users' mobility within the coverage of wireless networks, an accepted call that has not been completed in the current cell has to be transferred (handed over) to another cell. The call may not be able to get a channel in the new cell to continue its service due to limited radio resources in wireless networks. Eventually, it may be dropped. However, wireless network subscribers are more intolerant to dropping a handoff call than blocking a new call. Therefore, in order to ensure that handoff call dropping probability is kept below a certain level, handoff calls are usually admitted with a higher priority compared with new calls. Handoff call prioritization has an advantage of low handoff call dropping probability. However, the advantage of low handoff call probability is at the expense of new call blocking probability, which is high. Several handoff-priority-based schemes have been proposed in the literature. Some of these schemes are briefly reviewed as follows:

Guard Channel

In Guard Channel scheme, some channels (referred to as guard channels) are specifically reserved in each cell to take care of handoff calls. For example, if the total number of available channels in a single cell is C and the number of guard channels is C - H, a new call is accepted if the total number of channels used by ongoing calls (i.e., busy channels) is less than the threshold H, whereas a handoff call is always accepted if there is an available channel (Hong & Rappaport, 1986). l. Guard channel (GC) scheme can be divided into two categories namely static and dynamic strategies. In static guard channel scheme, the value of H is constant whereas in dynamic guard channel scheme, H varied with the arrival rates of new and handoff calls. Figure 9 shows the state transition diagram for a single-class service using guard bandwidth scheme.

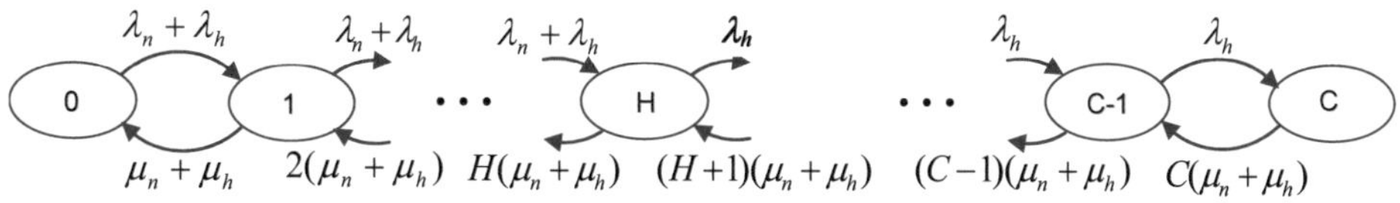

Fig. 9. State transition diagram for guard bandwidth scheme.

Fractional Guard Channel

In fractional guard channel scheme, handoff calls are prioritized over new calls by accepting an incoming new call with a certain probability that depends on the number of busy channels. In other words, when the number of busy channels becomes larger, the acceptance probability for a new call becomes smaller, and vice versa. This approach helps to reduce the handoff call dropping probability. The policy has a threshold, H for limiting the acceptance of new calls. A handoff is accepted as long as there is a channel available. Before the wireless system gets to threshold, H, new calls are accepted with a probability of 1. After threshold, H, a new call is accepted with a probability of α_p where $0 \le \alpha_p \le 1$ and H<p<C. New calls are rejected when the system reaches the maximum capacity. Figure 10 is the state transition diagram for fractional guard bandwidth policy.

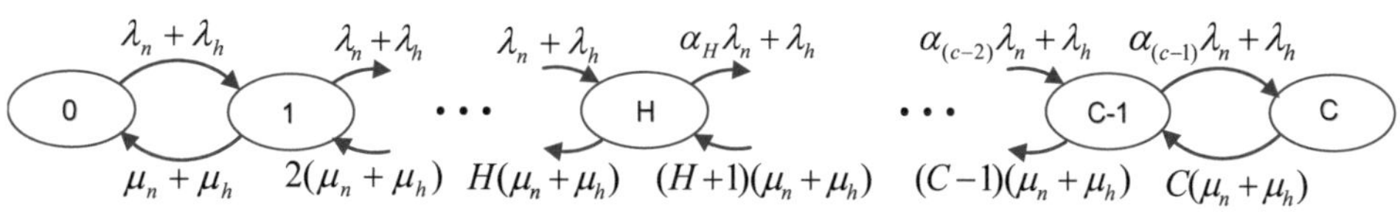

Fig. 10. State transition diagram for fractional guard bandwidth policy.

Queuing Priority Scheme

Queuing priority scheme accepts calls (new and handoff) whenever there are free channels. When all the channels are occupied, handoff calls are queued while new calls are blocked or all incoming calls are queued with certain rearrangement in the queue. When radio resource becomes available, one or some of the calls in the handoff queue are served until there is no more resource. The remaining calls are queued until resource becomes available again. However, a call is only queued for a certain period of time. If radio resource is not available within this period, the call will be dropped.

The main disadvantage of queuing priority scheme is that is needs a lot of buffers to deal with real-time multimedia traffic. It also needs a sophisticated scheduling mechanism in order to meet the QoS requirements of delay-sensitive calls, i.e. to guarantee that the queued data will be transmitted without excessive delay. (Chen, *et al*, 2002).

QoS Degradation Scheme

QoS degradation can either be bandwidth degradation or delay degradation. In bandwidth degradation method, calls are categorized as adaptive (degradable) and non-adaptive (non-degradable) calls. Degradable calls have flexible QoS requirements (e.g., minimum and maximum data rates). For most multimedia applications, e.g., voice over IP or video conferencing, service can be degraded temporarily as long as it is still within the pre-defined range. Bandwidth degradation reduces handoff call dropping by reducing the bandwidth of the ongoing adaptive calls during network congestion. When a handoff call arrives and there is network congestion, the system is able to free some radio resource to admit the handoff calls by degrading some of the ongoing adaptive calls. In delay degradation method, the amount of radio resources allocated to non-real-time (delay-tolerant) services is reduced during network congestion. When a handoff call arrives and there is no radio resource to accommodate the handoff call. Some non-real-time services are degraded to free some bandwidth, which is used to accommodate the incoming handoff call.

2.2.4 Service-class prioritization

In wireless systems which support multiple service classes, the limited bandwidth has to be shared among the multiple traffic classes. Complete sharing scheme allows the network radio resource to be shared among the various service classes without preference for any class. However, one major challenge in the design of CAC policy is to provide preferential treatment among users of different service classes while still utilizing the system resources efficiently. Preferential treatments are given to certain classes of calls for the following reasons: (1) some calls (such as voice call) have stringent QoS requirements and therefore require preferential treatment. (2) Some subscribers in a particular service class are willing to pay more for better QoS. Service class prioritization scheme is more complicated than complete sharing and complete partitioning schemes.

Figure 11 shows the prioritization scheme used in this paper. As shown in the figure, the two-class J-RAT heterogeneous wireless network (where J is the total number of RATs in the network) has different thresholds for prioritizing the two classes of calls. Th_{1j} and T_{2j} are the thresholds for rejecting class-1 and class-2 handoff calls in RAT j, respectively whereas Tn_{1j} and Tn_{2j} are the thresholds for rejecting class-1 and class-2 new calls in RAT j, respectively. It can be seen that handoff calls are prioritized over new call by using higher thresholds for handoff call. It can also be seen that class-1 calls are prioriotized over class-2 calls. The rejection thresholds can be static or dynamic. Static thresholds are very simple to implement but are less efficient whereas dynamic threshold are more efficient but are more complicated.

Bandwidth allocation to individual calls in cellular networks can be static or adaptive. In static bandwidth allocation, a fixed unit of radio resource is allocated to each call, and the allocated unit is fixed during the entire duration of the call. In adaptive bandwidth allocation, resource allocated to each call varies between a minimum value and a maximum value. When the network is underutilized, maximum amount of radio resources are allocated to certain type of calls (adaptive calls). However, when the network is being over subscribed, minimum amount of radio resources are allocated to adaptive calls in order to free up some amount of radio resources to accommodate more calls. Adaptive bandwidth allocation improves radio resource allocation efficiency but it is more complicated. They also incur more signalling overhead. Figure 12 shows adaptive bandwidth allocation allocation for class-i calls, where $b_{i,min}$ and $b_{i,max}$ are the minimum and maximum bandwith units that can be allocated to class-i calls respectively. For fixed bandwidth allocation, $b_{i,min} = b_{i,max}$.

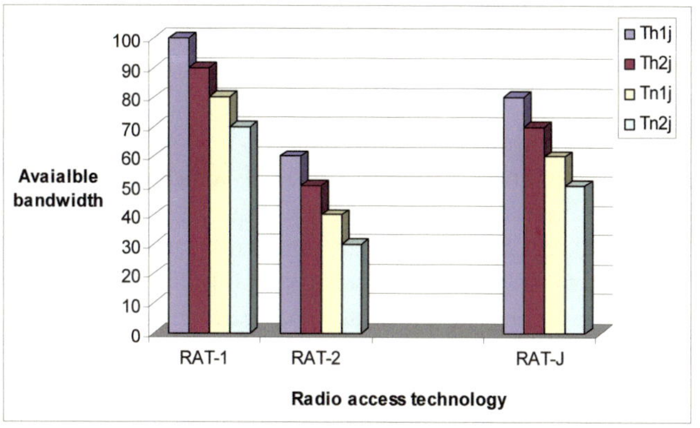

Fig. 11. Call prioritization in a two-class J-RAT heterogeneous wireless network.

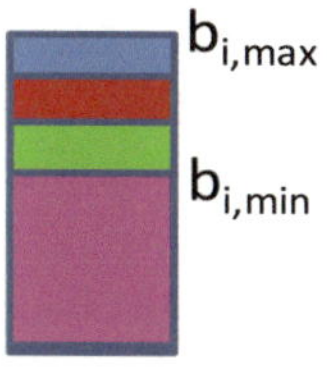

Fig. 12. Bandwidth allocation for adaptive calls.

3. Modelling of joint call admission control algorithm in heterogeneous cellular networks

We present a model of a load-based JCAC algorithm in a two-RAT heterogeneous cellular network supporting two classes of calls: class-1 call (voice) and class-2 call (video). The load based-JCAC algorithm admits an incoming call into the least loaded RAT in the heterogeneous wireless networks (scenarios 1 and 2 in Table 2). We also consider independent call admission control (ICAC) where radio resources are independently managed in the two RATs (scenarios 3 and 4 in Table 2). The four scenarios considered in the simulations are summarized in Table 2.

Scenario	Bandwidth allocation for class-1 and class-2 calls	Resource management	Acronym
1	Adaptive allocation	joint	AJCAC
2	Fixed allocation	joint	FJCAC
3	Adaptive allocation	independent	AICAC
4	Fixed allocation	independent	FICAC

Table 2. Scenarios Considered in the Simulations.

Scenarios 1 and 3 use adaptive bandwidth allocation where full rate bandwidth is allocated to class-1 calls when the network is underutilized whereas half rate bandwidth is allocated to class-1 calls when the networks is over subscribed. Similarly, class-2 calls are allocated a maximum amount of bandwidth when the network is underutilized whereas they are allocated a minimum amount of bandwidth when the network is oversubscribed.

Scenarios 2 and 4 uses fixed bandwidth allocation where fixed amount of bandwidth (full rate) is allocated to class-1 calls and fixed among of bandwidth (maximum rate) is allocated to class-2 calls at all times.

3.1 System model and assumptions

We consider a generic heterogeneous cellular network, which consists of J number of RATs with co-located cells, similar to (Zhang, 2005). Cellular networks such as GSM, GPRS, UMTS, EV-DO, LTE, etc, can have the same and fully overlapped coverage, which is technically feasible, and may also save installation cost (Holma & Toskala, 2001).

We consider cases where radio resources are independently or jointly managed in the heterogeneous network and each cell in RAT j (j =1,…,J) has a total of B_j basic bandwidth units (bbu). The physical meaning of a unit of radio resources (such as time slots, code sequence, etc) is dependent on the specific technological implementation of the radio interface. However, no matter which multiple access technology (FDMA, TDMA, CDMA, or OFDMA) is used, we could interpret system capacity in terms of effective or equivalent bandwidth. Therefore, whenever we refer to the bandwidth of a call, we mean the number of bbu that is adequate for guaranteeing the desired QoS for this call, which is similar to the approach used for wireless networks in (Falowo & Chan, 2007).

Our approach is based on decomposing a heterogeneous cellular network into groups of co-located cells. As shown in Fig. 3, overlapping cells form a group of co-located cells. A newly arriving call will be admitted into one of the cells in the group of co-located cells where the call is located. If the call cannot be admitted into any of the cells it will be blocked. Following the general assumption in cellular networks, new and handoff class-i calls arrive in the group of co-located cells according to Poisson process with rate λ_i^n and λ_i^h respectively. Note that the arrival rates of a split Poisson process are also Poisson (Bertsekas & Tsitsiklis, 2002). The channel holding time for class-i calls is exponentially distributed with mean $1/\mu_i$ (Orlik & Rappaport, 2001).

3.2 Markov model

The load-based based JCAC algorithm can be modeled as a multi-dimensional Markov chain. The state space of the group of co-located cells can be represented by a (2*K*J)-dimensional vector given as:

$$\begin{aligned}
&S = \{\Omega = (m_{i,j}, n_{i,j} : i = 1, \dots, k,\ \ j = 1, \dots, J): \\
&\sum_{c=1}^{m_{i,j}} b_{i,\,assigned_c} \le Tn_{i,j} \ \ \forall\, i,j \ \wedge \\
&\sum_{c=1}^{n_{i,j}} b_{i,\,assigned_c} \le Th_{i,j} \ \ \forall\, i,j \ \wedge \\
&\sum_{i=1}^{k}\sum_{c=1}^{m_{i,j}} b_{i,\,assigned_c} + \sum_{i=1}^{k}\sum_{c=1}^{n_{i,j}} b_{i,\,assigned_c} \le B_j \ \ \forall\, j\}
\end{aligned} \tag{1}$$

The non-negative integer $m_{i,j}$ denotes the number of ongoing new class-i calls in RAT j, and the non-negative integer $n_{i,j}$ denotes the number of ongoing handoff class-i calls in RAT j. S denotes the state space of all admissible states of the group of collocated cells. $B_{i,assigned}$ is the number of bbu allocated to an incoming class-i call, and the values varies between $b_{i,min}$ and $b_{i,\,max}$.

Let $\rho_{new_{i,j}}$ and $\rho_{han_{i,j}}$ denote the load generated by new class-*i* calls and handoff class-*i* calls, respectively, in RAT-*j*. Let $1/\mu_i^n$ and $1/\mu_i^h$ denote the channel holding time of new class-i call and handoff class-i call respectively, and let $\lambda_{i,j}^n$ and $\lambda_{i,j}^h$ denote the arrival rates of new class-i call and handoff class-i call in RAT j , respectively, then,

$$\rho_{new_{i,j}} = \frac{\lambda_{i,j}^n}{\mu_i^n} \quad \forall\, i,j, \tag{2}$$

$$\rho_{han_{i,j}} = \frac{\lambda_{i,j}^h}{\mu_i^h} \quad \forall\, i,j \tag{3}$$

From the steady state solution of the Markov model, performance measures of interest can be determined by summing up appropriate state probabilities. Let *P(s)* denotes the steady state probability that system is in state **s (s∈S)**. From the detailed balance equation, *P(s)* is obtained as:

$$P(s) = \frac{1}{G}\prod_{i=1}^{k}\prod_{j=1}^{J}\frac{(\rho_{new_{i,j}})^{m_{i,j}}}{m_{i,j}!}\frac{(\rho_{han_{i,j}})^{n_{i,j}}}{n_{i,j}!} \quad \forall\, s \in S \tag{4}$$

where G is a normalization constant given by:

$$G = \sum_{s\in S}\prod_{i=1}^{k}\prod_{j=1}^{J}\frac{(\rho_{new_{i,j}})^{m_{i,j}}}{m_{i,j}!}\frac{(\rho_{han_{i,j}})^{n_{i,j}}}{n_{i,j}!} \tag{5}$$

A new class-*i* call is blocked in the group of co-located cells if none of the RATs in the group of co-located cells has enough bbu to accommodate the new call. Let $S_{bi} \subset S$ denote the set of states in which a new class-*i* call is blocked in the group of collocated cells. Thus the new call blocking probability (NCBP), P_{b_i} , for a class-*i* call in the group of co-located cells is given by:

$$P_{b_i} = \sum_{s\,\in S_{b_i}} P(s) \tag{6}$$

A handoff class-*i* call is dropped in the group of co-located cells if none of the RATs in the group of collocated cells has enough bbu to accommodate the handoff call. Let $S_{d_i} \subset S$ denote the set of states in which a handoff class-*i* call is dropped in the group of co-located cells. Thus the handoff call dropping probability (HCDP) for a class-*i* call, P_{d_i} , in the group of co-located cells is given by:

$$P_{d_i} = \sum_{s \in S_{d_i}} P(s) \tag{7}$$

4. Numerical results

In this section, the performance of the JCAC scheme is evaluated through simulations. Results for both class-1 calls and class-2 calls are presented for the four scenarios shown in Table 2. The parameters used in the simulations are B_1=20, B_2=40, $Tn_{1,1}$=$Tn_{2,1}$=12, $Th_{1,1}$=$Th_{2,1}$=20, $Tn_{1,2}$=$Tn_{2,2}$=24, $Th_{2,1}$=$Th_{2,2}$=40, μ_1=μ_2=0.5. Some other parameters used are shown in Table 3.

Scenario	Bandwidth allocation
1	$b_{1,min}$=1bbu, $b_{2,\,min}$=3bbu, $b_{1,max}$=2bbu, $b_{2,\,max}$=7bbu
2	$b_{1,min}$= $b_{1,max}$= 2bbu, $b_{2,min}$=$b_{2,\,max}$=7bbu
3	$b_{1,min}$=1bbu, $b_{2,\,min}$=3bbu, $b_{1,max}$=2bbu, $b_{2,\,max}$=7bbu
4	$b_{1,min}$= $b_{1,max}$= 2bbu, $b_{2,min}$=$b_{2,\,max}$=7bbu

Table 3. Simulation Parameters.

4.1 Comparison of new call blocking probabilities for the four scenarios

Figure 13 shows the variation of new class-1 call blocking probability (Pb1) with call arrival rates for the four scenarios. Pb1 increases with increase in arrival rates for each of the four scenarios. However, the AJCAC scheme has the lowest call blocking probability whereas the FICAC scheme has the highest call blocking probability. Thus joint radio resource management and bandwidth adaptation reduces new call blocking probability in heterogeneous cellular networks.

Figure 14 shows the variation of new class-2 call blocking probability (Pb2) with call arrival rates for the four scenarios. Pb2 increases with increase in arrival rates for each of the four scenarios. Moreover Pb2 in each of the scenarios is higher than the corresponding Pb1 because class-2 calls require more bbu than class-1 calls. Thus, it is possible to block a class-2 call when it is still possible to admit a class-1 call into the network. However, the AJCAC scheme has the lowest call blocking probability for class 2 calls whereas the FICAC scheme

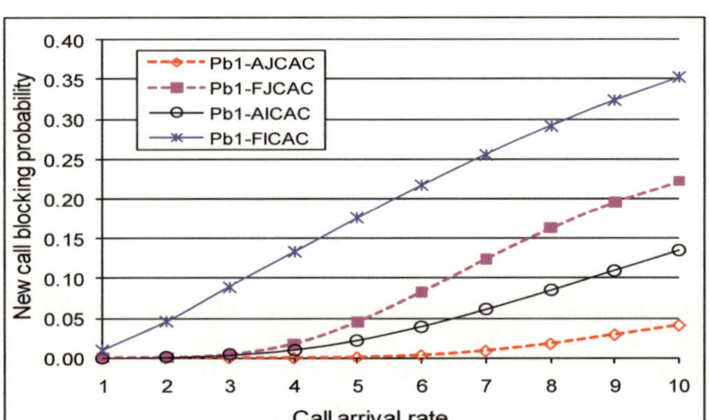

Fig. 13. New class-1 call blocking probability against call arrival rate.

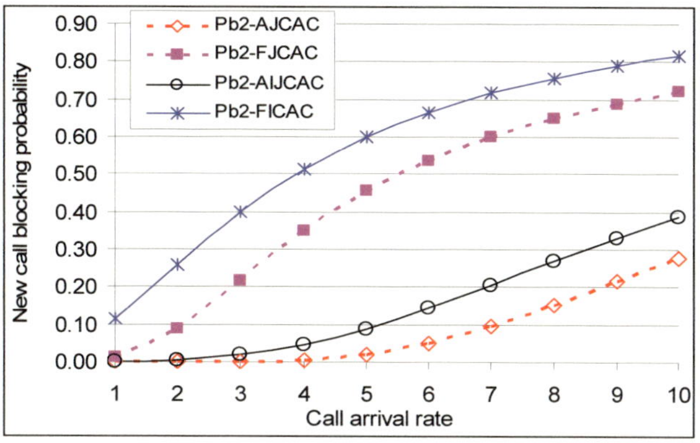

Fig. 14. New class-2 call blocking probability against call arrival rate.

has the highest call blocking probability. Thus joint radio resource management and bandwidth adaptation reduces new call blocking probability in heterogeneous cellular networks.

4.2 Comparison of handoff call dropping probabilities for the four scenarios

Figure 15 shows the variation of handoff class-1 call dropping probability (Pd1) with call arrival rates for the four scenarios. Pd1 increases with increase in arrival rates for each of the four scenarios. However, the AJCAC scheme has the lowest call dropping probability whereas the FICAC scheme has the highest call dropping probability. Thus joint radio resource management and bandwidth adaptation reduces handoff call dropping probability in heterogeneous cellular networks.

Figure 16 shows a similar trend to Figure 15. The AJCAC scheme has the lowest call dropping probability for class 2 calls whereas the FICAC scheme has the highest call dropping probability. Thus joint radio resource management and bandwidth adaptation reduces handoff call dropping probability in heterogeneous cellular networks.

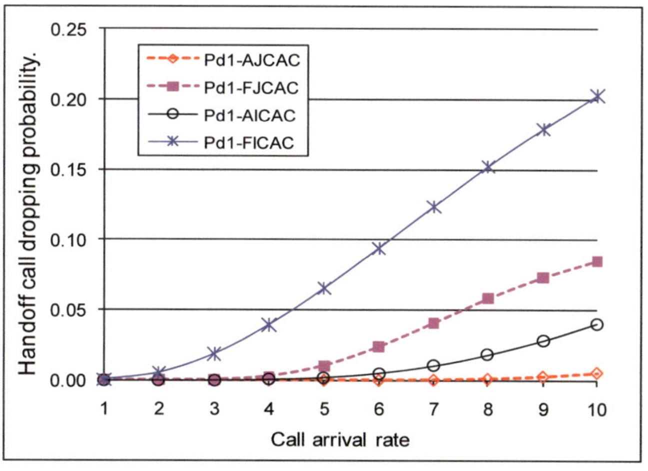

Fig. 15. Handoff class-1 call dropping probability against call arrival rate.

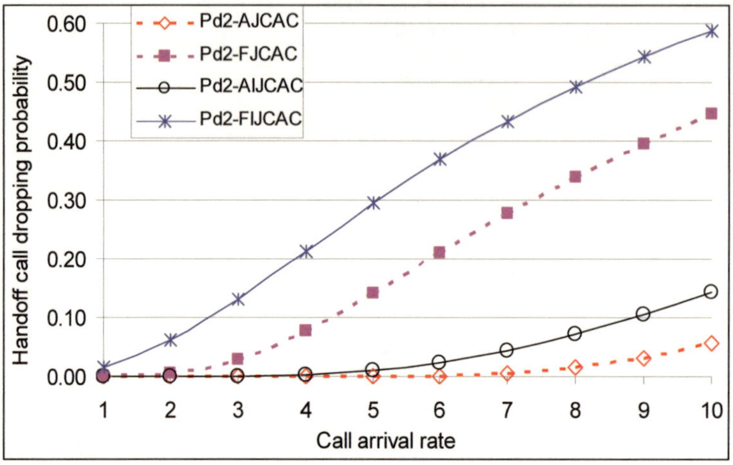

Fig. 16. Handoff class-2 call dropping probability against call arrival rate.

4.3 Comparison of call blocking/dropping probabilities for scenarios 1 and 2

Figure 17 compares the new class-1 call blocking probability and handoff class-1 call dropping probability for the Fixed and adaptive JCAC schemes. It can be seen that the Pd1 of AJCAC is less than the Pb1 of AJCAC. Similarly, the Pd1 of FJCAC is less than the Pb1 of FJCAC. Thus, handoff calls are prioritized over new calls by using the threshold based prioritization scheme shown in Figure 11.

Figure 18 compares the new class-2 call blocking probability and handoff class-2 call dropping probability for the Fixed and adaptive JCAC schemes. It can be seen that the Pd2 of AJCAC is less than the Pb2 of AJCAC. Similarly, the Pd2 of FJCAC is less than the Pb2 of FJCAC. Thus, handoff calls are prioritized over new calls by using the threshold based prioritization scheme shown in Figure 11.

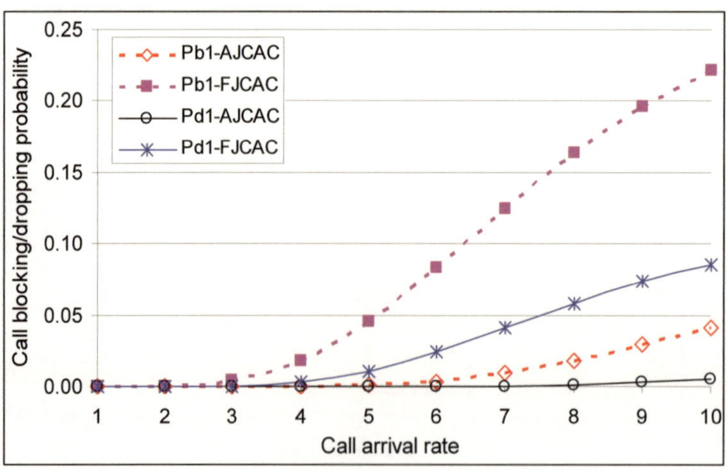

Fig. 17. Class-1 call blocking/dropping probability for JCAC schemes.

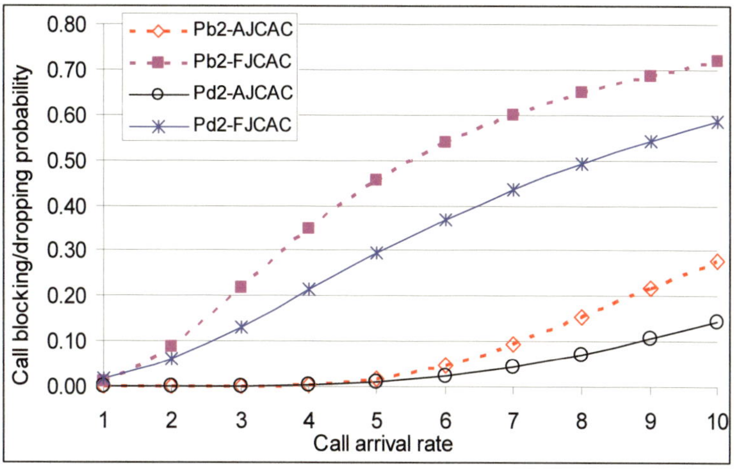

Fig. 18. Class-2 call blocking/dropping probability for JCAC schemes.

4.4 Comparison of call blocking/dropping probabilities for scenarios 3 and 4

Figure 19 compares the new class-1 call blocking probability and handoff class-1 call dropping probability for the Fixed and adaptive ICAC schemes. It can be seen that the Pd1 of AICAC is less than the Pb1 of AICAC. Similarly, the Pd1 of FICAC is less than the Pb1 of FICAC. Thus, handoff calls are prioritized over new calls by using the threshold based prioritization scheme shown in Figure 11.

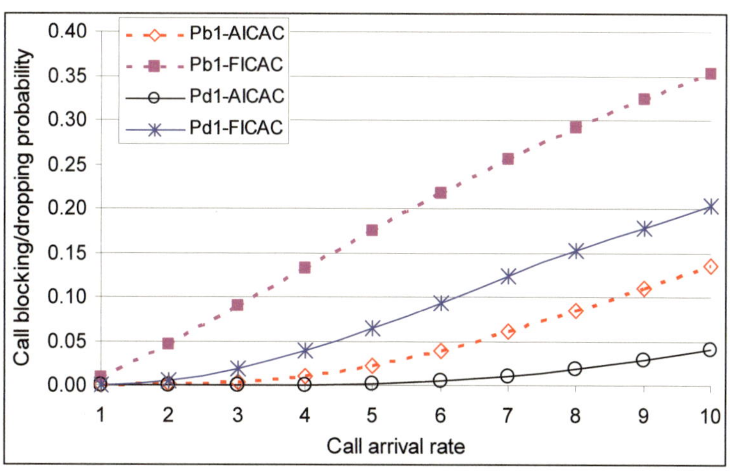

Fig. 19. Class-1 call blocking/dropping probability for ICAC schemes.

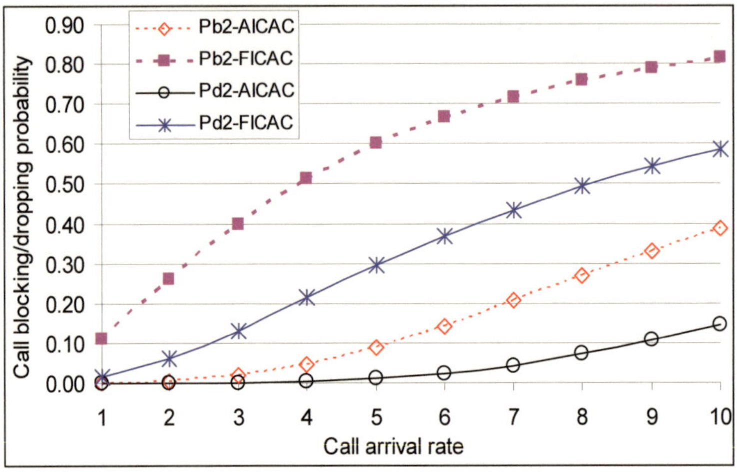

Fig. 20. Class-2 call blocking/dropping probability for ICAC schemes.

Figure 20 compares the new class-2 call blocking probability and handoff class-2 call dropping probability for the Fixed and adaptive ICAC schemes. It can be seen that the Pd2 of AICAC is less than the Pb2 of AICAC. Similarly, the Pd2 of FICAC is less than the Pb2 of FICAC. Thus, handoff calls are prioritized over new calls by using the threshold based prioritization scheme shown in Figure 6.

7. Conclusion

The coexistence of multiple cellular networks in the same geographical area has enabled more efficient utilization of radio resources and enhanced quality of service provisioning through joint radio resource management. An overview of joint call admission control in heterogeneous cellular networks has been given in this chapter. Different approaches for selecting RATs in heterogeneous cellular networks namely: random-selection, network load, service-cost, service-class, path-loss, layer, terminal modality, computational intelligence, and non computational intelligence techniques have been itemized. A Markov model for a load-based JCAC algorithm has been presented. Considering four different scenarios, simulation results are obtained and compared. Results show that joint management of radio resources and bandwidth adaptation reduce call blocking/dropping probability in heterogeneous cellular networks.

8. References

Bertsekas, D. P. & and Tsitsiklis, J. N. (2002). *Introduction to Probability,* Athena Scientific, Belmont, Mass, USA, 2002.

Chen, H.; Kumar, S. & Jay Kuo, C.-C. (2002). Dynamic Call Admission Control Scheme for QoS Priority Handoff in Multimedia Cellular Systems, Proceedings of IEEE Wireless Communications and Networking Conference, Orlando, vol. 1, March, 2002, pp. 114-118.

Falowo, O. E.; & Chan, H. A. (2007). Adaptive Bandwidth Management and Joint Call Admission Control to Enhance System Utilization and QoS in Heterogeneous Wireless Networks. *EURASIP Journal on Wireless Communications and Networking*, Vol. 2007, Article ID 34378, 11 Pages, 2007, DOI:10.1155/2007/34378, 2007.

Falowo, O. E. & Chan, H. A. (2010). Joint Call Admission Control Algorithm for Fair Radio Resource Allocation in Heterogeneous Wireless Networks Supporting Heterogeneous Mobile Terminals, *7th Annual IEEE Consumer and Communication & Networking Conference (IEEE CCNC)*, Las Vegas, Nevada, USA, 9-12 January, 2010.

Fettweis, G. (2009). Current Frontiers in Wireless Communications: Fast & Green & Dirty, IEEE Wireless Communications & Networking Conference (WCNC), Budapest, Hungary, April 5-8, 2009.

Gelabert, X.; Pe´ rez-Romero, J.; Sallent, O.; & Agustı´, R. (2008). A Markovian Approach to Radio Access Technology Selection in Heterogeneous Multiaccess/Multiservice Wireless Networks. *IEEE Transactions on Mobile Computing*, Vol. 7, No. 10, October 2008.

Ho, C. & Lea, C. (1999). Improving Call Admission Policies in Wireless Networks. *Wireless Networks*, Vol. 5, issue 4, 1999, pp. 257-265.

Holma H. & Toskala, A. (2001). WCDMA for UMTS. *John Wiley & Sons*, New York, NY, USA, 2nd Edition, 2001.

Hong and, D. & Rappaport, S. S. (1986). Traffic Model and Performance Analysis of Cellular Radio Telephony Systems with Prioritized and Non-prioritized Handoff Procedures. *IEEE Transaction of Vehicular Technology*, Vol. 35, August. 1986.

Lee, S.; Sriram, K,; Kim, K.; Kim, Y,; & Golmie, N.; (2009). Vertical Handoff Decision Algorithms for Providing Optimized Performance in Heterogeneous Wireless Networks. *IEEE Transactions on Vehicular Technology*, Vol. 58, No. 2, February 2009.

Niyato, D.; & Hossain, E. (2008). Noncooperative Game-Theoretic Framework for Radio Resource Management in 4G Heterogeneous Wireless Access Networks. *IEEE Transactions on Mobile Computing, Vol. 7, No. 3, March 2008.*

Orlik, P. V. & Rappaport, S. S. (2001). On the handover arrival process in cellular communications, ACM/Baltzer Wireless Networks, Vol. 7, No. 2, pp. 147-157, Mar./Apr. 2001.

Pérez-Romero, J; Sallent, O.; Agustí, R. & Díaz-Guerra, M. A (2005). *Radio Resource Management strategies in UMTS*, John Wiley & Sons.

Vidales, P.; Baliosian, J.; Serrat, J.; Mapp, G.; Stajano, F.; & Hopper, A. (2005). Autonomic System for Mobility Support in 4G Networks. *IEEE Journal on Selected Areas in Communications*, Vol. 23, No. 12, December 2005.

Zhang, W. (2005). Performance of real-time and data traffic in heterogeneous overlay wireless networks, *Proceedings of the 19th International Teletraffic Congress (ITC 19)*, Beijing, 2005.

12

Providing Emergency Services in Public Cellular Networks

Jiazhen Zhou[1] and Cory Beard[2]
[1]*Howard University, Washington DC*
[2]*University of Missouri - Kansas City*
United States

1. Introduction

When emergency situations like natural disasters or terrorist attacks happen, demand in telecommunication networks will go up drastically, causing congestion in the networks. Due to the local nature of most disaster events, this kind of congestion is usually most serious at access networks, which is of special concern for cellular networks. With serious congestion in the cellular networks, it is very difficult for customers to obtain access to services. It might be possible to use reserved spectrum for national security and emergency preparedness (NS/EP) customers, such as with police and fire radio systems. However, capacity may be limited. Also, the deployment of additional equipment takes time and could not address the urgent need for communications quick enough.

On the other hand, publicly available wireless communication capabilities are pervasive and always ready to use. It would be very beneficial if NS/EP customers could use the commercially available wireless systems to respond to natural and man-made disasters (Carlberg et. al., 2005).

For several years, but especially in response to the events of September 11, 2001, the U.S. government and the wireless telecommunications industry have worked together to specify a technically and politically feasible solution to the needs of homeland security for priority access and enhanced session completion. This has resulted in definition of an end-to-end solution for national security and emergency preparedness sessions called the Wireless Priority Service (WPS) defined in the Wireless Priority Service Full Operating Capability (WPS FOC) by the FCC (FCC, 2000), (National Communications System, 2002), (National Communications System, 2003). First-responders, NS/EP leadership, and key staff are able to use this capability in public cellular networks.

To support emergency services in public cellular networks, NS/EP users should be identified and provided better guaranteed services than general customers. When NS/EP customers present access codes and have been authorized to use the emergency service, *special admission control policies* are employed in the base station to make sure their session requests get better admission. Proper methods are also deployed in the core network to provide end-to-end service for NS/EP users.

The basic requirement on the *special admission control policies* is that better admission of NS/EP customers, including both high admission probability and quick access, should be guaranteed. However, as the main purpose of public cellular networks is also to provide services for public customers, certain resources should be ensured for public users even when there are high demands of spectrum resources from NS/EP users at the same time. For this reason, we can say that NS/EP traffic has "conditional priority" — the resources allocated for the NS/EP customers must be balanced with the demand of public users. If NS/EP traffic is light, low blocking probability for them should be guaranteed; if NS/EP traffic becomes unexpectedly heavy, then public traffic should be protected through guaranteeing a certain amount of resources for public use.

In this Chapter, important parameters for NS/EP and public customers are first identified. Three candidate admission control policies are then analyzed , with two types that provide high system utilization evaluated and compared in detail. Each control policy has significant benefits and drawbacks with neither one clearly superior, so guidelines are provided for choosing best schemes that are suitable to each operator and social context.

2. Important performance metrics

There are different possible admission control strategies for providing emergency services in public cellular networks. The important performance metrics that should be considered in evaluating these strategies include: system utilization, admission probability of public and NS/EP customers, access waiting time, and termination probability.

2.1 System utilization

System utilization is a measurement of the usage of system resources, like the spectrum resources in a cellular network. In normal situations, system utilization directly determines the revenues of the operators. The higher the system utilization, the higher the revenues the operators will obtain.

Since disasters are mostly unexpected and only happen occasionally, when we are considering possible admission control strategies, we need to make sure they can help get maximum system utilization while providing priority services to NS/EP users.

2.2 Admission of NS/EP and public traffic

Priority treatment for NS/EP traffic gives high probability of admission when demand is not extremely high (for example, less than 25% of a cell's capacity). Operators will set a threshold for expected NS/EP load, and will admit sessions with high probability if demand stays within those bounds.

As an effective measurement on whether public customers are well protected in case of high NS/EP traffic, channel occupancy of public traffic is used, which measures the average amount of channel resources utilized. For example, (Nyquetek, 2002) points out that it has been agreed by government and operators that at least 75% of channel resources should be guaranteed for public use. To achieve this goal, flexible admission control strategies should be employed to cope with different load cases.

2.3 Waiting time

Queueing based methods, where session requests are put into queues so that they will be served when system resources are available later, are very common in admission control strategies since they are effective in increasing the utilization of systems. However, this means that some customers will inevitably wait some time before being allocated a channel to start their sessions. The access waiting time experienced by customers, which decides the customer satisfaction, is another important metric used to evaluate the quality of service in telecommunication systems.

For NS/EP customers, long waiting for admission is unreasonable due to the urgent need of saving life and property. An ideal admission control policy should cause minimum or even no waiting for NS/EP customers.

2.4 Session termination probability

Session termination probability is concerned with reasons why a session might be terminated before it is able to be completed. This might occur because of a failed handoff or a hard preemption.

In a cellular network, a commonly agreed upon standard is that terminating an ongoing session is much worse than blocking a new attempt. This is why channels are often reserved for handoff traffic in cellular networks. However, the cost is that the system utilization can be sacrificed. This is a dilemma also encountered in emergency situations: should we try to guarantee few ongoing public sessions are terminated, or should we make higher system utilization more important so that we can admit more NS/EP sessions? Different operators might have different opinions on this issue. However, this also reminds us that some termination of sessions, especially if this is rare, might be acceptable during specific situations like when disasters happen.

3. Candidate admission control strategies

The admission control policies discussed in this paper are all load based. This means that admission is based on whether the new session will make the load on the system too high. The appropriateness of this load based admission control model for a 3G/4G network is discussed in Section 6.

3.1 Reservation based strategies

To guarantee certain resources for a special class of traffic, reservation strategies (Guerin, 1988) are often used in cellular networks; one example is guard channel policies to hold back resources for handoffs. The main idea of a reservation based strategy is that it limits the amount of sessions that can be admitted for some classes to hold back resources in case other classes need them. The benefit is that the high priority traffic could use specially reserved resources, thus achieving better admission. Yet the disadvantage is that the reserved resources can be wasted if the high priority traffic is not as high as expected, while the low priority traffic is probably suffering blocking. As discussed previously in section 2.1, the spectrum resources are especially valuable when disasters happen, so reserving channels for NS/EP traffic will probably cause waste of resources since NS/EP traffic volume is hard to predict. This is why the strategies employing reservation schemes (including guard channel policy (Guerin, 1988) and upper limit strategy (Beard & Frost, 2001)) are not as useful in public cellular networks that support emergency services.

3.2 Pure queueing based strategies

For a pure queueing based policy, all classes of traffic can have their own queues or shared queues. Session requests that cannot get immediate service will be put into queues. When system resources become available, session requests in the queues will be scheduled according to some specific rule to determine which queue gets served next. Since pure queueing based strategies will try to serve customers waiting in the queues whenever system resources become available, the newly released resources will be immediately taken and thus *guarantee no waste of a system's capability*.

In a pure queueing based scheme, both NS/EP traffic and public originating traffic can be put into separate queues. When channels become available, these two queues will be served according to a certain probability (Zhou & Beard, 2010), or using round-robin style scheduling (Nyquetek, 2002). For the former work, the scheme is called Adaptive Probabilistic Scheduling (APS). The latter work by Nyquetek Inc. evaluated a series of pure queueing based methods, with a representative one being the Public Use Reservation with Queuing All Calls (PURQ-AC).

3.3 Preemption based strategies

As opposed to pure queueing based strategies, preemption based strategies allow high priority customers to take resources away from ongoing low priority sessions. Preempted sessions can be put into a queue so that they can be resumed later, hopefully after a very short time. It can be shown that preemption strategies can even obtain slightly higher system utilization than pure queueing strategies.

The largest benefit of preemption based strategies is that they can guarantee immediate access and assure the admission of NS/EP traffic. However, an uncontrolled preemption strategy tends to use up all channel resources and will be against the goal to protect public traffic. Furthermore, it can be annoying to preempted users. Due to these side effects, in some places like the United States it is not currently allowed, even though allowed in other places.

In (Zhou & Beard, 2009), a controlled preemption strategy was presented to suppress these side effects while exploiting the unique benefits. When the resources occupied by the NS/EP sessions surpass a threshold, preemption will be prohibited. By tuning the preemption threshold, the channel occupancy for each class can be adjusted as we prefer.

4. Analysis for the admission control schemes

The adaptive probabilistic scheduling (APS) scheme and the preemption threshold based scheduling (PTS) scheme can be both analyzed using multiple dimensional Markov chains. The main performance metrics, including admission and success probability, waiting time, and termination probability can be computed.

The main types of sessions considered are emergency sessions, public handoff sessions and public originating sessions. As shown in (Nyquetek, 2002), the current WPS is provided only for leadership and key staff, so it is reasonable to assume that most emergency users will be stationary within a disaster area. Handoff for emergency sessions is not considered here, but this current framework can be readily extended to deal with emergency handoff traffic when necessary.

Strictly speaking, the distributions of session durations, inter-arrival time, and the length of customer's patience are probably not exponentially distributed. As shown in (Jedrzycki & Leung, 1996), the channel holding times in cellular networks can be modeled much more accurately using the lognormal distribution. Consult (Mitchell et. al., 2001) where Matrix-Exponential distributions can be added to standard Markov chains like the one here to model virtually any arrival or service process. However, in reality the exponential distribution assumption for sessions is still mostly used, both in analysis-based study like (Tang & Li, 2006), and simulation-based study like Nyquetek's report (Nyquetek, 2002). For this work, similar to what is used widely and also assumed in (Nyquetek, 2002), all session durations and inter-arrival times are independently, identically, and exponentially distributed.

4.1 Adaptive probabilistic scheduling

In (FCC, 2000), the FCC provided recommendations and rules regarding the provision of the Priority Access Service to public safety personnel by commercial providers. It required that "at all times a reasonable amount of CMRS (Commercial Mobile Radio Service) spectrum is made available for public use." To meet the FCC's requirements, when emergency traffic demand is under a certain "protection threshold", high priority should be given to emergency traffic; when the emergency traffic is extremely high so that it can take most or all of the radio resources, a corresponding strategy should be taken to avoid the starvation of public traffic by guaranteeing a certain amount of radio resources will be used by public. The "protection threshold" can be decided by each operator and thus is changeable. So our strategy should be able to deal with the above requirement for any "protection threshold" value; this is why we introduce an *adaptive probabilistic scheduling* strategy instead of fixed scheduling like in (Nyquetek, 2002).

4.1.1 Description and modeling of APS

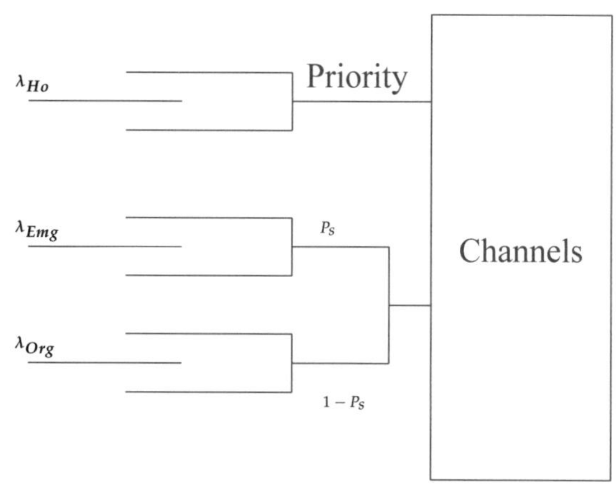

Fig. 1. Probabilistic scheduling

The basic APS scheme is illustrated in Fig. 1. In the figure, $\lambda_{Emg}, \lambda_{Ho}, \lambda_{Org}$ represent the arrival rate of emergency, public handoff, and public originating sessions respectively. When an incoming session fails to find free channels, it is put into a corresponding queue if the queue is not full. To reduce the probability of dropped sessions, the handoff sessions are assigned a non-preemptive priority over the other two classes of traffic. Note that when a disaster happens, it is uncommon for general people (who generate public traffic) to move into a disaster area. Thus, the handoff traffic into a disaster area will not be high, so setting the public handoff traffic as the highest priority will not make emergency traffic starve. If

Fig. 2. State Diagram for Probabilistic Scheduling Scheme

there is no session waiting in the handoff queue when a channel is freed, a session will be randomly chosen from either the emergency queue or public originating queue according to the scheduling probability already set. The scheduling probability for emergency traffic is denoted as P_s. The algorithm to decide P_s for different cases will be introduced in Section 4.1.3. The queues are finite and customers can be impatient when waiting in the queue, so blocking and expiration are possible.

A 3-D Markov chain can be built to model the behavior of the three classes of traffic as shown in Fig. 2. Here the total number of channels is C, and the queue lengths are L_1, L_2, L_3 individually. Each state is identified as (n, i, j, k), where n is the number of channels used, i, j, k is the number of sessions in handoff queue, emergency queue, and public originating queue respectively. The arrival rate for handoff, emergency and originating sessions is λ_{Ho}, λ_{Emg}, λ_{Org}, and the service rate for all sessions is μ. The expiration times of all three classes of sessions are exponentially distributed with rates $\mu_{exp,ho}$, $\mu_{exp,emg}$ and $\mu_{exp,org}$ respectively.

The probabilistic scheduling policy is implemented using a parameter P_s, which is the probability that an emergency session will be scheduled when a channel becomes free. This can be seen in Fig. 2, for example, where the part of the departure rates from state (C,0,L2,L3) that relate to scheduling are either $P_sC\mu$ or $(1 - P_s)C\mu$ for choosing to service an emergency or public session respectively.

We can see that the Markov chain in Fig. 2 does not have a product form solution. This is because the boundary (first layer) is not product form due to the probabilistic scheduling. So state probabilities will be obtained by solving the global balance equations from this Markov chain directly.

It is worthy to mention that the Markov chain size in Fig. 2 is $L_1L_2L_3$ states, thus it is not affected by the number of channels. Since the queue size employed does not need to be large (like 5) due to the reneging effect of customers waiting in the queue, the computational complexity of this Markov chain is lower than the preemption threshold strategy in (Zhou & Beard, 2009), which has a Markov chain of size CL_1L_2. Note that a typical value of C is around 50.

4.1.2 Performance evaluation

With the state probabilities solved, performance metrics can be computed for blocking, expiration and total loss probability, admission probability, average waiting time, and channel occupancy for each class.

In this system, the loss for each class of traffic consists of two parts: those sessions that are blocked when the arrivals find the queue full; and those sessions reneged (also called expired) when waiting too long in each queue. So we have: $P_{Loss} = P_B + P_{Exp}$ for each class of traffic.

(1) Blocking probability

Blocking for each class of traffic happens when the corresponding queue is full. Thus,

$$P_{B,ho} = \sum_{j=0}^{L_2} \sum_{k=0}^{L_3} P(C, L_1, j, k) \tag{1}$$

$$P_{B,emg} = \sum_{i=0}^{L_1} \sum_{k=0}^{L_3} P(C, i, L_2, k) \tag{2}$$

$$P_{B,org} = \sum_{i=0}^{L_1} \sum_{j=0}^{L_2} P(C, i, j, L_3) \tag{3}$$

(2) Expiration Probability

At a state (C, i, j, k), the arrival rates for each class are $\lambda_{Ho}, \lambda_{Emg}, \lambda_{Org}$, and the expiration rates for each class are $i\mu_{exp,ho}$, $j\mu_{exp,emg}$, $k\mu_{exp,org}$ independently. The probability of expiration is the ratio of departures due to expiration per unit time (expiration rate) over arrivals per unit time (arrival rate). Thus, we can find the overall expiration probability just based on the steady state probability, the expiration rate, and the arrival rate at each state as follows:

$$P_{Exp,ho} = \sum_{i=1}^{L_1} \sum_{j=0}^{L_2} \sum_{k=0}^{L_3} P(C, i, j, k) \frac{i\mu_{exp,ho}}{\lambda_{Ho}} \tag{4}$$

$$P_{Exp,emg} = \sum_{i=0}^{L_1} \sum_{j=1}^{L_2} \sum_{k=0}^{L_3} P(C, i, j, k) \frac{j\mu_{exp,emg}}{\lambda_{Emg}} \tag{5}$$

$$P_{Exp,org} = \sum_{i=0}^{L_1} \sum_{j=1}^{L_2} \sum_{k=0}^{L_3} P(C, i, j, k) \frac{k\mu_{exp,org}}{\lambda_{Org}} \tag{6}$$

(3) System utilization and channel occupancy

The channels are not fully used when there are still free channels available. When there are n channels being used, that means $C - n$ channels are idle, and the total portion of channels unused is thus $\frac{C-n}{C}$. So the system utilization can be calculated by considering those portion of channels unused at those possible states:

$$SysUtil = 1 - \sum_{n=0}^{C-1} \frac{(C-n)P(n,0,0,0)}{C} \tag{7}$$

We define "channel occupancy" as the proportion of channels occupied by each class of traffic. *Channel occupancy* is an important metric to measure whether the public traffic is well protected when emergency traffic is heavy. After the system utilization is obtained, with the assumption that each class of session has the same average session duration, the channel occupancy of each class can be calculated by comparing the admitted traffic of each class. The total admitted traffic rate is: $\lambda_{adm,tot} = \lambda_{Ho}(1 - P_{B,ho}) + \lambda_{Emg}(1 - P_{B,emg}) + \lambda_{Org}(1 - P_{B,org})$. Thus, we have:

$$ChOcp_{ho} = \frac{\lambda_{Ho}(1 - P_{B,ho})}{\lambda_{adm,tot}} SysUtil \tag{8}$$

$$ChOcp_{emg} = \frac{\lambda_{Emg}(1 - P_{B,emg})}{\lambda_{adm,tot}} SysUtil \tag{9}$$

$$ChOcp_{org} = \frac{\lambda_{Org}(1 - P_{B,org})}{\lambda_{adm,tot}} SysUtil \tag{10}$$

(4) Waiting time

Sessions waiting in the queue could be patient enough to wait until the next channel becomes available, or become impatient and leave the queue before being served.

If a customer needs to be put into a queue before being served or reneging, the average time staying in the queue (irrespective of being eventually served or not) can be calculated using Little's law: $\overline{T} = \frac{N_q}{\lambda(1-P_B)}$. Note that the effective arrival rate at each queue is $\lambda(1 - P_B)$, and the average queue length for each class is N_q. The mean queue length for each class can be calculated based on the steady states we compute, so we have:

$$\overline{T_{ho}} = \sum_{i=0}^{L_1} \sum_{j=0}^{L_2} \sum_{k=0}^{L_3} \frac{iP(C,i,j,k)}{\lambda_{Ho}(1 - P_{B,ho})} \tag{11}$$

$$\overline{T_{emg}} = \sum_{i=0}^{L_1} \sum_{j=0}^{L_2} \sum_{k=0}^{L_3} \frac{jP(C,i,j,k)}{\lambda_{Emg}(1 - P_{B,emg})} \tag{12}$$

$$\overline{T_{org}} = \sum_{i=0}^{L_1} \sum_{j=0}^{L_2} \sum_{k=0}^{L_3} \frac{kP(C,i,j,k)}{\lambda_{Org}(1 - P_{B,org})} \tag{13}$$

4.1.3 The adaptive probabilistic scheduling algorithm

With main performance metrics like channel occupancies computed, an algorithm for searching the best value of P_S can be obtained.

Denote the "system capacity" as the largest possible throughput of the cell. In other words, it is the total service rate of the cell - $C\mu$.

Using dynamic probabilistic scheduling, the scheduling probability for emergency traffic when there is a channel available can be adjusted according to the different arrival rates for each class of traffic. The algorithm to find the scheduling probability for emergency traffic P_s is:

Algorithm 1: Determining the scheduling probability

Step 1: Set the initial value of P_s to 1, which means giving absolute priority to emergency traffic as opposed to public originating traffic.

Step 2: Solving the Markov chain, get the general representation about channel occupancies using equations (8) - (10) . With the current P_S value applied, if the channel occupancy of public traffic is already higher than 75%, that means the emergency traffic obviously does not affect the performance of public traffic, and thus can be accepted, stop here. Otherwise go to step 3 to search for the suitable weighting parameter.

Step 3: Use a binary search method to search for the best weighting parameter: Let $P_s = 1/2$, calculate the channel occupancy of public traffic using the general representation obtained in step 2. If it is larger than the required value, search the right half space [1/2, 1]; otherwise search the left half space [0, 1/2]. Repeat step 3 until the suitable P_s that meets the channel occupancy requirement of public traffic is found.

4.2 Preemption threshold based scheduling

4.2.1 Description and modeling of PTS

The preemption threshold based scheduling scheme (PTS) is illustrated in Fig. 3. When an incoming emergency session fails to find free capacity, and if the number of active emergency sessions is less than the preemption threshold, it will preempt resources from a randomly picked ongoing public session. The preempted session will be put into the handoff/preempted session queue. For an arriving public handoff session, it will also be buffered in the handoff/preempted session queue when no capacity is immediately available. Correspondingly, there is also an originating session queue, which is further helpful for preventing starvation of public traffic. If an incoming emergency session fails to find free resources to preempt, it will simply be dropped.

We suggest not to have a buffer for emergency users for two reasons: (1) Make sure there is no access delay for emergency sessions; (2) Guarantee the public traffic has enough system resources when emergency traffic is very heavy. If emergency traffic is queued in this case, public traffic could not be well protected even if preemption is not allowed. The reason that we use the same buffer for handoff and preempted sessions is that both of these two types of sessions are broken sessions, so they have the same urgency to be resumed. More precise configuration like using two different buffers is possible, but will not be obviously beneficial. In fact, it will make the implementation and analysis more time consuming, because it will have a much larger Markov chain state space.

When capacity becomes available later, one session from the queues is served. A priority queue based scheduling policy will be used, and it is reasonable to assume that handoff/preempted sessions have higher priority over the originating sessions. The queues are finite and customers can be impatient when waiting in the queue, so blocking and expiration are possible.

Since customers have different patience, it is reasonable to assume their impatience behavior to be random rather than deterministic like assumed in Nyquetek's study. We assume that the expiration times of traffic in the same queue are exponentially and identically distributed, and the patience of a customer is the same after each preemption.

If session durations are memoryless (i.e., exponentially distributed), this means that if at any point a session is interrupted, the remaining service time is still exponential with the same

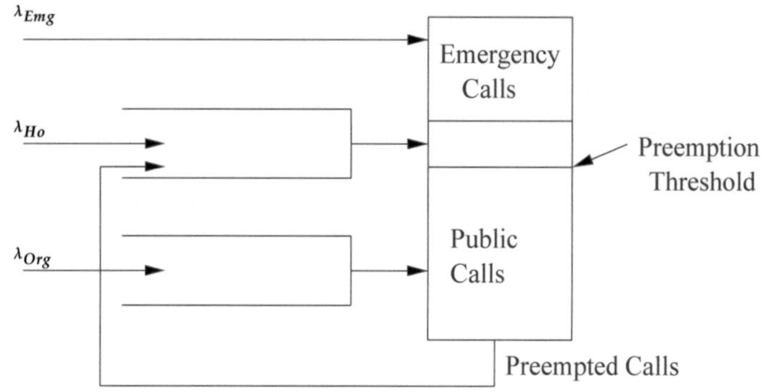

Fig. 3. Preemption threshold based scheduling

average service time as when it began. It is, therefore, reasonable to model a restarted session as a renewal process. In other words, the preempted session will be restarted with re-sampling of the exponential random variable (Conway et. al., 1967).

Same as for the APS scheme, the total number of channels is denoted as C. The length of the handoff/preempted queue is L_1 and the length of originating queue is L_2, and the *preemption threshold* is R. Each state is identified as (i, j, m, n), where i, j is the number of channels occupied by emergency and public sessions respectively, m, n represents the number of sessions in the handoff/preempted session queue and the public originating session queue individually. The arrival rates for emergency, handoff, and originating sessions are λ_{Emg}, λ_{Ho}, λ_{Org} respectively. The mean expiration rates for sessions waiting in the handoff/preempted queue and originating queue are denoted as $\mu_{exp}^{ho/prm}$ and μ_{exp}^{org}. To facilitate analysis, the average service rate for each class is assumed to be the same and denoted as μ. This also means that the session duration in a single cell is exponentially distributed with mean $1/\mu$, whether the session ends in this cell or is handed off to another cell. A Markov chain can be formed, and the state probabilities can be obtained by solving the following categories of balance equations:

(1) When the channels are not full, the typical state transition is shown in Fig. 4. Since the queues are empty in this case, in the notation we replace $P(i, j, 0, 0)$ with $P(i, j)$ for simplicity. The corresponding balance equation is:

$$\begin{aligned} &P(i,j)(\lambda_{Emg} + \lambda_{Ho} + \lambda_{Org} + (i+j)\mu) \\ &= P(i-1,j)\lambda_{Emg} + P(i,j-1)(\lambda_{Ho} + \lambda_{Org}) \\ &+P(i,j+1)(j+1)\mu + P(i+1,j)(i+1)\mu. \end{aligned} \tag{14}$$

For the states on the edge, some terms of this equation will disappear.

(2) When the channels are full, queueing is involved, the typical state transition is shown in Fig. 5. The corresponding balance equation is:

$$\begin{aligned} &P(i,C-i,m,n)(\lambda_{Emg} + \lambda_{Ho} + \lambda_{Org} + C\mu + m\mu_{exp}^{ho/prm} + n\mu_{exp}^{org}) \\ &= P(i-1,C-i+1,m-1,n)\lambda_{Emg} + P(i,C-i,m-1,n)\lambda_{Ho} \\ &+P(i,C-i,m,n-1)\lambda_{Org} + P(i+1,C-i-1,m+1,n)(i+1)\mu \\ &+P(i,C-i,m+1,n)((C-i)\mu + (m+1)\mu_{exp}^{ho/prm}) + P(i,C-i,m,n+1)(n+1)\mu_{exp}^{org}) \end{aligned} \tag{15}$$

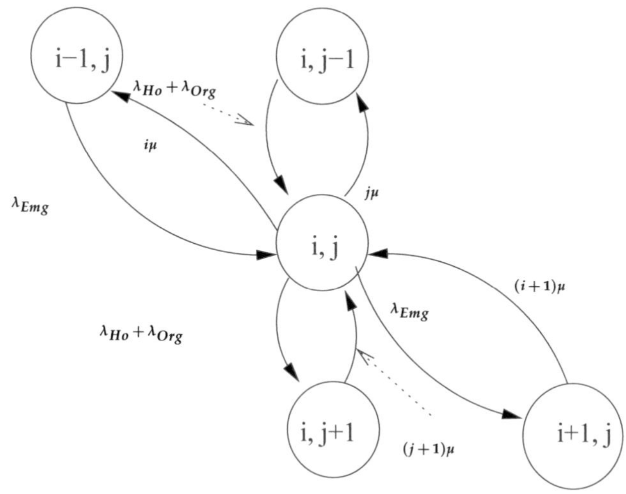

Fig. 4. The typical state change when channels are non-full

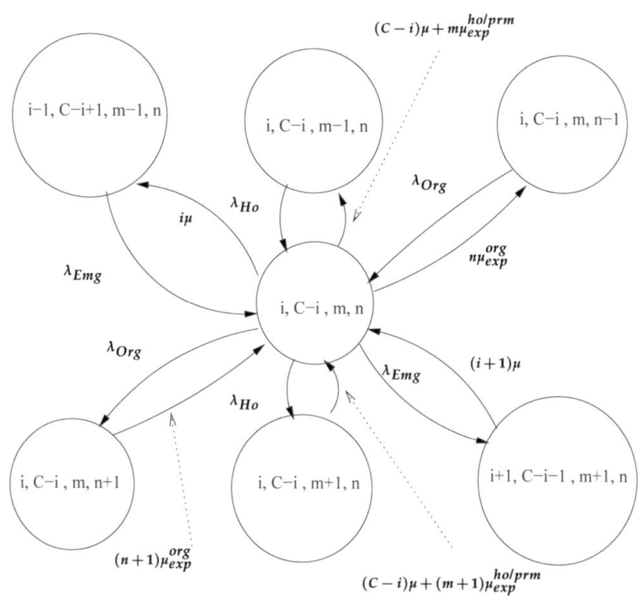

Fig. 5. The typical state change when channels are full and $i < R$

Note that when $i \geq R$, no preemption will be allowed, which will make the terms involving λ_{Emg} disappear.

With the practical consideration of expiration and preemption threshold, a product form solution for the equilibrium equations has not been found. Since we have limited the system to one buffer for handoff and preempted sessions, the computation requires operations on a matrix with size CL_1L_2, which means it depends on the number of channels and the size of the two buffers. Note that as pointed out in (Nyquetek, 2002), the buffer size need not be long (=5) because the effect will not be obvious after a certain point. Due to this fact, the computation is feasible.

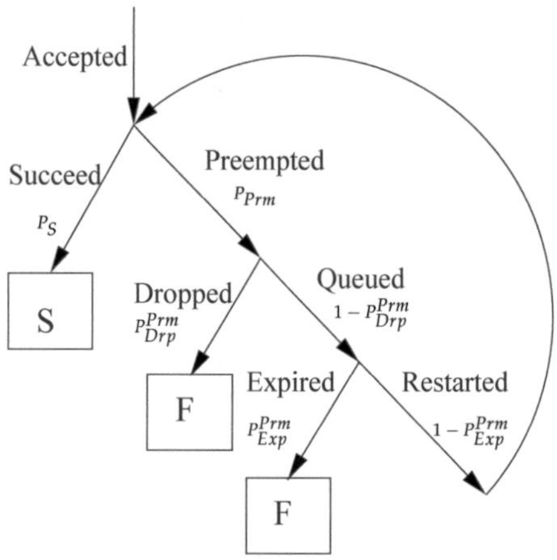

Fig. 6. Probability flow for low priority sessions

4.2.2 Performance evaluation

With the state probabilities solved, performance metrics, including average channel occupancy and the success probability, i.e., probability of finishing normally without expiring or dropping for each class can be obtained and will be shown in this subsection. Computation of related parameters, like admission probability, blocking probability of each class, the expiration probability of sessions in each queue, and preemption probability for a low priority session given that it is admitted, has been provided by (Zhou & Beard, 2006).

(1) System utilization and channel occupancy

Similar to the APS scheme, the system utilization can be computed by considering those portion of unused channels at all possible states:

$$SysUtil = 1 - \sum_{n=1}^{C-1} \sum_{i=0}^{n} \frac{(C-n)P(i, n-i, 0, 0)}{C} \tag{16}$$

The channel occupancies for emergency traffic and public traffic can also be computed based on steady states:

$$ChOcp^{Emg} = \sum_{n=1}^{C} \sum_{i=1}^{n} \sum_{k=0}^{L_1} \sum_{l=0}^{L_2} \frac{iP(i,n-i,k,l)}{C} \tag{17}$$

$$ChOcp^{Pub} = \sum_{n=1}^{C} \sum_{j=1}^{n} \sum_{k=0}^{L_1} \sum_{l=0}^{L_2} \frac{jP(n-j,j,k,l)}{C} \tag{18}$$

(2) Probability flow of low priority sessions

In Fig. 6, the probability flow of low priority sessions is shown. In the frame, "F" means failed, "S" means successful.

A session can be preempted multiple times, and with the renewal process assumption on resumed sessions, the number of preemption times will not affect the preemption probability of a session. Thus the number of preemption times is geometrically distributed with:

$$Pr(\text{Preempted } n \text{ times}) = P_{Prm}(1-A)A^{n-1}, n = 1, 2, ... \tag{19}$$

Here $A = P_{Prm}(1 - P^{Prm}_{Drp})(1 - P^{Prm}_{Exp})$ is the probability for a session to stay active; $(1 - A)$ is the probability that the session ends (succeeds, expires or is blocked after being preempted).

P_{Drp}^{Prm} is the probability for a preempted session to be dropped (due to a full queue) after being preempted, and P_{Exp}^{Prm} is the expiration probability for sessions waiting in the preempted session queue.
Thus the expected value of preempted times is $\frac{P_{Prm}}{1-A}$, or expressed in the form of preemption and expiration probability:

$$\overline{PrmTimes} = \frac{P_{Prm}}{1 - P_{Prm}(1 - P_{Exp}^{Prm})P_{Drp}^{Prm}} \tag{20}$$

(3) Success probability
For emergency sessions, all of the admitted sessions will be successfully finished, thus providing high dependability since they cannot be pre-empted. This kind of dependability cannot be assured for low priority sessions.
According to Fig. 6 we can compute the success probability given a session is admitted, which is denoted as P_{SGA}: for an admitted session, it will succeed only if it does not expire and is not blocked after being preempted. Note that $P_S = 1 - P_{Prm}$, we have:

$$\begin{aligned} P_{SGA} &= P_S \sum_{i=0}^{\infty} (P_{Prm}(1 - P_{Drp}^{Prm})(1 - P_{Exp}^{Prm}))^i \\ &= \frac{(1 - P_{Prm})}{1 - P_{Prm}(1 - P_{Drp}^{Prm})(1 - P_{Exp}^{Prm})} \end{aligned} \tag{21}$$

The successful finishing probabilities are decided by P_{SGA} and the corresponding admission probabilities :

$$P_{Succ}^{Ho} = P_{Adm}^{Ho} P_{SGA} \tag{22}$$

$$P_{Succ}^{Org} = P_{Adm}^{Org} P_{SGA} \tag{23}$$

5. Comparison of main admission control policies

In this Section, comparisons among the APS scheme (Zhou & Beard, 2010), the PURQ-AC scheme (Nyquetek, 2002), and the PTS scheme (Zhou & Beard, 2009) are provided. The main performance metrics considered include the achievable channel occupancy of public traffic, success probability, waiting time, and termination probability for each class of traffic.
The main parameters are: the number of channels in a cell is 50, and the average duration for each session is 100 seconds, so the maximum load that the system can process, called *system capacity*, is $C\mu = 0.5$ sessions/second $= 30$ sessions/minute. The load of public handoff traffic is 6 sessions/minute (20% of system capacity). Same as the parameters used in Nyquetek's report (Nyquetek, 2002), the average impatience times for handoff and originating traffic are both 5 seconds, and for emergency traffic it is 28 seconds, while the buffer sizes for all three queues are 5.

5.1 Comparison of achievable channel occupancy

As pointed out in (Nyquetek, 2002), the resource guarantee for public users is implemented through achieving at least 75% channel occupancy for public traffic. To evaluate whether all three schemes can achieve this goal, two different loads of emergency traffic are studied.

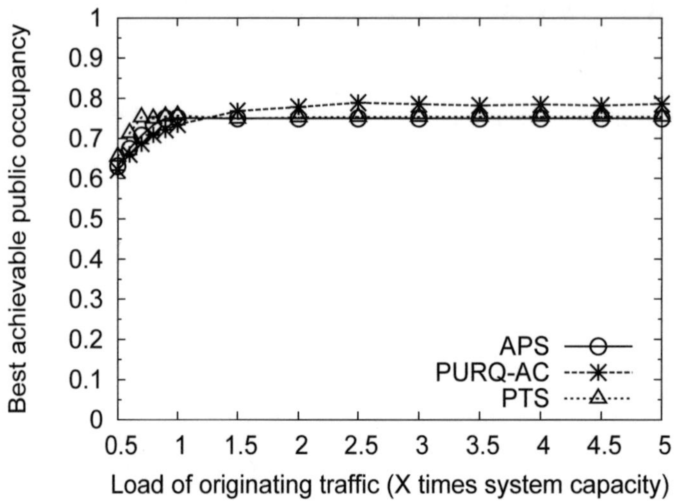

Fig. 7. Comparison of achievable channel occupancy for public traffic - Emergency traffic = 30% system capacity

When the load of emergency traffic is at 30% of the system capacity as shown in Fig. 7, all three schemes can guarantee at least 75% for public use if the load of the originating traffic is higher than the engineered system capacity. However, it can be seen that PURQ-AC will achieve even more than 75% when the public traffic is even higher. Although this protects the benefit of public traffic well, it does not achieve the guaranteed goal (25%) for emergency traffic.

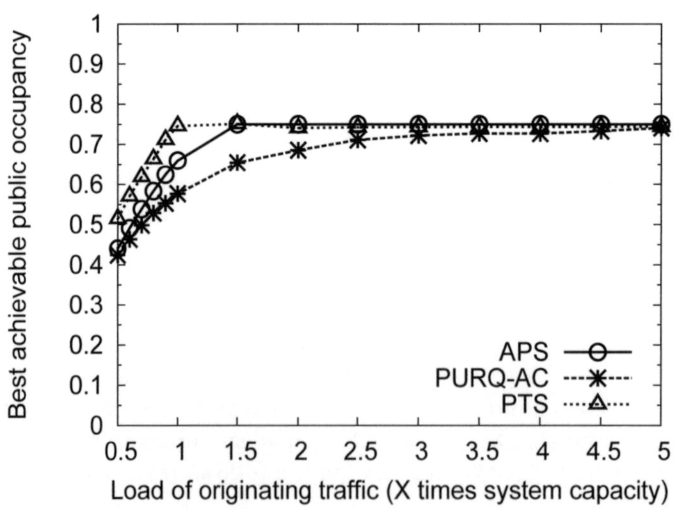

Fig. 8. Comparison of achievable channel occupancy for public traffic - Emergency traffic = 160% system capacity

Fig. 8 shows an extreme case: the emergency traffic is unexpectedly high at 160% of the system load. When the public traffic is not heavy enough, the PURQ-AC policy cannot guarantee 75% of channel occupancy for public traffic (also shown in Nyquetek (2002), Fig. 3-7). Actually, the

75% goal is not achieved until the originating traffic is 5 times of the system capacity. When the load of public originating traffic is at 100% of the system load, only 58% of channel resources are used for public traffic. In contrast, the APS scheme can achieve the goal with much lower load, at 1.5 times of the system capacity. If the originating traffic is even lower, the APS scheme can still guarantee higher channel occupancy for public traffic than what PURQ AC can do. The PTS strategy is even better for lower load. In summary, *the APS scheme can can protect both public and emergency traffic more effectively than the PURQ-AC strategy*. The PTS scheme could be even better than the APS in this aspect.

It is worthy to note that Nyquetek also recommended a "super count" scheme (similar to a leaky bucket scheme) to give the low load traffic better guarantees to be scheduled according to the 1/4 rule, which has been incorporated in the simulations that generate the results in Figs. 7 and 8. The main reason behind the above difference is that, with 1/4 scheduling, 75% of channel occupancy for public traffic is hard to be guaranteed with PURQ AC because some factors, such as different impatience times for each class of customers, are not considered in (Nyquetek, 2002). For example, very short impatience times (5 seconds) in the originating traffic queue will cause a lot of customers to drop their sessions before a channel is available, thus leading to much smaller effective amounts of originating traffic to compete with emergency traffic. In contrast, the adaptive probabilistic scheduling and the preemption threshold methods are dynamic and can consider the effects of these factors and still achieve the desired channel occupancies.

5.2 Success probability for each class

As another main goal, admission of emergency traffic should be guaranteed when its volume is not unexpectedly high. To compare the effectiveness of the preemption threshold strategy and PURQ AC in this aspect, the achieved success probability of emergency traffic and handoff traffic is shown in Fig. 9. It can be seen that the APS scheme is better than the PURQ AC policy, and the PTS scheme is even better. APS is better than PURQ AC because the emergency traffic can be given higher priority through a higher P_S value when the public traffic is comparatively high.

5.3 Waiting time

When the PTS scheme is applied, emergency traffic need not wait before being admitted. But for APS and PURQ AC, emergency customers have to wait before being admitted. As shown in Fig. 10, the APS scheme can cause obviously lower waiting time than the PURQ AC scheme when the emergency traffic load is moderate.

For the originating traffic, the waiting time in the APS scheme is almost the same as the PTS scheme, and is obviously lower than PURQ AC (Fig. 11).

From the comparisons in Fig. 10 11 it can be concluded that the APS scheme is obviously better than the PURQ AC method in terms of waiting time since it has lower access time for both emergency users and public users. The PTS scheme is again better than any of other two strategies.

5.4 Termination probability

As can be seen from the above comparisons, the PTS strategy is almost always the best in terms of achievable channel occupancy for public traffic, success probability for each class, and the waiting time. However, it has a significant downside: public sessions in this scheme can be

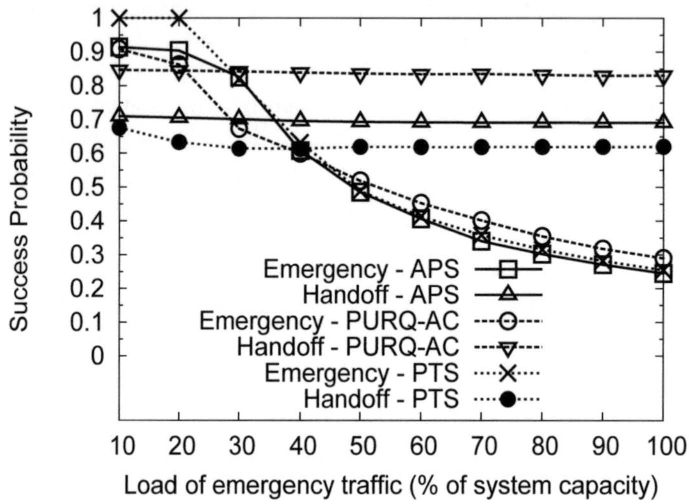

Fig. 9. Comparison of success probability

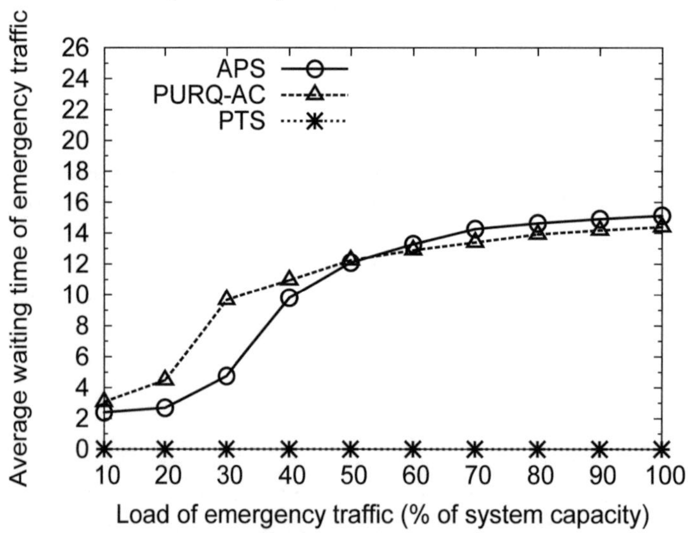

Fig. 10. Comparison of waiting time for emergency traffic

terminated due to preemption. In Fig. 12 the termination probability of public originating traffic is shown for the same scenarios considered above. For most load cases, the termination probability of public traffic for the PTS scheme is around 10%. In contrast, the queueing and scheduling based strategies APS and PURQ-AC obviously do not have such a problem. This shortcoming of the preemption based strategy is the main reason that it is not allowed in the current WPS in the United States, regardless of its other benefits.

6. Extension of the load based model to 3G/4G systems

The admission control policies discussed in this chapter are assumed to be load based. This means that admission is based on whether the new session will make the load surpass the

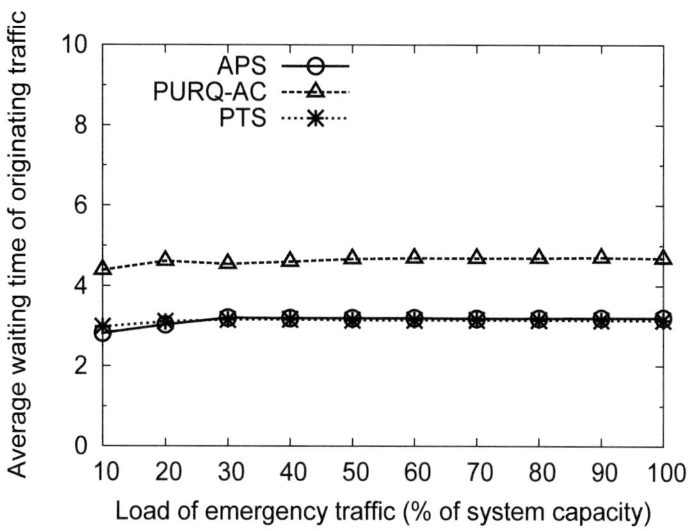

Fig. 11. Comparison of waiting time for originating traffic

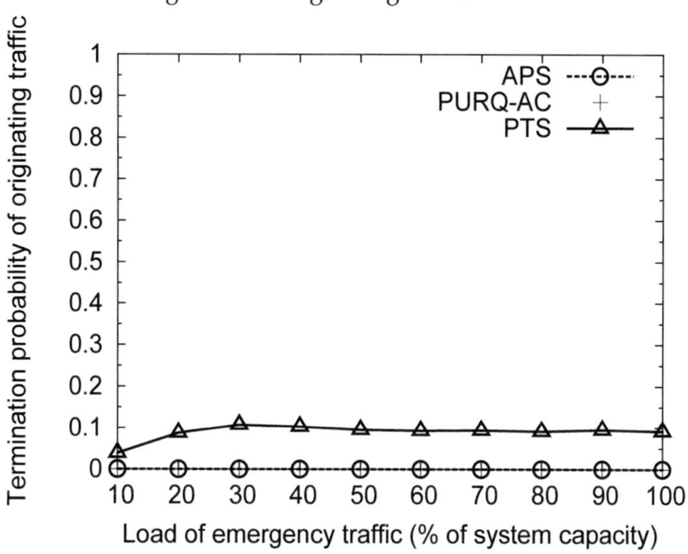

Fig. 12. Comparison of termination probabilities

capacity of the system. The load is usually measured by the number of users in a 2G system. With multiple access schemes like CDMA, WCDMA, OFDMA applied, one main difference is that interference rather than the number of users can be the main factor to be considered for the admission control problem in a 3G/4G system.

With a CDMA based access scheme, admission can be done indirectly by setting an interference based criteria, for example a limit on CDMA Rise over Thermal (RoT), then determining ahead of time the load where a new session would cause the system to exceed the interference limit. In fact, as pointed out in (Ishikawa & Umeda, 1997), load based admission control is still suitable. In their analysis for what they call number based CAC, the interference threshold is transferred into the maximum acceptable number of users. Then the blocking rate (measured grade of service) and the outage probability of communication quality (measured quality of service) are evaluated. The numerical results show that the number-based CAC and the interference-based CAC agree well with each other. They concluded that load-based admission is preferred because of its simplicity and ease of implementation, although interference based admission has the advantage that the threshold value has less sensitivity to other system parameters like the propagation model, traffic distribution, or the transmission rate.

As opposed to balancing blocking rate and outage probability of communication quality like in (Ishikawa & Umeda, 1997), we are mainly considering the fairness in resource use between emergency users and public users. When an emergency happens, there is much more demand than the system can handle. No matter how we try to balance capacity and quality of service, there is still blocking. So the capacity of the system, in terms of the maximum number of admitted users, can be determined according to the requirements on quality of service (QoS) only. With the capacity of the system known, the probabilistic scheduling can be tuned to achieve ideal channel occupancies for both emergency and public traffic. Note here that we assume the capacity is static for a period of time, but it can be recomputed if the SIR threshold needs to be changed, for instance, due to increased interference from neighboring cells or due to cell breathing to shift users to neighboring cells.

Another important difference is that data applications are much more common in a 3G/4G network. How would load based admission control be accomplished with both voice sessions and data sessions (emergency and public) in the same cell? Admission of voice sessions can easily be controlled based on whether a new session would go beyond the voice loading limit. Data sessions, however, can be handled in two distinctly different ways. On the one hand, if data sessions need some level of guaranteed QoS, they can be admitted similarly to voice sessions, by equating a data session to a certain number of voice sessions or a certain amount of needed bandwidth. On the other hand, service providers may treat data sessions differently by expecting them to use whatever is left over after the voice sessions are satisfied. For example, in the 3G EV-DO, Rev. A standard, "HiCap" data sessions are given different power levels and Hybrid ARQ termination targets as compared to "LoLat" voice traffic. HiCap data traffic is expected to be able to tolerate longer packet delays and to probably use TCP to adapt to the network congestion.

In conclusion, interference-based admission control can be converted into a load based admission control problem. Furthermore, the elastic property of data sessions makes it possible for us to use the same model as that of a 2G scenario. This is why we can conclude that the schemes and modeling methods shown in this chapter is suitable for all 2G, 3G, and 4G systems.

7. Summary

Due to the special requirements of emergency traffic, the reservation based admission control strategy is inappropriate due to its possible waste of resources when emergency situations do not occur. Among the schemes that can guarantee high system utilizations, the dynamic schemes like the preemption based PTS scheme and the queueing and scheduling based APS scheme demonstrate their privileges over the static schemes like PURQ-AC. The PTS scheme is almost always the best in the guaranteed protection of both public and emergency traffic and in much shorter access waiting time. However, its disadvantage is possible high termination probability. In contrast, the APS scheme is also quite good in protecting both public and emergency traffic, and it can still guarantee low termination probability for public sessions. The operators can choose the strategy that suits their specific needs.

8. References

Beard, C. (2005). Preemptive and Delay-Based Mechanisms to Provide Preference to Emergency Traffic, In *Computer Networks Journal*, Volume 47, Issue 6, pp. 801-824, April 2005.

Beard, C. & Frost, V. (2001). Prioritized Resource Allocation for Stressed Networks, In *IEEE/ACM Transactions on Networking*, Volume. 6, Issue 5,October 2001, pp. 618-633.

Carlberg, K., Brown, I., & Beard, C. (2005). Framework for Supporting Emergency Telecommunications Service (ETS) in IP Telephony, *Internet Engineering Task Force, Request for Comments 4190*, November 2005

Conway, R., Maxwell, W., & Miller L. (1967). Theory of Scheduling, published by *Addison Wesley*, 1967.

Federal Communications Commission Report and Order (2000). The Development of Operational, Technical and Spectrum Requirements for Meeting Federal, State and Local Public Safety Agency Communication Requirements Through the Year 2010, *FCC 00-242*, rel. July 13, 2000.

Guerin, R. (1988). Queueing-Blocking system with two arrival streams and guard channel, In *IEEE Transactions on Communications*, Volume 36, Issue 2, Feburary 1988, pp. 153-163.

Ishikawa, Y. & Umeda, N. (1997). Capacity Design and Performance of Call Admission Control in Cellular CDMA Systems, In *IEEE EEE Journal on Selected Areas in Communications*, Volume 15, Issue 8, 1997, pp.1627-1635

Jedrzycki, C. & Leung, V. (1996). Probability distribution of channel holding time in cellular telephony systems, In *Proc. IEEE Veh. Technol. Conf.*, May 1996, pp.247-251.

Mitchell, K., Sohraby, K., van de Liefvoort, A., & Place, J. (2001). Approximation Models of Wireless Cellular Networks Using Moment Matching, In *IEEE Journal of Selected Areas in Communications*, Volume 19, Issue 11, pp. 2177-2190, 2001.

National Communications System (2002). Wireless Priority Service (WPS) Industry Requirements for the Full Operating Capability (FOC) for GSM-Based Systems, *prepared by Telcordia*, Issue 1.0, Sept. 2002.

National Communications System (2003). Wireless Priority Service (WPS) Industry Requirements for the Full Operating Capability (FOC) for CDMA-Based Systems, *prepared by Telcordia*, Issue 1.0, Mar. 2003.

Nyquetek Inc. (2002). Wireless Priority Service for National Security / Emergency Preparedness: Algorithms for Public Use Reservation and Network Performance, August 30, 2002. Avaiable at http://wireless.fcc.gov/releases/da051650PublicUse.pdf.

Tang, S. & Li, W. (2006). An adaptive bandwidth allocation scheme with preemptive priority for integrated voice/data mobile networks, In *IEEE Transactions on Wireless Communications*, Volume 5, Issue 9, September 2006.

Zhou, J. & Beard, C. (2006). Comparison of Combined Preemption and Queuing Schemes for Admission Control in a Cellular Emergency Network, In *IEEE Wireless Communications and Networking Conference (WCNC)*, Las Vegas, NV, April 3-5, 2006.

Zhou, J. & Beard, C. (2009). A Controlled Preemption Scheme for Emergency Applications in Cellular Networks, In *IEEE Transactions on Vehicular Technology*, Volume 58, Issue 7, Sept. 2009 pp.3753-3764.

Zhou, J. & Beard, C. (2010). Balancing Competing Resource Allocation Demands in a Public Cellular Network that Supports Emergency Services, In *IEEE Journal on Selected Areas in Communications*, Volume 28, Issue 5, June. 2010.

13

Performance Analysis of Seamless Handover in Mobile IPv6-based Cellular Networks

Liyan Zhang[1], Li Jun Zhang[2] and Samuel Pierre[3]
[1]*School of Electronics and Information Engineering, Dalian Jiaotong University*
[2]*Division R&D, Geninov Inc.*
[3]*Department of Computer Engineering, Ecole Polytechnique de Montreal*
[1]*China*
[2,3]*Canada*

1. Introduction

The commercial proliferation of cellular voice and data service has placed a new challenge for mobile communication systems. Next-generation wireless systems are envisioned to have an all-IP-based infrastructure with the support of heterogeneous access technologies (Akyildiz et al., 2004). Under the circumstance, the Internet Protocol (IP) is selected as the common interconnection protocol to integrate disparate wireless systems, so that mobile users can roam among multiple wireless networks, regardless of the underlying different radio access technologies (Akyildiz et al., 2005; Makaya & Pierre, 2008; Mohanty & Xie, 2007). However, with the advent of new value-added services (video-conference, multimedia streaming, etc.) and novel concepts introduced into Long Term Evolution (LTE) architecture of the 4th Generation (4G) networks, provisioning efficient mobility management with quality of service guarantees and seamless handoff feature become even more important for next-generation wireless network design.

Generally, *mobility management* allows mobile communication systems to locate roaming terminals for voice/data delivery as well as maintaining network connectivity when the terminal moves into a new service area (Akyildiz et al., 1999). Typically, such process contains two aspects: location management and handoff management (Quintero et al., 2004; Zhang et al., 2010).

Location management enables telecommunication systems to find out the network attachment points of roaming nodes for call/data delivery. It usually contains two components: *location update* and call delivery (or data delivery). The former requires mobile nodes to provide the system with their location information, while the latter indicates that the system is queried for the location information of specific mobile nodes, and then services are delivered to them while they are away from their home network (Zhang et al., 2010).

Handoff management aims to maintain network connectivity when mobile nodes change their network attachment points or access points. Obviously, handoff protocols need to preserve mobile users' network connectivity as they move from one network to another, while simultaneously reducing disruption to the ongoing call/data sessions. Therefore, reducing handoff delay and maximizing session continuity are always the primary goals of handoff management (Dimopoulou et al., 2005). Generally, *handoff seamlessness* means lower packet losses, minimal handoff latencies, lower signaling overheads and limited handoff failures (Makaya & Pierre, 2008).

Handoff can be classified into: horizontal (or intra-system) and vertical (or inter-system) handover due to the coexistence of various radio access technologies in the next-generation wireless networks. *Horizontal handoff* takes place when mobile nodes move between access points supporting the same network technology while *vertical handoff* happens when mobile terminals move among access points supporting different network technologies (Nasser et al., 2006). This chapter proposes IP-layer-based mobility management solutions, which are suitable for both intra-system and inter-system handoff.

Handoff latency is defined as the time taken for a mobile node to obtain a new IP address from a visiting network and register itself with its home network (Haseeb & Ismail, 2007), during which the mobile node cannot send or receive any data packets. The handoff latency is the primary cause of packet losses in a network, and needs to be minimized as much as possible, particularly for supporting real-time applications.

According to handled object, mobility can be classified into: *network mobility* (NEMO) (Devarapalli et al., 2005) and *host mobility*. NEMO-based schemes aim to manage the mobility of an entire network (Ernst & Lach, 2007). Such protocols allow a mobile network to change its point of attachment to the Internet, and ensure its reachability in the topology, without interrupting packet delivery to/from that mobile network (Manner & Kojo, 2004).

The NEMO basic support protocol (Devarapalli et al., 2005) enables mobile networks to attach to different access points in the Internet. Such a protocol is an extension of mobile IPv6 (MIPv6) (Johnson et al., 2004) and allows session continuity for every node in the mobile network as the network moves. It also enables every roaming node in such a network to be reachable (Devarapalli et al., 2005). A number of solutions are proposed for network mobility. For example, a novel architecture is recently proposed to provide NEMO support in proxy MIPv6 domain, namely *N-PMIPv6* (Soto et al., 2009). Such a protocol handles mobile networks' connectivity in network-based localized mobility domain. To improve handover performance, issues that combine network mobility and host mobility are discussed in this proposal.

Host mobility management allows a mobile node to change its point of attachment to the network, without interrupting IP packet delivery to/from that node (Manner & Kojo, 2004). Numerous protocols are designed within the Internet Engineering Task Force (IETF) working groups for host mobility, such as MIPv6 (Johnson et al., 2004), hierarchical mobile IPv6 (HMIPv6) (Soliman et al., 2008), mobile IPv6 fast handovers (FMIPv6) (Koodli, 2008), fast handovers for HMIPv6 (F-HMIPv6) (Jung et al., 2005), and proxy mobile IPv6 (PMIPv6) (Gundavelli et al., 2008), etc. Moreover, some working groups are still striving to improve the performance of such specifications. And this chapter focuses on host mobility support issues.

The remainder of this chapter is organized as follows. We provide an overview of the related work pertaining to mobility management in IPv6-based wireless networks in Section II. Then, we elaborate the proposed seamless mobility management schemes, namely seamless mobile IPv6 (SMIPv6) in Sections III. To assess its efficiency, we design analytical models and present the analysis of numerical results in Section IV. Finally, we draw our conclusion marks in the last section.

2. Background and related work

In all-IP-based wireless networks, mobile nodes can freely change their network attachment points while communicating with correspondent nodes. Accordingly, *mobility management* becomes a critical issue to track mobile users' current location and to efficiently deliver services to them when they are away from their home network.

Generally, IP mobility includes *macromobility* and *micromobility*. Macromobility designates mobility over a large area; this refers to situations where mobile nodes move between different IP domains (Manner & Kojo, 2004). Typically, protocols such as mobile IPv4 (MIPv4) (Perkins, 2002), MIPv6 (Johnson et al., 2004) and PMIPv6 (Gundavelli et al., 2008) are best suited for macromobility management. This chapter only addresses mobility management in IPv6-based wireless networks.

Micromobility refers to mobility over a small area, i.e. within an IP domain. Usually, micromobility protocols maintain a location database that maps mobile host identifiers to location information, and they complement IP mobility by offering fast and seamless handoff control in limited geographical areas and IP paging in support of scalability and power conservation (Campbell et al., 2002). Typical micromobility management protocols are Cellular IP (Valko, 1999), handoff-aware wireless access internet infrastructure (HAWAII) (Ramjee et al., 2002), HMIPv6 (Soliman et al., 2008), FMIPv6 (Koodli, 2008) and F-HMIPv6 (Jung et al., 2005).

Cellular IP and HAWAII use two approaches to optimize handoff performance: bicasting, buffering & forwarding techniques (Campbell et al., 2002; Ramjee et al., 2002; Valko, 1999). HMIPv6, FMIPv6 and F-HMIPv6 confine mobility related signaling within a local domain. Therefore registrations with distant home agent and correspondent nodes are eliminated as long as mobile nodes remain inside their local domain. Accordingly, micromobility protocols yield better performance than macromobility solutions for roaming within a local domain, namely *intra-domain movement*.

MIPv6 (Johnson et al., 2004) is the baseline host-based mobility management protocol, which provides mobile users with unbroken network connectivity while they move around the Internet. Regardless of its current location, a mobile node is always identified by its home address. While away from the home network, the mobile node configures a new IP address (care-of-address), which indicates its current location within the Internet topology. The uniqueness of such a new address must be verified before utilization through neighbor discovery procedure defined in (Narten et al., 2007). Such verification is called *duplicate address detection* (Thomson et al., 2007). Each time when this mobile moves, it has to inform a router called home agent of its new IP address. However, this results in triangular routing problem. To fix this, route optimization is designed to allow mobile and correspondent nodes to communicate via a direct routing path (Arkko et al., 2007).

Usually, handover takes place when a mobile node changes its network attachment point. After acquiring a new IP address from the visiting network and successfully executing the duplicate address detection, the mobile node sends a *binding update* (BU) message to its home agent, which is a default router at the home network. The home agent then binds the mobile's home address to the new care-of-address, and replies the mobile with a *binding acknowledgement* (BA) message. Subsequently, the home agent intercepts the packets addressed to the mobile and tunnels them to the mobile's new location.

Following by successful home registration, return routability tests are carried out to ensure communication security between the mobile and each correspondent node. Upon completion, corresponding registration is done by exchanging BU and BA messages between the mobile and correspondent nodes. As a result, correspondent nodes can directly communicate with mobile node without bypassing the home agent.

Generally, the overall handoff process includes *link layer switching* (or layer two handoff), *movement detection* to discover new access networks, new care-of *address configuration*, *duplicate address detection*, *home registration*, *return routability tests*, and *correspondent registration*. Eventually, handoff latency results in packet loss and degrades network performance, which is unacceptable and detrimental to real-time traffic causing user perceptible service deterioration (Kempf et al., 2003). Thus improving the performance of MIPv6 is one major challenge for wireless networks to provide mobile users with seamless mobility, session continuity and guaranteed quality of service.

As MIPv6 handles local mobility and global mobility in the same fashion (Haseeb & Ismail, 2007), mobility management induces lengthy registration delays and unavoidable packet losses. Thereby separating local and global mobility domain is necessary. Under such circumstances, protocols such as Cellular IP (Valko, 1999), HAWAII (Ramjee et al., 2002), HMIPv6 (Soliman et al., 2008), FMIPv6 (Koodli, 2008), F-HMIPv6 (Jung et al., 2005) are designed to improve handoff performance.

Cellular IP (Valko, 1999) is a lightweight and robust protocol to support local mobility, it uses distributed caching techniques for location management and routing. Distributed paging cache coarsely maintains the position of idle mobile nodes in a service area while distributed routing cache maintains the position of active mobile nodes in the service area and dynamically updates the routing state of mobile nodes when they move to other service areas (Valko, 1999). Handoff is initiated by a mobile node through sending out a message to the old access point, which then modifies its routing cache, configures a new routing path for the mobile along with a timer. Before time out, packets destined to the mobile are delivered at both the old and new access points. Such delivery technique is called *bicasting*, which helps to reduce packet losses caused by handoff. Cellular IP shows great benefit for environments where mobile nodes migrate frequently. However, it requires the capability of mobile node to listen simultaneously two logical channels because of bicasting, this limits its applicability for mobiles with only one radio device.

HAWAII (Ramjee et al., 2002) is a protocol designed for micromobility management. It segregates the network into a hierarchy of domains, and each domain is controlled by a domain root router. Such router maintains a forwarding table with host-based entries. When a mobile node enters into a foreign domain, traditional Mobile IP mechanisms (Perkins, 1996) are executed. The mobile node is assigned a new care-of-address by a DHCP server. Such procedure is called *stateful address configuration*. The mobile node then executes duplicate address detection. Upon success, it carries out home registration with the home agent. As a result, packets addressed to the mobile are intercepted by its home agent, which then tunnels them to the mobile's new location, identified by the new care-of-address. When moving within the same domain, the mobile retains its care-of-address unchanged, and IP connectivity is maintained using dynamically established paths configured by the protocol HAWAII (Ramjee et al., 2002). Therefore, disruption to ongoing sessions is minimized during handoff.

HMIPv6 (Soliman et al., 2008) is designed to reduce the signaling cost and location update delay outside a local mobility domain. Like HAWAII, this protocol also divides the network into a hierarchy of domains, and each domain is managed by a mobility anchor point (MAP). While entering a MAP domain, a mobile node configures two IP addresses: an on-link local care-of-address (LCoA) and a regional care-of-address (RCoA). The mobile node then

verifies the uniqueness of the LCoA through duplicate address detection. Upon success, it sends a *local binding update* (LBU) message to the MAP, which then verifies the uniqueness of the RCoA, binds the mobile's LCoA with the RCoA, and replies the mobile with an acknowledgment message. As a result, a bidirectional tunnel is established between the mobile node and MAP (Soliman et al., 2008). Afterward, the mobile informs its home agent and each correspondent node of the RCoA. Accordingly, they bind the mobile's RCoA with its home address. Packets destined to the mobile are intercepted by the MAP, encapsulated and forwarded to the mobile's LCoA. A movement within the MAP domain merely incurs LBUs to the MAP without further propagation to home agent and correspondent nodes, thus significantly reducing signaling load and handoff latency for local movements (Zhang, 2008).
FMIPv6 (Koodli, 2008) is known as low latency address configuration protocol, which enables mobile nodes to rapidly detect their movements and to obtain a prospective IP address with a new access router before disconnecting with the current access router. It also offers mobile nodes the opportunity to utilize available link layer event notification (triggers) to accelerate network layer handoff (Kempf et al., 2003). Hence, delays pertaining to access network discovery and new IP address generation are completely removed from handoff latency. Moreover, a bidirectional tunnel is setup between the previous access router (PAR) and new access router (NAR) to avoid packet losses. The PAR binds the mobile's previous care-of-address with the new care-of-address. Therefore, packets addressed to the mobile are intercepted by the PAR, tunneled to the NAR, which then decapsulates and forwards them to the mobile node. During handoff, no registration is necessary with either the home agent or any correspondent node. However, because of the utilization of pre-handover triggers, the performance of FMIPv6 largely depends on the trigger time. In case where the pre-handoff trigger is delivered too closely to the actual link switching, the communication using FMIPv6 becomes unreliable (Kempf et al., 2003).
Both FMIPv6 and HMIPv6 are designed in their own fashion to improve the MIPv6 performance, it is necessary to combine them together. However, simple superimposition of FMIPv6 over HMIPv6 induces unnecessary processing overhead for re-tunneling at the PAR and inefficient usage of network bandwidth (Jung et al., 2005). To resolve such problems, F-HMIPv6 (Jung et al., 2005) enables mobile nodes to exchange handoff signaling messages with a MAP and to establish a bidirectional tunnel between the MAP and NAR, instead of between the PAR and NAR. However, the performance of this protocol is largely dependent on various wireless system parameters such as user mobility model, user density, domain size, session-to-mobility ratio, etc (Zhang, 2008).
The aforementioned protocols present typical solutions in the literature for mobility management in wireless networks. However, these protocols all have pros and cons. Cellular IP requires mobile nodes to be equipped with multiple radio interfaces. HAWAII needs the capability of working together with Mobile IP mechanisms, and its forwarding technique requires more buffer space at access routers. MIPv6 is suitable for macromobility management with some drawbacks such as high signaling overheads, unacceptable packet losses and lengthy handoff latencies, thus cannot support real-time traffic. HMIPv6 cannot meet the requirements for delay-sensitive traffic, such as voice over IP (VoIP), due to high packet loss rate and long handoff delay (Makaya & Pierre, 2008). FMIPv6 is hindered by the problems of supporting quality of service and scalability. Additionally, neither signaling overheads nor packet losses are effectively reduced using FMIPv6, thus supporting seamless mobility becomes impossible. F-HMIPv6 allows mobile users to benefit from both FMIPv6 and HMIPv6, but the handoff latency for intra-domain roaming lasts about 90*ms* whereas the

handover delay for inter-domain roaming rises to about 240*ms* (Jung et al., 2005), making this protocol unsuitable for multimedia streaming traffic (Zhang & Pierre, 2008). As a result, none of these protocols provides a perfect solution for seamless mobility management.
Under the circumstance, we propose a new protocol called seamless mobile IPv6 (SMIPv6) (Zhang et al., 2005; Zhang & Marchand, 2009; Zhang & Pierre, 2008; 2009). Compared with current alternatives, the key advantage of our proposal is that (1) mobile nodes don't have to be equipped with multiple radio interfaces; (2) forwarding IP packets from the PAR to NAR is carried out much earlier than FMIPv6 and F-HMIPv6; (3) Ongoing real-time session is resumed on the new link much earlier than FMIPv6 and F-HMIPv6; (4) it is flexible to work with MIPv6 and HMIPv6; (5) SMIPv6 can support multimedia streaming traffic during handoff.

3. Proposed seamless mobile IPv6

The main idea of SMIPv6 (Zhang et al., 2005; Zhang & Marchand, 2009; Zhang & Pierre, 2008; 2009) is to pre-configure bidirectional secure tunnels among access routers before actual handover and to utilize such tunnels to accelerate mobility management procedure of FMIPv6 (Koodli, 2008). The quality of service related parameters (e.g. delay, jitter, packet loss), and security aspects (e.g. authentication methods, tunneling keys) are specified for each unidirectional tunnel through negotiation between radio access networks (Zhang & Marchand, 2006; Zhang & Pierre, 2009). Such access networks can be managed by either the same or different operators. The utilization of pre-established tunnels enables network operators to serve their own mobile users and those from their competitors. As a result of negotiation, a set of specific mobile users is given the opportunity to exploit pre-configured tunnels during handoff. Additionally, using pre-established bidirectional tunnels allows mobile nodes to retain their previous valid IP addresses unchanged in a new visiting network or domain (Zhang & Marchand, 2009; Zhang & Pierre, 2009). This minimizes interruption of ongoing multimedia sessions during handoff. And new routing policy is added to access routers, this enables the delivery of packets to mobile nodes that use a topologically invalid address within an access network.
The proposed mobility management procedure in SMIPv6 comprises two stages: configuring bidirectional secure tunnels between radio access networks prior to actual handoff and using such tunnels to accelerate mobility management procedure during handoff.

3.1 Tunnel establishment

The first stage of SMIPv6 consists of using *tunnel establishment* method (Zhang & Marchand, 2006; Zhang & Pierre, 2009) to set up bidirectional secure tunnels among access routers. This method allows dynamically establishing a tunnel with a set of minimal characteristics between two tunnel endpoints. Such tunnels enable radio access networks to establish business and security relationship with their neighborhood. Consequently, communication services are offered to a list of subscribers from either the same or various mobile operators.
Tunnel establishment method requires that each node (access router in our case) comprises a tunneling protocol module. The source node determines a first set of desired characteristics of the tunnel. Its tunneling protocol module sends a *tunnel request* message to a destination node. Such request comprises the specific conditions of the unidirectional tunnel and a shared secret key with an index value thereof. The destination node then determines a second set of desired characteristics, and replies with a *tunnel reply* message. Upon reception of this message, the source node verifies if the second set of characteristics is at least equal to the

set of minimal characteristics. If so, it replies a *tunnel acknowledgment* message. Otherwise, negotiation keeps on between the involved nodes until time out. The shared secret is used to encrypt data and the index value indicates which shared secret is used during subsequent communication. Usually, such negotiation is done before mobile users handoff from one access network to another. With the determined characteristics, both nodes configure their *Forwarding* and *Reverse Tunnels Lists*, respectively (Zhang & Marchand, 2006; Zhang & Pierre, 2009).

3.2 Seamless mobility management

The seamless mobility management procedure in SMIPv6 allows mobile nodes to utilize pre-configured bidirectional secure tunnels during handoff (Zhang & Marchand, 2009). To realize such functionality, we introduce a new network entity, called intelligent access router (iAR) with novel routing policy, which allows it to handle the traffic using topologically invalid addresses within the access network, and to handle tunneled packets, of which the ultimate destination node is not yet attached to the network.

The following example shows the way of an iAR handling tunneled packets, of which the final destination is not within its network. The iAR receives a tunneled packet from another access router, which format is shown in Table I. Upon receiving such packet, the iAR (NAR in our case) usually removes the outer header, verifies the IP address of the destination node, and finds out that the destination (MN in our case) is not in its subnet. Normal IP routing policy requires the destination router to forward such packet to the source router (PAR in our case) with which the destination node is supposed to be attached. This induces the routing loop problem, which still exists in FMIPv6. SMIPv6 fixes such problem by allowing the iAR to buffer such packet for a certain of time, waiting for the attachment of the mobile node. Upon timeout and the absence of the MN, the destination router (iAR) simply discards the received tunneled packet.

source-addr 1	dest-addr 1	source-addr 2	dest-addr 2	token	data
PAR-addr	NAR-addr	CN-addr	MN-PCoA	TK1	data

Table 1. Example of a tunneled packet

The subsequent example shows the way of an intelligent access router (iAR) handling a packet from a topologically invalid address within the access network. The iAR (NAR in our case) receives from a mobile node (MN) a packet, which format is shown in Table II. Generally, such kind of packets are dropped by access router due to ingress filtering. However, SMIPv6 allows the iAR to verify the IP address of the source node (MN in our case), finds out the IP address of the associated router, from which the source node obtained its valid IP address, namely previous care-of-address (PCoA). The iAR (or NAR) then checks out if there are pre-established tunnels between the two involved access networks. If so, it tunnels the packets to the previously associated router (PAR). The PAR then removes the outer header of the tunneled packets, carries out ingress filtering, and decrypts the data using a pre-shared key with the MN. Upon success, the PAR then sends the packets with the decrypted data to the correspondent node (CN).

source-addr	dest-addr	encrypted data
MN-PCoA	CN-addr	encrypted data using pre-shared key between MN-PAR

Table 2. Example of a packet from MN to NAR

Since SMIPv6 empowers mobile nodes to use their valid previous care-of-address (PCoA), the context information of mobile nodes can be kept intact at previous access router. Hence, delay pertaining to context transfer process (Loughney et al., 2005) is eliminated completely from handoff latency. Additionally, mobile nodes can resume and initiate communication on the new link using their valid PCoAs due to pre-configured tunnels. Compared with the bidirectional edge tunnel handover for IPv6 (BETH) (Kempf et al., 2001), SMIPv6 does not need to exchange *handover request* and *handover reply* messages to establish bidirectional tunnels during handoff; neither does it exploit link layer pre-triggers to facilitate IP layer handoff. Given that both FMIPv6 (Koodli, 2008) and BETH (Kempf et al., 2001) protocols utilize pre-handover triggers, their performance, in terms of packet loss and handoff latency, depends greatly on the pre-handoff trigger time, thus becoming unreliable when such trigger is delivered too closely to the actual link switching (Gwon & Yegin, 2004; Kempf et al., 2003).

3.2.1 Predictive SMIPv6

We assume that mobile nodes (MNs) roam in the IPv6-based wireless networks, and each MN acquires a valid care-of-address (CoA) from its previous access router (PAR). Additionally, the MN has established a security association with the PAR before actual handoff. As a result, both of them configure a pre-shared key (PSK). In an overlap zone covered by the PAR and its neighbors, a mobile node receives beacons from nearby access points (APs). Such beacons contain APs' identifiers (AP-ID). Horizontal handoff requires the mobile to select the most suitable AP by analyzing the received signal strength, while vertical handoff asks for the mobile to use techniques such as score function (McNair & Zhu, 2004). Upon selection the best AP, the mobile node sends a *seamless binding update* (SBU) message to the PAR before breaking their connection. Such message contains the new AP's identifier (NAP-ID) and a session token generated by the mobile. Such token will be used to avoid replay attack.

Like FMIPv6 (Koodli, 2008), we assume that the PAR has some knowledge about its neighbors, such as their IP address, the associated APs' identifiers, etc. Upon receiving the SBU message, the PAR maps the NAP's ID to the IP address of the corresponding router (NAR in our case) and starts intercepting packets destined to the mobile. The PAR caches one copy of the intercepted packets, and then tunnels them to the NAR. An example of such tunneled packets is shown in Table I. The session token is inserted into the tunneled packet. Afterwards, the PAR adds an entry to its *Forwarding Tunnels List*. Such list is used to track the state of the tunnel from the PAR to NAR. Note that packets buffered by the PAR will be forwarded to the mobile in case of *ping pong* and *erroneous* movements. The former implies that mobile nodes move between the same two access routers rapidly while the latter connotes that mobile nodes think entering into a new network, but they are actually either moving to a different access network or aborting their movements by returning to the old access network (Zhang & Pierre, 2009).

Upon receipt of the tunneled packets from the PAR, the NAR removes the outer header, verifies the presence of the destination node in its subnet. In case of absence, the NAR puts the inner packets into a buffer and starts a timer. Subsequently, the NAR extracts the session token from the inner packets and puts it into the *Token List*. Note that each intelligent AR (iAR) manages a token list, which is indexed by mobile's IP address to facilitate information retrieval. The NAR also creates a *host route entry* for the mobile's previous care-of-address (PCoA), and allocates a unique new care-of-address (NCoA) to the pending mobile node. Here we advocate that each iAR manages a private address pool and guarantees the uniqueness of

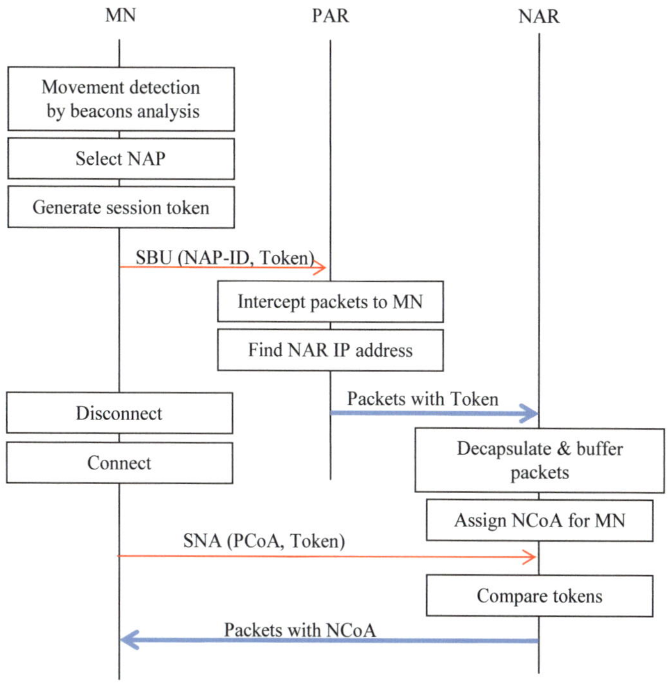

Fig. 1. Mobility management with predictive SMIPv6

each address in the pool. By this means, duplicate address detection can be removed from the overall handoff, thus improving handover performance (Zhang & Pierre, 2009).

Once attached to the new link, the MN sends a *seamless neighbor advertisement* (SNA) message to the NAR immediately. Such a message includes all the fields of *unsolicited neighbor advertisement* (UNA) (Narten et al., 2007), the IP address of the PAR, and a session token that is the same as the one sent to the PAR at the beginning of handoff. The latter two fields will be added as new options into the UNA. The IP source address of the SNA is the mobile's PCoA, and IP destination address is typically the all-nodes multicast address. The source link layer address (LLA) is the mobile's MAC address and the destination LLA is the new AP's link layer address.

Upon receipt of the SNA message, the NAR retrieves the token from its *Token List* with the assistance of the mobile's PCoA. The NAR also verifies whether the received token from the mobile node is the same as the one from the PAR. If they are identical, the NAR retrieves those buffered packets addressed to the mobile, and forwards them to the mobile node, along with the assigned NCoA. Figure 1 illustrates mobility management using predictive SMIPv6.

In case where those two tokens are different, the NAR obtains the IP address of the PAR from the SNA, and sends a *fast binding update* (FBU) message to the PAR on behalf of the mobile node. Such a message contains the mobile's MAC address and its PCoA. The PAR then verifies the mobile's identities and replies with a *fast binding acknowledgment* (FBack) message. Soon afterward, the PAR adds an host entry into its *Forwarding Tunnels List* and *Reverse Tunnels List*, respectively. Upon receiving the FBack message, the NAR forwards the buffered packets and the NCoA to the mobile. Consequently, the mobile becomes reachable on the new link under both CoAs: PCoA and NCoA (Zhang & Pierre, 2009).

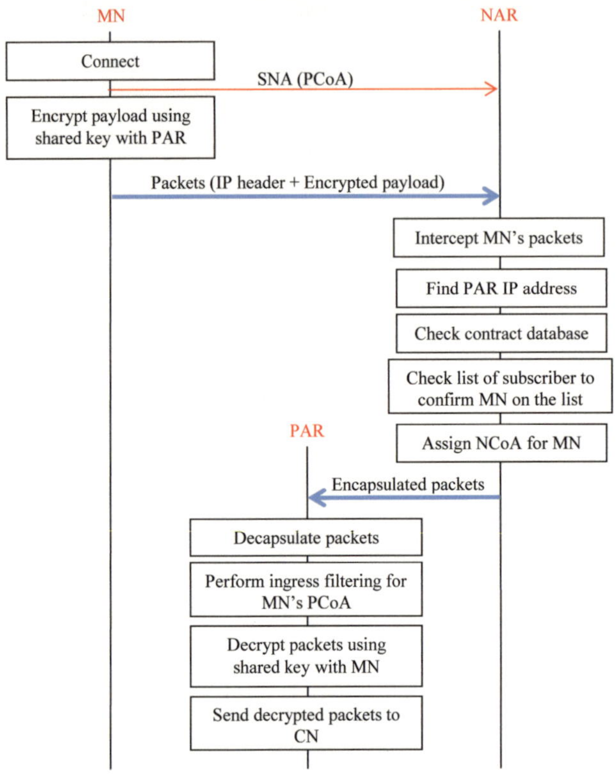

Fig. 2. Mobility management with reactive SMIPv6

3.2.2 Reactive SMIPv6

Typically, a *session* is identified by a group of information such as session ID, source address, destination address, source port number, destination port number, etc. When moving from one network to another, an mobile node loses its network connectivity and becomes unreachable because its previous IP address is invalid in the visiting network. Under the circumstances, the mobile node has to acquire a new IP address and registers the new address with its home agent and all active correspondent nodes. Prior to successful registration, the mobile cannot receive and send packets in the foreign network, thus the ongoing session is disrupted during handoff. In case where the mobile executes multimedia applications such as video-streaming, it cannot tolerate the degraded quality of the session. SMIPv6 resolves such problem by allowing mobile nodes to utilize their previous valid IP addresses on the new link via pre-configured bidirectional secure tunnels, thus guarantee seamless roaming with ongoing real-time sessions.

Reactive mobility management takes place when a mobile node (MN) initiates a new communication session with a correspondent node (CN) on a new link using a topologically invalid IP address. The mobile sends *seamless neighbor advertisement* (SNA) message to the NAR immediately after attachment (Zhang & Pierre, 2009). Figure 2 illustrates mobility management using reactive SMIPv6 (Zhang & Pierre, 2009).

For the sake of security, we advocate that mobile nodes encrypt the outgoing packets using the pre-shared key with the previous access router (PAR) before transmitting them over a visiting network. Note that instead of using the pre-shared key, the encapsulating security payload (ESP) protocol (Kent, 2005) may also be applicable to provide confidentiality, data origin authentication, connectionless integrity, and anti-replay service.

The new access router (NAR) then intercepts these outgoing packets, of which the source node utilizes its previous care-of-address (PCoA). An example of such packets is shown in Table II. Normally, the NAR should drop these packets because of ingress filtering, this induces high packet losses during handoff. To resolve such problem, SMIPv6 allows the NAR to extract the subnet prefix information from the mobile's PCoA, and obtains the IP address of the previously associated router (PAR). The NAR then checks out if there are any pre-established tunnels to the PAR. Such information is stored in a *contract database* (CD). The NAR also verifies if the mobile node is given the priority to use those pre-configured tunnels. If so, the NAR tunnels the intercepted packets to the PAR, and adds an entry into the *Reverse Tunnels List* for further tunnel maintenance and billing issues. If there is no pre-established tunnels, the NAR uses the *tunnel establishment* method to set up tunnels to the PAR. Upon success, the NAR tunnels the outgoing packets of the mobile to the PAR. If the mobile is not on the list of pre-configured subscribers who can benefit from the value-added service, the NAR will simply drop the packets, the same way as FMIPv6.

Upon receipt of the tunneled packets from the NAR, the PAR removes the outer header, performs ingress filtering for the mobile node's PCoA. Upon success, the PAR decrypts the inner packets using a pre-shared key with the mobile. On completion, the PAR forwards the decrypted packets to the destination node, a correspondent node (CN). The PAR also adds a host entry into its *Reverse Tunnels List*. Once terminating its ongoing session using the PCoA on the NAR's link, the MN can follow the legacy MIPv6 (Johnson et al., 2004) or HMIPv6 (Soliman et al., 2008) registration procedures.

3.3 Tunnel maintenance

Tunnel maintenance usually takes place after handoff, during which a mobile node may send a *Tunnel Bye* message to the new access router (NAR), which then releases the reserved bandwidth for the specific mobile, and forwards the same message to the previous access router (PAR) (Zhang & Marchand, 2006; Zhang & Pierre, 2009). As a consequence, entries in *Forwarding Tunnels List* and *Reverse Tunnels List* are removed or refreshed. However, SMIPv6 requires bidirectional tunnel remains active until mobile nodes complete the binding update procedures with their correspondents, same as the way of FMIPv6.

3.4 Summary

Using FMIPv6 (Koodli, 2008), even though a mobile node is IP-capable on the new link, it cannot use the new care-of-address (NCoA) directly with a correspondent node (CN) before the CN binds the mobile's NCoA with its home address, neither can the mobile use its previous care-of-address (PCoA) on the new link because of ingress filtering. In other words, FMIPv6 delivers better performance for downlink traffic. However, our proposed SMIPv6 allows mobile nodes to utilize their valid PCoAs immediately after attaching to a new link. Hence, the new proposal provides not only expedited forwarding packets to mobile nodes but also accelerated sending packets to their correspondents via a direct routing path, thus optimizes handoff performance.

On the other hand, the protocol SMIPv6 (Zhang et al., 2005; Zhang & Marchand, 2009) is independent of any network architecture. As a result, bidirectional secure tunnels can be pre-configured between any network entities acting as tunnel end-points. When such

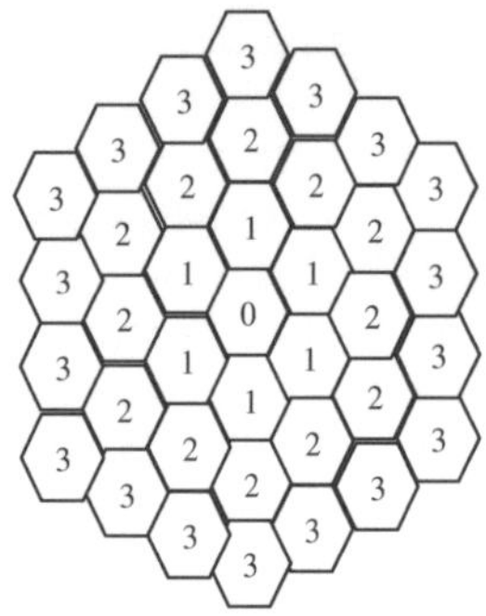

Fig. 3. An example of a MAP domain with 3 rings

tunnels are established between mobility anchor points, handoff delays and packet losses are reduced for both intra-domain and inter-domain movements, thus improves the handover performance of HMIPv6 (Soliman et al., 2008) and F-HMIPv6 (Jung et al., 2005) protocols. Furthermore, SMIPv6 can be freely implemented at any access network, this solves the problem of scalability in FMIPv6. When SMIPv6 mobility management mechanism is unavailable, mobile nodes can rely on the FMIPv6 (Koodli, 2008) protocol. In addition, intermediate routers are not involved in the tunnel setup and tunneling procedures, thus no extra overhead is added to them, this optimizes network resource usage.

4. Evaluation using analytical models

Performance evaluation of mobility management schemes is usually based on simulation and test-bed approaches (Gwon et al., 2004; Perez-Costa & Hartenstein, 2002; Perez-Costa et al., 2003). However, network scenarios for simulations vary greatly, the handoff performance comparison of the aforementioned mobility management protocols is rarely viable. Under the circumstance, analytical models are designed to evaluate system performance for users roaming in IPv6-based wireless cellular networks.

We assume that mobile service areas are partitioned into cells of equal size. Each cell is surrounded by rings of cells, except for cells in the outermost ring. Each domain is composed of n rings of the same size. We name the inmost cell "0", the central cell. Cells labeled "1" constitute the first ring around cell "0", and so on. Each ring is labeled in accordance with the distance to the cell "0". We assume that each cell is managed by one access router. Figure 3 shows an example of a MAP domain with three rings (Zhang & Pierre, 2008).

4.1 Mobility models

There are two mobility models proposed in the literature: the fluid-flow and random-walk models (Akyildiz & Wang, 2002). The former is more suitable for mobile users with high mobility, sporadic speed and direction changes. The latter is often used for pedestrian mobility, which is mostly limited to small geographical areas such as residential sites or premises.

4.1.1 The random-walk model

Under the random-walk model, the next position of a mobile node is determined by its previous position plus the value of a random variable with an arbitrary distribution. Assuming that a mobile node is located in a cell of ring r, the probability for the mobile to

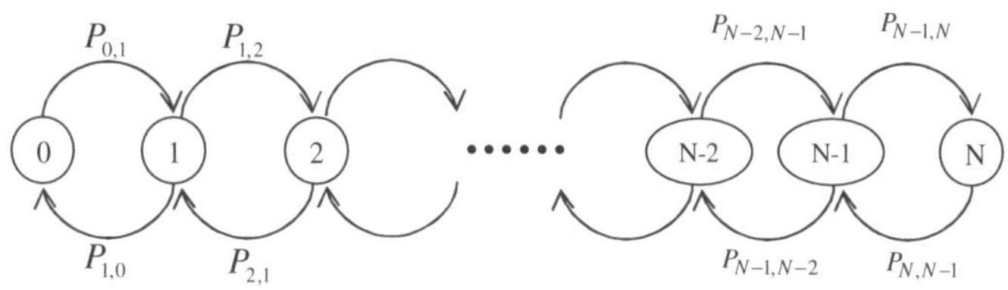

Fig. 4. State diagram for the random-walk model

move forward to a cell of ring $r+1$ ($p^+(r)$) and backward to a cell of ring $r-1$ ($p^-(r)$) are shown as follows (Zhang & Pierre, 2008):

$$p^+(r) = \frac{1}{3} + \frac{1}{6r} \tag{1}$$

$$p^-(r) = \frac{1}{3} - \frac{1}{6r} \tag{2}$$

We present the random-walk model with a one-dimensional Markov chain in which the state is defined as the distance between the current cell located the mobile node and central cell. Thus a mobile node is in state r if and only if it is now residing in a cell of ring r. Figure 4 shows the state transition diagram of this Markov chain (Zhang & Pierre, 2008).

Assuming that the probability for a mobile node to stay in the current cell is q, the probability for the mobile node to move to another cell is $1-q$. The transition probability $P_{r,r+1}$ and $P_{r,r-1}$ represent the probabilities that a mobile node moves from its current state r to the state $r+1$ and $r-1$, shown as follows (Zhang & Pierre, 2008):

$$P_{r,r+1} = \begin{cases} 1-q \; if \; r=0 \\ (1-q) \times (\frac{1}{3} + \frac{1}{6r}) \; if \, 1 \le r \le n \end{cases} \tag{3}$$

$$P_{r,r-1} = (1-q) \times (\frac{1}{3} - \frac{1}{6r}) \tag{4}$$

Let $\Phi_{r,n}$ be the steady-state probability of state r within a mobility anchor point (MAP) domain of n rings. Using the transition probabilities in Equations (3) and (4), $\Phi_{r,n}$ is shown as follows (Zhang & Pierre, 2008):

$$\Phi_{r,n} = \Phi_{0,n} \prod_{i=0}^{r-1} \frac{P_{i,i+1}}{P_{i+1,i}} \tag{5}$$

As $\sum_{r=0}^{n} \Phi_{r,n} = 1$, the expression of $\Phi_{0,n}$ is also given as follows (Zhang & Pierre, 2008):

$$\Phi_{0,n} = \frac{1}{1 + \sum_{r=1}^{n} \prod_{i=0}^{r-1} \frac{P_{i,i+1}}{P_{i+1,i}}} \tag{6}$$

Assuming that a mobility anchor point (MAP) domain is composed of n rings, and each cell is controlled by an access point integrating the functionality of an access router. The probability for a mobile node to perform an inter-domain mobility P is given as (Zhang & Pierre, 2008):

$$P = \Phi_{n,n} \times P_{n,n+1} \tag{7}$$

Where the $\Phi_{n,n}$ is the steady-state probability of the state n, $P_{n,n+1}$ is the probability that a mobile node moves from a cell in ring n to a cell in ring $(n+1)$.

4.1.2 The fluid-flow model

Using the fluid-flow model, the movement direction of a mobile node (MN) within a mobility anchor point (MAP) domain is distributed uniformly in the range of $(0, 2\pi)$. Let v be the average speed of an MN (m/s); R the cell radius (m); L_c and L_d the perimeters of a cell and a MAP domain with n rings (m); S_c and S_d the areas of a cell and a MAP domain with n rings (m^2); R_c and R_d be the cell and domain crossing rates, which denote the average number of crossings of the boundary of a cell and a domain per unit of time $(/s)$, shown as follows (Zhang & Pierre, 2008):

$$R_c = \frac{v \times L_c}{\pi \times S_c} = \frac{v \times 6R}{\pi \times 2.6R^2} = \frac{6v}{\pi \times 2.6R} \tag{8}$$

$$R_d = \frac{v \times L_d}{\pi \times S_d} = \frac{v \times (12n+6)}{\pi \times [3n \times (n+1)+1] \times 2.6R} \tag{9}$$

4.2 Cost functions

To analyze the performance of SMIPv6, we define the total cost as the sum of the mobility signaling cost and the packet delivery cost (Zhang & Pierre, 2008; Zhang et al., 2010).

4.2.1 Mobility signaling cost

Generally, mobile nodes perform two types of movements: intra-domain and inter-domain. The former are movements within an administrative domain while the latter implies movements between domains. Accordingly, two mobility management procedures are carried out for HMIPv6 and F-HMIPv6: the intra-domain and inter-domain cases. The latter includes the intra-domain and legacy MIPv6 mobility management procedures. However, FMIPv6 and SMIPv6 only address the problem of inter-cell handoff, because their domain is defined as a set of access routers.

We assume that mobility management protocols such as HMIPv6 (Soliman et al., 2008), F-HMIPv6 (Jung et al., 2005), FMIPv6 (Koodli, 2008) and SMIPv6 all support route optimization (RO) and only a pair of messages (*neighbor solicitation* and *neighbor advertisement*) exchanged for duplicate address detection. In addition, we assume that the distance between the previous access router (PAR) and MAP equals the one between the new access router (NAR) and MAP. And processing costs at the mobile node and correspondent node are ignored during analysis.

The mobility signaling overhead functions for MIPv6 (Johnson et al., 2004) with tunnel and RO modes, intra- and inter-domain HMIPv6, predictive and reactive FMIPv6, intra- and inter-domain F-HMIPv6 are given in (Zhang, 2008; Zhang & Pierre, 2008). The signaling overhead functions for predictive SMIPv6 (P-SMIPv6) and reactive SMIPv6 (R-SMIPv6) are expressed as follows (Zhang & Pierre, 2008; Zhang et al., 2010):

$$S_{P-SMIPv6} = 2\kappa \tag{10}$$

$$S_{R-SMIPv6} = \kappa \tag{11}$$

Where κ represents the unit transmission cost in a wireless link. Equation (10) implies that for predictive SMIPv6, 2 messages (SBU and SNA) are exchanged between a mobile node and

intelligent access routers (iARs) via radio link during handover, and the signaling cost for each message is represented by κ. The same principle applies to Equation (11).
Under the random-walk model, the mobility signaling cost functions for MIPv6 with tunnel and route optimization (RO) modes, HMIPv6, predictive FMIPv6 (P-FMIPv6), reactive FMIPv6 (R-FMIPv6), F-HMIPv6 are given in (Zhang & Pierre, 2008). The mobility signaling cost functions for predictive SMIPv6 (P-SMIPv6) and reactive SMIPv6 (R-SMIPv6) are expressed as follows (Zhang & Pierre, 2008; Zhang et al., 2010):

$$C^{s}_{P-SMIPv6} = \frac{S_{P-SMIPv6} \times (1-q)}{E(T)} \quad (12)$$

$$C^{s}_{R-SMIPv6} = \frac{S_{R-SMIPv6} \times (1-q)}{E(T)} \quad (13)$$

Where q is the probability that a mobile node remains in its current cell, $E(T)$ is the average cell residence time (s), $S_{P-SMIPv6}$ and $S_{R-SMIPv6}$ represent the mobility signaling overheads obtained from Equations (10) and (11).
Using the fluid-flow model, the mobility signaling cost functions for MIPv6 (Johnson et al., 2004) with tunnel and RO modes, HMIPv6 (Soliman et al., 2008), predictive and reactive FMIPv6 (Koodli, 2008), F-HMIPv6 (Jung et al., 2005) are given in (Zhang & Pierre, 2008). The mobility signaling cost functions for predictive SMIPv6 (P-SMIPv6) and reactive SMIPv6 (R-SMIPv6) are expressed as follows (Zhang & Pierre, 2008; Zhang et al., 2010):

$$C^{s}_{P-SMIPv6} = R_c \times S_{P-SMIPv6} \times (1-q) \quad (14)$$

$$C^{s}_{R-SMIPv6} = R_c \times S_{R-SMIPv6} \times (1-q) \quad (15)$$

Where R_c is the cell crossing rate, i.e. the average number of crossings of the boundary of a cell per unit of time $(/s)$, q is the probability that a mobile node remains in its current cell, $S_{P-SMIPv6}$ and $S_{R-SMIPv6}$ represent the mobility signaling overheads obtained from Equations (10) and (11).

4.2.2 Packet delivery cost

Packet delivery cost per session are defined as the cost of delivering a session from a source node to a destination node, which includes all nodes' processing costs and link transmission costs from the source to the destination.
We assume that HMIPv6 (Soliman et al., 2008), FMIPv6 (Koodli, 2008), F-HMIPv6 (Jung et al., 2005) and SMIPv6 (Zhang et al., 2005; Zhang & Marchand, 2009; Zhang & Pierre, 2008) support route optimization (RO). Under this mode, only the first packet of a session is transmitted to a home agent (HA) to detect whether a mobile node is away from its home network or not. All successive packets of the session are routed directly to the mobile's new location. Under the circumstance, the processing cost at a home agent is expressed as (Zhang & Pierre, 2008):

$$P_{HA} = \lambda_p \times \theta_{HA} \quad (16)$$

Where λ_p denotes the arrival rate of the first packet of a session, which is assumed to be the average packet arrival rate (packets per second). θ_{HA} indicates the unit cost for processing packets at the home agent (HA), which is assumed to be identical for all nodes' home agents.

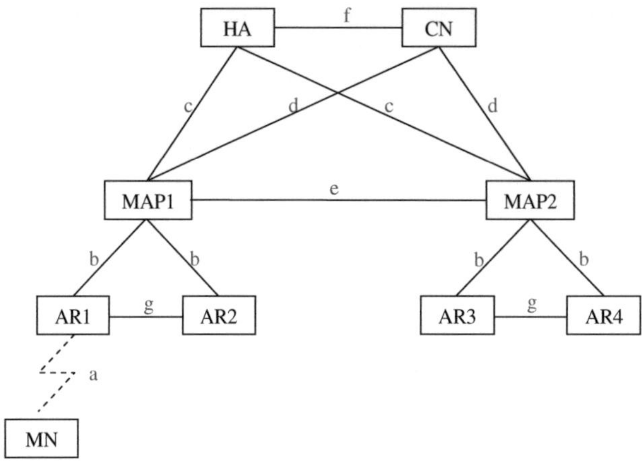

Fig. 5. Network topology for performance analysis

The packet delivery cost functions for MIPv6 with tunnel and RO modes, HMIPv6, FMIPv6 and F-HMIPv6 are given in (Zhang, 2008; Zhang & Pierre, 2008). The packet delivery cost for SMIPv6 is expressed as follows (Zhang & Pierre, 2008; Zhang et al., 2010):

$$C^p_{SMIPv6} = P_{AR} + C^p_{MIPv6-RO} + \tau \times \lambda_s \times d_{PAR-NAR} \tag{17}$$

Where λ_s denotes the session arrival rate (packets per second), P_{AR} the processing cost at access router (AR), d_{x-y} the hop distance between network entities x and y, τ is the unit transmission cost in a wired link, and $C^p_{MIPv6-RO}$ represents the packet delivery cost for MIPv6 with route optimization (RO) mode.

Using SMIPv6 (Zhang et al., 2005; Zhang & Marchand, 2009; Zhang & Pierre, 2008), intelligent access routers manage Forwarding and Reverse Tunnels Lists, so the processing cost at an access router mainly comprises the lookup costs for searching such lists. We assume that such cost is proportional to the number of mobile nodes served by the access router, and identical for each access router. Accordingly, the processing costs at an access router can be expressed as follows (Zhang & Pierre, 2008):

$$P_{AR} = \lambda_s \times (\epsilon \times E_{MN}) \tag{18}$$

Where λ_s is the session arrival rate (packets per second), ϵ is a weighting factor showing the relationship between the lookup cost and size of the tunneling lists, and E_{MN} the average number of mobile nodes in a cell.

4.3 Numerical results

This section analyzes the impact of various wireless system parameters on the above-mentioned costs. The parameter values are taken from (Pack & Choi, 2003; Woo, 2003; Zhang et al., 2002), i.e. $\alpha = 0.1$ and $\beta = 0.2$, $\lambda_s = 1$, $\lambda_p = 0.1$, $\theta_{HA} = 20$, $\tau = 1$, $\kappa = 2$, $N_{CN} = 2$, $L_c = 120m$. The network topology is shown in Figure 5 (Zhang & Pierre, 2008). In addition, we fix the value of $\epsilon = 0.1$, $R = 20m$. The hop distance between different domains is assumed to be identical, i.e. $d_{HA-CN} = f = 6$, $d_{CN-MAP} = d = 4$, $d_{HA-MAP} = c = 6$, $d_{AR-MAP} = b = 2$, $d_{AR1-AR2} = d_{PAR-NAR} = 2$. And all links are assumed to be full-duplex in terms of capacity and delay.

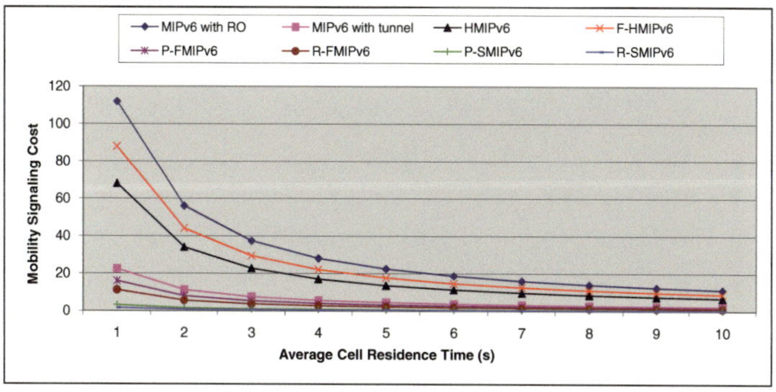

(a) $q = 0.2$

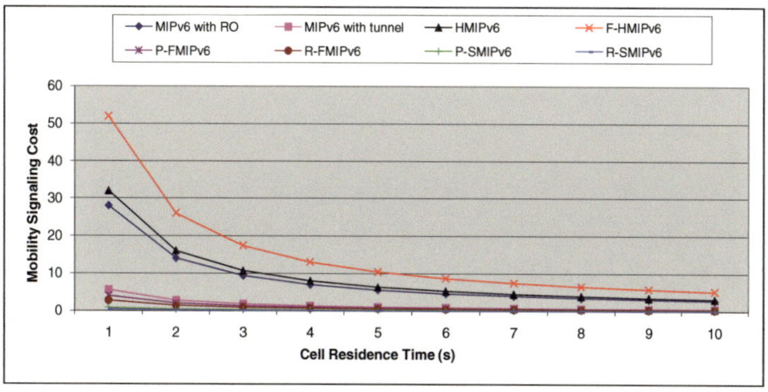

(b) $q = 0.8$

Fig. 6. Signaling cost vs. cell residence time

4.3.1 Signaling cost versus cell residence time

Figures 6.a and 6.b show the relationship between the mobility signaling cost and average cell residence time for $q = 0.2$ and $q = 0.8$, using the random-walk model. Mobile nodes are roaming in a mobility anchor point (MAP) domain with one ring. Note that q represents the probability that a mobile node remains in its current cell. Figure 6.a shows dynamic mobile users, who are eager to move to other cells, while Figure 6.b illustrates the mobility signaling costs for static mobile nodes. The longer a mobile node remains in a current cell, the lower the mobility signaling cost. We explain this as the mobile node is less likely to move between subnets, so fewer handoffs are required when the mobile stays longer in its current cell. In addition, both predictive and reactive SMIPv6 deliver better performance than MIPv6 and its extensions. On the other hand, MIPv6 (Johnson et al., 2004) with route optimization (RO) mode requires the most signaling cost when $q = 0.2$, and F-HMIPv6 (Jung et al., 2005) demonstrates the highest signaling cost when $q = 0.8$.

Compared with MIPv6 with RO mode, predictive SMIPv6 presents 97.13% less signaling cost for $q = 0.2$ and 97.20% less for $q = 0.8$; reactive SMIPv6 presents 98.57% less signaling cost for $q = 0.2$ and 98.54% less for $q = 0.8$. Compared with MIPv6 with tunnel mode, predictive SMIPv6 needs 85.67% less signaling cost for $q = 0.2$ and 85.98% less for $q = 0.8$; reactive SMIPv6 needs 92.84% less signaling cost for $q = 0.2$ and 92.68% less for $q = 0.8$.

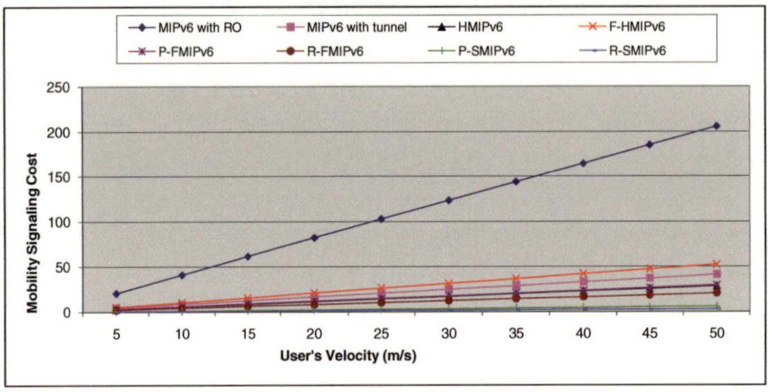

(a) $n = 1$

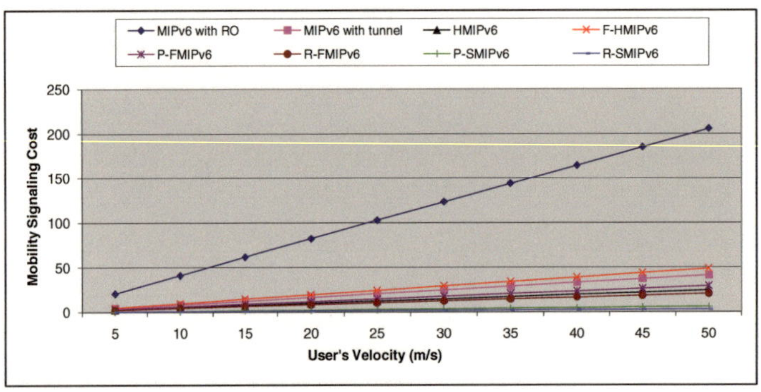

(b) $n = 4$

Fig. 7. Signaling cost vs. user's velocity

Compared with HMIPv6, predictive SMIPv6 requires 95.28% less signaling cost for $q = 0.2$ and 97.55% less for $q = 0.8$; reactive SMIPv6 requires 97.64% less signaling cost for $q = 0.2$ and 98.72% less for $q = 0.8$.

Compared with predictive FMIPv6, predictive SMIPv6 presents 79.96% less signaling cost for $q = 0.2$ and 80.34% less for $q = 0.8$; reactive SMIPv6 presents 89.98% less signaling cost for $q = 0.2$ and 89.74% less for $q = 0.8$. Compared with reactive FMIPv6, predictive SMIPv6 needs 71.34% less signaling cost for $q = 0.2$ and 71.95% less for $q = 0.8$; reactive SMIPv6 needs 85.67% less signaling cost for $q = 0.2$ and 85.37% less for $q = 0.8$.

Compared with F-HMIPv6, predictive SMIPv6 requires 96.35% less signaling cost for $q = 0.2$ and 98.49% less for $q = 0.8$; reactive SMIPv6 requires 98.18% less signaling cost for $q = 0.2$ and 99.21% less for $q = 0.8$.

Comparing the two figures, we find that increasing the probability that mobile nodes remain in their current cells leads to significant reduction of mobility signaling over the network. This is because mobile nodes are less likely to perform handoffs.

4.3.2 Signaling cost versus user velocity

Figures 7.a and 7.b demonstrate the relationship between the mobility signaling cost and user's average velocity for MAP domains of one ring and four rings, using the fluid-flow model (Zhang & Pierre, 2008). The probability that a mobile node remains at its current cell

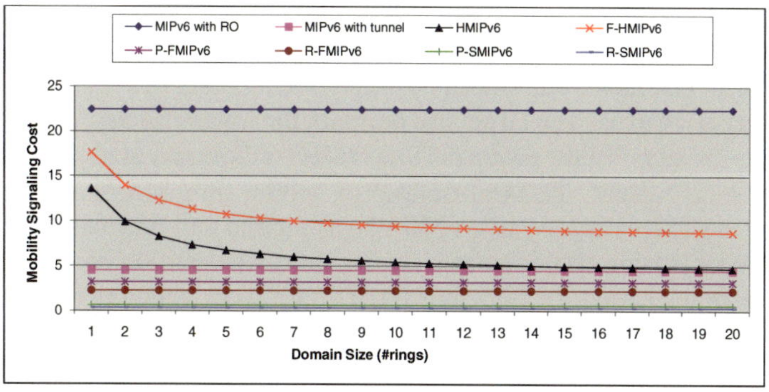

(a) $q = 0.2$

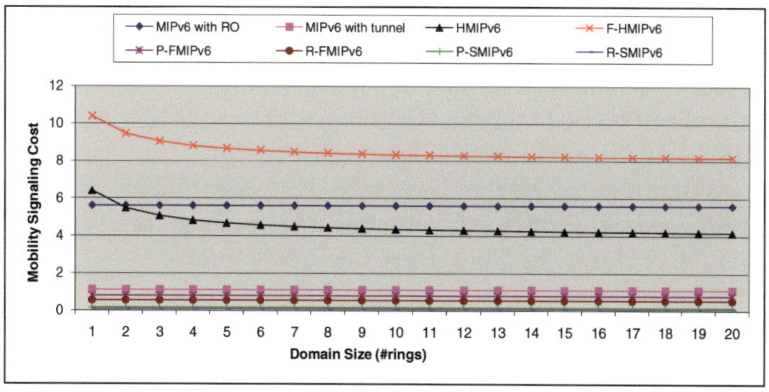

(b) $q = 0.8$

Fig. 8. Signaling cost vs. domain size

is set to 0.2. A lower velocity leads to a lower cell and domain crossing rate and results in less signaling cost. In addition, we find that predictive and reactive SMIPv6 (Zhang & Pierre, 2008) deliver better performance than MIPv6 (Johnson et al., 2004) and its extensions.

For $n = 1$, shown in Figure 7.a, MIPv6 with route optimization (RO) mode engenders the most exorbitant cost, which rises to 113.12, on average. In comparison, F-HMIPv6 (Jung et al., 2005) climbs to 28.74; MIPv6 with tunnel mode needs 22.62; predictive FMIPv6 (P-FMIPv6) rises to 16.16, HMIPv6 (Soliman et al., 2008) requires 15.85, reactive FMIPv6 (R-FMIPv6) is about 11.31. However, the average signaling cost for predictive SMIPv6 (P-SMIPv6) is 3.23, and 1.62 for reactive SMIPv6 (R-SMIPv6).

Comparing the two figures, we find that increasing the MAP domain size leads to significant reduction of mobility signaling cost for localized domain-based mobility management schemes, such as HMIPv6 (Soliman et al., 2008) and F-HMIPv6 (Jung et al., 2005). We explain this as a mobile node roaming in a domain with larger size is less likely to perform inter-domain movements. As a result, Figure 7.b shows that F-HMIPv6 descends to 26.64, which presents 7.31% less signaling cost than that in Figure 7.a. At the same time, HMIPv6 descends to 13.38, on average. This presents 15.58% less signaling cost than that in Figure 7.a. However, signaling costs for other protocols remain unchanged while increasing the MAP domain size.

4.3.3 Signaling cost versus domain size

Figures 8.a and 8.b show the relationship between the mobility signaling cost and domain size for $q = 0.2$ and $q = 0.8$, using the random-walk model (Zhang & Pierre, 2008). The average cell residence time is set to $5s$. The larger the domain, the lower the mobility signaling cost for localized domain-based mobility protocols like HMIPv6 (Soliman et al., 2008) and F-HMIPv6 (Jung et al., 2005). However, the performance of MIPv6 (Johnson et al., 2004) with tunnel and RO modes, predictive and reactive FMIPv6, predictive and reactive and SMIPv6 remain unchanged while increasing the domain size; the same observation as that from Figures 7.a and 7.b. On the other hand, we find that SMIPv6 delivers better performance than other protocols.

For $q = 0.2$, the average signaling cost for MIPv6 with RO mode is 22.40; 10.22 for F-HMIPv6, 6.22 for HMIPv6, 4.48 for MIPv6 with tunnel mode, 3.20 for predictive FMIPv6 (P-FMIPv6) and 2.24 for reactive FMIPv6 (R-FMIPv6), 0.64 for predictive SMIPv6 (P-SMIPv6) and 0.32 for reactive SMIPv6 (R-SMIPv6). These values are shown in Figure 8.a.

For $q = 0.8$, the average signaling cost for F-HMIPv6 is 8.56, 5.60 for MIPv6 with RO mode; 4.56 for HMIPv6, 1.12 for MIPv6 with tunnel mode, 0.80 for predictive FMIPv6 (P-FMIPv6) and 0.56 for reactive FMIPv6 (R-FMIPv6), 0.16 for predictive SMIPv6 (P-SMIPv6) and 0.08 for reactive SMIPv6 (R-SMIPv6), as shown in Figure 8.b.

Comparing the two figures, we find that increasing the probability that mobile nodes remain in their current cells leads to significant reduction of signaling cost. This is because mobile nodes are less likely to perform handover from one cell to another.

4.3.4 Packet delivery cost versus session arrival rate

Figures 9.a and 9.b show the relationship between the packet delivery cost and session arrival rate for MAP domains with one ring and four rings (Zhang & Pierre, 2008). The average number of mobile nodes in a cell is set to 10. Generally, the higher the session arrival rate, the higher the packet delivery cost.

For MAP domains with 1 ring, MIPv6 with tunnel mode requires the highest costs amongst all schemes. We explain this as all of the session packets must cross a triangular path via a home agent, whose steep processing costs are detrimental. On the other hand, MIPv6 with route optimization (RO) mode delivers better performance than other approaches, since all the packets (except the first one) in a session are delivered to mobile nodes via a direct path, and there is no additional processing cost at the MAP neither at the access router. HMIPv6 (Soliman et al., 2008) and F-HMIPv6 (Jung et al., 2005) deliver identical performance, as do FMIPv6 (Koodli, 2008) and SMIPv6 (Zhang et al., 2005; Zhang & Marchand, 2009; Zhang & Pierre, 2008; 2009).

For MAP domains with 1 ring, shown in Figure 9.a, the mean packet delivery cost is 198.00 for MIPv6 with tunnel mode, 100.99 for F-HMIPv6 and HMIPv6, and 75.90 for FMIPv6 and SMIPv6, 59.40 for MIPv6 with RO mode.

For MAP domains with 4 ring, shown in Figure 9.b, the mean packet delivery cost is 401.42 for F-HMIPv6 and HMIPv6, which present 297.48% more cost for delivering packets. However, the performance of MIPv6, FMIPv6 and SMIPv6 remain unchanged while increasing the domain size; the same observation as that from Figures 7.a, 7.b, 8.a and 8.b.

The two figures also show that increasing the MAP domain size leads to a rapid augmentation of packet delivery cost for domain-based localized mobility management protocols, like F-HMIPv6 and HMIPv6; this is due to the processing cost at the MAP, especially the routing

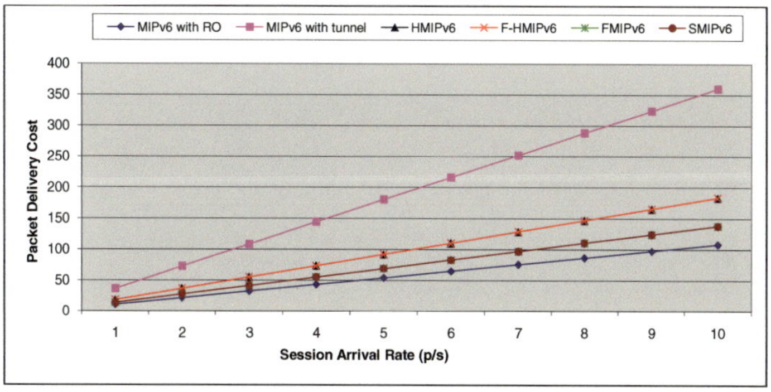

(a) $n = 1$

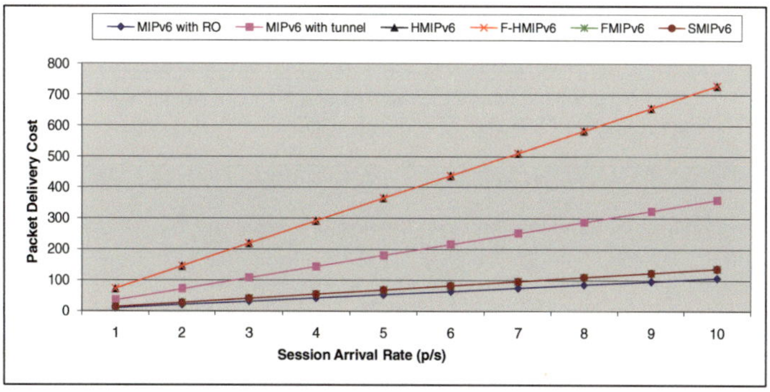

(b) $n = 4$

Fig. 9. Packet delivery cost vs. session arrival rate

cost, which is proportional to the logarithm of the number of access routers in a MAP domain (Zhang & Pierre, 2008).

4.3.5 Total cost versus session-to-mobility ratio

Figures 10.a and 10.b show the relationship between the total cost and average session-to-mobility ratio for MAP domains with one ring, using the random-walk model (Zhang & Pierre, 2008). The *session-to-mobility ratio* (SMR) is defined as the ratio of the session arrival rate to the user mobility ratio, it is analogous to the call-to-mobility ratio (CMR) used in cellular networks.

Under the random-walk model, $SMR = \frac{\lambda_s}{\frac{1}{E(T)}} = \lambda_s \times E(T)$, i.e. the session arrival rate divided by the cell crossing rate. E(T) denotes the average cell residence time. As the value of λ_s is fixed to 0.5, the augmentation of the SMR implies an increase of the cell residence time. as a result, reducing the total cost.

In case of $SMR \leq 1$, i.e. $\lambda_s \leq \frac{1}{E(T)}$, the mobility signaling cost is more dominant than packet delivery cost over the total cost, shown in Figure 10.a. Under this circumstance, MIPv6 with RO mode has the highest total cost amongst all schemes. The total cost in descent order is MIPv6 with RO mode (171.02, on average), F-HMIPv6 (137.56), HMIPv6 (108.27), MIPv6 with

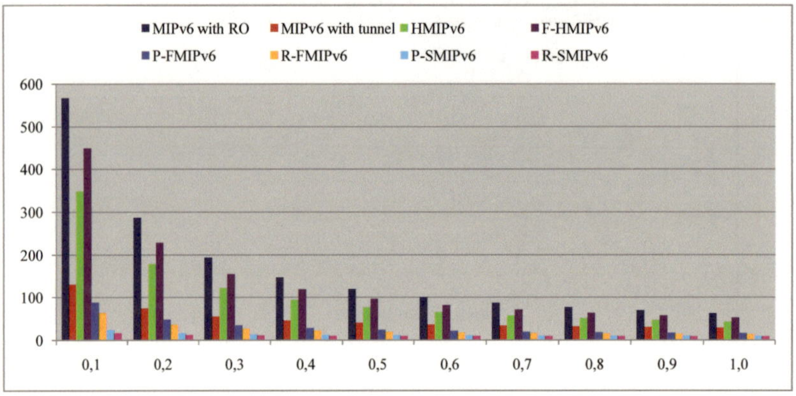

(a) $SMR \leq 1$

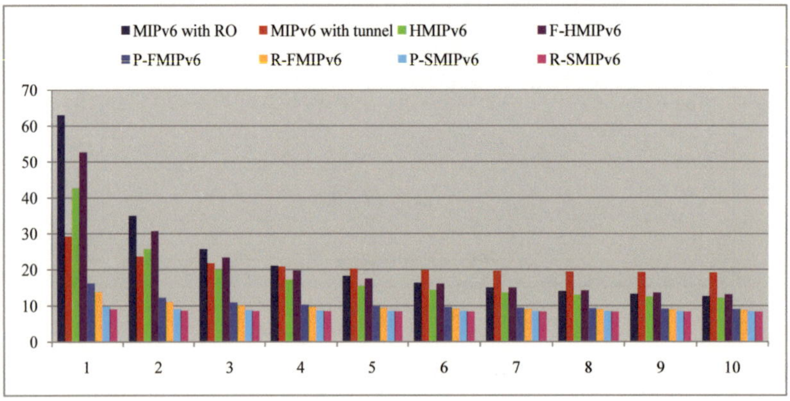

(b) $1 \leq SMR \leq 10$

Fig. 10. Total cost vs. SMR for $n = 1$

tunnel mode (50.80), predictive FMIPv6 (31.63), reactive FMIPv6 (24.60), predictive SMIPv6 (12.89), and reactive SMIPv6 (10.54).

In addition, as $SMR \geq 1$, the impact of mobility signaling cost on the total cost reduces while packet delivery cost becomes more important over the total cost. The higher the SMR, the more important is the packet delivery cost over the total cost. As a result, when $SMR \geq 5$, MIPv6 with tunnel mode requires the highest cost than other protocols. The total cost on average in descent order is MIPv6 with RO mode (23.40), F-HMIPv6 (21.57), MIPv6 with tunnel mode (21.28), HMIPv6 (18.64), predictive FMIPv6 (10.54), reactive FMIPv6 (9.84), predictive SMIPv6 (8.67), and reactive SMIPv6 (8.43). Such values are shown in Figure 10.b. Besides, SMIPv6 yields the best performance amongst all schemes, due to lower signaling cost and no additional processing cost at the MAP.

Figures 11.a and 11.b also illustrate the variation of total cost as the average session-to-mobility ratio changes for MAP domains with four rings, using the random-walk model. The total cost decreases as the SMR augments, the same observation applies to Figures 10.a and 10.b. Besides, increasing the MAP domain size leads to a reduction of total cost for HMIPv6 and F-HMIPv6, yet no impact on MIPv6, FMIPv6 and SMIPv6 protocols.

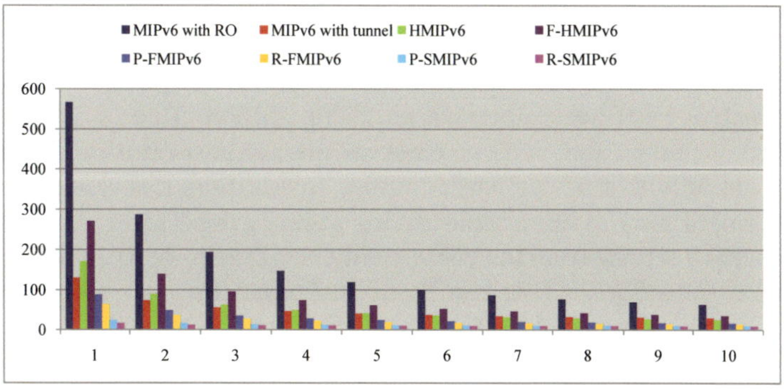

(a) $SMR \leq 1$

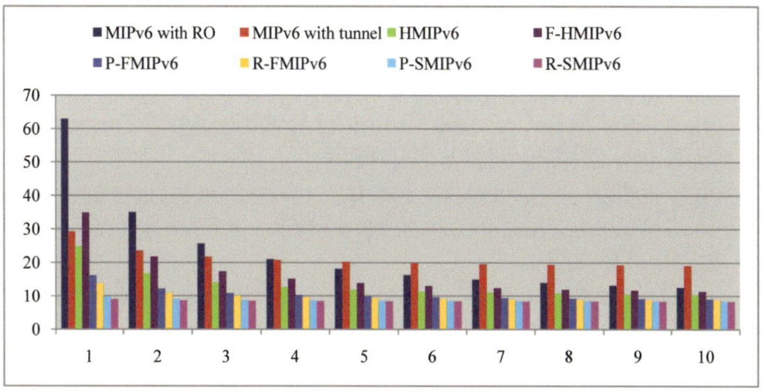

(b) $1 \leq SMR \leq 10$

Fig. 11. Total cost vs. SMR for $n = 4$

In case of $SMR \leq 1$, the total cost in descent order is MIPv6 with RO mode (171.02, on average), F-HMIPv6 (85.41), HMIPv6 (56.12), MIPv6 with tunnel mode (50.80), predictive FMIPv6 (31.63), reactive FMIPv6 (24.60), predictive SMIPv6 (12.89), and reactive SMIPv6 (10.54). We find that F-HMIPv6 presents 37.91% less total cost than that shown in Figure 10.a and HMIPv6 presents 48.17% less total cost than that shown in Figure 10.a.
However, with $SMR \geq 1$, the total cost in descent order is MIPv6 with RO mode (23.40), MIPv6 with tunnel mode (21.28), F-HMIPv6 (16.35), HMIPv6 (13.42), predictive FMIPv6 (10.54), reactive FMIPv6 (9.84), predictive SMIPv6 (8.67), and reactive SMIPv6 (8.43). Such values are shown in Figure 11.b. This is because the impact of packet delivery cost over total cost increases as SMR augments. When $SMR \geq 5$, MIPv6 with tunnel mode requires the highest cost than other protocols. We also observe that predictive FMIPv6 tends to deliver the same performance as reactive FMIPv6, and predictive SMIPv6 tends to provide the same performance than reactive SMIPv6, shown in Figure 11.b.

5. Conclusion

This chapter proposes a new seamless mobility management protocol, called SMIPv6. The novelty of this protocol consists of pre-configuring bidirectional secure tunnels before handoff and utilizing such tunnels to accelerate mobility management procedure during handoff. To

evaluate the efficiency of the proposal, we employ analytical models, numerical results show that SMIPv6 delivers better performance than MIPv6 and its extensions.
Even though SMIPv6 delivers better performance than MIPv6 (Johnson et al., 2004) and its enhancements such as HMIPv6 (Soliman et al., 2008), FMIPv6 (Koodli, 2008) and F-HMIPv6 (Jung et al., 2005), we notice that such schemes are always host-centric. They require mobile nodes to signal mobility to other network entities. In addition, this chapter only focuses on mobility management issue without considering security aspect. In fact, each time before mobile users obtains a service from the visiting network, they have to undergo authentication and authorization procedure. This results in additional delays. Accordingly, new fast authentication protocol is required for seamless mobility management.

6. References

Akyildiz, I.F., McNair, J., Ho, J.S.M., Uzunalioglu, H. & Wang, W. (1999). Mobility management in next-generation wireless systems, *Proceedings of the IEEE*, Vol. 87, No. 8, pp. 1347-1384, ISSN: 0018-9219.

Akyildiz, I.F., Mohanty, S. & Xie, J. (2005). Ubiquitous mobile communication architecture for next-generation heterogeneous wireless systems, *IEEE Communications Magazine*, Vol. 43, No. 6, pp. 529-536, ISSN: 0163-6804.

Akyildiz, I.F. & Wang, W. (2002). A dynamic location management scheme for next-generation multitier PCS systems, *IEEE Transactions on Wireless Communications*, Vol. 1, No. 1, pp. 178-189, ISSN: 1536-1276.

Akyildiz, I.F., Xie, J. & Mohanty, S. (2004). A survey of mobility management in nextgeneration all-IP-based wireless systems, *IEEE Wireless Communications*, Vol. 11, No. 4, pp. 16-28, ISSN: 1536-1284.

Arkko, J., Vogt, C. & Haddad, W. (2007). Enhanced route optimization for mobile IPv6, RFC 4866, Internet Engineering Task Force. URL: http://tools.ietf.org/rfc/rfc4866.txt.

Campbell, A.T., Gomez, J., Kim, S., Wan, C.-Y., Turanyi, Z.R. & Valko, A.G. (2002). Comparison of IP micro-mobility protocols, *IEEE Wireless Communications*, Vol. 9, No. 1, pp. 72-82, ISSN: 1536-1284.

Devarapalli, V., Wakikawa, R., Petrescu, A. & Thubert, P. (2005). Network mobility (NEMO) basic support protocol, RFC 3963, Internet Engineering Task Force. URL: http://tools.ietf.org/rfc/rfc3963.txt.

Dimopoulou, L., Leoleis, G. & Venieris, I. S. (2005). Fast handover support in a WLAN environment: challenges and perspectives, *IEEE Network*, Vol. 19, No. 3, pp. 14-20, ISSN: 0890-8044.

Ernst, T. & Lach, H.-Y. (2007). Network mobility support terminology, RFC 4885, Internet Engineering Task Force. URL: http://tools.ietf.org/rfc/rfc4885.txt.

Gundavelli, S., Leung, K., Devarapalli, V., Chowdhury, K. & Patil, B. (2008). Proxy mobile IPv6, RFC 5213, Internet Engineering Task Force. URL: http://tools.ietf.org/rfc/rfc5213.txt.

Gwon, Y. & Yegin, A. (2004). Enhanced forwarding from the previous care-of address (EFWD)for fast handovers in mobile IPv6, *Proceedings of 2004 IEEE Wireless Communications and Networking* (IEEE WCNC 2004), pp. 861-866, ISBN: 0-7803-8344-3, Atlanta, Georgia, USA, 21-25 March 2004, IEEE.

Gwon, Y., Kempf, J. & Yegin, A. (2004). Scalability and robustness analysis of mobile IPv6, fast mobile IPv6, hierarchical mobile IPv6, and hybrid IPv6 mobility protocols using a large-scale simulation, *Proceedings of 2004 IEEE International Conference on*

Communications (ICC 2004), pp. 4087-4091, ISBN: 0-7803-8533-0, Paris, France, 20-24 June 2004, IEEE.

Haseeb, S. & Ismail, A.F. (2007). Handoff latency analysis of mobile IPv6 protocol variations, *Computer Communications*, Vol. 30, No. 4, pp. 849-855, ISSN: 0140-3664.

Johnson, D., Perkins, C. & Arkko, J. (2004). Mobility support in IPv6, RFC 3775, Internet Engineering Task Force. URL: http://tools.ietf.org/rfc/rfc3775.txt.

Jung, H.Y., Kim, E.A., Yi, J.W. & Lee, H.H. (2005). A scheme for supporting fast handover in hierarchical mobile IPv6 networks, *ETRI Journal*, Vol. 27, No. 6, pp. 798-801.

Kempf, J., Calhoun, P., Dommety, G., Thalanany, S., Singh, A., McCann, P.J. & Hiller, T. (2001). Bidirectional edge tunnel handover for IPv6, draft, Internet Engineering Task Force. URL: http://tools.ietf.org/id/draft-kempf-beth-ipv6-02.txt.

Kempf, J., Wood, J. & Fu, G. (2003). Fast mobile IPv6 handover packet loss performance: measurements for emulated real time traffic, *Proceedings of 2003 IEEE Wireless Communications and Networking* (WCNC 2003), pp. 1230-1235, ISBN: 0-7803-7700-1, New Orleans, Louisiana, USA, 20-20 March 2003, IEEE.

Kent, S. (2005). IP encapsulating security payload (ESP), RFC 4303, Internet Engineering Task Force. URL: http://tools.ietf.org/rfc/rfc4303.txt.

Koodli, R. (2008). Mobile IPv6 fast handovers, RFC 5268, Internet Engineering Task Force. URL: http://tools.ietf.org/rfc/rfc5268.txt.

Loughney, J., Nakhjiri, M., Perkins, C. & Koodli, R. (2005). IP mobility support, RFC 4067, Internet Engineering Task Force. URL: http://tools.ietf.org/rfc/rfc4067.txt.

Makaya, C. & Pierre, S. (2008). An architecture for seamless mobility support in IP-based nextgeneration wireless networks, *IEEE Transactions on Vehicular Technology*, Vol. 57, No. 2, pp. 1209-1225, ISSN: 0018-9545.

Manner, J. & Kojo, M. (2004). Mobility related terminology, RFC 3753, Internet Engineering Task Force. URL: http://tools.ietf.org/rfc/rfc3753.txt.

McNair, J. & Zhu, F. (2004). Vertical handoffs in fourth-generation multinetwork environments, *IEEE Wireless Communications*, Vol. 11, No. 3, pp. 8-15, ISSN: 1536-1284.

Mohanty, S. & Xie, J. (2007). Performance analysis of a novel architecture to integrate heterogeneous wireless systems, *Computer Networks*, Vol. 51, No. 4, pp. 1095-1105, ISSN: 1389-1286.

Narten, T., Nordmark, E., Simpson, W. & Soliman, H. (2007). Neighbor discovery for IP version 6 (IPv6), RFC 4861, Internet Engineering Task Force. URL: http://tools.ietf.org/rfc/rfc4861.txt.

Nasser, N., Hasswa, A. & Hassanein, H. (2006). Handoffs in fourth generation heterogeneous networks, *IEEE Communications Magazine*, Vol. 44, No. 10, pp. 96-103, ISSN: 0163-6804.

Pack, S. & Choi, Y. (2003). Performance analysis of hierarchical mobile IPv6 in IP-based cellular networks, *Proceedings of 2003 IEEE Conference on Personal, Indoor and Mobile Radio Communications* (PIMRC 2003), pp. 2818-2822, ISBN: 0-7803-7822-9, Beijing, China, 7-10 September 2003, IEEE.

Perez-Costa, X. & Hartenstein, H. (2002). A simulation study on the performance of mobile IPv6 in a WLAN-based cellular network, *Computer Networks*, Vol. 40, No. 1, pp. 191-204, ISSN: 1389-1286.

Perez-Costa, X., Torrent-Moreno, M. & Hartenstein, H. (2003). A performance comparison of mobile IPv6, hierarchical mobile IPv6, fast handovers for mobile IPv6 and their combination, *ACM SIGMOBILE Mobile Computing and Communications Review*, Vol. 7, No. 4, pp. 5-19, ISSN: 1559-1662.

Perkins, C. (1996). IP mobility support, RFC 2002, Internet Engineering Task Force. URL: http://tools.ietf.org/rfc/rfc2002.txt.

Perkins, C. (2002). IP mobility support for IPv4, RFC 3344, Internet Engineering Task Force. URL: http://tools.ietf.org/rfc/rfc3344.txt.

Quintero, A., Garcia, O. & Pierre, S. (2004). An alternative strategy for location update and paging in mobile networks, *Computer Communications*, Vol. 27, No. 15, pp. 1509-1523.

Ramjee, R., Varadhan, K., Salgarelli, L., Thuel, S.R., Wang, S.-Y. & La-Porta, T. (2002). HAWAII: a domain-based approach for supporting mobility in wide-area wireless networks, *IEEE/ACM Transactions on Networking*, Vol. 10, No. 3, pp. 396-410, ISSN: 1063-6692.

Soliman, H., Castelluccia, C., El-Malki, K. & Bellier, L. (2008). Hierarchical mobile IPv6 (HMIPv6) mobility management, RFC 5380, Internet Engineering Task Force. URL: http://tools.ietf.org/rfc/rfc5380.txt.

Soto, I., Bernardos, C., Calderon, M., Banchs, A. & Azcorra, A. (2009). NEMO-enabled localized mobility support for Internet access in automotive scenarios, *IEEE Communications Magazine*, Vol. 47, No. 5, pp. 152-159, ISSN: 0163-6804.

Thomson, S., Narten, T. & Jinmei, T. (2007). IPv6 stateless address autoconfiguration, RFC 4862, Internet Engineering Task Force. URL: http://tools.ietf.org/rfc/rfc4862.txt.

Valko, A.G. (1999). Cellular IP : a new approach to Internet host mobility, *ACM SIGCOMM Computer Communication Review*, Vol. 29, No. 1, pp. 50-65, ISSN: 0146-4833.

Woo, M. (2003). Performance analysis of mobile IP regional registration, *IEICE Transactions on Communications*, Vol. E86-B, No. 2, pp. 472-478, ISSN: 0916-8516.

Zhang, L.J. (2008). Fast and seamless mobility management in IPv6-based next-generation wireless networks, *PhD thesis*, Ecole Polytechnique de Montreal, Montreal, Canada.

Zhang, L.J. & Marchand, L. (2006). Tunnel establishment, *US Patent Application*, US 11/410,205. Filed on April 25, 2006.

Zhang, L.J., Marchand, L. & Pierre, S. (2005). Optimized seamless handover in mobile IPv6 networks, *US Patent*, US 60/674,356 . Published on April 25, 2005.

Zhang, L.J. & Marchand, L. (2009). Handover enabler, *US Patent*, US 7,606,201 B2. Published on October 20, 2009.

Zhang, L.J. & Pierre, S. (2008). Evaluating the performance of fast handover for hierarchical MIPv6 in cellular networks, *Journal of Networks*, Vol. 3, No. 6, pp. 36-43, ISSN: 1796-2056.

Zhang, L.J. & Pierre, S. (2009). *Next-Generation Wireless Networks: Protocols, Architectures, Standards, Mobility and Performance*, LAP LAMBERT Academic Publishing, ISBN: 3-8383-1906-0, Cologne, Germany.

Zhang, L.J., Zhang, L., Marchand, L. & Pierre, S. (2010a). A survey of IP-layer mobility management protocols in next-generation wireless networks, in *Next Generation Mobile Networks and Ubiquitous Computing*, Samuel Pierre (ed.), chapter 9, Information Science Publishing, ISBN: 1-6056-6250-X, Hershey, PA, USA.

Zhang, L.J., Zhang, L., Marchand, L. & Pierre, S. (2010b). Mobility management protocols design for IPv6-based wireless and mobile networks, in *Fixed Mobile Convergence Handbook*, Syed A. Ahson & Mohammad Ilyas, (Ed.), chapter 9, CRC Press, Taylor & Francis Group, ISBN: 1-4200-9170-0, New York, NY, USA.

Zhang, X., Castellanos, J.G. & Campbell, A.T. (2002). P-MIP: paging extensions for mobile IP, *Mobile Networks and Applications*, Vol. 7, No. 2, pp. 127-141, ISSN: 1383-469X.

Part 3

Reliabilty Issuses in Cellular Networks

14

Automation of Cellular Network Faults

Okuthe P. Kogeda and Johnson I. Agbinya[1]
Computer Science Department, University of Fort Hare, Alice 5700, South Africa.
+Center for Real-time Information Network(CRIN), Faculty of Engineering, University of Technology, Sydney, NSW 2007, Australia

1. Introduction

The internet explosion and increasing number of services on offer and subscribers has placed a lot of pressure on cellular network service providers. Cellular network subscribers have different requirements and needs. This requires that the operation of the network be optimal at all times, to attract and retain subscribers. This can happen with proper operation and maintenance of the network itself. The automation of cellular network faults, where these faults are reported before they occur is the approach for avoiding the catastrophic failures that may cause network blackout.

An application of Mobile Intelligent Agents (MIA) in monitoring the network elements for any potential failure of these core objects of the network to be avoided is explored in this chapter. The main concern is the prediction of possible cellular network faults using scenarios extracted from correlation of certain cellular network parameters that may not be evident to human operators. These could be solved using an advanced automated solution. This chapter proposes and discusses the development of a MIA system for computer-aided analysis, simulation and diagnosis based on mobile intelligent software agents (Wooldridge & Jennings, 1995). We propose a framework that utilizes different Artificial Intelligent (AI) techniques and probabilistic methods. Neural networks, fuzzy logic, genetic algorithms, among others, are some of the established artificial intelligent techniques used into software agents (Thottan & Ji, 1999).

In this work we combine a Bayesian Network Model (BNM) with mobile intelligent agents for automating fault prediction in cellular network service providers, in a project called Modelling of Reliable Service-Based Operations Support System (MORSBOSS). The major advantage of using Bayesian network model is that the cellular network faults can be automatically detected based on a similar fault occurrence that the system has experienced previously. The information about the previous fault occurrences can be stored and retrieved from a database. This information shows the causal relation between network elements, network faults and services. It also shows the belief or likelihood of a fault at a particular network element. Fault prediction is therefore based on the historical memory of the system about known faults.

This Chapter is organized as follows: In Section 2, we give a detailed overview of Cellular network faults. Definition, characteristics, causes and classification of cellular network faults are provided in this Section. Methods and algorithms of cellular network faults modeling are provided in this Section. Bayesian network, cellular network modeling process and assumptions are also provided in this Section. In Section 3, we provide Mobile Intelligent Agent (MIA) and reasons for choosing MIA. In Section 4, we present Mobile Intelligent Agent model for cellular network faults prediction. We give reasons for cellular network faults prediction and provide such models in this Section. In Section 5, we present the implementation architecture, stating reasons for the choice of JADE architecture. The experimental results are provided in Section 6 and then we draw conclusion in Section 7.

2. Cellular network faults

2.1 Definition

Cellular network fault can be defined as an abnormal operation or defect at the component, equipment, or sub-system level, which significantly degrades performance of an active entity in the network or disrupts communication. All errors are not faults as protocols can mostly handle them. Generally faults may be indicated by an abnormally high error rate.
A fault can be defined as an inability of an item to perform a required *function* (a set of processes defined for purpose of achieving a specified objective), excluding that inability due to preventive maintenance, lack of external resources, or planned actions (ETSI Guide, 2001).

2.2 Characteristics of cellular network faults

There is lack of a generally accepted definition of what constitutes a behaviour of a normal cellular network fault (Hajji et al., 2001; Hajji & Far, 2001; Lin & Druzdzel, 1997). Therefore, it is very difficult to characterize the cellular network faults accurately. However, there are estimations (based on statistics of the network traffic) as to what characterize a cellular network fault. Cellular network faults are characterized by transient performance degradation, high error rates, loss of service provision to the customers (i.e., loss of signal, loss of connection, etc), poor quality of service provision, delay in delivery of services and getting connectivity, etc.

2.3 Causes of cellular network faults

The main causes of network faults differ from network to network. Managing complex hardware and software systems has always been a difficult task. The Internet and the proliferation of web-based services have increased the importance of this task, while aggravating the problem (faults) in at least four ways (Meira, 1997; Thottan & Ji, 1998; Hood & Ji, 1997a; Lazar et al., 1992):

- The speed of software development and release means less reliable and more frequently updated software.
- Multi-tier and distributed software architectures increase the complexity of the cellular network environment and obscure causes of both functional and performance problems.
- Internet style service construction implies more dynamic dependencies among the distributed software elements of the overall services making it difficult to construct and maintain accurate system models.
- Internet scale deployments increase the number of service elements under a particular administrator's responsibility.
- Many heterogeneous networks
- New innovations means interoperation of different networks must be kept to some level leading to faults.
- Overloading of power supply gadgets, natural disasters, etc.

2.4 Classification of cellular network faults

In addition to the definitions given in Section 2.1, a fault is regarded as an abnormal and/or an accidental condition that is either caused by a defective network element, problem in the network layer or at sub-system level. Such problems often cause a previously functional unit to fail. A cellular network fault can also be viewed as a defect that causes a reproducible or catastrophic malfunction. A reproducible malfunction is one that occurs consistently under the same circumstances.

Cellular network faults may also cause malfunctions and outages. Malfunctions are mostly experienced when the software and hardware are working with some errors. However, outages are often manifested when the software and hardware are completely knocked out; they will not be working at all. When this occurs, ESPs will not only lose revenues but also the customers may shun away. However, in case of outages, ESPs may device contingent plans that are meant to improve services by ensuring that the duration of each outage is kept to minimum time possible.

Cellular network faults are classified as *malfunctions* and *outages*. The model gives an overview of cellular network faults types that are commonly experienced by a cellular network service provider under study. Fig.1 depicts the classification of cellular network faults.

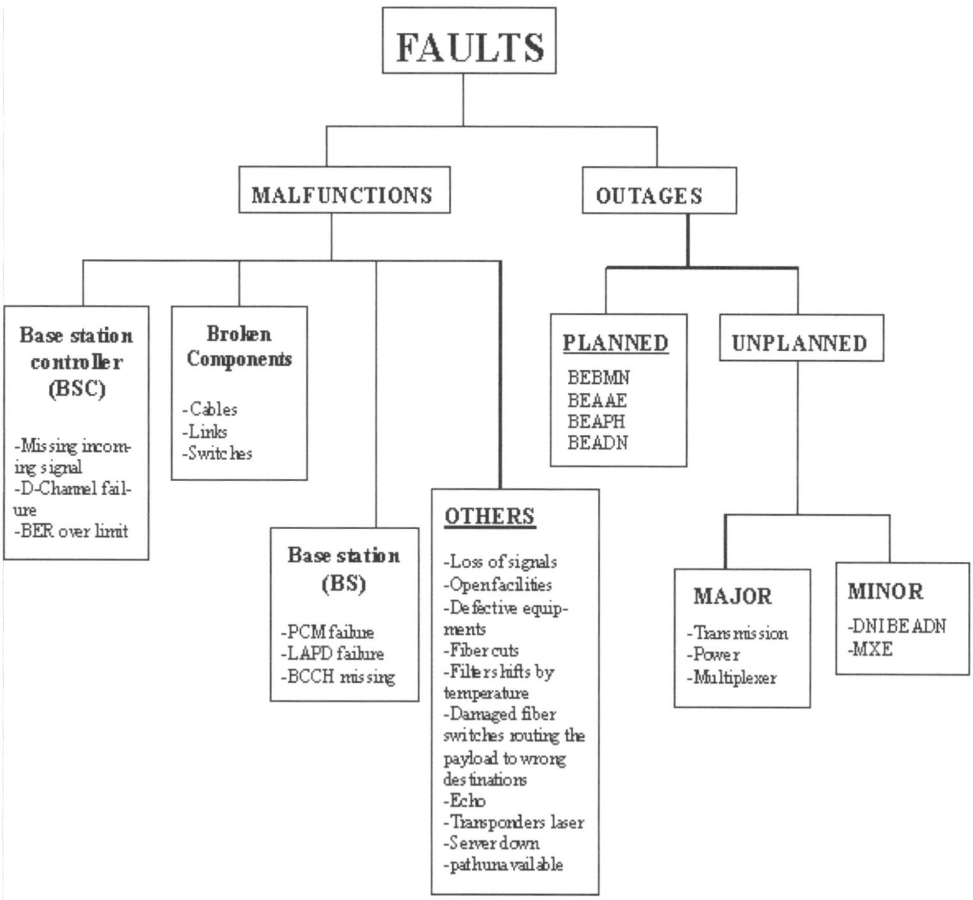

Fig. 1. Classes of Cellular network faults

2.5 Methods and algorithms for cellular network faults modelling

A detailed review of methods and algorithms for faults modeling is provided by Meira (Meira, 1997) and Holst (Holst, 1997) in their respective works. While Meira (Meira, 1997) gives the methods and algorithms for alarm correlation, a summary of some of these methods under machine learning are provided by Holst (Holst, 1997). Most of these methods and algorithms can be utilized in the faults prediction process. Probabilistic approaches and the approaches in which the network entities are modeled as finite state machines are identified (Meira, 1997) for faults identification, detection and prediction in cellular network service providers. Others apply principles defined in non-conventional logics and others adopt *ad hoc* methods to deal with faults modeling.

Among the methods and algorithms include: Fuzzy Logic (Zadeh, 1962, 1965), Artificial Neural Networks (Duda et al., 2001; Widrow, 1989), Decision tress (Mitchel, 1997; Winston, 1992), Model Based Reasoning (Khardon & Roth, 1997; Khardon et al., 1999), Case Based Reasoning (Watson, 1997; Agnar & Plaza, 1994; Klaus-Dieter et al., 1999) Rule Based Systems (Langley & Simon, 1995), Blackboard (Corkill, 1991) and Bayesian networks (Heckerman et al., 1997) among others.

There is no single model or method in terms of complexity, cost, precision, time, and robustness, which can be regarded as best method to be used in network faults modeling. Recent approaches lean towards a combination of two or more methods together to solve complex problems (like cellular networks systems) (Walrand et al., 2000), (Heckerman et al., 1997), (Frey et al., 1997). The choice of a method would depend on a specific problem case. However, the following factors may be considered: (1) Implementation complexity, (2) Facility for construction of a theoretical model of the *object network*, (3) Performance, (4) Facility to adapt to change in the object network, and (5) Precision. Network faults prediction must take into account the object network characteristics, which can be looked at from technology viewpoint i.e., CDMA, FDMA, GSM, UMTS, WCDMA, CDMA2000 (Frey at al., 1997), GPRS, EDGE, DECT, etc.

The nature of application area will also dictate the choice of a method to adopt. The high cost of implementation and adaptation to changes in the object network make it difficult to apply rule-based approaches in large cellular networks and hence are better used in network elements whose configuration is rarely altered. The other approaches that are less sensitive to changes in the object network i.e., case-based approaches still lack a theoretical basis, which would allow their utilization in large, size commercial cellular service network systems.

The complexity of the problem to be solved sometimes brings the *exceptions*, which can be effectively treated by mechanisms that are used for the implementation of *non-monotonic* reasoning (Williamson et al., 2005). This means each identified exception requires the reformulation of the already established rules or the creation of new ones at the development phase. The complexity of the solution increases, which reduces the performance and robustness. Rule-based systems provide additional structuring which facilitates the development of applications but because of the tendency to reduce performance, they are not attractive for the implementation of more complex cellular network systems.

The complexity of cellular network faults prediction problem makes it extremely difficult to obtain exact solutions (Donini et al., 1990). This brings uncertainty as a factor in fault prediction process that needs to be considered. Fuzzy logic, Bayesian networks (cf. Section 2.6), case-based reasoning and artificial neural networks are some of the approaches that can deal with uncertainty. Each of these alternatives have got advantages and disadvantages, for example, defenders of fuzzy logic based approaches argue that they simplify applications development and result in working products with excellent performance; on the other hand it is tuning (like membership functions) is very hard and it lacks a solid mathematical support which hinders its adoption in a larger number of applications.

Bayesian network based methods, which were first utilized in 1921 in the analysis of harvesting results (Katzela et al., 1996) count on a solid mathematical support. They win more acceptances in the community of computing scientists as a suitable option to the solution of problems involving uncertainty (Katzela et al., 1996). These factors contributed to the adoption of Bayesian networks in this work.

2.6 Bayesian networks

Bayesian networks model is named after Thomas Bayes, who proved a special case of what is called Bayes' theorem. The term Bayesian, however, came into use only around 1950 (Heckerman, 1997). It provides an approach to the treatment of uncertainty with incomplete and inaccurate available data to produce inferences.

A Bayesian network is a Directed Acyclic Graph (DAG) in which each node represents a random variable to which conditional probabilities are associated, given all the possible combinations of values of the variables represented by the directly preceding nodes; an edge in this graph represents conditional probabilities between the variables corresponding to the interconnected nodes (Meira, 1997; Heckerman, 1997).

The terms subjective probability, personal probability, epistemic probability and logical probability describe some of the schools of thought, which are customarily called "Bayesian". These terms overlap but there are differences of emphasis. A subjective probability expresses the degree of belief of an expert related to the occurrence of a given event, based on the information this person has available up to the moment. The use of subjective probabilities is very often the only resource in situations where analytical or experimental data is very hard or even impossible to obtain.

It is possible sometimes to evaluate conditional probabilities from empirical data obtained from the past behaviour of the network service provider under study. Given a Bayesian network and a set of evidences it is possible to evaluate the network, that is, to calculate the conditional probability associated with each node, given the evidences observed up to the moment. Generally speaking, this is a NP-hard problem (Chickering et al., 2004) but with the use of appropriate heuristics and depending on the problem dealt with; networks containing thousands of nodes may be evaluated in an acceptable time. An example of a Bayesian network of faults prediction is shown in Fig. 2.

Why Bayesian Networks?

The main reasons why Bayesian networks was chosen for cellular network faults prediction include (Hood et al., 1997a), (Hood et al., 1997b), (Hood et al., 1998):

- *Mathematical support*: the Bayesian networks count on a solid mathematical support, which allows the analysis of the model in view of the knowledge of its performance and precision before an implementation is carried out.
- *Robustness*: approximate answers can be obtained, even when the existing information are incomplete or imprecise whenever new information become available, the Bayesian networks allow a corresponding improvement in the precision of the correlation results.
- The facilities are readily available for the construction of the Bayesian network.

- Bayesian networks have the capacity to identify, in polynomial time, all the conditional independence relationships that are extracted from the information gained by the Bayesian network structure.
- The capacity for non-monotonic reasoning, through which previously obtained conclusions may be withdrawn as a consequence of the knowledge of new information.
- The capacity to carry out inferences on the present state of telecommunication networks from the combination of: a) statistical data empirically surveyed during the network functioning, b) subjective probabilities supplied by specialists, and c) information (that is, "evidences" or "alarms") received from the telecommunications network, in real time.
- It is simple and effective.

Concerns about Bayesian Networks

Although the use of Bayesian network for knowledge generation from data produces good results on some benchmark data sets, there are still some concerns. These include (Heckerman, 1996, 1999):

- *Node ordering requirement* - many Bayesian network learning algorithms require additional information, which is mostly an ordering of the nodes to reduce the search space. Unfortunately, this information is not always available.
- *Computational Complexity* - practically all Bayesian network learners are slow, both in theory and in practice. For example, most dependency-analysis based algorithms require an exponential numbers of "conditional independence" tests.
- *Lack of publicly available learning tools* - although there are many algorithms for this learning task, very few systems for learning Bayesian networks systems are publicly available. Even fewer systems can be applied to real-world data-mining applications where the data sets often have hundreds of variables and millions of records.

2.7 Cellular network faults modelling process

The cellular network fault modeling process consists of five independent processes. These processes are indefinitely repeated and take into account the existence of the assumptions outlined in Section 2.8. These include:

- Network fault alarms acquisition by the network management system
- Classification of the cellular network fault alarms received, according to time windows and originating network element.
- Correlation at network element level, from network fault alarms originally generated by the network elements or obtained through other correlation processes, depending on the correlation topology adopted. The existence of this process is not mandatory at a first moment, as it can be gradually implemented in each network element, according to necessity and taking into consideration peculiarities of each one of them.
- Random variables corresponding to each network element are updated, according to the state of these network elements, which is given by the network fault alarms received and by the result of the correlations carried out on these fault alarms.
- Network fault alarm correlation at the cellular network level, through the evaluation of the new probabilities associated to the fault states defined for each network element, in view of the evidences available at each moment.

2.8 Assumptions

For the models to perform some of the functions there were assumptions made as a condition for the model application in this work. These include:

- It was assumed that it is possible to define a discrete random variable values representative of the state for each network element corresponding to a node in the graph of the cellular network model.
- There is an integral network management system that collects information, from which values are attributed, in real time, to the variables mentioned at element level.
- The managed cellular network service provider is modeled in conformity with the ITU-T rules.
- The prior probabilities related to the variables of each root node, and the local probabilities related to the variables of the other nodes, may as well be alternatively attributed to:
  - through the *relative frequencies* of the corresponding events which are calculated from the data collected by the network management system
  - by using *relative likeliness*. This consists of estimating the probabilities from the subjective judgment of an expert. This method will be useful whenever there is not enough data to permit the estimation of the relative frequencies, which may occur due to the low frequency of the phenomena observed, or even due to the nonexistence of sufficient network management resources.

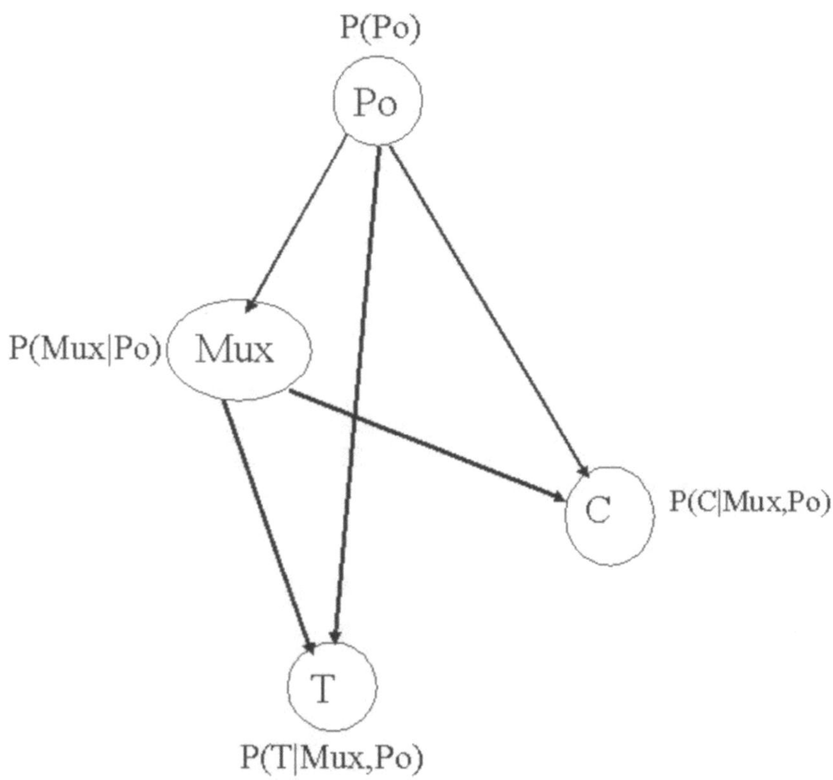

Fig. 2. A Bayesian Network of faults prediction

3. Mobile intelligent agents

3.1 Definition and related work

An Intelligent Agent (IA) is defined as "software that assists people and acts on their behalf. Intelligent agents work by allowing people to delegate work that they could have done, to the software agent. Agents can, just as assistants can, automate repetitive tasks, remember things you forgot, intelligently summarize complex data, learn from you, and even make recommendations to you" (Gilbert, 1997). In this work, we define an IA as an autonomous program with the capability of controlling its own actions and decision-making based on prior knowledge, past experience and on its perception of its environment in pursuit of predefined goals. However, these kinds of agents cannot migrate from one host to another making them undesirable for this work. Mobile Intelligent Agent (MIA), on the other hand, have ignited interest among researchers in Information Communication Technology (ICT) with applications as diverse as e-commerce, computer games, interface design, personalised information management and management of complex cellular networks.
Proactive anomaly detection using distributed intelligent agents was proposed in (Thottan & Ji, 1998) and faults prediction at the network layer using intelligent agents was proposed in (Thottan & Ji, 1999). While these works explored ways of applying intelligent agents and Bayesian Belief Network (BNN) in cellular network faults prediction, they did not address the relationship between cellular network faults and service provision, and only explored the network layer. Pissinou et al (Pissinou et al., 2000) apply mobile agents to automate the fault management in wireless and mobile networks. While this work delves into detection of faults and automated recovery from faults, it does not deal with reporting such faults before they occur. Andrzej Bieszczad et al (Bieszczad et al., 1998) discussed the potential uses of mobile agents in network management and in Elerftheriou and Galis (Eleftheriou & Galis, 2000) MIAs are explored for network management systems. While both papers discuss the application of mobile agents in fault analysis, they fall short of predicting such faults and relating them to cellular network services. Thottan and Ji (Thottan & Ji, 1998, 1999) presents a proactive anomaly detection using distributed intelligent agents and faults prediction at the network layer using intelligent agents. They used intelligent agents, throbbing technique and Bayesian Belief Network (BNN) in network faults detection and prediction. The deployed agents obtain relevant MIA data, provide temporally and spatially correlated predictive alarms, and time correlated abnormal changes in the individual MIA variables. Their testing showed successful prediction rate of seven faults out of nine faults with a prediction time in the order of minutes. However, the authors failed to relate faults to network services. In the present work we address this relationship.
We define Mobile Intelligent Agent (MIA) as a software program that acts on behalf of a user or another program and is able to migrate from one network node to another in a network system under its own control. The MIA chooses when and where it will migrate and may interrupt its own execution and continue elsewhere on the network. The agent returns results and messages in an asynchronous fashion. We exploit the intelligence and ability to cooperate features of MIA in this work.

3.2 Why Mobile Intelligent Agents?

Almost every task that can be performed by MIA can be done by stationary intelligent agents. However, the use of MIA brings certain benefits over other technologies such as stationary intelligent agents, remote objects, etc, including (Chess et al., 1998; Bieszczad et al., 1998; Eleftheriou & Galis, 2000):

- Efficiency savings - CPU consumption is limited, because a mobile agent executes only on one node at a time. Other nodes do not run an agent until needed.
- Space savings - Resource consumption is limited, because a mobile agent resides only on one node at a time. In contrast, static multiple servers require duplication of functionality at every location. Mobile agents carry the functionality with them, so it does not have to be duplicated.
- Reduction in network traffic - Code is very often smaller than data that it processes, so the transfer of mobile agents to the sources of data creates less traffic than transferring the data. Remote objects can help in some cases, but they also involve marshalling of parameters, which may be large.
- Asynchronous autonomous interaction - Mobile agents can be delegated to perform certain tasks even if the delegating entity does not remain active.
- Interaction with real-time systems - Installing a mobile agent close to a real-time system may prevent delays caused by network congestion.
- Robustness and fault tolerance - If a distributed system starts to malfunction, then mobile agents can be used to increase availability of certain services in the concerned areas. For example, the density of fault detecting or repairing agents can be increased. Some kind of meta-level management of agents is required to ensure that the agent-based system fulfils its purpose.
- Supports for heterogeneous environments - Mobile agents are separated from the hosts by the mobility framework. If the framework is in place, agents can target any system. The costs of running a Java Virtual Machine (JVM) on a device are affordable.
- On-line extensibility of services - Mobile agents can be used to extend capabilities of applications, for example, providing services. This allows for building systems that are extremely flexible.
- Convenient development paradigm - Creating distributed systems based on mobile agents is relatively easy.
- Easy software upgrades - A mobile agent can be exchanged virtually at will.
- Dynamic adaptation - mobile agents can sense their environment and react autonomously to changes and can distribute themselves among hosts in the network to maintain optimal configuration. In the case of a mobile agent moving across a number of host nodes, it can adapt its future behaviour according to information that it has already collected and stored in its state.

However, MIA still suffer from certain cost constraints, which include (Chess et al., 1998; Eleftheriou & Galis, 2000): migration and machine load overhead; high costs in speed for Remote Method Invocation (RMI); resource management; standardization and interoperability; and lastly the directory service used by the agents seemed to slow down communication when the number of agents increases.

4. The mobile intelligent agent model

4.1 Proposed mobile architecture

The proposed MORSBOSS mobile intelligent model is designed into three tiers conducive to real-time processing. The first tier is specific network elements, which are within the cellular network environment. The mobile agent monitors the network elements/nodes in this environment. The next tier where MORSBOSS agent operates is the mobile agency. It mediates between the managed network elements and the data tier. The MORSBOSS agent analyses and then logs the fault alarms into the database. Data is the last tier where generated alarms (evidence of fault occurrence) are stored. The alarms variables are aggregated and computed based on BNM and then these alarms are combined using a priori information about the relationships between the variables to produce network element alarms, which are the indicators of the network health as shown in Fig. 3.

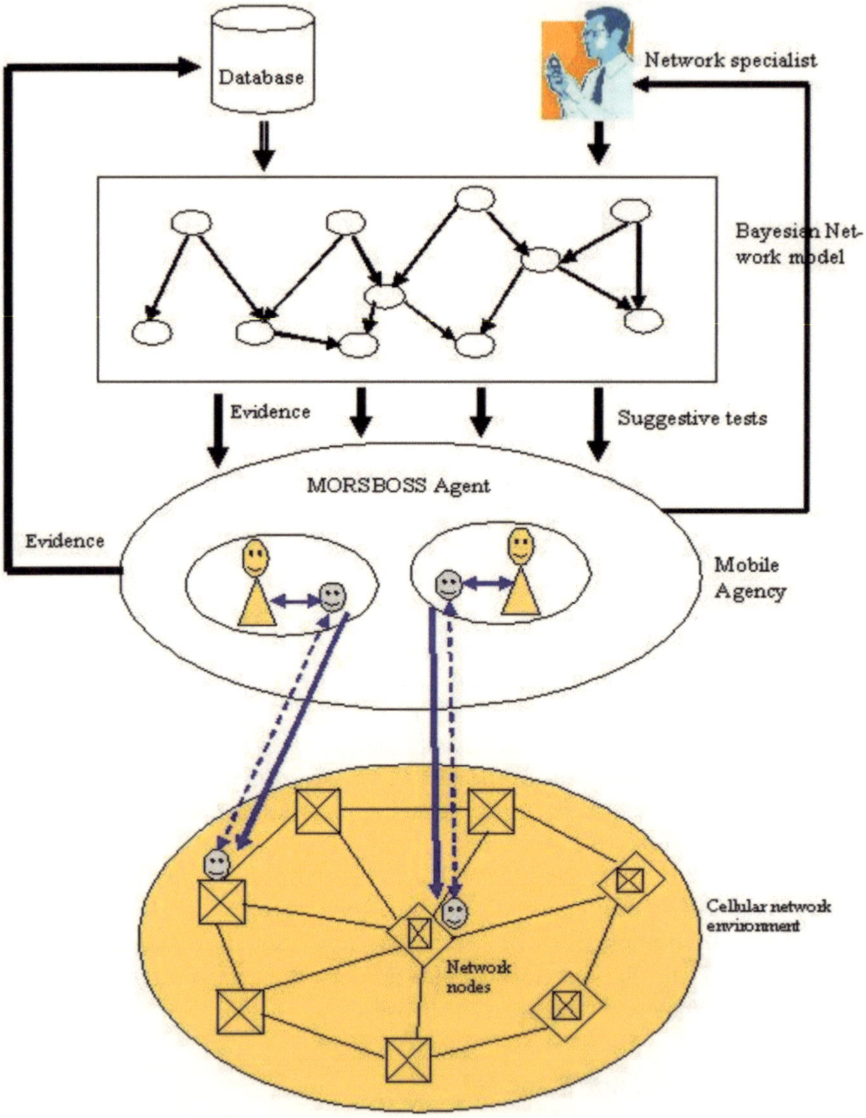

Fig. 3. Proposed MORSBOSS Architecture

4.2 Why cellular network faults prediction?

Faults prediction brings a number of benefits to the cellular network service providers. Some of these include (Hood & Ji, 1997a, 1997b, 1998; Danyluk & Provost, 2002):

1. Accurate fault prediction models support project planning and steering.
2. Faults prediction helps network managers in re-routing of network traffic. Network managers can take corrective action before the faults occur, thereby ensuring services reliability and availability over the network.
3. Decision making, i.e., whether to buy from a particular vendor or not, whether to buy a particular hardware or not, etc. It may also be decided that for elements with a high-predicted fault-proneness, say, above 25%, the element design shall undergo quality assurance (QA) activities such as inspections, extensive unit testing, etc. QA tests help to improve the quality of system with each fault that is discovered and corrected.
4. Operations cost will be minimized if network faults are found as soon as they occur. Faults which are discovered early are cheaper to repair and hence such a scenario leads to offering of cheap and reliable services.
5. The fault prediction models provide a mapping from a hard to interpret design measurement data to easily interpreted external quality data.
6. Fault prediction models provide a sound method to combine multiple factors into one cohesive model i.e., take into account the various factors that make certain network elements fault-prone.
7. Highly accurate fault prediction models can be beneficial when highlighting trouble areas in cellular network system.

Uncertainty Causing Factors

Cellular networks being dynamic in nature and as has been demonstrated in this chapter, uncertainty is inherently associated with the cellular network faults prediction process. This uncertainty is largely driven by the possibility of including erroneous data. There are basically four main sources of data errors. These are:

- The influence of factors not captured by the managed system model
- The imprecision in the calculation of the probabilities distribution values
- The imprecision in the capture and transference of the alarms.
- Imprecision in the information obtained from other correlating processes.

4.3 Cellular network faults prediction models

The cellular network under study has power (Po), cell (C), transmission (T), and multiplexer (Mux) faults as the network variables to be estimated as shown in Fig. 4. Each variable has observations, which are stored in the database. The arrows indicate cause and effect of the cellular fault. The structure was chosen from the natural grouping of the network dynamics into four variables mentioned above. The data from different database variables were combined through the probabilistic framework defined by the Bayesian belief network, where the probabilities were estimated from the network variables as in (Kogeda & Agbinya, 2006; Kogeda et al., 2006). These include the conditional probability, for example, assuming multiplexer (Mux) is ok then the only reason it may not perform its functions is when power (Po) fails. This can be calculated by equation (1) and the joint probability distribution using equation (2).

$$p(Mux \mid Po) = \frac{p(Mux) \times p(Po \mid Mux)}{p(Po)} \tag{1}$$

$$\mathrm{p(Po,Mux,C,T) = p(Po) \times p(Mux) \times p(C \mid Po,Mux) \times p(T \mid Mux)} \tag{2}$$

The prediction factor, 'belief' that a variable $X_k \notin \{X_m,...,X_p\}$ assumes the value x_k is computed using equation (3) when one knows a set of evidences $e = \{X_m = x_m,...,X_p = x_p\}$, constituted by all the known values of the random variables of Bayesian network, where $\{X_m,...,X_p\} \subset X = \{X_1, X_2,...,X_n\}$:

$$p(X_k = x_k \mid e) = \frac{p(X_k = x_k) \times p(e \mid X_k = x_k)}{p(e)} \tag{3}$$

The above equations (1, 2 & 3) form part of the MIA engine, which is able to foresee and move to a node likely to be faulty. The proposed model in this paper utilizes different artificial intelligent techniques. The development of the proposed application (software) requires knowledge extraction from the alarms stored in the database (MORSBOSS database), which must be updated with the new knowledge that might have been discovered from the alarm data and the belief of foreseen fault occurrence.

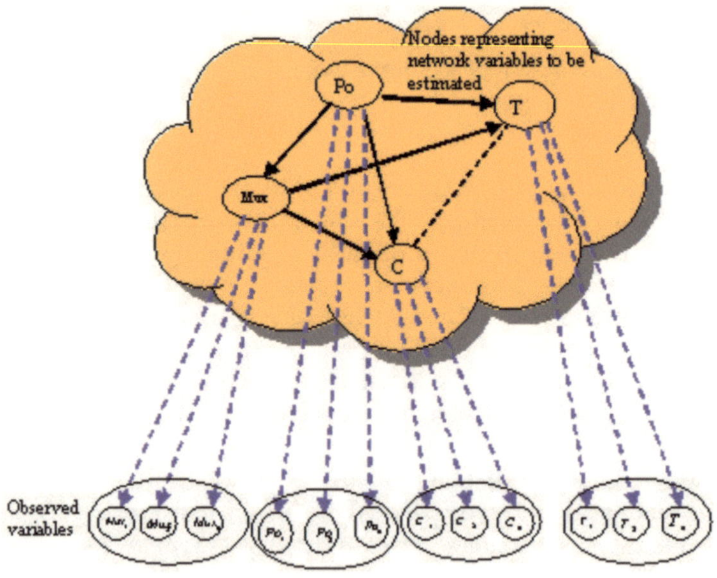

Fig. 4. A Bayesian network for fault prediction

The MORSBOSS database developed in MySQL stores the variable alarms, which are grouped into four distinct fault variables using probabilistic framework defined by Bayesian belief network. Computation of conditional and joint probabilities based on the information at hand is done within MORSBOSS agent using equations (1) and (2). It then interprets the computed figures, which may lead to evidence of an eminent fault and suggestive tests to be done in order to prevent the eminent failure. It is at this point that MORSBOSS agent sends alarm message to the engineering assistant agent about the health of the network. This message can be in the form of excellent, good, fair or critical. Whenever a critical alarm is sent to the engineering assistant agent who in turn informs selected customers (based on profile) about the foreseen fault, the network is at a high risk of failure. Depending on time to fault, the MORSBOSS agent can decide to inform the selected customers and engineering assistant agent at the same time.

Besides informing the engineering assistant agent, the MORSBOSS Agent moves with speed to the node, which is deemed to become faulty. Since it is a mobile agent, it clones itself in order to cooperate and do the task at hand. In this manner, the agent is able not only to predict but also to be ready to resolve the error; for example, the traffic can be re-routed to other routes avoiding the faulty node. Once a fault occurs then the agent communicates the message (i.e., start time, end time, node id, service id, etc) as evidence, which is sent to the database as shown in Fig. 3.

The interpretation of the computed values may show no change, positive change or negative change. The agent will determine this change within a given time window. We took average time to fault as our time window for repeating this computation in order to determine the node with high frequency failure.

MIA operates in an environment, which is cellular network node. The nodes often go into different states of operation. These states can be normal or abnormal. When a state of a node is normal, then such a node is assumed to be operating well without errors and abnormal state implies that the node is faulty. Let us assume that the node (N) may be in finite state set E of discrete states: $N = \{e, e', ...\}$ Where e is normal state and e' is abnormal state. The MIA is assumed to have an array of possible actions available to them, which they can perform to transform the state of the node. The state of the node also dictates which action the MIA will perform. The action of MIA is also dependent on the history of the node. State transformer function (JADE-LEAP website) is used to represent the effect that MIA's action have on the node:

$$\tau : R^a \rightarrow p(N) \tag{4}$$

Where R is a run of MIA in an environment; a actions.

We compute the likelihood value of a fault occurring using Bayesian network model. Using MIA, M and our modeling techniques, we define the likelihood of an environment being in abnormal state by:

$$M = \begin{cases} p(e_0, e_1, e_2, ... e_n) & \\ 0 & otherwise \end{cases} \tag{5}$$

Where $e_0 \in N$ is the initial state of the node.

The Utility (U) function of the MIA can be measured on some particular runs (r) using:

$$U(r) \cong \frac{fr}{fd} \tag{6}$$

Where fr is number of faults fixed in r; fd is number of faults that appeared in r.

Let us write $p(r \mid M, E)$ to denote the probability that run r occurs when MIA m is placed in environment E, clearly as:

$$\sum_{r \in R(M,E)} p(r \mid M, E) = 1. \tag{7}$$

Then the optimal MIA $Mopt$ in an environment E is defined as the one that maximizes expected utility:

$$M_{opt} = \arg\max_{M \in m} \sum_{r \in R(M,E)} U(r)p(r \mid M,E) \quad (8)$$

5. Implementation architecture

This work has been implemented using Java Agent Development Framework and Lightweight Extensible Agent Platform (JADE-LEAP) (Moreno et al., 2002; JADE-LEAP website, 2007), a mobile agent platform distributed by Telecom Italia (Moreno et al., 2002; JADE website, 2007).

5.1 Why JADE architecture?

JADE (JADE website) was chosen for a combination of benefits offered including (Altmann et al., 2001):

- Simplicity in usage and agent programming;
- good online community support and documentation;
- support for the Foundation for Intelligent Physical Agents (FIPA) (FIPA Specification, 2007) standards;
- efficient and tolerant of faulty programming and
- it is open source (free), among others.

However, it is worth noting that JADE does not provide support for migration between different execution environments as a drawback.

JADE has already been integrated into different major architectures, such as J2EE and .NET allowing JADE to execute multi-platform proactive applications. JADE can take J2EE for servers, J2SE (Fagin et al., 1995) for PCs and desktop computers, Personal Java (Pjava) for wireless mobile devices supporting Pjava (i.e., Personal Digital Assistant (PDA)) and J2ME with Connected Limited Device Configuration (CLDC) and Mobile Information Device Profile (MIDP) for mobile devices supporting MIDP (i.e., cell phones) as shown in Fig. 5.

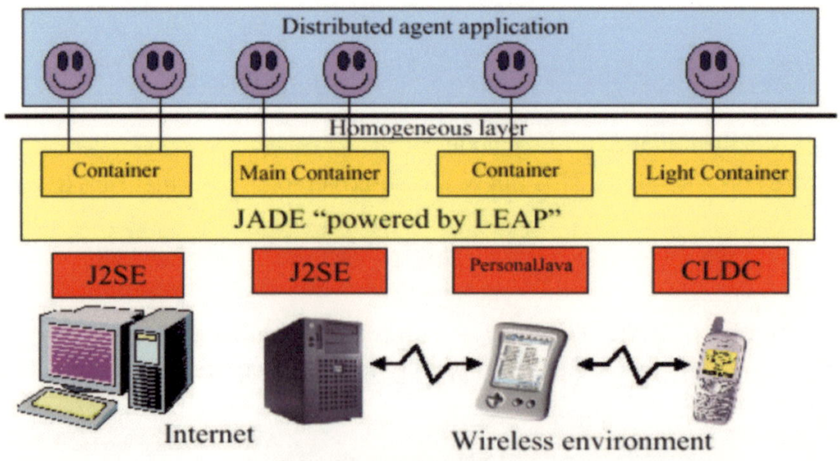

Fig. 5. JADE-LEAP agents' architecture, from JADE-LEAP user's guide

A MIA agent is a JADE-LEAP agent, which is a mobile agent with autonomy and high degree of collaboration. The system automatically creates a new Client Agent to act on behalf of the user once a new user logs onto the system. The agent name is unique, and based on the user ID. The Client Agent is able to obtain and maintain the user's personal information and preferences, and to react to various incoming and outgoing messages and requests intended for the user. The Client Agent is automatically removed from the system once the user log outs.

6. Experimental results

We set up a simple wireless network of eight devices. 2 iPAQs, Router, 2 desktop PCs, 2 laptops and an access point are the devices that act as nodes in our simple wireless network. The implemented MIA is then executed from the desktop-1, which acts like host to the MIA. The PCs, laptops and iPAQs are connected to the network wirelessly via the wireless access point as shown in Fig.6. We decided to send simple messages, which arrived in time as any normal healthy cellular network service provider.

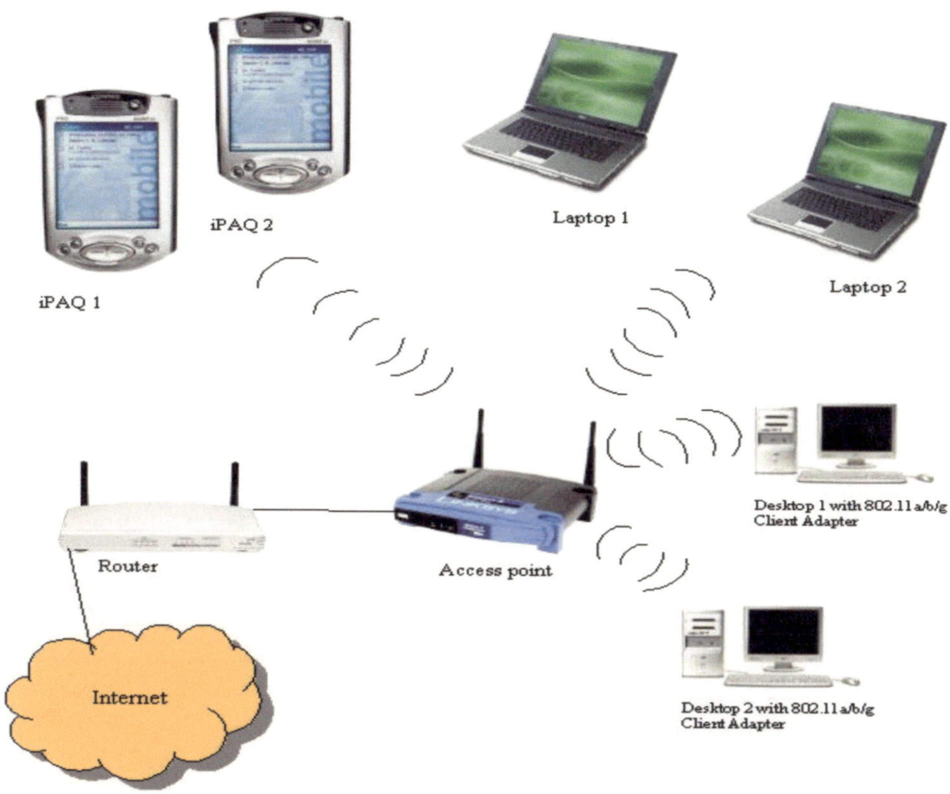

Fig. 6. Wireless network for Experiment

In our testing of the models, we injected faults to various nodes after a random number of minutes. The injection was done with a view to see if the MIA could respond to not only detecting the faults but also in predicting the faults before they occurred. We made 400 runs over a period of three months. A run took an average of 30 minutes. During each run the MIA would update the 'belief' after a time window of 5 minutes, by logging in the faults and computing the new 'belief' of each and every variable. The node status is stored in the

database and bears a typical format (for example, faultID: 1 serviceID: 1 faultName: Cell prob: 0.0898438 faultState: VERY UNLIKELY).
The customers consume services, whose performances are affected by the fault. In case of a fault, the MIA may inform the customer of a fault as well as network engineer. The typical message to the customer who consumes the service affected is (Dear Customer, Fault ID: 1 has been created for service ID: 1 MORSBOSS will endeavour to resolve the problem ASAP. Please do not reply this SMS).
We injected faults to the network during each run of the MIA. Most of the network faults were injected automatically. Ms Visual Basic for application program was written to automatically put off the devices at random time intervals, VBScript for injecting transmission fault, and manually moving portable devices away from the wireless access point thereby leading to loss of signal fault, among others. In our experiments of 400 runs, our model could report 316 faults before they occurred, 38 faults were reported after they had occurred, 26 faults could not be detected, 20 faults alarms were false, giving us a success rate of about 79% prediction rate, 9.5% delayed results, 6.5% no results, and 5% false alarms. The probability of the network being faulty in each of the runs we made is shown in Fig. 7. The utility of the MIAs is shown in Fig. 8.

7. Conclusion

The model utilizing different artificial intelligent techniques was proposed in this Chapter. Cellular network faults prediction using mobile intelligent agent technology and Bayesian belief networks was presented. The experimental results show a prediction rate of 79% with 5% false alarms. The research will strive to improve on the success rate and incorporate swarm intelligence in furthering the cellular network fault prediction.

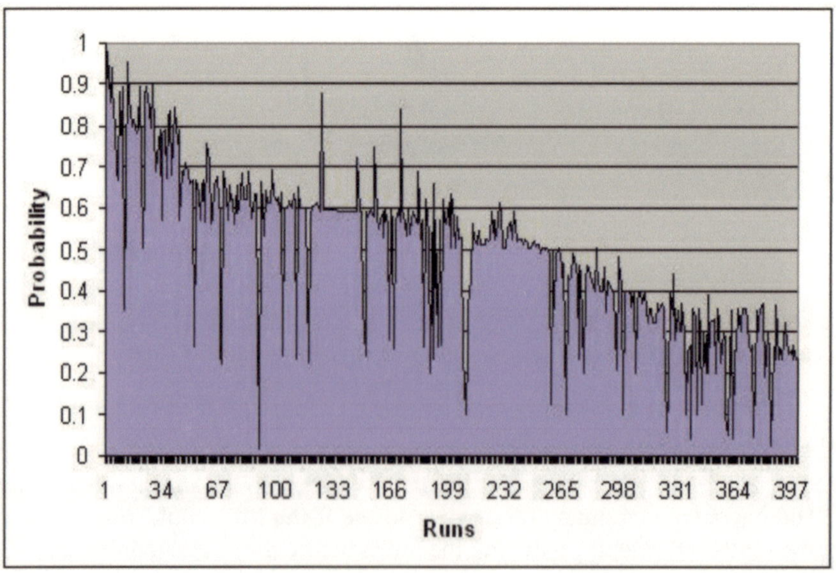

Fig. 7. Probability of faults in each run

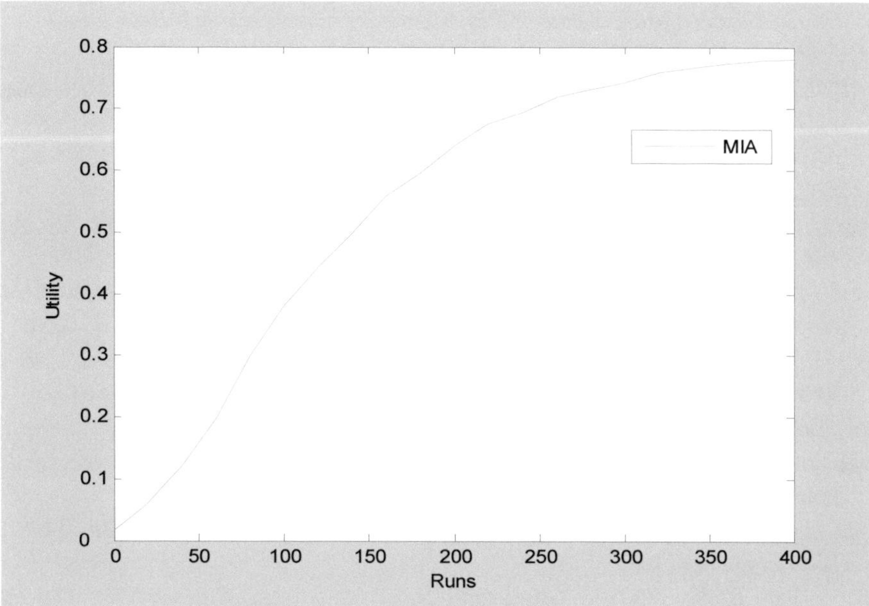

Fig. 8. MIAs Utility

8. Acknowledgement

The authors would like to thank the National Research Foundation, South Africa for the financial support that has made this work possible

9. References

Agnar, A. & Plaza, E. (1994). Case-Based Reasoning: Foundational Issues, Methodological Variations, and System Approaches, *Artificial Intelligence Communications* 7, no. 1, pp. 39-52, 1994.

Altmann, J.; Gruber, F.; Klug, L; Stockner, W. & Weippl, E. (2001). Using Mobile Agents in Real World: A Survey and Evaluation of Agent Platforms, *Workshop on Infrastructure for Agents, MAS and Scalable MAS, at the 5th International Conference on Autonomous Agents*. ACM Press, Montreal, Canada, pp. 33-39, June 2001.

Bieszczad, A.; Pagurek, N. & White, T. (1998). Mobile Agents for Network Management, *IEEE Communications Surveys and Tutorials*, Vol. 1(1), 1998.

Chess, D. M.; Harrison, C. G. & Kershenbaum, A. (1998). Mobile Agents: Are They a Good Idea?, IBM Research Report, April 27-30, 1998.

Chickering, D.; Heckerman, D. & Meek, C. (2004). Large-Sample Learning of Bayesian Networks is NP-Hard, *Journal of Machine Learning Research.* V5:pp.1287-1330, 2004.

Corkill, D. D. (1991). Blackboard Systems, *AI Expert* 6(9): pp.40 – 47, Sept.1991.

Danyluk, A. & Provost, F. (2002). Telecommunications Network Diagnosis, *In W. Kloesgen and J. Zytkow* (eds.), *Handbook of Knowledge Discovery and Data Mining,* Oxford University Press, 2002.

Donini, F. M.; Lenzerini, M.; Nardi, D.; Pirri, F. & Schaerf, M. (1990). Non-monotonic reasoning, *Artificial Intelligence Review*, Vol. 4(3), pp.163-210, 1990.

Duda, R. O.; Hart, P. E. & Stork, D. G. (2001). Pattern classification (2nd edition), Wiley, 2001, ISBN 0471056693.

Eleftheriou, G. & Galis, A.(2000). Mobile Intelligent Agents for Network Management Systems, *Proceedings London Communication Symposium*, 2000.

ETSI Guide (2001), Final drafts ETSI EG 202 009-3 V1.1.0 (2001-12) User Group; Quality of telecom services; Part 3: Template for Service Level Agreements (SLA). 2001.

Fagin, R.; Halpern, J. Y.; Moses, Y. & Vardi, M. Y. (1995). Reasoning about Knowledge, MIT Press, Cambridge, MA, 1995.

FIPA Specifications., Available from: http://www.fipa.org/

Frey, J. & Lewis, L. (1997). Multi-level reasoning for managing distributed enterprises and their networks, *In Integrated Network Management V*, pp.5-16, 1997.

Gilbert, D. (1997). Intelligent Agents: The Right Information at the Right Time, White Paper, *IBM Corporation*, 1997.

Hajji, B. & Far, B. H. (2001b). Continous Network Monitoring for Fast Detection of Performance Problems, *Proceedings of 2001 International Symposium on Performance Evaluation of Computer and Telecommunication Systems*, July 2001.

Hajji, B.; Far, B. H. & Cheng, J. (2001a). Detection of Network Faults and Performance Problems, *Proceedings of the Internet Conference*, Osaka, Japan, Nov. 2001.

Heckerman, D. (1996). Bayesian Networks for Knowledge Discovery, *Advances in Knowledge Discovery and Data Mining*, pp.273-305, 1996.

Heckerman, D. (1999). A Tutorial on Learning with Bayesian Networks, *In Learning in Graphical Models*, pp. 301-354, MIT Press, 1999.

Heckerman, D.; Meek, C. & Cooper, G. (1997). A Bayesian Approach to Causal Discovery, *Technical Report*, MSR-TR-97-05, Microsoft Research, Feb. 1997.

Holst, A. (1997). The Use of a Bayesian Neural Network Model for Classification Tasks, *Thesis, Studies of Artificial Neural Systems*, Department of Numerical Analysis and Computing Science, Royal Institute of Technology, S-100 44 Stockholm, Sweden, September 1997.

Hood, C. S. & Ji, C. (1997a). Proactive Network Fault Detection, *Proceedings of the IEEE INFOCOM*, pp. 1139-1146, Kobe, Japan, April 1997.

Hood, C. S. & Ji, C. (1997b). Automated proactive anomaly detection, *In Integrated Network Management V*, California, USA, Vol.86, pp.688-699, 1997.

Hood, C. S. & Ji, C. (1998). Intelligent Agents for Proactive Fault Detection, *IEEE Internet Computing*, Vol. 2(2), pp. 65-72, March/April 1998.

JADE website., Available from: http://jade.tilab.com/

JADE-LEAP website., Available from: http://leap.crm-paris.com/

Katzela, I.; Bouloutas, A. T. & Calo, S. (1996). Comparison of distributed fault identification schemes in communication networks, *Technical report, IBM Corp.*, T.J. Watson Research Center, Yorktown Heights, NY, USA, January 1996.

Khardon, R. & Roth, D. (1997). Defaults and Relevance in Model Based Reasoning, *Artificial Intelligence*, Vol. 97, Number 1-2, pp. 169-193, 1997.

Khardon, R.; Mannila, H. & Roth, D. (1999). Reasoning with Examples: Propositional Formulae and Database Dependencies, *Acta Inf*, 36(4), pp.267–286, 1999.

Klaus-Dieter, A.; Bergmann, R. & Branting, L. K. (1999). Case-Based Reasoning Research and Development, *Proceedings of the Third International Conference on Case-Based Reasoning*, Berlin, Germany, July 27-30,1999.

Kogeda, O. P. & Agbinya, J. I. (2006a). Prediction of Faults in Cellular Networks Using Bayesian Network Model, *Proceedings of 1st IEEE International Conference on Wireless Broadband and Ultra Wideband Communications (AUSWIRELESS 2006)*, pp. ISBN: 0-9775200-0-5, Sydney Australia, March 13-16, 2006.

Kogeda, O. P.; Agbinya, J. I. & Omlin, C. W. (2006b). A probabilistic Approach to Faults Prediction in Cellular Networks, *Proceedings of the 5th International Conference on Networking (ICN2006)*, ISBN: 0-7695-2570-9, Mauritius, April 23-28, 2006.

Langley, P. & Simon, H. A. (1995). Applications of Machine Learning and Rule Induction, *Communication ACM* Vol. 38(11): pp. 54 - 64, November 1995.

Lazar, A.; Wang, W. & Deng, R. (1992). Models and algorithms for network fault detection and identification: A review, *Proceedings of IEEE ICC*, Singapore, pp.999-1003, November 1992.

Lin, Y. & Druzdzel, M. J. (1997). Computational Advantages of Relevance Reasoning in Bayesian Belief Networks, *Proceedings of the Thirteenth Annual Conference in Uncertainty in Artificial Intelligence (UAI-97)*, pp. 342-350, Morgan Kaufmann Publishers, Inc., San Francisco, CA, 1997.

Meira, D. M. (1997). A Model for Alarm Correlation in Telecommunications Networks, *PhD Thesis*, Federal University of Minas Gerais, Belo Horizonte, Brazil, Nov. 1997.

Mitchell, T. (1997). Decision Tree Learning, in T. Mitchell, *Machine Learning*, The McGraw-Hill Companies, Inc., pp. 52-78, 1997.

Moreno, A.; Valls, A. & Viejo, A. (2002). Using JADE-LEAP to implement agents in mobile devices, 2002.

Pissinou, N.; Bhagyavati & Makki, K. (2000). Mobile Agents to Automate Fault Management in Wirelss and Mobile Networks, *IPDPS Workshops*, Lecture Notes in Computer Science, Vol. 1800, 2000, pp. 1296-1300, Springer Verlag.

Thottan, M. & Ji, C. (1998). Proactive Anomaly Detection Using Distributed Intelligent Agents, *IEEE Network*, Sept./Oct. 1998.

Thottan, M. & Ji, C. (1999). Fault Prediction at the Network Layer Using Intelligent Agents, *Proceedings of IEEE/IFIP Integrated Network Management*, May 1999, Boston, MA.

Walrand, J. & Varaiya, P. (2000). High-performance Communication Networks, *Second edition*, Morgan Kaufmann, 2000.

Watson, I. (1997). Applying Case-Based Reasoning: Techniques for Enterprise Systems. *Morgan Kaufmann*, July 1997, ISBN: 9781558604629.

Widrow, B. (1989). DARPA Neural Network Study, *AFCEA International Press*, 1989.

Williamson, C.; Halepovic, E.; Sun, H. & Wu, Y. (2005). Characterization of CDMA2000 Cellular Data Network Traffic, *LCN*, pp.712–719, Nov. 2005.

Winston, P. (1992). Learning by Building Identification Trees, in P. Winston, *Artificial Intelligence*, Addison-Wesley Publishing Company, pp. 423-442, 1992.

Wooldridge, M. J. & Jennings, N. R. (1995). Intelligent Agents: Theory and Practice. *The Knowledge Engineering Review,* Vol. 2, 1995, pp. 115-152, ISSN

Zadeh, L. A. (1962). In the engineering journal, *Proceedings of the IRE,* 1962.

Zadeh, L. A. (1965). Fuzzy sets, *Information and Control*, pp. 8:338-353, 1965.

15

Forward Error Correction for Reliable e-MBMS Transmissions in LTE Networks

Antonios Alexiou[2], Christos Bouras[1,2], Vasileios Kokkinos[1,2]
Andreas Papazois[1,2] and Georgia Tseliou[1,2]
[1]*Research Academic Computer Technology Institute,*
[2]*Computer Engineering and Informatics Department, University of Patras,*
Greece

1. Introduction

The Long Term Evolution (LTE) project focuses on enhancing the Universal Terrestrial Radio Access (UTRA) and optimizing 3rd Generation Partnership Project (3GPP) radio access architecture. A key new feature of LTE is the possibility to exploit the Orthogonal Frequency-Division Multiplexing (OFDM) radio interface to transmit multicast or broadcast data as a multicell transmission over a synchronized Single Frequency Network (SFN): this is known as Multimedia Broadcast and Multicast Service (MBMS) over Single Frequency Network (MBSFN) operation. MBSFN transmission enables a more efficient operation of the MBMS (3GPP, 2008a), allowing over-the-air combining of multi-cell transmissions towards the User Equipments (UEs). This fact makes the MBSFN transmission appear to the UE as a transmission from a single larger cell. Transmission on a dedicated carrier for MBSFN with the possibility to use a longer Cyclic Prefix (CP) with a sub-carrier bandwidth of 7.5 kHz is supported as well as transmission of MBSFN on a carrier with both MBMS transmissions and point-to-point (PTP) transmissions using time division multiplexing. MBMS service defines two delivery methods: the download and the streaming delivery.

There are many ways to provide reliability in multicast transmission. The best-known method that operates efficiently for unicast transmission is the Automatic Repeat re-Quest (ARQ). When ARQ is applied in a multicast session, receivers send requests for retransmission of lost packets over a back channel towards the sender. Although ARQ is an effective and reliable tool for point-to-multipoint (PTM) transmission, when the number of receivers increases, it reveals its limitations. One major limitation is the feedback implosion problem which occurs when too many receivers are transmitting back to the sender. A second problem of ARQ is that for a given packet loss rate and a set of receivers experiencing losses, the probability that every single data packet needs to be retransmitted quickly approaches unity as the number of receivers increases. In other words, a high average number of transmissions are needed per packet. In wireless environments, ARQ has another major disadvantage. On most wired networks the feedback channel comes for free, but on wireless networks the transmission of feedback from the receiver can be expensive, either in terms of power consumption, or due to limitations of the communication infrastructure. Thus, due to its requirement for a bidirectional communication link, the application of ARQ over wireless networks may be too costly or, in some cases, not possible. Forward Error Correction (FEC) is an error control method that can be used to augment or replace other methods for reliable data transmission. The main attribute of FEC schemes is that the sender adds redundant information in the messages transmitted to the receiver. This information allows the receiver to reconstruct the source data. Such schemes inevitably add a constant overhead in the transmitted data and are computationally expensive. In multicast protocols however, the use of FEC techniques has very strong motivations. The encoding eliminates the effect of independent losses at different receivers. This makes these schemes able to scale irrespectively of the actual loss pattern at each receiver. Additionally, the dramatic reduction in the packet loss rate largely reduces the need to send feedback to the sender. FEC schemes are therefore so simple as to meet a prime objective for mobile multicast services, which is scalability to applications with thousands of receivers. MBMS service for multicast transmission uses MBSFN. This is the reason why 3GPP recommends the use of FEC for MBMS and, more specifically, adopts the use of systematic Raptor FEC code (3GPP, 2008b). The Raptor codes belong to the class of fountain codes and are very popular due to their high probability for error recovery and their efficiency during encoding and decoding. In this chapter, we study the application of FEC for MBSFN transmissions over LTE cellular networks. First, we make a cost analysis and define a model for the calculation of the total telecommunication cost that is required for the transmission of the MBSFN data to end users. Then, we propose an innovative error recovery scheme for the transmission of the FEC redundant information during MBMS download delivery. This scheme takes advantage of the MBSFN properties and performs an adaptive generation of redundant symbols for efficient error recovery. The redundant encoding symbols are produced continuously until all the multicast receivers have acknowledged the complete file recovery. Then, we investigate the performance of the proposed scheme against the existing approaches under different MBSFN deployments, user populations and error rates. In this framework, we evaluate the performance of our scheme and we examine whether the use of FEC is beneficial, how the optimal FEC code dimension varies based on the network conditions, which parameters affect the optimal FEC code selection and how they do it.

This work is structured as follows: in Section 2 we present the study related to this scientific domain. In Section 3 we provide an overview of MBMS architecture and we describe the key concepts that our study deals with. The telecommunication cost analysis of the MBSFN delivery scheme is described in Section 4. In Section 5 we describe some approaches for transmission as well as our proposed scheme and in Section 6 the evaluation results of the conducted experiments. Finally, in Section 7 the conclusions are briefly described and in Section 8 all the planned next steps of this work are listed. For the reader's convenience, Appendix A presents an alphabetical list of the acronyms used in the chapter.

2. Related work

The research over FEC for broadcast and multicast transmission has recently moved from the domain of fixed networks to the wireless communication field. The standardization of MBMS by 3GPP triggered the research on the use of FEC for multicasting in the domain of mobile networks. Even though this research area is relatively new, a lot of solutions have been proposed so far.

In (Luby et al., 2006) an introduction in the Raptor code structure is presented. The Raptor codes are described through simple linear algebra notation. Several guidelines for the practical implementation of the relevant encoders and decoders are presented and the good performance of file broadcasting with Raptor codes is verified. The simulation results verify the efficient performance of the whole process. The same authors in (Luby et al., 2007) present an investigation on MBMS download delivery services in Universal Mobile Telecommunications System (UMTS) considering a comprehensive analysis by applying a detailed and complex channel model and simulation setup. It is concluded that the optimal operating point in this trade-off uses low transmission power and a modest amount of Turbo FEC coding that results in relatively large radio packet loss rates.

The study presented in (Alexiou et al., 2010a) investigates the impact of FEC use for MBMS and examines whether it is beneficial or not and how the optimal FEC code dimensioning varies based on the network conditions, elaborating the parameters which affect the optimal FEC code selection. The simulation results show the behaviour of the standardized FEC scheme evaluated against parameters such as multicast user density and multicast user population. In (Alexiou et al., 2010b), the applicability of FEC via Raptor code in the multicast data transmission is studied while focusing on power control in the Radio Access Network (RAN). The evaluation considers the properties of PTP, PTM as well as hybrid transmission mode that combine both PTP and PTM bearers in RAN. The main assertion that came out is the fact that increasing the power in order to succeed a better Block Error Rate (BLER) is cheaper from power perspective than increasing the power to send the redundant symbols added by FEC decoder.

The study in (Lohmar et al., 2006) focuses particularly on the file repair procedure. The trade-off between FEC protection and successive file repair is discussed extensively. The authors propose a novel file repair scheme that combines PTM filer repair transmission with a PTP file repair procedure. After the analysis, it is proved that the new scheme can achieve better performance than a PTP-only file repair procedure. The overall goal is the optimization of 3G resource usage by balancing the FEC transmission overhead with file repair procedures after the MBMS transmission.

The adoption of FEC is examined from another aspect in (Wang & Zhang, 2008). A potential bottleneck of the radio network is taken into consideration and the authors investigate which are the optimal operation points in order to save radio resources and use the available spectrum more efficiently. The conducted simulation experiments and the corresponding numerical results demonstrate the performance gain that Raptor code FEC offers in MBMS coverage. In more detail, the spectrum efficiency is significantly improved and resource savings are achieved in the radio network.

The reliability and efficiency in download delivery with Raptor codes are examined in (Gasiba et al., 2007). The authors propose two algorithms; one allowing to find a minimum set of source symbols to be requested in the post delivery and one allowing to find a sufficient number of consecutive repair symbols. Both algorithms guarantee successful recovery. These post-repair methods are combined with the regular Raptor decoding process and fully exploit the properties of these codes. Selected simulations verify the efficient performance of file distribution with Raptor codes as well as the algorithms for file repair in case of file distribution to more than one user. Despite the extraordinary performance of Raptor codes, reliable delivery cannot be guaranteed, especially in heterogeneous receiver environments.

Generally, it should be noted that all the existing related work covers research either on the application layer FEC for prior to LTE cellular networks or FEC for the LTE physical layer. It is important to mention that the use of FEC for the multicast transmission over LTE networks has not been studied yet. Any related work, as the works presented above, is dedicated to the previous generations of mobile networks. Therefore, it is our belief and the motivation behind our work that the impact of FEC in MBSFN transmissions should constitute a new domain where the LTE research community should focus on. The contribution of this work includes the review of the current error recovery methods, an extensive cost analysis of the data delivery during MBSFN transmissions in LTE cellular networks and the proposal of a new error recovery scheme which the simulation experiments prove to be more cost effective than the existing ones.

3. Overview of MBMS

3.1 LTE Architecture for MBMS

The LTE architecture for MBMS, or as it is commonly referred to, evolved MBMS (e-MBMS) architecture is illustrated in Fig. 1.

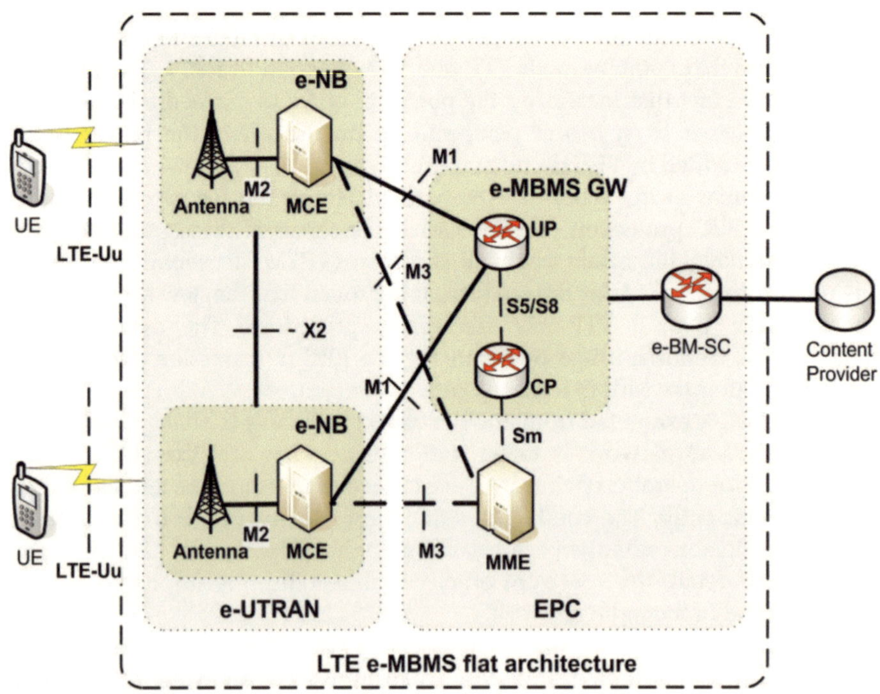

Fig. 1. e-MBMS flat architecture

Within evolved UTRA Network (e-UTRAN) the evolved Node Bs (e-NBs) or base stations are the collectors of the information that has to be transmitted to users over the air-interface. The Multicell/multicast Coordination Entity (MCE) coordinates the transmission of synchronized signals from different cells (e-NBs). MCE is responsible for the allocation of the same radio resources, used by all e-NBs in the MBSFN area for multi-cell MBMS transmissions. Besides allocation of the time / frequency radio resources, MCE is also responsible for the radio configuration, e.g., the selection of modulation and coding scheme. The e-MBMS Gateway (e-MBMS GW) is physically located between the evolved Broadcast Multicast Service Centre (e-BM-SC) and e-NBs and its principal functionality is to forward the e-MBMS packets to each e-NB transmitting the service. Furthermore, e-MBMS GW performs MBMS Session Control Signalling (Session start/stop) towards the e-UTRAN via the Mobility Management Entity (MME). The e-MBMS GW is logically split into two domains. The first one is related to control plane, while the other one is related to user plane. Likewise, two distinct interfaces have been defined between e-MBMS GW and e-UTRAN namely M1 for user plane and M3 for control plane. M1 interface makes use of IP multicast protocol for the delivery of packets to e-NBs. M3 interface supports the e-MBMS session control signalling, e.g., for session initiation and termination (3GPP, 2009; Holma & Toskala, 2009).

The e-BM-SC is the entity in charge of introducing multimedia content into the LTE network. For this purpose, the e-BM-SC serves as an entry point for content providers or any other broadcast/multicast source which is external to the network. An e-BM-SC serves all the e-MBMS GWs in a network.

3.2 Application layer FEC

3GPP has standardized Turbo codes as the physical layer FEC codes and Raptor codes as the application layer FEC codes for MBMS aiming to improve service reliability (3GPP, 2008a). The use of Raptor codes in the application layer of MBMS has been introduced to 3GPP by Digital Fountain (3GPP, 2005). Generally in the literature, FEC refers to the ability to overcome both erasures (losses) and bit-level corruption. However, in the case of an IP multicast protocol, the network layers will detect corrupted packets and discard them or the transport layers can use packet authentication to discard corrupted packets. Therefore the primary use of application layer FEC to IP multicast protocols is as an erasure code. The payloads are generated and processed using a FEC erasure encoder and objects are reassembled from reception of packets containing the generated encoding using the corresponding FEC erasure decoder.

Raptor codes belong to the class of the fountain codes. Fountain codes are record-breaking, sparse-graph codes for channels with erasures, where files are transmitted in multiple small packets, each of which is either received without error or not received. The conventional file transfer protocols usually split a file up into k packet sized pieces and then repeatedly transmit each packet until it is successfully received. A back channel is required for the transmitter to find out which packets need retransmitting. In contrast, fountain codes make packets that are random functions of the whole file. The transmitter sprays packets at the receiver without any knowledge of which packets are received. Once the receiver has received any m packets - where m is just slightly greater than the original file size k - the whole file can be recovered. The computational costs of the best fountain codes are astonishingly small, scaling linearly with the file size.

The Raptor decoder is therefore able to recover the whole source block from any set of FEC encoding symbols only slightly more in number than the number of source symbols. The Raptor code specified for MBMS is a systematic fountain code producing n encoding symbols E from $k < n$ source symbols C. This code can be viewed as the concatenation of several codes. The most-inner code is a non-systematic Luby-Transform (LT) code with l input symbols F, which provides the fountain property of the Raptor codes. This non-systematic Raptor code does not use the source symbols as input, but it encodes a set F of intermediate symbols generated by some outer high-rate block code. This means that the outer high-rate block code generates the F intermediate symbols using k input symbols D.

Finally, a systematic realization of the code is obtained by applying some pre-processing to the k source symbols C such that the input symbols D to the non-systematic Raptor code are obtained. The description of each step and the details on specific parameters can be found in (3GPP, 2008a).
The study presented in (Luby et al., 2006) shows that Raptor codes have a performance very close to ideal, i.e., the failure probability of the code is such that in case that only slightly more than k encoding symbols are received, the code can recover the source block. In fact, for $k > 200$ the small inefficiency of the Raptor code can accurately be modelled by the following equation (Luby et al., 2007):

$$p_f(m,k) = \begin{cases} 1 & \text{if } m < k, \\ 0.85 \times 0.567^{m-k} & \text{if } m \geq k. \end{cases} \quad (1)$$

In (1), $p_f(m,k)$ denotes the failure probability of the code with k source symbols if m symbols have been received. It has been observed that for different k, the equation almost perfectly emulates the code performance. While an ideal fountain code would decode with zero failure probability when $m = k$, the failure for Raptor code is still about 85%. However, the failure probability decreases exponentially when number of received encoding symbols increases.

3.3 File repair procedure

The purpose of file repair procedure is to repair lost or corrupted file segments that appeared during the MBMS download data transmission (3GPP, 2008b). At the end of the MBMS download data transmission each multicast user identifies the missing segments of the transmitted file and sends a file repair request message to the file repair server. This message determines which exactly the missing data are. Then, the file repair server responds with a repair response message. The repair response message may contain the requested data, redirect the client to an MBMS download session or to another server, or alternatively, describe an error case.
The file repair procedure has significant disadvantages since it may lead to feedback implosion in the file repair server due to a potential large number of MBMS clients requesting simultaneous file repairs. Another possible problem is that downlink network channel congestion may be occurred due to the simultaneous transmission of the repair data towards multiple MBMS clients. Last but not least, the file repair server overload, caused by bursty incoming and outgoing traffic, should be avoided. The principle to protect network resources is to spread the file repair request load in time and across multiple servers. The resulting random distribution of repair request messages in time enhances system scalability.

4. Cost analysis of MBSFN

4.1 Introduction

In this section, we present a performance evaluation of MBSFN delivery scheme. As performance metric for the evaluation, we consider the total telecommunication cost for both packet delivery and control signals transmission (Ho & Akyildiz, 1996). In our analysis, the cost for MBSFN polling is differentiated from the cost for packet deliveries. Furthermore, in accordance with (Ho & Akyildiz, 1996), we make a further distinction between the processing costs at nodes and the transmission costs on links. For the analysis, we apply the notations presented in Table 1:

Symbol	Explanation
D_{Uu}	Transmission cost of single packet over Uu interface
C_{Uu}	Total transmission cost over Uu (air) interface
D_{M1}	Transmission cost of single packet over M1 interface
C_{M1}	Total transmission cost over M1 interface
$C_{polling}$	Total transmission cost for polling
C_{SYNC}	Total processing cost for synchronization at eBM-SC
D_{p_eNB}	Cost of polling procedure at each e-NB
D_{M2}	Transmission cost of single packet over M2 interface
N_p	Total number of packets of the MBSFN session
N_{eNB}	Number of e-NBs that participate in MBSFN
N_{cell}	Total number of e-NBs in the topology
N_{p_burst}	Mean number of packets in each packet burst
C_{MBSFN}	Total telecommunication cost of the MBSFN delivery

Table 1. Notations

Before presenting in detail the parameters introduced in Table 1, some general assumptions of our analysis and the topology under examination are presented.

4.2 General assumptions and topology

We assume that the topology is scalable and has the possibility to consist of an infinite number of cells according to Fig. 2. Moreover, in order to calculate the total cost, we assume that the users can be located in a constantly increasing area of cells in the topology, called "UE drop location cells". Therefore, in the case when UE drop location cells are equal to 1, all users are located in the centre cell (see Fig. 2). The six cells around the centre cell constitute the inner 1 ring. Likewise, the inner 2 ring consists of the 12 cells around the first ring. Following this reasoning, we can define the "inner 3 ring", the "inner 4 ring" etc.

In this chapter the following user distributions are examined:

- All MBSFN users reside in the centre cell (UE drop location cells = 1).
- All MBSFN users reside in the area included by the inner 1 ring (UE drop location cells = 7).
- All MBSFN users reside in the area included by the inner 2 ring (UE drop location cells = 19).
- And so forth…
- All the infinite cells of the topology contain MBSFN users (UE drop location cells = infinite, i.e., number of cells >> 721 or number of cell rings >> 15).

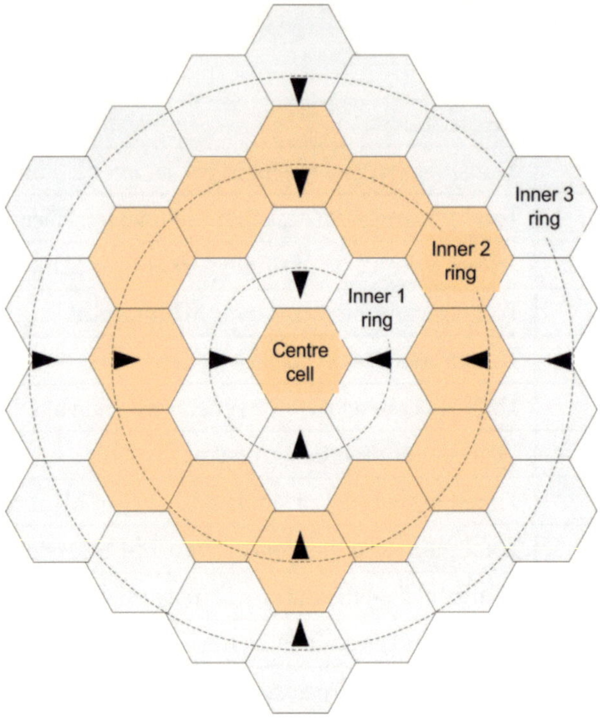

Fig. 2. Topology under examination

The performance of the MBSFN increases rapidly when rings of neighbouring cells outside the "UE drop location cells" area assist the MBSFN service and transmit the same MBSFN data. More specifically according to (3GPP, 2008a; Rong et al., 2008), even the presence of one assisting ring can significantly increase the overall spectral efficiency. Moreover, we assume that a maximum of 3 neighbouring rings outside the "UE drop location cells" can transmit in the same frequency and broadcast the same MBSFN data (assisting rings), since additional rings do not offer any significant additional gain in the MBSFN transmission (3GPP, 2008a; Rong et al., 2008). Our goal is to examine the number of neighbouring rings that should be transmitting simultaneously to the UE drop location cells in order to achieve the highest possible gain, in terms of overall packet delivery cost. For this purpose, we define the following three MBSFN deployments (where "A" stands for an Assisting ring and "I" for an Interference ring, i.e.: a ring that does not participate in the MBSFN transmission):

- AII: The first ring around the UE drop location cells, contributes to the MBSFN transmission, the second and third rings act as interference.
- AAI: The first and the second ring around the UE drop location cells assist in the MBSFN transmission, the third ring acts as interference.
- AAA: indicates that each of the 3 surrounding rings of the UE drop location cells assists in the MBSFN transmission

The system simulation parameters that were taken into account for our simulations are presented in Table 2. The typical evaluation scenario used for LTE is macro Case 1 with 10 MHz bandwidth and low UE mobility. The propagation models for macro cell scenario are based on the Okamura-Hata model (3GPP, 2008a; Holma & Toskala, 2009).

Parameter	Units	Case 1
Inter Site Distance (ISD)	m	500
Carrier Frequency	MHz	2000
Bandwidth	MHz	10
Penetration Loss (PL)	dB	20
Path Loss	dB	Okumura-Hata
Cell Layout		Hexagonal grid, 3 sectors per site, infinite rings
Channel Model		3GPP Typical Urban (TU)
# UE Rx Antennas		2
UE speed	Km/h	3
BS transmit power	dBm	46
BS # Antennas		1
BS Ant. Gain	dBi	14

Table 2. Simulation parameters

4.3 Air interface cost

In this section the transmission cost over the air interface is defined for different network topologies, user distributions and MBSFN deployments. Fig. 3 depicts the resource efficiency of SFN transmission mode (i.e., the spectral efficiency of the SFN transmission normalized by the fraction of cells in the SFN area containing UEs) as the number of UE drop location cells increases, for the 3 different MBSFN deployments (AII, AAI, AAA) presented in the previous paragraph. More specifically, Fig. 3 presents the way the resource efficiency changes with the number of UE drop location cells for a macrocellular Case 1 environment (3GPP, 2008a).

In Fig. 3, we observe that when all users are distributed in the centre cell, the resource efficiency for AAA is 0.06, for AAI 0.12 and for AII 0.19. As a result, when all the MBSFN users reside in the centre cell, AII is the best deployment in terms of resource efficiency. However, we have to mention that in the specific example; the best deployment was selected based only on the air interface performance. Next in our analysis, we will present an alternative/improved approach that selects the best MBSFN deployment based on the overall cost.

To define the telecommunication cost over the air interface, we define as resource efficiency percentage (*RE_percentage*) the fraction of current deployment resource efficiency to the maximum SFN resource efficiency. This percentage indicates the quality of the resource efficiency our current deployment achieves for the macrocellular Case 1, compared to the maximum resource efficiency that can be achieved in Case 1. Then, we define the cost of packet delivery over the air interface (D_{Uu}) as the inverse of *RE_percentage*. This means that as the resource efficiency of a cell increases, the *RE_percentage* increases too, which in turn means that the cost of packet delivery over the air interface decreases.

Finally, the total telecommunication cost for the transmission of the data packets over Uu interface is derived from (2), where N_{eNB} represents the number of e-NBs that participate in MBSFN transmission, N_p the total number of packets of the MBSFN session, and D_{Uu} is the cost of the delivery of a single packet over the Uu interface.

$$C_{Uu} = D_{Uu} \cdot N_p \cdot N_{eNB} \tag{2}$$

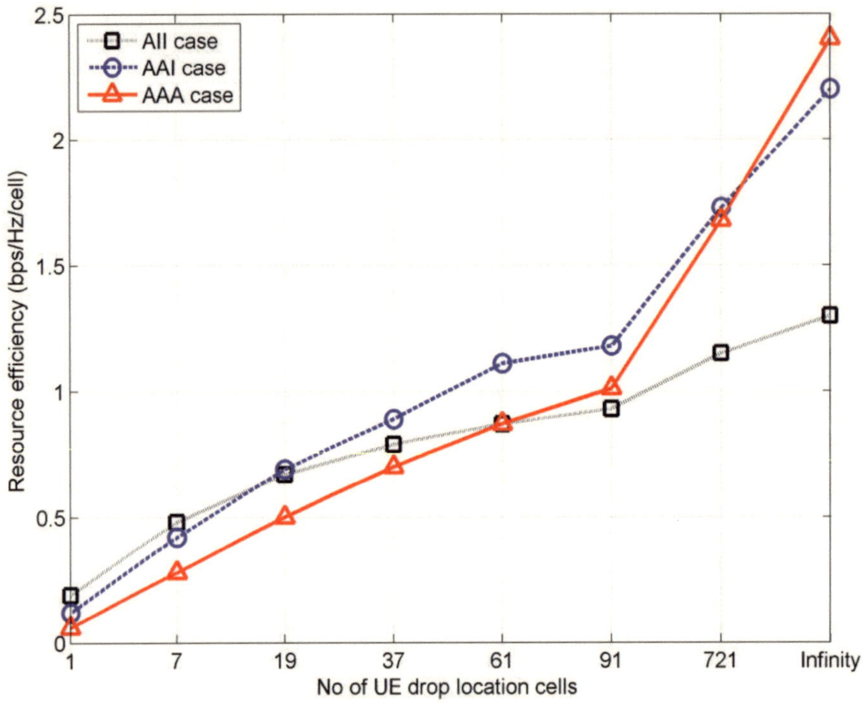

Fig. 3. Resource efficiency vs. number of UE drop location cells for ISD = 500m

4.4 Cost over M1 interface

M1 interface uses IP multicast protocol for the delivery of packets to e-NBs. In multicast, the e-MBMS GW forwards a single copy of each multicast packet to those e-NBs that participate in MBSFN transmission. After the correct multicast packet reception at the e-NBs that serve multicast users, the e-NBs transmit the multicast packets to the multicast users via Multicast Traffic Channel (MTCH) transport channels. The total telecommunication cost for the transmission of the data packets over M1 interface is derived from (3), where D_{M1} is the cost of the delivery of a single packet over the M1 interface.

$$C_{M1} = D_{M1} \cdot N_P \cdot N_{eNB} \tag{3}$$

More specifically, D_{M1} depends on the number of hops between the nodes connected by M1 interface and the profile of the M1 interface in terms of link capacity (Alexiou et al., 2007). In general, a high link capacity corresponds to a low packet delivery cost over M1 and a small number of hops, corresponds to a low packet delivery cost.

4.5 Synchronization cost

In order to implement a SFN, each of the transmitting cells should be tightly time-synchronized and use the same time-frequency resources for transmitting the bit-identical content. The overall user plane architecture for content synchronization is depicted in Fig. 4.

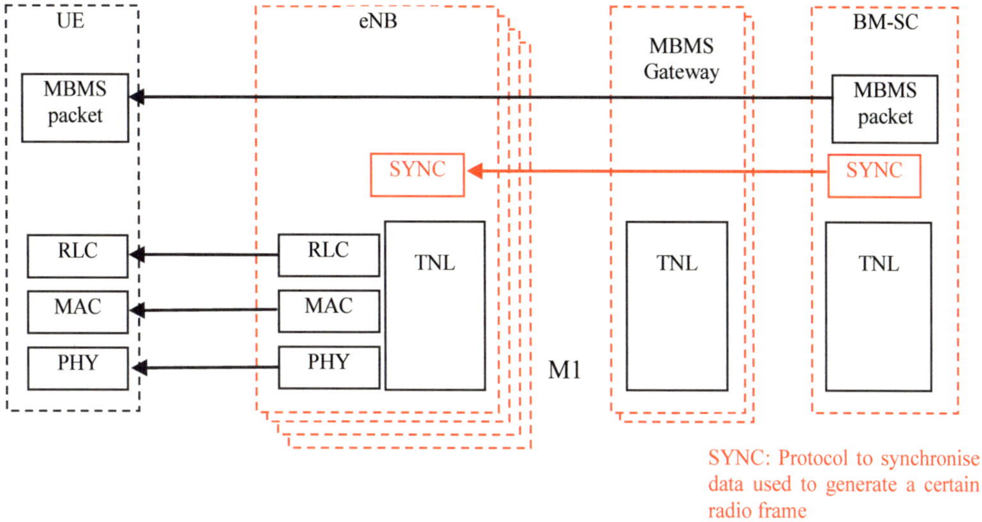

Fig. 4. Content synchronization in MBSFN

The SYNC protocol layer is defined on transport network layer to support content synchronization. It carries additional information that enables e-NBs to identify the timing for radio frame transmission and detect packet loss. Every e-MBMS service uses its own SYNC entity. The SYNC protocol operates between e-BM-SC and e-NB. As a result of synchronization, it is ensured that the same content is sent over the air to all UEs (3GPP, 2009). The e-BM-SC should indicate the timestamp (*T*) of the transmission of the first packet of a burst of data (block of packets) by all e-NBs and the interval between the radio transmissions of the subsequent packets of the burst as well. Since the synchronization protocol has not yet been standardized and many alternative protocols have been proposed (3GPP, 2007a), we assume that the transmission timestamp of the first packet of a burst of data is sent before the actual burst in a separate Packet Data Unit (PDU). When time *T* is reached, the e-NB buffer receives another value of *T* and new packet data which correspond to the next burst. All in all, in this case the transmission timing for subsequent bursts is implicitly determined by the size and the number of previous packets (3GPP, 2007a). This in turn means that the synchronization cost depends on the total numbers of multicast bursts/packets per MBSFN session. The total telecommunication cost for the transmission of the synchronization packets is derived from the following equation where D_{M1} is the cost of the delivery of a single packet over the M1 interface and N_{p_burst} is the mean value of the number of packets transmitted each time in the sequential bursts of the MBSFN session.

$$C_{SYNC} = \frac{N_P}{N_{p\_burst}} \cdot D_{M1} \cdot N_{eNB} \tag{4}$$

4.6 Polling cost

To determine which cells contain users interested in receiving a MBSFN service, we assume that a polling procedure is taking place. In contrast to counting procedure used in UMTS MBMS, where the exact number of MBMS users was determined, with polling we just determine if the cell contain at least one user interested for the given service.

The e-NBs initiate the detection procedure by sending a UE feedback request message on Multicast Control Channel (MCCH). The cost of sending this request message corresponds to the cost of polling procedure at e-NB (D_{p_eNB}). The message includes the MBMS service ID that requires the user feedback and a "dedicated access information" (in the form of a particular signature sequence) that is to be used for the user feedback by the UEs. After receiving the feedback request message, the UEs which are interested in receiving the particular e-MBMS service, respond to the request by sending a feedback message using the allocated "dedicated access resources" over non-synchronous Random Access Channel (RACH).
The e-NB receives the feedback from the UEs in the form of signature sequence. If energy is detected corresponding to the known signature sequence, this indicates that at least one user in the coverage area of the e-NB is interested in or activated the particular e-MBMS service. This information (packet) is sent to the MCE over M2 interface which in turn estimates which cells contain multicast users interested for the given e-MBMS service (3GPP, 2006).
The total cost associated to the polling procedure is derived from (5), where N_{eNB} represents the number of e-NBs that participate in MBSFN transmission, N_{cell} is the total number of e-NBs in the topology and D_{M2} is the cost of the delivery of a single packet over the M2 interface.

$$C_{Polling} = D_{p\_eNB} \cdot N_{cell} + D_{M2} \cdot N_{eNB} \tag{5}$$

4.7 Total telecommunication cost

Based on the analysis presented in the previous paragraphs, the total telecommunication cost of the MBSFN delivery scheme is derived from (6)

$$\begin{aligned} C_{MBSFN} &= C_{Uu} + C_{M1} + C_{SYNC} + C_{Polling} = \\ &\left(D_{Uu} + D_{M1} + \frac{D_{M1}}{N_{p\_burst}} \right) \cdot N_p \cdot N_{eNB} + \left(D_{p\_eNB} \cdot N_{cell} + D_{M2} \cdot N_{eNB} \right) \end{aligned} \tag{6}$$

5. Proposed scheme

The scheme that we propose introduces the exclusive sending of redundant encoding symbols instead of using the file repair procedure for the complete recovery of a transmitted file. It is important to clarify that the transmission of all the encoding symbols is performed over the MBSFN infrastructure. The scheme takes advantage of the fact that the Raptor FEC decoder, based on a fountain code, is able to recover the source blocks from any set of encoding symbols only slightly more in number than the number of source symbols. Therefore, it is proposed that the Raptor FEC encoder in the sender generates redundant symbols until it takes an acknowledgement from all the receivers that all the initial source symbols have been recovered. Our work investigates the application of FEC over the download delivery method, so the rest of our analysis focuses only on this MBMS delivery method.
In the rest of this section, we describe our proposed scheme in more detail and we present it against existing error recovery approaches specified by 3GPP for the MBMS download delivery method (3GPP, 2008b). In general, depending on the error recovery scheme used, the following three different approaches can be distinguished:

- Approach A1: Retransmission of the lost file's segments with MBSFN.

- Approach A2: Prefixed FEC overhead during the e-MBMS service transmission combined with retransmission of lost file's segments.
- Approach A3: Exclusive transmission of redundant symbols for file recovery (proposed scheme).

Assuming that an MBMS download delivery of a file is performed using MBSFN operation, then based on the error recovery approach used (A1, A2 or A3), the transmission process proceeds as illustrated in Fig. 5.

Initially, we examine the case where no FEC is used (Fig. 5, A1). In this case, the single error recovery scheme used is the file repair procedure and thus the receivers request the retransmission of the lost file's segments at the end of the process. Since MBSFN operation is used, the lost segments are transmitted to all the users in the area irrespectively of whether they have requested them or not. On the other hand, in case FEC is used (Fig. 5, A2 and A3), then the file to be downloaded is partitioned into one or several so-called source blocks. As mentioned above, for each source block, additional repair symbols can be generated by applying Raptor FEC encoding.

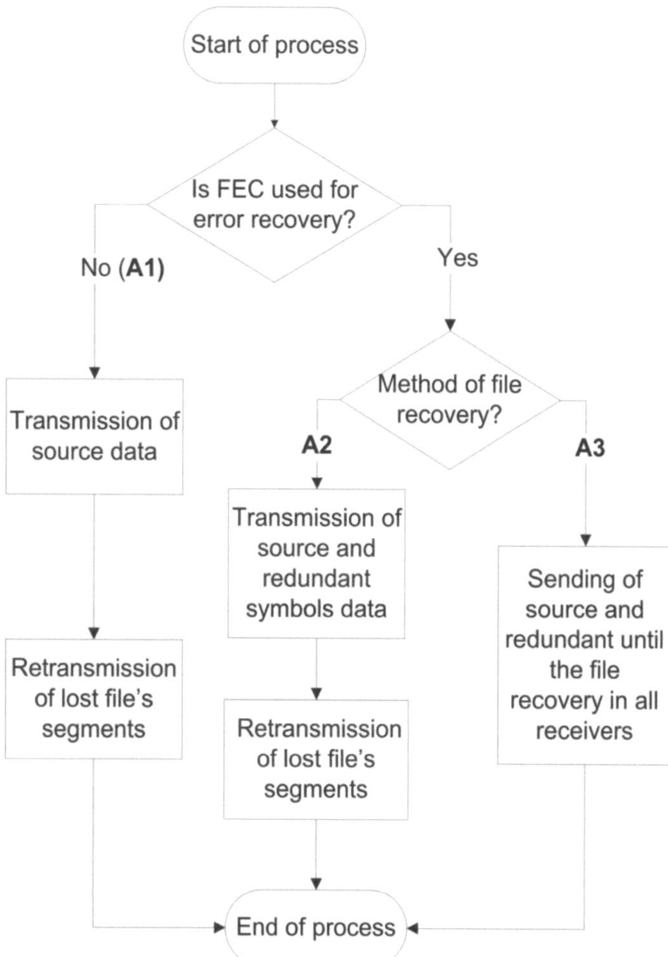

Fig. 5. Flowchart of error recovery approaches

The ideal situation in an MBMS session is that all the multicast receivers have collected the source blocks from the file and therefore the complete file recovery is possible. Nevertheless, the above occasion rarely happens. In most of cases, due to miscellaneous network conditions, receivers cannot recover all the source blocks since some of the received blocks are corrupted and they are rejected. In order to solve this situation and repair lost or corrupted file segments, we can use the standardized method defined by 3GPP in (3GPP, 2008b) (Fig. 5, A2). According to this approach, the complete error recovery may be achieved through the retransmission of source and redundant data through the file repair procedure, i.e., the selective retransmission of lost file's segments that takes place at the end of the transmission. On the other hand, the scheme that we propose introduces the exclusive use of FEC for efficient error recovery during MBMS transmission over MBSFN. In more detail, redundant symbols are produced continuously by the sender until the sender has received acknowledgment messages from all the receivers participating in the multicast group (Fig. 5, A3). On the MBMS receiver's side, each receiver sends back to the sender an acknowledgment message upon collection of the encoding symbols that are sufficient for the complete file recovery. The sender keeps track of which receivers have acknowledged and continues to send redundant encoding symbols until all receivers have acknowledged the complete file reception.

6. Performance evaluation

6.1 Simulation model

During our simulation experiments, we compare the proposed approach (A3) with the existing error recovery approaches (A1 and A2) presented above. The performances of the above approaches are evaluated through a realistic simulation model that incorporates all the network parameters and is consistent with the corresponding 3GPP specifications. In this framework, we consider the performance of our approach under different error rates, user populations and FEC configurations.

As already mentioned, the evaluation of the above approaches is performed from telecommunication cost perspective. The estimation of each factor of the cost is based on the telecommunication cost for MBSFN transmission given by equation (6). It should be noted that the above recovery processes are provided via MBSFN transmissions. Our simulation model incorporates all the properties of a typical Raptor code defined for data delivery over e-MBMS as they are defined by 3GPP in (3GPP, 2009). The total telecommunication cost for a complete file reception is the sum of the cost for the initial file transmission, the cost for the transmission of the additional packets due to FEC encoding and the cost for the selective retransmission of lost packets. The estimation of each of the above three terms is based on the telecommunication cost for MBSFN transmission given by (6).

It is worth clarifying that since n encoding symbols are produced from $k < n$ source symbols, then the overhead added due to the Raptor encoding, i.e., the number of repair symbols divided by the number of source symbols, is equal to the fraction $(n-k)/k$. Given that the packet size is fixed, the FEC overhead that is needed for the transmission of a file of given size is also equal to the same fraction. Thus, it is obvious that, in terms of percentage over the initial file size, the overhead of the additional packets that are needed for the download delivery of a given file is $(n-k)/k$. This packet overhead creates additional cost which is taken into account by our scheme. During the decoding procedure in each UE, there is a decoding failure probability represented by (1). When a packet loss rate $p_{loss} > 0$ is applied over the e-MBMS bearer, the number of the received symbols m may become less than the n symbols initially transmitted. As a result of the packet loss, the failure probability $p_f(m,k)$ increases. If the recovery of the k source symbols through decoding procedure fails in a UE and selective retransmission is invoked by the UE for the recovery of the lost packets, then this procedure creates an additional cost which is also taken into account by our scheme.

The system simulation parameters that were taken into account for our simulations are presented in Table 2. The typical evaluation scenario used for LTE is macro Case 1 with 1.4 MHz bandwidth and low UE mobility. The propagation models for macro cell scenario are based on the Okamura-Hata model (3GPP, 2007b).

6.2 Cost vs. MBSFN deployment

Having analyzed the distinct costs of the MBSFN delivery scheme, we evaluate the total cost of each of the MBSFN deployments (AAA, AAI, AII) for different user distributions for the distinguished error recovery approaches (A1, A2, A3). The topology we use is the one described in Section 4.2. Through this experiment, our goal is to evaluate each MBSFN deployment for different user distribution and not to examine whether FEC use is beneficial or not.

Fig. 6 depicts the total cost of the SFN transmission without FEC, with a prefixed FEC overhead and using redundant symbols for the 3 different deployments (AII, AAI, AAA) as the number of UE drop location cells increases. We observe that for the first 3 user distributions (cases of 1, 7, 19 UE drop location cells), the AII deployment ensures the lowest cost for the delivery of the MBSFN data and therefore is the most efficient deployment for

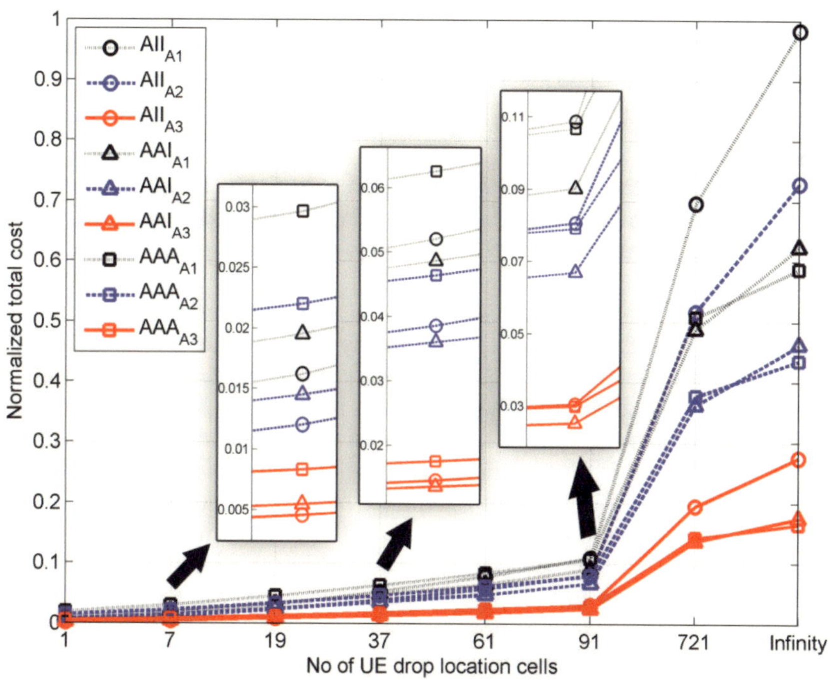

Fig. 6. Cost vs. MBSFN Deployment (Packet loss rate=5%, FEC overhead =5%, UE population=100)

the delivery of the MBSFN data. On the other hand, for UE drop location cells 37, 61, 91 and 721 cells, AAI is the most cost efficient deployment. Finally, for the case of the MBSFN transmission where the users are residing in infinite cells, the AAA deployment is more efficient than the other two deployments since it results in a lower overall cost.
Generally, it is necessary to switch between the 3 MBSFN deployments, when the number of UE drop location cells increases, so as to achieve the lowest possible transmission cost. As the number of UE drop location cells increases, the most efficient deployment for the delivery of the MBSFN data, switches from AII, to AAI and finally to AAA when the number of cells that have users interested in the MBSFN service approaches infinity (number of cells >> 721). This switching can save resources both in the core network and the air interface. For example, in the case of 721 UE drop location cells, we observe that the normalized total cost without FEC application is 0.6967 when AII is used. However, when AAI is used the total cost is 0.4879. Therefore, the deployment of AAI instead of AII can decrease the total telecommunication cost by (0.6967-0.4879) / 0.6967 = 29.96%.
At this point, it is important to clarify that for the rest of our analysis the number of the UE drop location cells is 7 so the deployment that we choose for the carried experiments is AII that results in the lowest telecommunication cost for the specific case. Table 3 lists all the additional simulation settings for the rest of our experiments.

Parameters	Units	Value
Cellular layout		Hexagonal grid, 19 cell sites
UE drop location cells		7
System bandwidth	MHz	1.4
UE # Rx Antennas		2

Table 3. Additional simulation settings for the experiment

6.3 Cost vs. packet loss

This section evaluates the total costs for different packet loss rates assuming: for the examined approaches. In the first instance of the experiment (Fig. 7), the fixed overhead used by the FEC encoding in approach A2 has been set to 5%. In Fig. 7, the normalized total telecommunication cost is plotted against the packet loss probability. As Fig. 7 presents, the conventional retransmission of lost segments (approach A1) is the most inefficient approach compared to the two other approaches that use FEC, irrespectively of the packet loss rate. Furthermore, in this figure, we observe that approach A2 has nearly the same total telecommunication cost with the proposed approach A3 until the packet loss rate reaches 3%. However, as the packet loss rate increases, the cost of approach A2 increases exponentially. On the other hand, an increase in the packet loss rate causes a linear increase in the cost of approach A3.
The first observation from Fig. 8 is that for higher fixed FEC overhead (15%) for approach A2, the approach A1 presents again the highest total telecommunication cost among the three approaches. Fig. 8 also reveals that approaches A2 and A3 show very close behaviour until packet loss approaches 10%. In approach A2, however, higher values of packet loss rate increase the total telecommunication cost drastically. Therefore, it is worth mentioning that a further increase in FEC overhead of A2 will just increase the total cost without actually improving the overall performance of the FEC scheme. To sum up, it has been shown that the proposed approach A3 ensures the lowest total cost irrespectively of the network conditions in terms of packet loss rate.

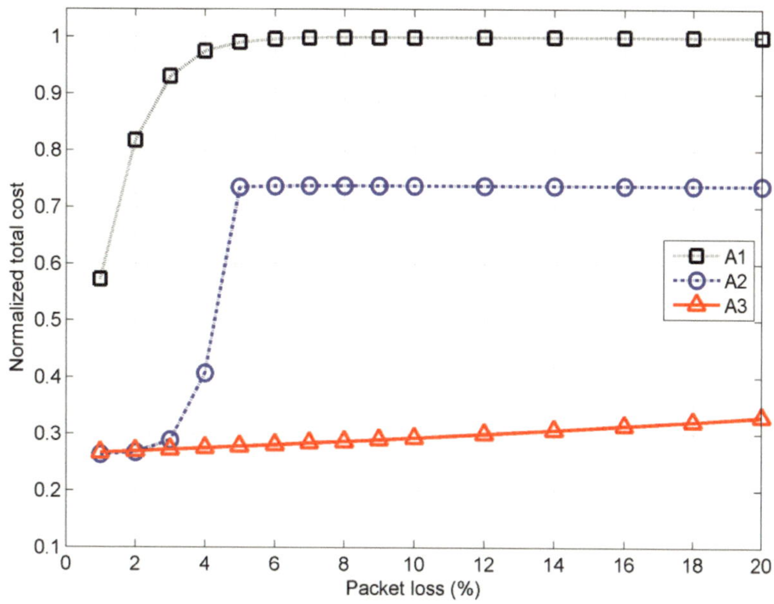

Fig. 7. Cost vs. packet loss rate (UE population = 100, fixed FEC overhead = 5%)

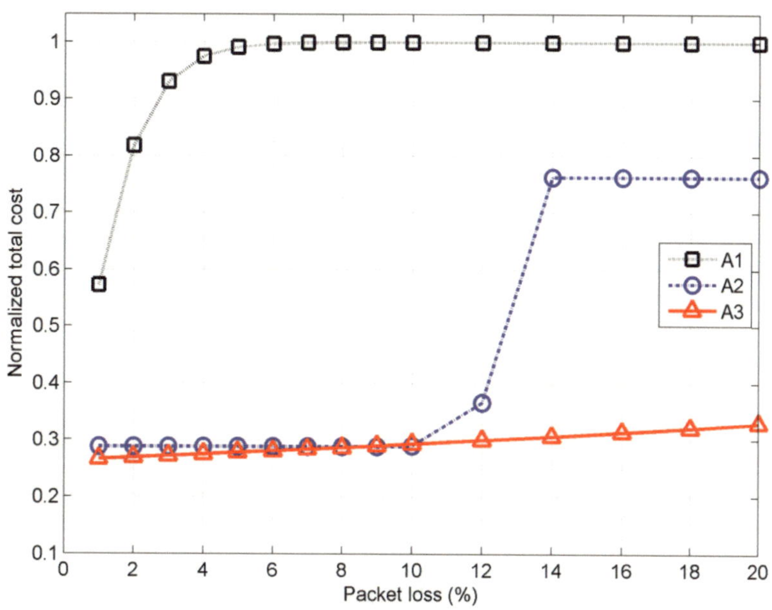

Fig. 8. Cost vs. packet loss rate (UE population = 100, fixed FEC overhead=15%)

6.4 Cost vs. FEC overhead

This paragraph presents the impact of the prefixed FEC overhead used on the approach A2 on the comparison of the three approaches under investigation. More specifically, Fig. 9 presents the normalized total cost of the three approaches as a function of the applied FEC overhead percentage, when the packet loss rate is equal to 5% and the total number of MBSFN users in the topology is 100. Obviously, the prefixed FEC overhead concerns only approach A2 and the total telecommunication cost for approaches A1 and A3 is constant and does not depend on this parameter (Fig. 9).

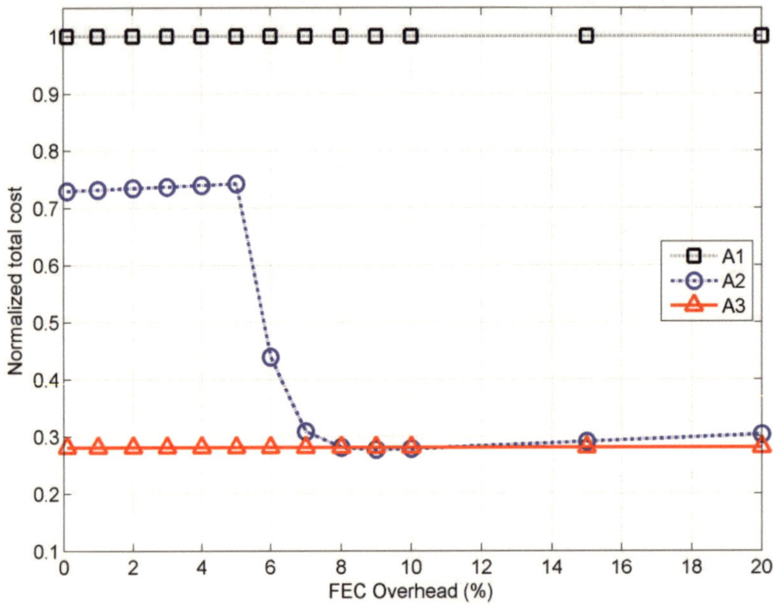

Fig. 9. Cost vs. fixed FEC overhead (packet loss rate = 5%, UE population =100)

On the other hand, the prefixed FEC overhead percentage has a direct impact on the performance of approach A2. Indeed, when approach A2 is applied and the additional information introduced by FEC remains low enough (0%-5%), the unreliable redundant retransmissions keep the total cost in unacceptable high levels. On the other hand, if the percentage of the applied FEC overhead is high enough (in the specific scenario higher than 10%) the total cost increases without actually improving the system's performance. The lowest values of total cost are achieved when the percentage of redundant information introduced by approach A2 is around 8%.

It is worth mentioning that the amount of the prefixed FEC overhead is a matter of argument in FEC schemes. Sometimes a small amount does not have any effect to the transmission and, consequently, the need for packets' retransmission and the total telecommunication cost increase. On the other hand, a large amount of a fixed FEC overhead may cause the same results. In any case, as depicted in Fig. 9, the proposed scheme (A3) ensures the lowest cost and proves a stable behaviour when network condition changes are often.

In order to further prove the efficiency and the stability of the proposed approach, we present an overview of how the value of the total telecommunication cost varies based on the FEC overhead used for approach A2 and the packet loss rate. The same experiment is conducted for the different MBSFN deployments (AAA, AII, AAI) with similar results and therefore we only present the results for the most efficient deployment (i.e., AII). It is should be mentioned that the term FEC overhead is only used for comparison purposes since the FEC overhead only affects the performance of approach A2 where this term actually represents the prefixed FEC overhead that is selected.

Fig. 10 summarizes the simulation results. It confirms the previous observations and reveals the efficiency of the proposed approach. More specifically, it can be noticed that the total cost introduced by the proposed approach (A3) increases linearly as the packet loss rate increases, ensuring in this way the system's stability. On the other hand, the increase in the packet loss rate causes an abrupt increment in the total cost of A1 and A2. However, the most important observation from Fig. 10 is that the proposed method ensures the lowest telecommunication cost irrespectively of the packet loss and the FEC overhead rate. This fact can relax the network in heavy load conditions.

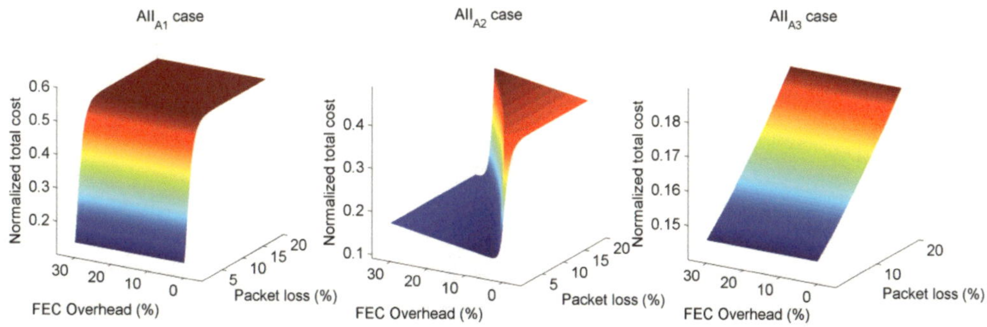

Fig. 10. Cost vs. packet loss rate vs. FEC overhead (UE population =100, deployment: AII)

6.4 Cost vs. multicast user population

One parameter that has a significant impact on the total telecommunication cost for the transmission of a multicast MBSFN service is the user population. Fig. 11 presents the normalized total cost of the three approaches as a function of the number of users in the

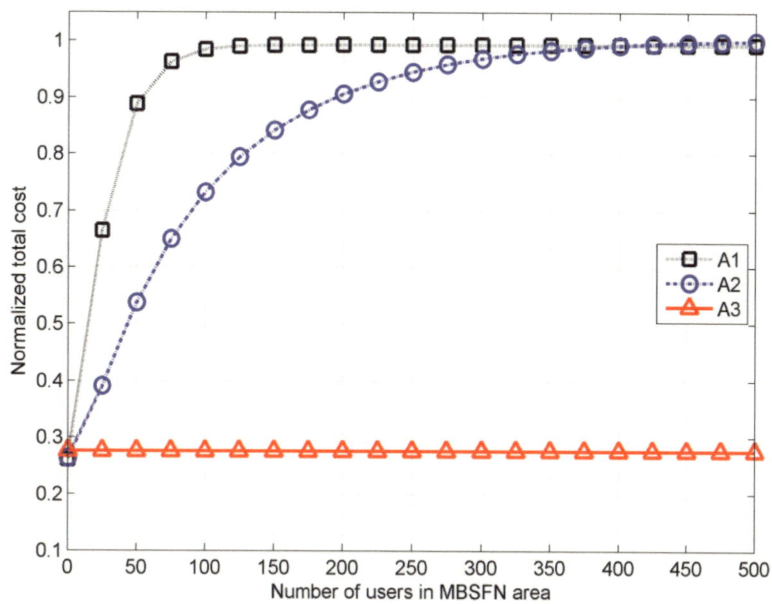

Fig. 11. Cost vs. multicast user population (packet loss rate=5%, fixed FEC overhead = 5%)

MBSFN area when the packet loss rate is equal to 5% and prefixed FEC overhead introduced by A2 is 5%. One important result is that the conventional retransmissions of lost segments (approach A1) and the application of a prefixed FEC overhead (approach A2) may keep the total cost in acceptable levels only for small number of users. As the number of users becomes large, it is evident that approaches A1 and A2 do not perform cost-efficiently. This occurs because an increase in the number of users results in an increase of failure probability. This in turn indicates that there is an extra need for retransmission of the lost segments.

On the other hand, Fig. 11 reveals that the normalized total cost of the proposed scheme is independent of the number of users and also remains in very low levels. Therefore sending redundant symbols is proven to be the most efficient way to ensure the reliable reception of MBSFN data among the three approaches.

7. Conclusion

In this chapter, we have presented a study on the application of FEC during MBSFN transmission over LTE cellular networks. We have investigated the performance of the file recovery approaches which are standardized by 3GPP for the multicast data delivery via e-MBMS and proposed an efficient new error recovery scheme for the MBSFN operation. The proposed scheme is based on the Raptor codes standardized by 3GPP for FEC use in cellular multicasting. It uses exclusively the FEC technique for the complete file recovery. The sender generates symbols, through a Raptor FEC encoder, and sends the redundant encoding symbols until it receives an acknowledgment message from all the receivers participating in the multicast group, that the file recovery has been completed. In order to evaluate our approach, we have conducted extensive simulation experiments. Also a direct comparison of our approach with the other existing approaches has been performed. Various MBSFN deployments, FEC code dimensions and error rates have been examined. Based on these parameters, we have calculated the total telecommunication cost that is required for the MBSFN transmission towards the mobile users for the various approaches. Our evaluation has been performed through a realistic simulation model that incorporates all the above parameters and is consistent with the relevant 3GPP specifications.

The simulation results have shown how the optimal FEC code dimension varies depending on the different network conditions. In more detail, we have concluded that parameters like the MBSFN deployment, the multicast user population and the packet loss rate affect the optimal FEC code dimension and we have investigated how they do it. It is important to mention that all the above results have been qualitatively assessed and explanations for the model behaviour have been provided.

The most important conclusion of our simulation experiment is that the proposed approach can offer improved performance during MBSFN operation in terms of total telecommunication cost. The main reason is that our approach can take advantage of the main property of MBSFN operation which specifies that MBMS data are broadcasted simultaneously over the air from multiple tightly time-synchronized cells. Therefore it transmits redundant information that is necessary to all receivers for the error recovery, instead of selectively retransmitting lost segments that are probably different among the receivers (due to different packet loss patterns). Based on the above procedure, the proposed approach can save resources both in the wired and more importantly in the wireless link, allowing the uses to experience more demanding applications and services.

8. Future work

The step that follows this work could be the investigation of the proposed scheme against a PTP file repair session. The reason is that, in some cases, the setup of multiple file repair procedures could be more efficient that the use of already setup MBSFN sessions. Another idea could be the modelling and the implementation of a mechanism that makes efficient Raptor code selection for LTE networks. This mechanism could monitor the network conditions, e.g., parameters like the multicast user population, the user distribution and the packet loss rate, and use them as input in order to forecast the appropriate amount of redundant symbols for FEC encoding. Finally, another possible field for future research may be the investigation of the FEC schemes from power control perspective. The work presented in this chapter could be the base for a scheme that combines FEC code selection with efficient power allocation in LTE cellular networks.

Appendix A. Acronyms

Acronym	Explanation
3GPP	3rd Generation Partnership Project
ARQ	Automatic Repeat re-Quest
BLER	Block Error Rate
CP	Cyclic Prefix
e-BM-SC	Evolved Broadcast Multicast Service Center
e-MBMS	Evolved MBMS
e-MBMS GW	E-MBMS Gateway
e-NBs	Evolved Node B
e-UTRAN	Evolved UTRA Network
FEC	Forward Error Correction
ISD	Inter Site Distance
LT	Luby-Transform
LTE	Long Term Evolution
MBMS	Multimedia Broadcast and Multicast Service
MBSFN	MBMS over Single Frequency Network
MCE	Multicell/multicast Coordination Entity
MME	Mobility Management Entity
OFDM	Orthogonal Frequency-Division Multiplexing
PL	Penetration Loss
PTM	Point-to-Multipoint
PTP	Point-to-Point
RACH	Random Access Channel
RAN	Radio Access Network
SFN	Single Frequency Network
TU	Typical Urban
UE	User Equipment
UMTS	Universal Mobile Telecommunications System
UTRA	Universal Terrestrial Radio Access

9. References

3GPP. (2005). *TSG SA WG4 S4-AHP205, Specification Text for Systematic Raptor Forward Error Correction*

3GPP. (2006). *R2-062271, Layer 1 signalling based user detection for LTE MBMS*

3GPP. (2007a). *R3-071453, Comparison of Robust E-MBMS Content Synchronization Protocols*

3GPP. (2007b). *TSG RAN WG1 R1-070051, Performance of MBMS Transmission Configurations*

3GPP. (2008a). *TS 23.246 V9.0.0, Technical Specification Group Services and System Aspects; MBMS; Architecture and functional description (Release 9)*

3GPP. (2008b). *TS 26.346 V7.8.0, Technical Specification Group Services and System Aspects; MBMS; Protocols and codes (Release 7)*

3GPP. (2009). *TS 36.300 V9.1.0, Technical Specification Group Radio Access Network; E-UTRA and E-UTRAN; Overall description; Stage 2 (Release 9)*

Alexiou, A.; Bouras, C. & Kokkinos, V. (2007). Evaluation of the Multicast Mode of MBMS, *Proceedings of the 18th Annual IEEE International Symposium on Personal, Indoor and Mobile Radio Communications (PIMRC'07)*, Athens, Greece, September 2007

Alexiou, A.; Bouras, C. & Papazois, A. (2010a). Adopting Forward Error Correction for Multicasting over Cellular Networks, *Proceedings of the 16th European Wireless Conference (EW 2010)*, Lucca, Italy, April 2010

Alexiou, A.; Bouras, C. & Papazois, A. (2010b). The Impact of FEC on Mobile Multicast Power Control, *Proceedings of the 6th Advanced International Conference on Telecommunications (AICT 2010)*, Barcelona, Spain, May 2010

Gasiba, T.; Stockhammer, T. & Xu, W. (2006). Reliable and Efficient Download Delivery with Raptor Codes, *Proceedings of the 4th International Symposium on Turbo Codes and Related Topics (ISTC'06)*, Munich, Germany, April 2006

Ho, J. & Akyildiz, I. (1996). Local Anchor Scheme for Reducing Signaling Costs in Personal Communications Networks. *IEEE/ACM Transactions on Networking,* Vol. 4, No. 5, 709–725, ISSN: 1063-6692

Holma, H. & Toskala, A. (2009). *LTE for UMTS – OFDMA and SC-FDMA Based Radio Access*, John Wiley & Sons, ISBN: 978-0-470-99401-6, Chichester, United Kingdom

Lohmar, T.; Peng, Z. & Mahonen, V. (2006). Performance Evaluation of a File Repair Procedure Based on a Combination of MBMS and Unicast Bearers, *Proceedings of the International Symposium on a World of Wireless Mobile and Multimedia Networks (WoWMoM 2006)*, pp. 349–357, Niagara-Falls, Buffalo-NY, USA, June 2006

Luby, M.; Gasiba, T.; Stockhammer, T. & Watson, M. (2007). Reliable Multimedia Download Delivery in Cellular Broadcast Networks. *IEEE Transactions on Broadcasting,* Vol. 53, No. 1, 235-246, ISSN: 0018-9316

Luby, M.; Watson, M.; Gasiba, T.; Stockhammer, T. & Xu, W. (2006). Raptor codes for reliable download delivery in wireless broadcast systems, *Proceedings of the Consumer Communications and Networking Conference (CCNC 2006)*, pp. 192-197, Las Vegas, NV, USA, January 2006

Rong, L.; Haddada, O. & Elayoubi, S. (2008). Analytical Analysis of the Coverage of a MBSFN OFDMA Network, *Proceedings of IEEE Global Communications Conference (GLOBECOM 2008),* New Orleans, LO, USA, November/December 2008

Wang, N. & Zhang, Z. (2008). The impact of application layer Raptor FEC on the coverage of MBMS, *Proceedings of the 2008 IEEE Radio and Wireless Symposium (RWS 2008)*, pp. 223–226, Orlando, FL, USA, January 2008

Part 4

Coordination of the Cellular Networks through Signaling

16

Metabolic Networking through Enzymatic Sensing, Signaling and Response to Homeostatic Fluctuations

Victoria Bunik
A.N.Belozersky Institute of Physico-Chemical Biology,
M.V.Lomonosov Moscow State University
Russia

1. Introduction

Metabolic networks include multiple pathways, where certain substrates are transformed to the pathway products through a chain of intermediates generated by sequential action of biological catalysts, i.e. enzymes. Stability of metabolic network and its performing biological functions are achieved through coordination of the network components. For this, critical enzymes of the network not only catalyze the substrate transformation, but also support the network function as a whole by sensing metabolic state, generating signals about the state changes and eventually adjusting the catalytic parameters of the network to abolish the change and signal. Stable function of the metabolic networks under variety of conditions is an important contributor to homeostasis of the living systems, which is defined as their ability to regulate internal environment so that it is maintained stably, also when external environment changes. Independent of the exact mechanism, a basic principle of the metabolic network regulation implies the catalyst ability to change its catalytic properties dependent on other network members. In this paper, pre-requisites and basic mechanisms of the coordination are formulated and exemplified in the data obtained on enzymes and multienzyme complexes. It will be shown that already primary, i.e. purely metabolic, networks without specific regulatory systems are able to coordinate their members. In this case, the same binding sites and members are used for both catalysis and regulation. Increasing complexity of networks may employ additional regulatory binding sites and/or enzymes to coordinate the network function. That is, the more complex networks may include special regulatory enzymes (in particular, those phosphorylating/ dephosphorylating other enzymes), or regulatory binding to the allosteric sites which are structurally separated from the site where the enzymatic catalysis takes place. However, the primary mechanisms of the coordination, i.e. those based on the chemistry of interactions between the enzyme active site with the network metabolites, are realized already in the basic metabolic networks formed by catalysts only. Moreover, they preserve their essentiality also in more sophisticated networks where additional regulatory elements, such as allosteric binding and specialized signal transduction systems, are added to "pure" self-regulated catalysts. The main focus of this paper is to consider the basic mechanisms of the metabolic network coordination and to show their applicability to both the primary and complex networks.

2. Pre-requisites for and mechanisms of the metabolic network coordination

Coordination of different enzymes in a metabolic network requires: (1) molecule(s) which concentration would be indicative of the network state/function; (2) binding of such indicator molecule(s) to the network enzyme(s), which results in changes of the enzyme catalytic properties to benefit the network function and stability. It is worth noting that impact of particular metabolic reactions on general metabolic functions is not equal (Sweetlove et al., 2008). For instance, only a limited amount of genes is essential for early embryonic development in zebrafish (Amsterdam et al., 2004). Also within central metabolic network there are enzymatic processes which perturbation greatly affects general performance, and those which changes are easily accommodated/compensated without an obvious effect on basic functions. This is in accord with the study of the global organization of metabolic fluxes, revealing that general metabolic activities are dominated by a limited number of reactions, so called "high-flux backbone" of metabolism (Almaas et al., 2004). Thus, regulation of different enzymes affects the network function to different extent, with the critical, or so called "key", enzymes having the greatest impact. The criticality may be defined by the degree of changes in the whole network upon perturbation of its single member. The regulation of enzymes by indicator molecules may occur through different mechanisms. In the simplest case, the enzymes responding to such indicator molecule(s) would have the same binding site for both catalysis and regulation. In such a case a structural similarity between the enzyme substrate(s) and regulator(s) is expected, if the enzyme complex with the regulator is formed by reversible binding. Alternatively, a regulator may irreversibly modify an enzyme. In this case the reactivities of the modificator and modified group on enzyme may enable the modification even in the absence of the modifier structural similarity to the enzyme substrate, with the enzyme-modificator complex stabilized covalently. In this section, it will be shown that the organization of metabolic networks through intercepting pathways transforming different substrates is well suited to satisfy the coordination requirements already in primary, i.e. consisting of pure catalysts, networks.

2.1 Structural similarity between the pathway intermediates.

2.1.1 Substrates and intermediates of their transformation in a pathway have certain similarity in structures.

A pathway within a metabolic network consists of a chain of consecutively working enzymes which transform the pathway substrate by a series of sequential steps. For instance, glucose is a substrate of the glycolytic pathway, where glucose is phosphorylated, isomerised and degraded to smaller molecules. Hence, many of the glycolytic intermediates have at least part of the glucose carbon skeleton and common functional groups, such as alcohol, aldehyde or phosphate ones. Owing to this, the pathway intermediates (e.g., glucose-6-phosphate, or its isomer fructose-6-phosphate) have a certain similarity to glucose and/or each other. Besides, a product of one enzyme in a pathway is the substrate of the enzyme catalyzing the next step. Thus, the consecutive transformation of a substrate molecule in a pathway implies structural similarity between the pathway intermediates. This could underlie partially overlapping binding to enzymes of not only their substrates, but other intermediates of the pathway as well (Fig.1). That is, the substrate of the complex (1) in Fig.1 may later in the pathway generate the breakdown products shown in (2) and (3), but the partial structural similarity of all the bound intermediates shown in Fig.1 still enables their interaction with the enzyme as shown. For instance, the depicted enzyme may catalyze the phosphorylation of glucose, whereas the breakdown products shown in the complexes (2) and (3) of Fig.1 may correspond to glyceraldehyde-3-phosphate and 3-phosphoglycerate arising on the later stages of the pathway. On the other hand, such breakdown products may interconnect different pathways within a network. For instance, 2-oxoglutarate dehydrogenase, which oxidatively decarboxylates 2-oxoglutarate to CO_2 and succinyl-CoA, may bind malonate (3C) and glyoxylic acid (2C) (Bunik & Pavlova 2006), which may be regarded as the breakdown derivatives of the 2-oxoglutarate (5C) molecule.

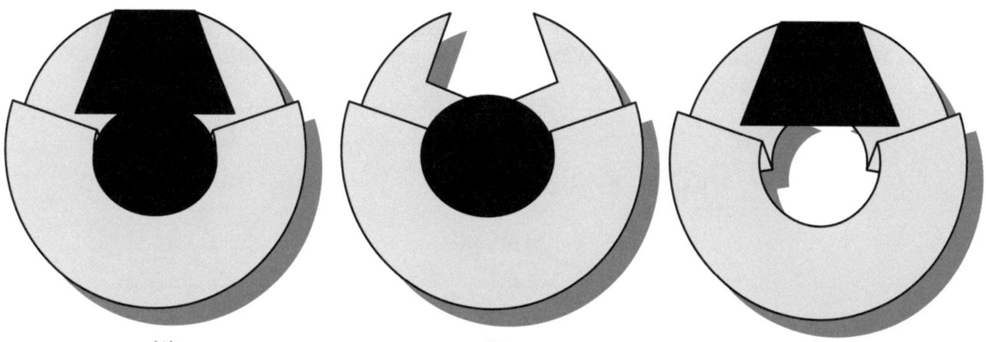

(1) (2) (3)

Fig. 1. Binding of a substrate (black in 1) and structurally relevant intermediates (black in 2 and 3), to an enzyme (grey). Certain flexibility of an enzyme structure, realized as a conformational change of its binding site in the presence of a ligand ("hand-glove" model), is schematically presented as a slight structural difference between the enzyme binding site in the free state and when ligand is bound.

Competition for the enzyme binding site between the substrate and intermediate will affect enzymatic transformation of substrate. This change in the enzyme catalytic properties enables the network coordination dependent on the relative concentrations of the intermediates. In the simplest case of the competitive inhibitor, the binding of a non-transformed structural analog of the enzyme substrate blocks the correct binding of the substrate and hence blocks catalysis (Fig.1). However, taking into account multiple binding points, relatively extended structures of both the enzyme ligands and binding sites, and different enzyme-substrate complexes along the reaction pathway, multiple binding modes of ligands are possible (Fig.2), and those may have different consequences for catalysis. A good example of this type of the regulation is provided by the 2-oxoglutarate dehydrogenase, which starts the overall reaction of a key system of the tricarboxylic acid cycle, the 2-oxoglutarate dehydrogenase multienzyme complex. The cycle and its auxiliary pathways metabolize different organic acids with up to three carboxylic groups. Extensive kinetic studies of the 2-oxoglutarate dehydrogenase reaction (Bunik et al., 2000; Bunik & Pavlova, 1997a, 1997b; 1996) showed that many of these organic acids and their structural analogs, such as oxalacetate, succinate, glutarate, malonate and 2-oxomalonate, are able to bind to an enzyme-substrate complex. The partial overlap with the binding of the 2-oxoglutarate dehydrogenase substrate, 2-oxoglutarate, occurs by means of either the dicarboxylate (D) group, or both the dicarboxylate and 2-oxo (O) groups. To a certain degree which depends on the structural similarity to 2-oxoglutarate, such binding imitates the pre-catalytic complex of 2-oxoglutarate with the 2-oxoglutarate dehydrogenase (Fig.2, SE). The catalytically essential interaction with the 2-oxoglutarate dehydrogenase coenzyme, thiamine diphosphate (ThDP) is possible only when the substrate binds to the active site. In this case the formation of a pre-catalytic complex SE is followed by a conformational transition to the catalytic ES complex, where the decarboxylation of 2-oxoglutarate takes place. Following the release of the two reaction products (p_1, p_2), the free enzyme E is ready for a new catalytic cycle. However, the ES complex is also able to bind another molecule, which may be a dicarboxylate (succinate, malonate, glutarate) or a 2-oxo dicarboxylate (oxalacetate), including the substrate 2-oxoglutarate itself. When bound to ES complex (Fig.2), such binding of a structural analog does not prevent catalysis. Moreover, 2-oxoglutarate may be bound to ES in either an inhibitory (S_iES) or an activatory (SES) mode. The activatory mode is associated with the slow compared to catalysis conformational

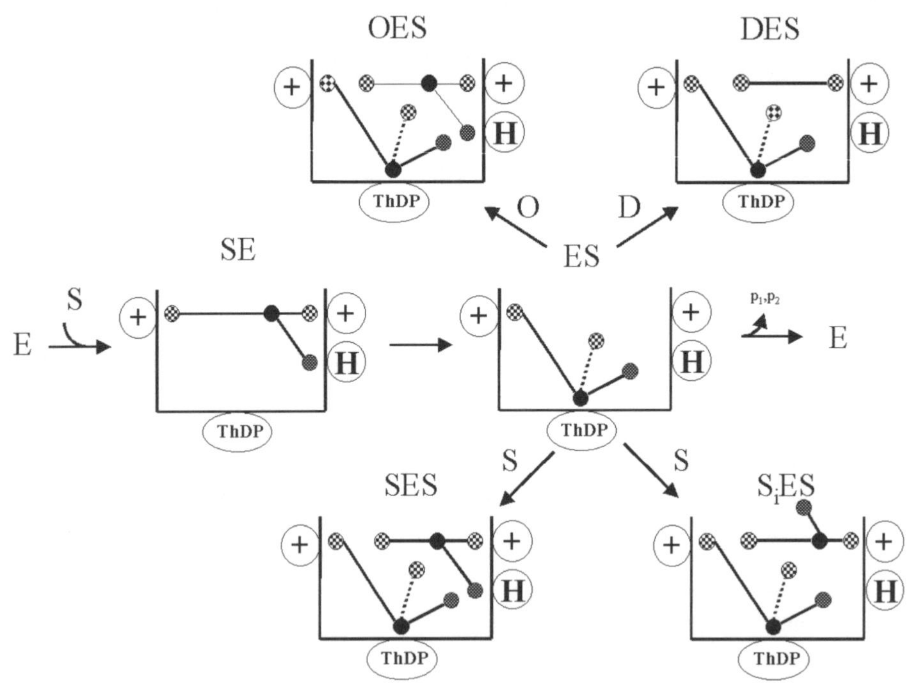

Fig. 2. Enzyme regulation through the multiple binding modes of a ligand, exemplified by the interaction of 2-oxoglutarate dehydrogenase with substrate and its structural analogs. The substrate 2-oxoglutarate (S) and its dicarboxylate structural analogs with (O) or without (D) 2-oxo group may bind to the active site of the 2-oxoglutarate dehydrogenase in different modes, which depend on the enzyme state and interactions realized. The two positively charged (+) and one histidine (H) residues of the enzyme active site interact, respectively, with the two carboxyl and 2-oxo groups of a 2-oxo dicarboxylate. The precatalytic complex SE is transformed into the catalytic complex ES by the formation of the substrate adduct with the coenzyme, thiamine diphosphate (ThDP). The middle part of the figure shows the catalytic cycle. The catalytic bond cleavage occurring in the ES complex is schematically shown by dotted line. The parts below and above the catalytic cycle demonstrate the complexes formed after ES complex binds another molecule of the substrate or its analogs, correspondingly. See the text for other explanations.

transition of the enzyme to a more active state. The inhibitory binding S_iES is similar to that of dicarboxylates (DES) and, like the bound dicarboxylates, prevents the activation through the conformational transition. Remarkably, only the 2-oxo dicarboxylates are able to block catalysis, obviously because they more closely imitate the precatalytic complex SE, forming the three bonds essential for the SE→ES transformation (Fig.2). Binding of dicarboxylates does not prevent catalysis, but blocks the activatory transition of the enzyme due to binding the second 2-oxoglutarate molecule (complex SES). Thus, dependent on the degree of the structural similarity with the substrate, the substrate structural analogs exhibit different abilities to block catalysis as competitive inhibitors at the stage of the precatalytic complex SE, or to modify catalytic properties as regulators bound to the complex ES (Bunik & Pavlova, 1997a, 1997b). In the latter case, the binding affects the kinetically slow catalysis-associated conformational transition of the protein (Bunik et al., 1991). This phenomenon represents the enzyme hysteretic properties, also called as "enzyme memory", being important for switching fluxes in the branch points of metabolic networks (Frieden, 1964).

Thus, the network state is encoded by the concentrations of certain intermediates. An enzyme in a network may respond to the network function as a whole through dependence of catalysis on the concentrations of not only the enzyme own substrate(s) and product(s), but also some other intermediates in the network. The response may go beyond the equilibrium binding of different ligands. Indeed, it may involve the reversible, but time-dependent (slow compared to the binding and catalysis) consequences of the binding of regulatory metabolites. Affecting the slow compared to catalysis conformational transition of the enzyme, such regulation does not simply prevent the catalytic transformation, but adjusts it in the time-dependent manner, enabling the enzyme hysteretic properties. This type of sensing mechanism may thus be employed for the time-dependent regulation of distribution of the substrate-dependent fluxes through different pathways. Obviously, it can contribute to the temporal network dynamics (Kholodenko et al., 2010), which is crucial for the network function.

It is remarkable that the mammalian 2-oxoglutarate dehydrogenase multienzyme complex which evolved to the very high catalytic selectivity, hardly allowing for the transformation of any natural 2-oxo acids other then 2-oxoglutarate or 2-oxoadipate, demonstrates a much less pronounced binding specificity at the level of the SE complex formation (Fig.2). That is, the enzyme does bind many of the substrate structural analogs for regulatory purposes. In this regard, it is interesting to note that the participation of the biological catalysts in the network coordination inevitably results in certain constraints to evolution of their catalytic properties to the highest binding specificity or catalytic efficiency. This may contribute to the notion that a higher level of the enzyme regulation, usually inherent in the enzymes functioning within more complex networks, is associated with a lower catalytic power compared to the less regulated orthologues (Hong et al., 1998). The view of enzymes losing catalytic power in order to satisfy regulatory requirements is also in accord with our knowledge from the directed enzyme evolution and protein engineering, which successfully create enzymes with a catalytic power higher compared to that achieved through the natural evolution. The lower catalytic power of the natural enzyme is often considered as a sub-optimal result of the evolution. However, such a view suffers from the lack of understanding that biological catalysts differ from chemical ones by the need to support not only the catalytic process per se, but also the metabolic network coordination/stability. This must impose the selection criteria additional to those increasing the catalytic efficiency and selectivity. For instance, binding of certain substrate structural analogs may interfere with the highest substrate specificity, but be required for regulation. It is also worth noting that the network coordination mechanisms and critical enzymes may differ, dependent on the network. That is, the evolution of different systems does not obligatory follow the same criteria. For instance, the high catalytic selectivity evolved in the mammalian 2-oxo acid dehydrogenase complexes, including the 2-oxoglutarate dehydrogenase one, is not an obligatory feature of these enzyme systems throughout different kingdoms of living organisms. In actinomicete Corynebacterium glutamicum a chimeric complex degrading both pyruvate and 2-oxoglutarate is present (Niebisch et al., 2006) instead of the two specific multienzyme complexes transforming either pyruvate or 2-oxoglutarate in many bacteria, plants and animals. Remarkably, the degradation of the two different substrates is performed by the two different gene products of the first component of the 2-oxo acid dehydrogenase multienzyme complex. That is, this chimeric complex uses the substrate-specific first components, like the other known 2-oxoglutarate dehydrogenase complexes. However, the overall control of the processes is changed in such a way that the oxidation of both substrates is regulated not separately, but together within one complex. The chimeric complex also has a very unusual regulation through posttranslational modification of a specific regulatory protein (Schultz et al., 2007, Niebisch et al., 2006). This example illustrates that the species-specific organization of metabolic networks and their coordination mechanisms greatly affect the catalysts evolved in living systems. Other examples illustrating this notion are also given below in Section 4. Thus, the evolution of enzymes and networks is interdependent, which creates an opportunity to change networks through changing their critical enzymes. Obviously, our understanding of the specific network coordination features and how they are met by evolutionary developments will underlie our ability to engineer cells with the designed network coordination. For instance, cells with new signaling circuits may be designed for medical and biotechnological applications (Lim, 2010).

2.1.2 Common metabolites interconnect pathways

In biological networks, there are a number of substrates or coenzymes, which are common not only within a pathway, but also between different pathways. For instance, the universal energy store and donor ATP is used in many reactions of a pathway. That is, in glycolysis ATP is the energy donor in the two reactions and is synthesized from ADP in the other two reactions. Other pathways (e.g., fatty acid oxidation) also use this molecule for the same purposes. Universal substrate and coenzyme for many oxidation-reduction steps are $NAD(P)^+/NAD(P)H$ and $FAD/FADH_2$, respectively. Remarkably, all these ubiquitous components of metabolic networks, as well as some important coenzymes or their derivatives, such as coenzyme A (CoA) or recently discovered thiamine adenine nucleotide (Bettendorff et al., 2007) comprise a common structural block of adenosine (adenine heterocycle bound to ribose) which connects to the catalytic part of the substrate or coenzyme (nicotinamide ring in NAD(P)H, isoalloxasine ring in FAD or thiazole ring in adenylated thiamine triphosphate) via the 5′-phosphorylated end of ribose. Moreover, special regulatory systems evolving for the coordination of more complex networks, employ regulators based on the same ubiquitous intermediates. For instance, apart from functioning as the universal redox substrate, interconnecting many pathways, nicotinamide adenine dinucleotide (NAD^+) is also the substrate for the regulatory NAD^+-consuming enzymes, such as ADP-ribose transferases and poly(ADP-ribose) polymerases (Belenky et al., 2007). The former enable regulatory post-translational modifications of specific enzymes, whereas poly(ADP-ribosylation) is a pluripotent cellular process important for maintenance of genomic integrity and RNA transcription in cells. Although all molecular mechanisms involved in the function of this regulatory system are not yet well characterized, the process basically depends on the depletion of the network NAD^+ and ATP (Du et al., 2003) and regulatory action of the intermediates accumulated upon the NAD^+ degradation, such as ADP-ribose (Perraud et al., 2005) and AMP (Formentini et al., 2009). Thus, the advanced regulation of the highly complex networks is based on the enzymatic sensing of the common key intermediates and their derivatives, occurring already in the primary metabolic networks. Other regulatory derivatives of nicotinamide adenine dinucleotides include the second messengers involved in calcium signaling, such as nicotinic acid-adenine dinucleotide phosphate (NAADP), which differs from $NADP^+$ by the presence of a nicotinic acid instead of a nicotinamide moiety (Rutter et al., 2008; Guse & Lee, 2008), and cyclic ADP-ribose (Graeff et al., 2009; Davis et al., 2008; Bai et al., 2005, Yue et al., 2009). As a result, common metabolites interconnecting different reactions of a pathway and different pathways of a network participate in the coordination of primary networks, whereas their derivatives regulate the metabolic networks evolved to a higher complexity. It thus appears that ubiquitous participation of these common intermediates in regulation of primary metabolic networks, associated with the wide representation of their protein binding sites, enabled evolution of the network coordination to further exploitation of these compounds in additional regulatory systems evolved.

Not all of the network indicator molecules are so widely used by the network enzymes as those considered above. Instead, the indicators may also be molecules involved in the processes crucial for the network function and stability. For instance, 2–oxoglutarate is an intermediate synthesized in the tricarboxylic acid cycle. It is irreversibly degraded by either the 2-oxoglutarate dehydrogenase functioning in the cycle or prolyl hydroxylase. The former enzyme exerts an essential control on mitochondrial oxygen consumption under increasing energy demands (Bunik, 2010; Cheshchevik et al., 2010; Cooney et al., 1981), whereas the latter is the hypoxia-inducible factor which controls cellular responses to hypoxia (McDonough et al., 2006; Ginouves et al., 2008). It is therefore obvious that the decrease in the mitochondrial oxidation of 2-oxoglutarate under limited oxygen may influence availability of 2-oxoglutarate for the hypoxia-inducible factor, thus contributing to regulation of the complex network dependent on oxygen sensitivity (Ginouves et al., 2008). In good accord with this assumption, cellular sensing of the 2-oxoglutarate level is involved in retrograde signaling of mitochondria to nucleus, leading to adaptations compensating mitochondrial impairment through the changed expression of key enzymes (Butow and Avadhani, 2004; Bunik & Fernie, 2009).

2.2 Biological network buffers

Some metabolites may be present in metabolic networks at very high concentrations. For instance, cells may contain 1-10 mM tripeptide glutathione (L-γ-glutamyl-L-cysteineyl-glycine) (Mieyal et al., 1995; Lopez-Mirabal & Winter, 2008) or 10-200 mM of dipeptides carnosine (beta-alanyl-L-histidine) and anserin (N2-methyl-carnosine) (Boldyrev, 2007). This creates an opportunity for the network coordination, which employs such highly abundant molecules as general intracellular sensors. For instance, the buffering or redox properties of these compounds may be used to integrate functional outcome of different enzymes which influence the pH or redox potential of intracellular milieu. The dipeptides are mostly known as general buffers and metal chelators. The latter property may be responsible for their

antioxidant action, which was also suggested to be due to the direct chemical reactions between the dipeptides and reactive oxygen species (ROS) and/or the secondary products arising due to ROS, such as the products of lipid peroxidation. That is, the conjugation of carnosine with α,β-unsaturated aldehydes formed from lipid peroxidation was shown as an important mechanism for the aldehydes detoxification (Aldini et al., 2002; Guiotto et al., 2005). It should be noted, however, that in case of bifunctional lipoxidation products such reactions may also damage the protein function. For instance, the bifunctional lipoxidation-derived aldehyde 4-oxo-2-nonenal can cross-link carnosine to proteins, causing an irreversible protein modification in vitro (Zhu et al., 2009), although occurrence of such processes in vivo is not known.
Significance of the very specific pattern of the dipeptide levels during development and of a number of species- and tissue-specific modifications of the basic carnosine structure is quite intriguing, but remains an enigma, as no specific protein/enzyme targets of the dipeptides have been revealed up to date (Boldyrev, 2007). Nevertheless, the exact structure of the dipeptides may be used as a chemical signature of different species, which may point to yet undefined function of such specificity for the network regulation. In contrast, the specific structure of glutathione is known to be required for its reduction by the NADPH-dependent glutathione reductase and also for the glutaredoxin-dependent deglutathionylation of proteins (Mieyal et al., 1995). Although the antioxidant function of glutathione employs its direct chemical reaction with ROS as well as that of the dipeptides, it goes beyond that, involving the bi-directional communication between the network enzymes. As will be shown in Section 2.3, the redox equilibrium between glutathione and its oxidized form, glutathione disulfide, may affect the enzymatic function of many enzymes in the network through the reversible chemical modification of the protein thiol groups. On the other hand, other enzymes affect the state of the glutathione redox buffer by generating ROS oxidizing glutathione to its disulfide and through the NADPH-dependent reduction of the glutathione disulfide (Gitler and Danon, 2003). Thus, the glutathione-dependent communication between the network enzymes is under control of the network function indicators $NAD(P)H/NAD(P)^+$ and ROS. Besides, the glutathione/glutathione disulfide redox buffer is also in redox equilibrium with other cellular thiols and disulfides of low molecular weight, which may control availability of some important SH-comprising substrates and coenzymes. Reactions between different cellular thiols and disulfides are represented by Equations 1-4, with the relative concentrations of different products dependent on the redox potentials of the participating redox couples ($^*RS^-/^*RS\text{-}SR^*$ and $^\#RS^-/^\#RS\text{-}SR^\#$) in the medium:

$$^*RS^- + {}^\#RS\text{-}SR^\# \leftrightarrow {}^*RS\text{-}SR^\# + {}^\#RS^- \quad (1)$$

$$^*RS^- + {}^*RS\text{-}SR^\# \leftrightarrow {}^*RS\text{-}SR^* + {}^\#RS^- \quad (2)$$

$$2{}^*RS^- + {}^\#RS\text{-}SR^\# \leftrightarrow {}^*RS\text{-}SR^* + 2{}^\#RS^- \quad (3)$$

$$^*RS\text{-}SR^* + {}^\#RS\text{-}SR^\# \xleftrightarrow{RS^-} 2\,{}^*RS\text{-}SR^\# \quad (4)$$

For instance, an important substrate of many enzymes, CoA, contains thiol group which enters the thiol-disulfide exchange reactions with the glutathione/glutathione disulfide (Gilbert, 1982). Furthermore, cystine or glutathione disulfide may oxidize dithiol group of reduced lipoic acid to its disulfide (Konishi et al., 1996). In living systems, lipoic acid is used as a covalently attached cofactor of the multienzyme complexes which oxidatively decarboxylate 2-oxo acids (Bunik & Strumilo, 2009). In the course of catalysis, the redox active disulfide of lipoic acid undergoes redox cycling between the dithiol and disulfide forms. Addition of low molecular weight disulfides of cysteine or glutathione inhibits the catalysis in accordance with the redox potentials of the thiols and disulfides participating in the reaction, pointing to the formation of their mixed disulfides with the complex-bound lipoate (Bunik, 2000). The mixed disulfide of the complex-bound lipoic acid with glutathione was also observed in situ (Applegate et al., 2008).

2.3 Common reactivities of the enzyme functional groups and structural elements

Complementary to the commonality of metabolites in the network, the enzyme structures and reactivities have much in common too. For instance, the common adenine nucleotide moiety discussed in Section 2.1.2 binds to the nucleotide binding domain of proteins, comprising an open twisted β sheet surrounded by α helices on both sides - so called Rossman fold (Branden and Tooze, 1999). The thiol/disulfide groups of the protein cysteine/cystine residues may undergo different thiol-disulfide exchange reactions (Equations 1-4) with the glutathione/glutathione disulfide or other SH/disulfide-comprising members of the network, including both the protein (thioredoxin, protein disulfide isomerase) and low molecular weight (cysteine/cystine; CoA/CoA-disulfide) components. Increasing network concentration of glutathione disulfide usually signals the prevalence of oxidative conditions and loss of the network reducing power. As a result, the thiol groups of proteins ($^{*}RS^{-}$) enter the reaction with glutathione disulfide ($^{\#}RS\text{-}SR^{\#}$), forming the mixed disulfide according to Equation 1 (Gilbert, 1984; Mieyal et al., 1995; Shelton et al., 2005). Similar to enzymes, the DNA-binding proteins regulating transcription may also change their function upon glutathionylation. For instance, the NFκB transcription factor is shown to be glutathionylated under hypoxic conditions in situ, with the loss of its DNA-binding activity causing the hypoxic cancer cell death (Qanungo et al., 2007). Currently, no enzyme has been shown to serve as a catalyst of this reaction (S-glutathionylation) in situ, although potential prototypes are reported, including human glutaredoxin 1 and the pi isoform of glutathione-S-transferase (Gallogly & Mieyal, 2007; Qanungo et al., 2007). The backward reaction of the protein deglutathionylation is catalyzed by glutaredoxins. The reversible post-translational modification of proteins through glutathionylation is important not only to protect the protein cysteine residues from irreversible oxidation under oxidative conditions. This modification also serves to transduce redox signals in order to either normalize homeostasis or, if this turns to be impossible, to start the death program. For instance, accumulation of hydrogen peroxide leads to depression of mitochondrial metabolism due to glutathionylation of 2-oxoglutarate dehydrogenase, with the peroxide consumption restoring both the metabolism and free thiols in the enzyme (Applegate et al., 2008). However, when the transcription factor p65-NFkappaB is S-glutathionylated, this potentiates the cell death through apoptosis (Qanungo et al., 2007). Such network switches between different states, which are dependent on the degree of homeostatic deviations, are widely spread in living systems. In particular, initial increase in the network hydrogen peroxide elicits antioxidant gene expression to reduce the peroxide, but if the peroxide level continues to increase, the prooxidant genes are expressed, inducing the cell death (Veal et al., 2007).

Another important network coordination system is based on the complete oxidation-reduction of protein vicinal thiols according to Equations 1-2, where $^{\#}RS\text{-}SR^{\#}$ depicts the protein disulfide. Usually the reductant *RS- is a small protein with the redox-active thiol/disulfide group, thioredoxin. In many enzymes this results in essential changes of the enzymatic activities, leading to metabolic switches (Gitler & Danon 2003). A classic example is the light-dark regulation of metabolic activity of chloroplasts, which is based on the light-induced reduction of the disulfide bonds in several key proteines, switching on the photosynthetic reactions of carbon fixation in plants through multiple mechanisms shown in Fig.3 (Buchanan, 1991; Danon & Mayfield, 1994; Jacquot et al., 1997; Gitler & Danon 2003). Not only metabolic enzymes, but also ion channels (Aon et al., 2007), transcription and translation factors (Chen PR, 2006; Hayashi, 1993; Jacquiersarlin & Polla, 1996; Ueno et al., 1999; Levings & Siedow, 1995; Danon & Mayfield, 1994), cytokines (Schenk et al., 1996), growth factors (Blum et al., 1996; Gasdaska et al., 1995), hormonal action (Boniface & Reichert, 1990; Makino et al., 1999) and even intercellular communication (Meng et al., 2010) may be regulated by the thioredoxin-dependent mechanism.

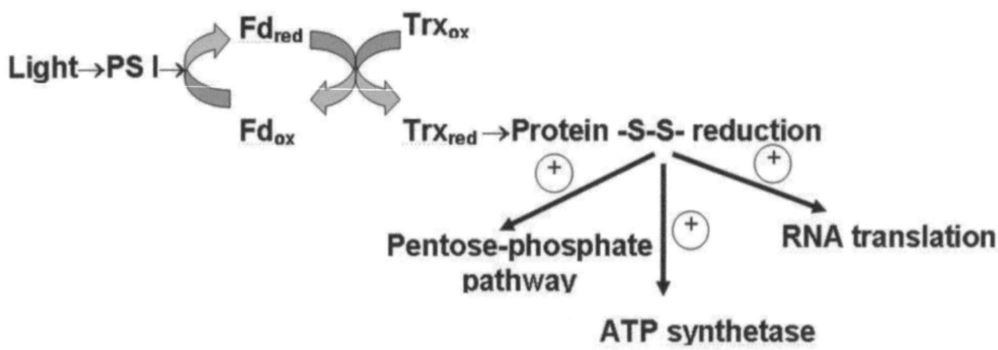

Fig. 3. Thioredoxin-mediated regulation of chloroplastic metabolism by light. Reduction of the chloroplastic photosystem I (PS I) by light enables electron flow to the redox-protein ferredoxin (Fd) which reduced form serves as the thioredoxin (Trx) reductant in the reaction catalyzed by the ferredoxin-dependent thioredoxin reductase. Thioredoxin reduces the regulatory disulfides in several chloroplastic proteins, which leads to their activation. The latter occurs in the key enzymes of the photosynthetic pentose-phosphate pathway, the energy producing ATP synthetase and the translation-regulating protein.

Many aspects of the redox regulation by reversible oxidation/reduction of protein thiols/disulfides are similar to another important posttranslational modification of proteins, involved in signal transduction, the phosphorylation/dephosphorylation (Fig.4). This modification usually occurs at the protein amino acid residues tyrosine, serine, threonin or histidine. Remarkably, the redox state of the nicotinamide adenine dinucleotides may be among the signals which are transduced by both the phosphorylation/dephosphorylation and thiol-disulfide-dependent systems. For instance, in response to increased NADH/NAD^+ ratio the pyruvate dehydrogenase complex is inactivated by phosphorylation (Roche et al., 2003), whereas the 2-oxoglutarate dehydrogenase complex is inactivated according to another mechanism controlled by thioredoxin (Bunik 2003). Thus, different coordination mechanisms may be involved in the transduction of the same signal through different enzymes. In general, the tight interplay between the redox states of the thiol/disulfides and nicotinamide adenine dinucleotides makes the pathways controlled by the coordination mechanism (1) in Fig.4 directly dependent on the network indicator molecules (i.e. SH/-S-S-; NAD(P)H/NAD(P)$^+$ and ROS). In contrast, the phosphorylation/dephosphorylation-dependent coordination mechanism (2) does not directly depend on the cellular ADP/ATP ratio, but rather switches off/on certain pathways in response to specific signals of different nature. That is, although such signals may still be generated by depletion in cellular ATP, this would not prevent the ATP-dependent protein phosphorylation under such conditions. However, depletion of NADPH directly decreases the NADPH-dependent reduction of the glutathione disulfide.

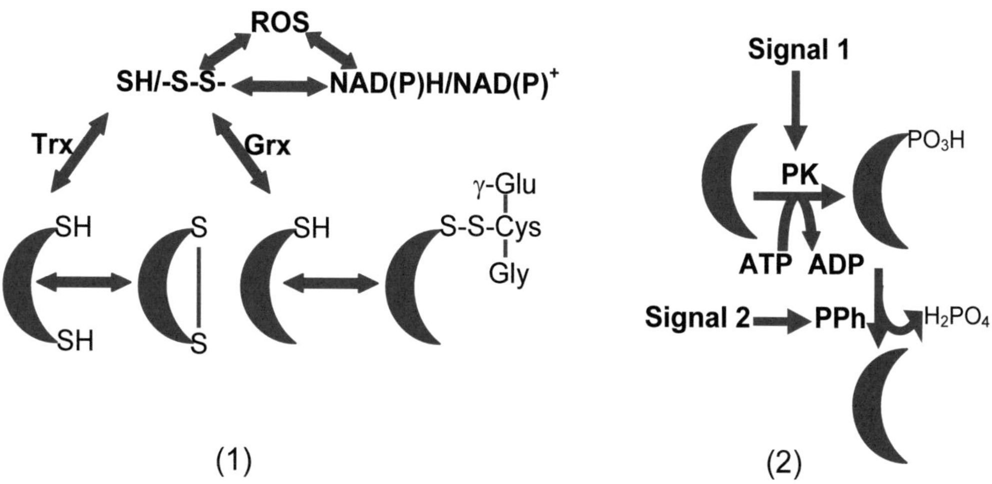

Fig. 4. Comparison of the two types of the protein regulation by post-translational modification. (1) - Modification of the protein cysteine residues in non-photosynthetic organisms; (2) – phosphorylation of the protein tyrosine, threonine, serine or histidine residues. The regulated protein is depicted as a crescent. Full oxidation/reduction of the neigbouring thiols in a protein is under the control of thioredoxin system (Trx), comprising thioredoxin and the thioredoxin reductase which in non-photosynthetic organisms is NADPH-dependent. S-glutathionylation of a protein thiol by glutathione (L-γ-glutamyl-L-cysteinyl-glycine) is controlled by glutaredoxin system (Grx), comprising glutaredoxin, glutathione and the NADPH-dependent glutathione reductase. Through catalysis by the thiol-disulfide oxidoreductases, the redox state of the nicotinamide adenine dinucleotides is related to that of cellular thiols. Both the redox states, i.e. those of the dinucleotides and thiols, are also directly related to the ROS production and scavenging. Thus, the coordination mechanism (1) may integrate signals from cellular thiols, ROS and nicotinamide adenine dinucleotides. In contrast, the coordination mechanism (2), involving the signal- and protein-specific protein kinases (PK) and protein phosphatases (PPh), does not directly depend on the overall ATP/ADP ratio.

It should be noted, however, that the network evolution to increased complexity and high differentiation precludes any of the regulation types to operate at the level of whole network. Indeed, in simpler bacterial organisms certain examples exist where the protein thiol oxidation into disulfide bonds or reduction of the disulfide bonds to thiols is determined by the redox status of the environment. That is, the oxidizing milieu of the bacterial periplasm stimulates the former, whereas reducing conditions of cytoplasm promotes the latter (Eser et al., 2009). Also in mammals under some experimental settings it is possible to observe the correlation between the global intracellular change in the cellular glutathione/gluthione disulfide ratio and function of the redox-dependent transcription factors (Haddad et al., 2000). However, in view of the known reducing nature of intracellular milieu, significance of the thiol-disulfide exchange-based regulation of proteins in such a milieu had been questioned (Ziegler, 1985). For instance, the redox potential of the transcription factor OxyR is -185 mV, while thiol/disulfide redox potential of cytoplasm was estimated between -260 and -280 mV, which raises the question of how OxyR may be activated by the disulfide formation at all (Aslund et al., 1999). Further studies showed that such simplified thermodynamic consideration is not plausible for predicting the coordination mechanisms in biological networks, because the local reaction kinetics cannot be neglected, even when the overall thermodynamics is unfavorable (Danon 2002; Toledano et al., 2004). In particular, transient activation of OxyR upon local accumulation of hydrogen peroxide was supposed to occur (Aslund et al., 1999). Indeed, it was shown that activation of OxyR due to specific disulfide bond formation occurs very fast and results in a metastable protein conformation that is locally strained. The rapid kinetic reaction path and conformational strain, respectively, are supposed to drive the oxidation and reduction of OxyR (Lee et al., 2004). Also Hsp33, a member of a newly discovered heat shock protein family, is a cytoplasmically localized protein, with its reactive cysteines responding quickly to oxidizing conditions by forming the disulfide bonds. The letter activates the chaperone function of Hsp33 (Jacob et al., 1999). Thus, it was shown that the disulfide bonds can indeed be formed within the reducing intracellular environment. Moreover, the disulfide bonds can also be reduced in the oxidizing extracellular space (Hogg 2003; Yang & Loscalzo, 2005; Yang et al., 2007).

Worth noting, the local accumulation of hydrogen peroxide necessary for cellular signaling through the thiol/disulfide-dependent processes is in some cases allowed by the phosphorylation-induced inactivation of peroxiredoxin I, which in its dephosphorylated state efficiently degrades hydrogen peroxide (Woo et al., 2010). This provides a good example of the cross-talk between the regulatory systems coordinating the network through the thiol-disulde exchange and phosphorylation reactions. Thus, significance of the local effects due to the thiol-disulfide exchange was established, owing to which most types of the network coordination in real systems should take into account local concentrations and kinetic effects. That is, most of the effects involving the network coordination through the posttranslational modification of the network enzymes occur transiently and in response to a local signal.

3. Coordination of the network through generation of secondary signal molecules

Binding of the indicator molecules as discussed in the section 2.1 may coordinate the network through the direct regulation of a critical enzyme with the indicator bound. However, the indicator binding may also coordinate network through generation of a secondary signal molecule, which would bind to other members of the network, not binding the indicator itself. Activation of an alternative catalytic pathway in the indicator-enzyme complex may induce generation of such secondary signal. This is exemplified in, but not limited to the widely known side reaction of partial dioxygen reduction with the formation of ROS. The latter are well suited for the signaling function because their reactivity allows for chemical modification of catalysts, whereas their concentration is tightly controlled by the cellular ROS scavenging systems, intimately linked to the network indicator molecules NAD(P)H/NADP+ and glutathione/glutathione disulfide. In multisubstrate enzymatic reactions, the partition between different catalytic routes, such as the main physiological and side reactions, may be controlled by the substrate ratio. Function of the 2-oxo acid dehydrogenase multienzyme complexes, each catalyzing the substrate-specific overall reaction of oxidative decarboxylation of pyruvate, 2-oxoglutarate or branched chain 2-oxo acids according to Equation 5, provides a good example of such regulation.

$$\underset{\displaystyle R}{\text{HOOC-}\overset{}{\underset{|}{\text{C}}}\text{=O}} + \text{HS-CoA} + NAD^+ \longrightarrow \underset{\displaystyle R}{\text{O=}\underset{|}{\text{C}}\text{\textasciitilde S-CoA}} + CO_2 + NADH + H^+ \qquad (5)$$

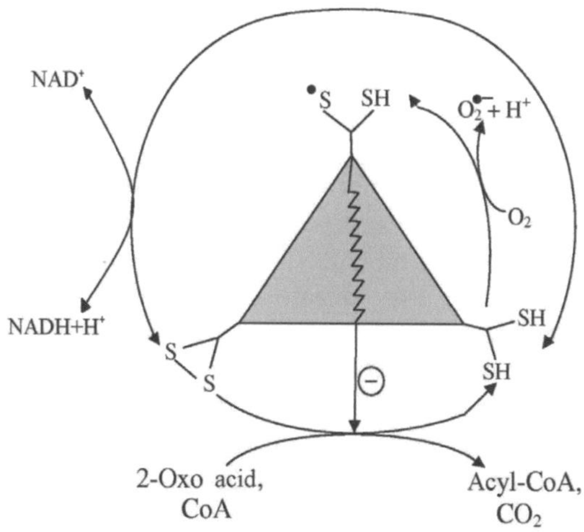

Fig. 5. Oxidative decarboxylation of a 2-oxo acid by the multienzyme 2-oxo acid dehydrogenase complexes. The overall physiological reaction occurring through the dihydrolipoyl intermediate, results in the formation of acyl-CoA, CO_2 and NADH. This comprises partial reactions at the bottom and at the outside left part of the figure. The catalytic disulfide/dithiol groups of the covalently bound lipoate residues of the complexes are schematically shown at the triangle corners. The side reaction of the one electron reduction of dioxygen is activated in the absence of the terminal substrate NAD^+, shown inside the circle at the right. The inactivation of the first enzymatic component of the complex by the thiyl radical of the complex-bound lipoate is indicated by the negative sign beside the arrow connecting the radical species and the component reaction at the bottom. See text for further discussion.

The complexes consist of multiple copies of the three catalytic components: the substrate specific 2-oxo acid dehydrogenase, dihydrolipoyl acyltransferase and dihydrolipoyl dehydrogenase. The physiological process (Equation 5) is defined in Fig.5 by the circled arrows and arrows outside the circle at the bottom and left sides. The process includes the intermediate acylation/reduction of lipoyl residues covalently bound to the second component, dihydrolipoyl acyltransferase. While this route uses the energy of the 2-oxo acid oxidative decarboxylation to reduce NAD^+ into NADH (reaction at the left of Fig. 5), dioxygen is reduced instead of NAD^+ in the side reaction (inside the circle at the right of Fig.5). As seen from the figure, the side reaction may be activated when the dihydrolipoyl intermediate arises in the presence of 2-oxo acid and CoA, but the terminal substrate NAD^+ is absent. Due to reversibility of the terminal dihydrolipoyl dehydrogenase reaction (reaction at the left of Fig. 5), the side reaction of dioxygen reduction may also be activated in the absence of 2-oxo acid and CoA, when the complex-bound lipoic acid is reduced by NADH. In so far, the activation of the side reaction involving dioxygen is controlled by availability of the substrates of physiological reaction, i.e. 2-oxo acid, CoA, NAD^+, the reaction product NADH and their ratio, which defines the steady-state level of the complex-bound dihydrolipoyl intermediate. The mechanism of the dioxygen reduction in the side reaction of the 2-oxo acid dehydrogenase complexes is such that only one electron is transferred from the dihydrolipoyl intermediate to dioxygen (Bunik & Sievers, 2002). As a result, the system operating through this route generates the two radical species: the superoxide anion radical and complex-bound thiyl radical (Fig.5). Both radical species may be considered as the secondary signal molecules. The superoxide anion radical belongs to ROS affecting the enzymatic function (i) directly, (ii) through shifting the thiol-disulfide redox status (Fig.4) to a higher oxidation state and (iii) through other signaling systems including those of phosphorylation/dephosphorylation (Woo et al., 2010). In particular, ROS inactivate aconitase and succinate dehydrogenase (Nulton-Persson & Szweda, 2001; Bulteau et al., 2003), which belong to the same pathway (TCA cycle) as the pyruvate and 2-oxoglutarate dehydrogenase complexes, all catalyzing the steps within the first half of the cycle. The second reaction product, the complex-bound thiyl radical, acts as affinity modificator inactivating the first component of the complexes. This is shown in Fig.5 by the negative regulation of the initial step of the overall process, exerted by this radical species (Bunik & Sievers, 2002; Bunik 2003). Thus, stimulation of the side route of catalysis by the 2-oxo acid dehydrogenase complexes is defined by the steady-state concentration of the dihydrolipoyl intermediate. This signal originates from the ratio of several network intermediates and indicators, such as 2-oxo acid, CoA, $NADH/NAD^+$, O_2 (Fig.5). The two types of secondary signaling molecules, i.e. the superoxide anion and thiyl radicals, are generated by the 2-oxo acid dehydrogenase complex as the complex response to sensing the ratio of its substrates and products. Importantly, that the complex senses not only a single metabolite, but an integral signal given by the ratio. One of the generated secondary signaling molecules, the superoxide anion radical, may affect the above mentioned enzymes and systems, i.e. act outside the active sites where it was generated. The other signaling molecule, thiyl radical, is bound to the complex and can therefore act only in its vicinity. In particular, it may prevent accumulation of the superoxide anion radical by inactivating the first component of the complex (Fig.5) so that the accumulation of the dihydrolipoyl intermediate is stopped. On the other hand, the thiyl radical of the complex-bound lipoate may interact with thioredoxin. The latter thiol-disulfide oxidoreductase was shown to be the potent thiyl radical scavenger (Hanine Lmoumene et al., 2000), owing to which it prevents the inactivation of the 2-oxo acid dehydrogenases by the thiyl radical as shown in Fig.6 (Bunik & Sievers, 2002; Bunik, 2000). Thus, generation of signaling molecules from the 2-oxo acid dehydrogenase complexes is coupled to the thioredoxin system. Dismutating the thiyl radicals of the complex-bound lipoate, thioredoxin prevents the inactivation of the first component of the complexes in the presence of 2-oxo acid and CoA (negative arrow in Fig.5, Fig.6), thus amplifying their generation of signaling molecules (side reaction inside the circle in Fig.5).

Fig. 6. Inactivation of the 2-oxo acid dehydrogenase complexes (depicted as a caterpillar) by their substrates 2-oxo acid and CoA (depicted as dragons) is prevented by a thiol-disulfide oxidoreductase thioredoxin (depicted as a fairy). Thioredoxin thus reconciles all the system components. The signaling implications of the reconciliation are discussed in the text. (The picture created by Dr. J. Schütz)

4. Contribution of biological context to chemical networking

Living systems widely use compartmentation of the medium, where metabolic networking occurs. Membrane-surrounded intracellular compartments, forming different organelles, represent an advanced compartmentation mechanism, separating different modules of the metabolic network in eukaryotic cells. Although this type of compartmentation is absent in prokaryotes, they do have a specific compartment of periplasm where the oxidizing conditions prevail, in contrast to the reducing conditions of cytoplasm. Development of intracellular imaging of high resolution strengthened the view that spatial organization is crucial for intracellular communication, because it provides for local interactions between and the density fluctuations of the system components (Dehmelt & Bastaniens, 2010). Indeed, as discussed in Section 2.3, such local changes are essential for the network coordination through the disulfide formation in signaling proteins, which may occur transiently under thermodynamically unfavorable conditions of generally reducing cytoplasmic milieu. However, there is a need to also have the stable structural protein disulfides. This is controlled by the enzymes oxidizing the neighboring cysteine residues of the nascent polypeptide into a disulfide and further isomerising this disulfide to its correct position in the final polypeptide after all cysteine residues have already been included. Function of this part of cellular network, which comprises many component proteins with the redox active thiols, is intensively studied, although not completely understood as yet (Thorpe & Coppock, 2007; Appenzeller-Herzog et al., 2010). However, it is well known that these reactions require separate cellular compartment where oxidizing conditions prevail to enable the disulfide formation. In bacteria this occurs in periplasm, while in eukaryotes the system is localized to the endoplasmic reticulum. Remarkably, that although the oxidized glutathione plays important role in the redox homeostasis of the endoplasmic reticulum, significant part of the redox equilibrium in this compartment is defined by the thiols and disulfides of the catalysts themselves (Appenzeller-Herzog et al., 2010). Obviously, this is required to make possible the dynamic equilibrium of the isomerization process, which includes both the reduction and oxidation of the disulfides under generally oxidizing conditions of the compartment. To achieve this, mammalian endoplasmic reticulum contains the protein disulfide isomerase at nearly millimolar concentration (Thorpe & Coppock, 2007). This example demonstrates benefits of comparable concentrations of the catalyst and its substrate, which may thus be employed by biological network, in contrast to the common chemical catalysis.

4.1 Protein-based compartmentation for the network coordination

Both pro- and eukaryotes employ still another compartmentation type, which does not require the membrane-separated compartments, but is created through the protein-protein interactions. This type of compartmentation may be called as microcompartmentation to distinguish it from the membrane-afforded organellar compartmentation. However, independent of the scale, already the homo- or heterooligomerization of catalysts, i.e. their assembly from several identical or different subunits, respectively, could create a compartment with certain advantages for the network function. The advantages are implied by the fact that many enzymes are built of at least two identical subunits, with the active sites formed at the interface between the subunits, even if each of the subunits possesses all the residues needed to perform catalysis (Branden & Tooze, 1999). One of the advantages of the enzyme function as dimers or tetramers is the ensuing opportunity of the cooperative interactions between the active sites which enables the information transfer (Fenton 2008; Branden & Tooze, 1999; Keleti 1986). In particular, this makes possible a steeper (positive cooperativity) or a slower (negative cooperativity) increase in the enzymatic activity in response to an increase in the substrate concentration, compared to the hyperbolic saturation of an enzyme with substrate in the absence of cooperative interactions. In case of heterooligomers with different active sites the cooperative effects may enable the information transfer between these sites so that the second active site would not be working if the first one lacks its substrate (Nagradova, 2003; Mouilleron & Golinelli-Pimpaneau, 2007), thus preventing a useless substrate consumption. Strong regulation of the catalytic process within the heterooligomeric multienzyme complexes is also evident from the fact that their component enzymes may even lose the capacity to catalyze the physiological reaction outside the complex. An example is provided by the 2-oxo acid dehydrogenase complexes. Their first components, the 2-oxo acid dehydrogenases, catalyze the 2-oxo acid oxidative decarboxylation reaction, reducing the second substrate, lipoamide, only when it is covalently bound to the second enzyme of the complex. Catalytic efficiency (kcat/Km) of the 2-oxo acid dehydrogenase with free lipoamide is 20000-fold less (Graham et al., 1989; Graham and Perham, 1990). The microcompartmentation may also serve protection to surrounding medium by creating channels through which the toxic intermediates are transported from one catalytic site to another (Doukov et al., 2008; Mouilleron & Golinelli-Pimpaneau, 2007). Another opportunity from the catalyst oligomerization is forming the reaction chambers. This may be needed for processing large protein substrates as in case of chaperones (Hofmann et al., 2010). However, reaction chambers are also useful to cage a common substrate so that its local concentration is increased only in a relevant part of a network, i.e. in the vicinity of the active sites where this substrate is required. This is true for the acyl carrier protein in the fatty acid synthetases and lipoic acid in the 2-oxo acid dehydrogenase complexes (Perham 2000; Jenni et al., 2007, Maier et al., 2006). Furthermore, the whole pathways may be compartmentalized. In this case, supramolecular structures, so called metabolones, are assembled, which comprise the catalysts belonging to the same pathway (Velot & Srere, 2000; Ovádi & Srere, 2000, Ovádi & Saks, 2004). It is reasonable to suggest that local concentration of the pathway intermediates within such metabolones may increase, because their diffusion into the bulk solution is hindered due to molecular crowding inherent in a cell containing many macromolecules and solutes. As a result of the hindered diffusion, the intermediates would rather bind to the vicinal enzymes belonging to the pathway metabolone, which enables a more efficient transformation of substrate and coordination of the pathway enzymes.

Why are some sequences of reactions catalyzed by relatively unstable metabolones, whereas others - by isolatable as a whole unit, i.e. rather stable, multienzyme complexes or even multifunctional enzymes? In the latter case, the genes for different enzymes catalyzing successive steps of a substrate transformation, merge to code for a single polypeptide with many active sites, but even in multienzyme complexes the overall reactions mostly occur only when the full structure is self-assembled. One may suggest that the degree of the evolutionary stabilization of such supramolecular structures depends, in particular, on the criticality of the metabolic process for the coordination of a given network and on the employed mechanisms to achieve such coordination.

4.2 Advantages for the network coordination of the random coupling vs channeling between the active sites of a supramolecular structure

It is important to note that the structure-function relationships in the multienzyme complexes established up to date do not support simplistic view that such complexes are created only to achieve efficient catalysis through intermediate channeling. Indeed, as mentioned above, the catalytic efficiency is greatly increased and the channeling is indeed important and occurs in case of toxic intermediates, such as, e.g., ammoniac, which is channeled between the active sites of the synthetase complexes using this intermediate (Nagradova, 2003; Mouilleron & Golinelli-Pimpaneau, 2007). However, in many multienzyme complexes muptiple active sites are randomly coupled to each other rather than channeling intermediates through the defined routes. It thus seems that - unless there is a need to protect from toxic intermediates - increasing the local concentration of intermediates in the enzymatic assemblies is preferred by the nature to the intermediate channeling. An obvious explanation would be that the most efficient catalysis employing channeling is unable to provide the equally efficient network coordination, because the channeling excludes the chemical information transfer between the channeled intermediate and the network. In contrast, the locally increased concentration of an intermediate could be a compromise between the needs to achieve both the efficient catalysis and network coordination. To demonstrate how this occurs, an example of the 2-oxo acid dehydrogenase multienzyme complexes may be considered. These complexes are formed around the core of a cubic or dodecahedral structure, which is built by 24 or 60 subunits of the dihydrolipoyl acyltransferase component of the complex, respectively (Izard et al., 1999). Like in other multienzyme complexes (Jenni et al., 2007), the microcompartment made by the complex core forms an internal cavity, connected with the surrounding medium by pores. In case of dodecahedral complexes the cavity diameter reaches 12 nm, and the pore size is 5 nm. The structure suggests that certain intermediates may be accumulated/compartmentalized in the cavity with the regulatory consequences. The core-forming component also has extensions outwards of the core, represented by the flexibly connected lipoyl-bearing domains of an extended structure (Fig.7). Dependent on the species, each subunit of the acyltransferase contains from one to three such domains which extrude from the core structure outwards. Thus, the 2-oxo acid dehydrogenase complexes are cellular microcompartments of the covalently bound lipoic acid. The number of the lipoic acid residues included in the complex varies between the different complexes (Perham, 2000). For instance, the cubic pyruvate dehydrogenase complex of Escherichia coli includes 72 molecules, with each of the 24 subunits of the core possessing the three lipoyl-bearing domains. The mammalian pyruvate dehydrogenase complex of dodecahedral structure contains more than 100 lipoate residues, dependent on the ratio of the isoforms of the acetyl transferase component including either two or one lipoate residues (Vijayakrishnan et al., 2010). As a result, increased local concentration of lipoic acid is available. However, the amount is excessive for the catalysis, as more than half of the lipoyl moieties of the dihydrolipoyl acyltransferase oligomer (Hackert et al., 1983; Danson et al., 1978; Collins J.H., Reed, 1978), or two of the three lipoyl-bearing domains of the acetyl transferase (Guest et al., 1990; 1997), may be removed without significant change in the overall activity of the complexes. In spite of no catalytic disadvantage, the physiological behaviour of such mutants is impaired. Indeed, bacteria with a decreased number of lipoyl domains in their pyruvate dehydrogenase complex have decreased growth rates and are washed-out from the mixed population of the mutant and wild type cells [Guest et al., 1997; Dave et al., 1995].

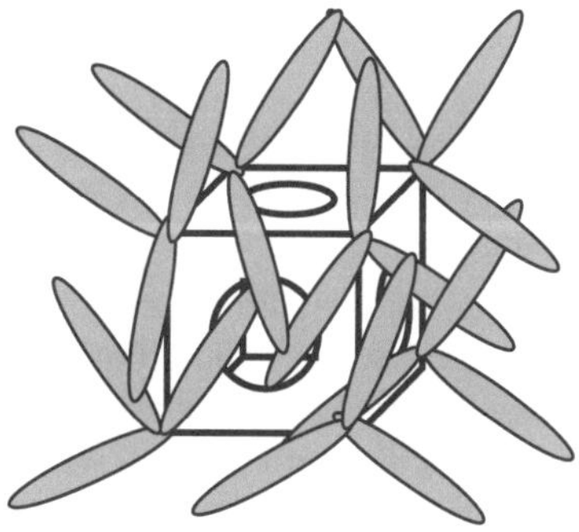

Fig. 7. Schematic presentation of the cubic core of a 2-oxo acid dehydrogenase complex (white) with the extruding lipoyl-bearing domains (grey). Catalytic domains of 24 subunits of the dihydrolipoyl acyltransferase forming the complex core are arranged in trimers at the eight cube corners. The lipoyl-bearing part of these 24 subunits shuttles between the active sites of the complex core and peripheral enzymes (not shown). This extended part may include up to three domains, with each carrying one lipoate residue.

Remarkably, the physiology is not perturbed, if the reduced number of lipoyl groups is not accompanied by the reduced length of the lipoyl-bearing protein molecule. That is, the mutant strain with the dihydrolipoyl acyltransferase bearing the three lipoyl domains with only the outermost one lipoylated behaved similar to the wild type under experimental conditions, although such a mutant complex catalyzes pyruvate oxidation 25% less efficiently than the complexes which are unable to provide the normal growth rates, i.e. those with dihydrolipoyl acyltransferase containing one or two fully lipoylated domains (Guest et al., 1997). Hence, the physiological advantage of the three lipoyl domains seems to depend on how far away the lipoyl group is able to protrude from the inner core rather than the catalytic requirements. This feature suggests that it is the interaction of the lipoyl residues with the surrounding medium beyond the peripheral enzymes of the complex, which provides the physiological advantage. Such interaction is experimentally supported by the thiol-disulfide exchange reactions between the complex-bound lipoate and cellular thiols (Bunik 2000; Applegate et al., 2008) and by other modifications of the complex-bound lipoate residues. For instance, both in vitro and in vivo the complex-bound lipoate is modified with 4-hydroxy-2-nonenal (Millar & Leaver, 2000; Humphries & Szweda, 1998; Moreau et al., 2003). Furthermore, such interaction is obvious from the thioredoxin-dependent scavenging of the complex-bound thiyl radicals of the dihydrolipoyl intermediate, discussed in the previous Section 3. Thus, apart serving the catalytic function, the complex-bound lipoate communicates with the surrounding medium, with the outcome of this communication significantly affecting the key reactions of energy metabolism, where lipoate participates as the cofactor. Another advantage of the high local concentration of the lipoate residues within the microcompartment created by the complexes (Fig.7) is the efficient and thioredoxin-dependent regulation of their superoxide anion radical production through the self-inactivation by the intrinsic thiyl radical (Fig.5). Enabling the electron migration between the neighboring lipoate residues, the lipoate microcompartment stabilizes the thyil radical, thus promoting its interaction with the surrounding medium including the first component of the complex and/or thioredoxin.

Thus, the protein structure-based compartmentation may be used not only for the catalytic, but also for the network coordination purposes, which may explain the wide occurrence of this type of compartmentation in the living systems. For instance, up to 16 stable and most abundant large multiprotein complexes with the mass more than 400 000 Da were characterized in a soil bacterium Desulfovibrio vulgaris (Han et al., 2009). Remarkably, the authors of this work conclude that the subunit composition and quaternary structure of these complexes vary to a great extent between different organisms. This view is supported by other data on the multienzyme complexes studied. For instance, even the limited number of organisms from which the 2-oxo acid dehydrogenase complexes were isolated, showed significant structural differences, which do not correlate with the catalytic performance. The variations include the different architectures known for the complexes (Izard et al., 1999; De Kok et al., 1998), their different lipoate content and stoichiometries of catalytic components and different localization of the lipoate residues. Those may be incorporated not only in the established lipoate holder, the core-forming dihydrolipoyl acyltransferase component of the complexes, but also in the peripheral components (De Kok et al., 1998; De la Sierra et al., 1997). As discussed in Section 2.1.1, the catalytic selectivity of the complexes may differ between the species as well (Niebisch et al., 2006). For the fatty acid synthase systems the three different architectures are known, from the dissociated system of individual monofunctional proteins in Escherichia coli, through the heterododecamer $\alpha_6\beta_6$ including multiple copies of the two multifunctional polypeptides in yeast to the dimer of the identical multifunctional polypeptides comprising all active sites plus acyl carrier protein in mammals (Maier et al., 2006; Jenni et al., 2007). Although evolution from the dissociated bacterial system comprising individual enzymes to the mammalian multifunctional dimer is clearly of catalytic importance, the difference between the fungal dodecamer and mammalian dimer, as well as the different structures known for the 2-oxo acid dehydrogenases are not interpretable from the catalytic viewpoint. In view of the role of the protein-based cellular microcompartments for the network coordination through the mechanisms discussed above, the absence of structural conservation in the multienzyme complexes is plausible to consider from the viewpoint of the species-specific significance of the microcompartmentation for the network coordination.

5. Conclusion

Already primary metabolic networks which consist of catalysts only, have all pre-requisites for participating in the network coordination. Independent of the network complexity, the network coordination mechanisms are based on the regulation of critical enzymes by the network intermediates of signalling importance. In addition, cellular networking widely uses benefits of the organellar and/or protein-based compartmentation, intracellular macromolecular crowding and in some cases comparable concentrations of the enzyme and its substrate. This allows for local subsets of the signal generation, transduction and response, which are different from those expected in a homogeneous medium. The coordinating function of catalysts may interfere with achievement of the highest catalytic efficiency and substrate specificity, because the network coordination requires the catalytic site to respond not only to its substrate, but also to the network regulator(s). A better compromise between the coordination and catalysis may be therefore found through development of specific regulatory binding sites and/or purely regulating catalysts, as observed in the networks of high complexity. The exact mechanisms of the coordination and the critical enzymes themselves may differ, dependent on the network. Remarkably, different coordination mechanisms may be involved in the transduction of the same signal through different enzymes, and each enzyme may respond to a variety of signals, some of them already integers of several critical parameters of the network function. This feature underlies a high interconnectivity of the network coordination mechanisms. Functional outcome of the catalytic activity adjustment (enzyme activation or inhibition) is less obvious in a network, than in a single pathway, owing to which genetic manipulations of living organisms often do not lead to desired changes in metabolism. To efficiently regulate metabolic network, we thus need to increase our knowledge on the metabolic state indicators, the enzymatically generated signaling molecules and molecular mechanisms of their regulatory action on enzymes catalyzing metabolic reactions.

6. Acknowledgement

V.I.Bunik greatly acknowledges the long-standing support of her work within the program RUS-1003594 by the Alexander von Humboldt Foundation (Bonn, Germany) and current financial support by grants from Russian Foundation of Basic Research (09-04-90473 and 10-04-90007)

7. References

Aldini, G., Granata, P. & Carini, M. (2002). Detoxification of cytotoxic alpha, beta - unsaturated aldehydes by carnosine: characterization of conjugated adducts by electrospray ionization tandem mass spectrometry and detection by liquid chromatography/mass spectrometry in rat skeletal muscle. *Journal of Mass Spectrometry*, Vol. 37, 1219–1228.

Almaas, E., Kovacs, B., Vicsek, T., Oltvai, Z.N. & Barabasi, A.L.. (2004). Global organization of metabolic fluxes in the bacterium Escherichia coli. *Nature*, Vol. 427, 839-843.

Amsterdam, A., Nissen, R.M., Sun, Z., Swindell, E.C., Farrington, S. & Hopkins, N. (2004). Identification of 315 genes essential for early zebrafish development. *Proc Natl Acad Sci USA*, Vol. 101, 12792-12797.

Aon, M.A., Cortassa, S., Maack, C. & O'Rourke, B. (2007). Sequential opening of mitochondrial ion channels as a function of glutathione redox thiol status. *J Biol Chem.*, Vol. 282, No. 30, 21889-21900.

Appenzeller-Herzog, C., Riemer, J., Zito, E., Chin, K.T., Ron, D., Spiess, M. & Ellgaard, L. (2010). Disulphide production by Ero1α-PDI relay is rapid and effectively regulated. *EMBO J.*, Vol. 29, No. 19, 3318-3329.

Applegate, M.A., Humphries, K.M. & Szweda, L.I. (2008). Reversible inhibition of alpha-ketoglutarate dehydrogenase by hydrogen peroxide: glutathionylation and protection of lipoic acid. *Biochemistry*, Vol. 47, 473-478.

Aslund, F., Zheng, M., Beckwith, J. & Storz, G. (1999). Regulation of the OxyR transcription factor by hydrogen peroxide and the cellular thiol-disulfide status. *Proc Natl Acad Sci U S A*, Vol. 96, No. 11, 6161-6165.

Bai, N., Lee, H.C. & Laher, I. (2004). Emerging role of cyclic ADP-ribose (cADPR) in smooth muscle. *Pharmacol Ther.*, Vol. 105, No. 2, 189-207.

Belenky, P., Bogan, K.L. & Brenner, C. (2007). NAD+ metabolism in health and disease. *Trends Biochem Sci.*, Vol. 32, No. 1, 12-19.

Bettendorff, L., Wirtzfeld, B., Makarchikov, A.F., Mazzucchelli, G., Frédérich, M., Gigliobianco, T., Gangolf, M., De Pauw, E., Angenot, L., Wins, P. (2007). Discovery of a natural thiamine adenine nucleotide. *Nat Chem Biol.*, Vol. 3, No. 4, pp. 211-212.

Blum, H., Rollinghoff, M. & Gessner, A. (1996). Expression and co-cytokine function of murine thioredoxin adult T cell leukemia-derived factor. *Cytokine,* Vol. 8, 6-13.

Boldyrev, A.A. (2007). 1901-1993 (Ed. Skulachev V.P.), *Nauka,* ISBN 978-5-02-036126-3, Moscow (Russian)

Boniface, J.J. & Reichert, L.E. Jr. (1990). Evidence for a novel thioredoxin-like catalytic property of gonadotropic hormones. *Science,* Vol. 247, No. 4938, 61-64.

Branden, C. & Tooze, J. (1999). Introduction to Protein Structure (second edition), Garland Publishing, Inc., ISBN 0-8153-2305-0, NY

Buchanan, B.B. (1991). Regulation of CO2 assimilation in oxygenic photosynthesis: the ferredoxin/thioredoxin system. Perspective on its discovery, present status, and future development. *Arch Biochem Biophys.*, Vol. 288, No. 1, 1-9.

Bulteau, A.L., Ikeda-Saito, M. & Szweda, L.I. (2003). Redox-dependent modulation of aconitase activity in intact mitochondria. *Biochemistry*, Vol. 42, No. 50, 14846-14855.

Bunik, V. (2010). Contribution of the 2-oxoglutarate dehydrogenase complex to the control of mitochondrial and cellular functions. *Proceedings of 7th conference on Mitochondrial Physiology,* ISBN 978-3-9502399-4-2, Obergurgl, Tyrol, Austria, 27.Sept-1.Oct 2010.

Bunik, V. (2003) 2-Oxo acid dehydrogenase complexes in redox regulation. Role of the lipoate residues and thioredoxin. *Eur. J. Biochem.* Vol. 270, pp. 1036-1042

Bunik, V. & Sievers, C. (2002) Inactivation of the 2-oxo acid dehydrogenase complexes upon generation of intrinsic radical species. *Eur. J. Biochem.*, Vol. 269, pp. 5004-5015

Bunik, V., Westphal, A.H. & DeKok, A. (2000) Kinetic properties of the 2-oxoglutarate dehydrogenase complex from *Azotobacter vinelandii.* Evidence for the formation of a precatalytic complex. *Eur.J.Biochem.,* Vol. 267, pp 3583-3591

Bunik, V. (2000). Increased catalytic performance of the 2-oxoacid dehydrogenase complexes in the presence of thioredoxin, a thiol-disulfide oxidoreductase. *Journal of Molecular Catalysis B:Enzymatic,* Vol. 8, 165-174.

Bunik, V.I. & Pavlova, O.G. (1997). Inactivation of α-ketoglutarate dehydrogenase in the course of enzymatic reaction. *Biochemistry (Mosc),* Vol. 62, No. 9, 973-982.

Bunik, V.I. & Pavlova, O.G. (1997). Inhibition of α-ketoglutarate dehydrogenase from pigeon breast muscle by the structural analogs of substarte. *Biochemistry (Mosc),* Vol. 62, No. 9, 1012-1020.

Bunik, V. & Pavlova, O. (1996). Interaction of 2-oxoglutarate dehydrogenase with 2-oxosubstrates and their structural analogs. in: Biochemistry and Physiology of Thiamin Diphosphate Enzymes, Eds. H.Bisswanger and A.Schellenberger, pp.367-373, A.u.C.Intemann, ISBN 3-926323-84-1, Prien

Chen, P.R., Bae, T., Williams, W.A., Duguid, E.M., Rice, P.A., Schneewind, O. & He, C. (2006). An oxidation-sensing mechanism is used by the global regulator MgrA in Staphylococcus aureus. *Nat Chem Biol.*, Vol. 2, No. 11, 591-595.

Cheshchevik, V., Janssen, A.J.M., Dremza, I.K., Zavodnik, I.B. & Bunik, V.I (2010). The OGDHC-exerted control of mitochondrial respiration is increased under energy demand. *Proceedings of 7th conference on Mitochondrial Physiology*, ISBN 978-3-9502399-4-2, Obergurgl, Tyrol, Austria, 27.Sept-1.Oct 2010.

Collins, J.H. & Reed, L.J. (1977). Acyl group and electron pair relay system: a network of interacting lipoyl moieties in the pyruvate and α-ketoglutarate dehydrogenase complexes from Escherichia coli. *Proc.Natl.Acad.Sci. USA*, Vol. 74, 4223-4227.

Cooney, G. J., Taegtmeyer, H. & Newsholme, E. A (1981). Tricarboxylic acid cycle flux and enzyme activities in the isolated working rat heart. *Biochem. J.*, Vol. 200, 701–703.

Danon, A. & Mayfield, S.P. (1994). Light-regulated translation of chloroplast messenger RNAs through redox potential. *Science*, Vol. 266, No. 5191, 1717-1719.

Danon, A. (2002). Redox reactions of regulatory proteins: do kinetics promote specificity? *Trends Biochem Sci.*, Vol. 27, No. 4, 197-203.

Danson, M.J., Ferscht A.R. & Perham R.N. (1978). Rapid intramolecular coupling of active sites in the pyruvate dehydrogenase complex of Escherichia coli: mechanism for rate enhancement in a multimeric structure. *Proc. Nat.Acad.Sci.USA*, Vol. 75, 5386-5390.

Dave, E., Guest, J.R. & Attwood, M.M. (1995). Metabolic engineering in Escherichia coli: lowering the lipoyl domain content of the pyruvate dehydrogenase complex adversely affects the growth rate and yield. *Microbiology*, Vol. 141, 1839-1849.

Davis, L.C., Morgan, A.J., Ruas, M., Wong, J.L., Graeff, R.M., Poustka, A.J., Lee, H.C., Wessel, G.M., Parrington, J. & Galione, A. (2008). Ca(2+) signaling occurs via second messenger release from intraorganelle synthesis sites. *Curr Biol.*, Vol. 18, No. 20, 1612-1618.

De Kok, A., Hengeveld, A.F., Martin, A. & Westphal, A.H. (1998). The pyruvate dehydrogenase complex from Gram-negative bacteria. *Biochim. Biophys. Acta*, Vol. 1385, 353–366.

De la Sierra, I.L., Pernot, L., Prange, T., Saludjian, P., Schiltz, M., Fourme, R. & Padron, G. (1997). Molecular structure of the lipoamide dehydrogenase domain of a surface antigen from Neisseria meningitidis. *J. Mol. Biol.*, Vol. 269, 129–141.

Dehmelt, L. & Bastiaens, P.I. (2010). Spatial organization of intracellular communication: insights from imaging. *Nat Rev Mol Cell Biol.*, Vol. 11, No. 6, 440-452.

Doukov, T.I., Blasiak, L.C., Seravalli, J., Ragsdale, S.W. & Drennan, C.L. (2008). Xenon in and at the end of the tunnel of bifunctional carbon monoxide dehydrogenase/acetyl-CoA synthase. *Biochemistry*, Vol. 47, No. 11, 3474-3483.

Du, L., Zhang, X., Han, Y.Y., Burke, N.A., Kochanek, P.M., Watkins, S.C., Graham, S.H., Carcillo, J.A., Szabó, C. & Clark, R.S. (2003). Intra-mitochondrial poly(ADP-ribosylation) contributes to NAD+ depletion and cell death induced by oxidative stress. *J Biol Chem.*, Vol. 278, No. 20, 18426-18433.

Fenton, A.W. (2008). Allostery: an illustrated definition for the 'second secret of life'. *Trends Biochem Sci.*, Vol. 33, No. 9, 420-425.

Formentini, L., Macchiarulo, A., Cipriani, G., Camaioni, E., Rapizzi, E., Pellicciari, R., Moroni, F. & Chiarugi, A. (2009). Poly(ADP-ribose) catabolism triggers AMP-dependent mitochondrial energy failure. *J Biol Chem.*, Vol. 284, No. 26, 17668-17676.

Gallogly, M.M. & Mieyal, J.J. (2007). Mechanisms of reversible protein glutathionylation in redox signaling and oxidative stress. *Curr Opin Pharmacol.*, Vol. 7, No. 4, 381-391.

Gasdaska, J.R., Kirkpatrick, D.L., Montfort, W., Kuperus, M., Hill, S.R., Berggren, M. & Powis G (1997). Oxidative inactivation of thioredoxin as a cellular growth factor and protection by a Cys(73)->Ser mutation. *Biochemical Pharmacology*, Vol. 52, 1741-1747.

Gilbert, H.F. (1982). Biological disulfides: the third messenger? Modulation of phosphofructokinase activity by thiol/disulfide exchange. *J Biol Chem.*, Vol. 257, No. 20, 12086-12091.

Gilbert, H.F. (1984). Redox control of enzyme activities by thiol/disulfide exchange. *Methods Enzymol.*, Vol. 107, 330-351.

Ginouvès, A., Ilc, K., Macías, N., Pouysségur, J. & Berra, E. (2008). PHDs overactivation during chronic hypoxia "desensitizes" HIFalpha and protects cells from necrosis. *Proc Natl Acad Sci U S A.*, Vol. 105, No. 12, 4745-4750.

Gitler, C. & Danon, A. (Eds.). (2003). Cellular implications of redox signaling, Imperial College Press, ISBN 1-86094-331-4, London

Graeff, R., Liu, Q., Kriksunov, I.A., Kotaka, M., Oppenheimer, N., Hao, Q. & Lee, H.C. (2009). Mechanism of cyclizing NAD to cyclic ADP-ribose by ADP-ribosyl cyclase and CD38. *J Biol Chem.*, Vol. 284, No. 40, 27629-27636.

Graham, L.D. & Perham R.N. (1990). Interaction of lipoyl domains with the E1p subunits of the pyruvate dehydrogenase multienzyme complex from Escherichia coli. *FEBS Lett.*, Vol. 262, 241-244.

Grahamm L.D., Packman L.C. & Perham R.N. (1989). Kinetics and specificity of reductive acylation of lipoyl domains from 2-oxo acid dehydrogenase multienzyme complexes *Biochemistry*, Vol. 28, 1574-1581.

Guest, J. R., Attwood, M.M., Machado, R.S., Matqi, K.Y., Shaw, J.E. & Turner, S.L. (1997). Enzymological and physiological consequences of restructuring the lipoyl domain content of the pyruvate dehydrogenase complex of Escherichia coli. *Microbiology*, Vol. 143, 457-466.

Guest, J.R., Ali, S.T., Artymiuk, P., Ford, G.C., Green, J. & Russel, G.C. (1990). Site-directed mutagenesis of dihydrolipoyl acetyltransferase and post-translational modification of its lipoyl domains. In: Biochemistry and physiology of thiamin diphosphate enzymes (Bisswanger h., Ullrich J. eds.), Chemie, Weinheim, 176-193.

Guiotto, A., Calderan, A., Ruzza, P. & Borin, G. (2005). Carnosine and carnosine-related antioxidants: A review. *Curr Med Chem*, Vol. 12, 2293–2315.

Guse, A.H., Lee, H.C. (2008). NAADP: a universal Ca2+ trigger. *Sci Signal.* , Vol. 1, No. 44, re10.

Hackert, M.L., Oliver, R.M. & Reed, L.J. (1983). A computer model analysis of the active-site coupling mechanism in the pyruvate dehydrogenase complex of Escherichia coli. *Proc.Natl.Acad.Sci. USA* , Vol. 80, 2907-2911.

Han, B.G., Dong, M., Liu, H., Camp, L., Geller, J., Singer, M., Hazen, T.C., Choi, M., Witkowska, H.E., Ball, D.A., Typke, D., Downing, K.H., Shatsky, M., Brenner, S.E., Chandonia, J.M., Biggin, M.D. & Glaeser, R.M. (2009). Survey of large protein complexes in D. vulgaris reveals great structural diversity. *Proc Natl Acad Sci U S A*, Vol. 106, No. 39, 16580-16585.

Hanine Lmoumene, E.LC., Conte, D., Jacquot, J.-P. & Houee-Levin, C. (2000). Redox properties of protein disulfide bond in oxidized thioredoxin and lysozyme: a pulse radiolysis study. *Biochemistry*, Vol. 39, 9295-9301.

Hayashi, T., Ueno, Y. & Okamoto, T. (1993). Oxidoreductive regulation of nuclear factor κB. Involvement of a cellular reducing catalyst thioredoxin. *J. Biol. Chem.*, Vol. 268, 11380-11388.

Hofmann, H., Hillger, F., Pfeil, S.H., Hoffmann, A., Streich, D., Haenni, D., Nettels, D., Lipman, E.A. & Schuler, B. (2010). Single-molecule spectroscopy of protein folding in a chaperonin cage. *Proc Natl Acad Sci U S A.*, Vol. 107, No. 26, 11793-11798.

Humphries, K.M. & Szweda, L.I. (1998). Selective inactivation of alpha-ketoglutarate dehydrogenase and pyruvate dehydrogenase: reaction of lipoic acid with 4-hydroxy-2-nonenal. *Biochemistry*, Vol. 37, 15835-15841.

Izard, T., Aevarsson, A., Allen, M.D., Westphal, A.H., Perham, R.N., de Kok, A. & Hol, W.G. (1999). Principles of quasi-equivalence and Euclidean geometry govern the assembly of cubic and dodecahedral cores of pyruvate dehydrogenase complexes. *Proc Natl Acad Sci U S A.*, Vol. 96, No. 4, 1240-1245.

Jacquiersarlin, M. R. & Polla, B. S. (1996). Dual regulation of heat-shock transcription factor activation and DNA-binding activity by H2O2: role of thioredoxin. *Biochem. J.*, Vol. 318, 187-193.

Jacquot, J.P., Lancelin, J.M. & Meyer, Y. (1997). Thioredoxins: structure and function in plant cells. *New Phytol*, Vol. 136, 543-570.

Jakob, U., Muse, W., Eser, M. & Bardwell, J.C. (1999). Chaperone activity with a redox switch. *Cell*, Vol. 96, No. 3, 341-352.

Jenni, S., Leibundgut, M., Maier, T. & Ban, N. (2006). Architecture of a fungal fatty acid synthase at 5 A resolution. *Science*, Vol. 311, No. 5765, 1263-1267.

Keleti, T. (1986) Basic Enzyme Kinetics, Akademiai Kiado, ISBN 963-05-4090-8, Budapest

Kholodenko, B.N., Hancock, J.F. & Kolch, W. (2010). Signalling ballet in space and time. *Nat Rev Mol Cell Biol.*, Vol. 11, No. 6, 414-426.

Lee, C., Lee, S.M., Mukhopadhyay, P., Kim, S.J., Lee, S.C., Ahn, W.S., Yu, M.H., Storz, G. & Ryu, S.E. (2004). Redox regulation of OxyR requires specific disulfide bond formation involving a rapid kinetic reaction path. *Nat Struct Mol Biol.*, Vol. 11, No. 12, 1179-1185.

Levings, C.S. 3rd & Siedow, J.N. (1995). Regulation by redox poise in chloroplasts. *Science*, Vol. 268, No. 5211, 695-696.

Lim, W.A. (2010). Designing customized cell signalling circuits. *Nat Rev Mol Cell Biol.*, Vol. 11, No. 6, 393-403.

López-Mirabal, H.R. & Winther, J.R. (2008). Redox characteristics of the eukaryotic cytosol. *Biochim Biophys Acta.*, Vol. 1783, No. 4, 629-640.

Maier, T., Jenni, S. & Ban, N. (2006). Architecture of mammalian fatty acid synthase at 4.5 A resolution. *Science*, Vol. 311, No. 5765, 1258-1262.

Makino, Y., Yoshikawa, N., Okamoto, K., Hirota, K., Yodoi, J., Makino, I. & Tanaka, H. (1999). Direct association with thioredoxin allows redox regulation of glucocorticoid receptor function. *J. Biol. Chem.*, Vol. 274, 3182-3188.

McDonough, M.A., Li, V., Flashman, E., Chowdhury, R., Mohr, C., Liénard, B.M., Zondlo, J., Oldham, N.J., Clifton, I.J., Lewis, J., McNeill, L.A., Kurzeja, R.J., Hewitson, K.S., Yang, E., Jordan, S., Syed, R.S. & Schofield, C.J. (2006). Cellular oxygen sensing: crystal structure of hypoxia-inducible factor prolyl hydroxylase (PHD2). *Proc. Natl.Acad. Sci. U.S.A.*, Vol. 103, 9814–9819.

Meng, L., Wong, J.H., Feldman, L.J., Lemaux, P.G. & Buchanan, B.B. (2010). A membrane-associated thioredoxin required for plant growth moves from cell to cell, suggestive of a role in intercellular communication. *Proc Natl Acad Sci U S A*, Vol. 107, No. 8, 3900-3905.

Mieyal, J.J.; Srinivasan, U. & Starke, D.W. (1995). Glutathionyl specificity of thioltransferases: mechanistic and physiological implications, In: Biothiols in Health and Disease, Packer L. and Cadenas E, (Eds), pp. 305-372, Marcel Dekker, Inc., ISBN 0-8247-9654-3, NY, USA

Millar, A.H. & Leaver, C.J. (2000). The cytotoxic lipid peroxidation product, 4-hydroxy-2-nonenal, specifically inhibits decarboxylating dehydrogenases in the matrix of plant mitochondria. *FEBS Lett*, Vol. 481, 117-121.

Milne, J.L., Shi, D., Rosenthal, P.B., Sunshine, J.S., Domingo, G.J., Wu, X., Brooks, B.R., Perham, R.N., Henderson, R. & Subramaniam, S. (2002). Molecular architecture and mechanism of an icosahedral pyruvate dehydrogenase complex: a multifunctional catalytic machine. *EMBO J.*, Vol. 21, No. 21, 5587-5598.

Moreau, R., Heath, S.H., Doneanu, C.E., Lindsay, J.G. & Hagen, T.M. (2003). Age-related increase in 4-hydroxynonenal adduction to rat heart alpha-ketoglutarate dehydrogenase does not cause loss of its catalytic activity. *Antioxid Redox Signal*, Vol. 5, 517-527.

Mouilleron, S. & Golinelli-Pimpaneau, B. (2007). Conformational changes in ammonia-channeling glutamine amidotransferases. *Curr Opin Struct Biol.*, Vol. 17, No. 6, 653-664.

Nagradova, N. (2003). Interdomain communication in bifunctional enzymes: how are different activities coordinated? *JUBMBV Life*, Vol. 55, 459-466.

Niebisch, A., Kabus, A., Schultz, C., Weil, B. & Bott, M. (2006). Corynebacterial protein kinase G controls 2-oxoglutarate dehydrogenase activity via the phosphorylation status of the OdhI protein. *J Biol Chem.*, Vol. 281, No. 18, 12300-12307.

Nulton-Persson, A.C. & Szweda, L.I. (2001). Modulation of mitochondrial function by hydrogen peroxide. *J Biol Chem.*, Vol. 276, No. 26, 23357-23361.

Ovádi, J. & Saks, V. (2004). On the origin of intracellular compartmentation and organized metabolic systems. *Mol Cell Biochem.*, Vol. 256-257, No. 1-2, 5-12.

Ovádi, J. & Srere, P.A. (2000). Macromolecular compartmentation and channeling. *Int Rev Cytol.*, Vol. 192, 255-280.

Perham, R.N. (2000). Swinging arms and swinging domains in multifunctional enzymes: catalytic machines for multistep reactions. *Annu Rev Biochem.*, Vol. 69, 961-1004.

Perraud, A.L., Takanishi, C.L., Shen, B., Kang, S., Smith, M.K., Schmitz, C., Knowles, H.M., Ferraris, D., Li, W., Zhang, J., Stoddard, B.L. & Scharenberg, A.M. (2005). Accumulation of free ADP-ribose from mitochondria mediates oxidative stress-induced gating of TRPM2 cation channels. *J Biol Chem.*, Vol. 280, No. 7, 6138-6148.

Qanungo, S., Starke, D.W., Pai, H.V., Mieyal, J.J. & Nieminen, A.L. (2007). Glutathione supplementation potentiates hypoxic apoptosis by S-glutathionylation of p65-NFkappaB. *J Biol Chem.*, Vol. 282, No. 25, 18427-18436.

Roche, T.E., Hiromasa, Y., Turkan, A., Gong, X., Peng, T., Yan, X., Kasten, S.A., Bao, H. & Dong, J. (2003). Essential roles of lipoyl domains in the activated function and control of pyruvate dehydrogenase kinases and phosphatase isoform 1. *Eur J Biochem.*, Vol. 270, No. 6, 1050-1056.

Rutter, G.A. & Bellomo, E.A. (2008). Ca2+ signalling: a new route to NAADP. *Biochem J.*, Vol. 411, No. 1, e1-3.

Schenk, H., Vogt, M., Droge, W. & Schulzeosthoff, K (1996). Thioredoxin as a potent costimulus of cytokine expression. *J. Immunol.*, Vol. 156, 765-771.

Schultz, C., Niebisch, A., Gebel & L., Bott, M. (2007). Glutamate production by Corynebacterium glutamicum: dependence on the oxoglutarate dehydrogenase inhibitor protein OdhI and protein kinase PknG. *Appl Microbiol Biotechnol.*, Vol. 76, No. 3, 691-700.

Shelton, M.D., Chock, P.B. & Mieyal, J.J. (2005). Glutaredoxin: role in reversible protein s-glutathionylation and regulation of redox signal transduction and protein translocation. *Antioxid Redox Signal.*, Vol. 7, No. 3-4, 348-366.

Sweetlove, L.J., Fell, D.A. & Fernie, A.R. (2008). Getting to grips with the plant metabolic network. *Biochem J*, Vol. 409, 27-41.

Toledano, M.B., Delaunay, A., Monceau, L. & Tacnet, F. (2004). Microbial H2O2 sensors as archetypical redox signaling modules. *Trends Biochem Sci.*, Vol 27, No. 7, pp.351-357

Thorpe, C. & Coppock, D.L. (2007) Generating disulfides in multicellular organisms: emerging roles for a new flavoprotein family. *J Biol Chem*, Vol 282, No. 19, pp. 13929-13933

Ueno, M., Masutani, H., Arai, R.J., Yamauchi, A., Hirota, K., Sakai, T., Inamoto, T., Yamaoka, Y., Yodoi, J. & Nikaido, T. (1999). Thioredoxin-dependent redox regulation of p53-mediated p21 activation. *J. Biol. Chem.*, Vol. 274, No. 50, 35809-35815.

Veal, E. A., Day, A. M. & Morgan, B. A. (2007). Hydrogen peroxide sensing and signalling. *Mol. Cell*, Vol. 26, 1–13.

Vijayakrishnan, S., Kelly, S.M., Gilbert, R.J., Callow, P., Bhella, D., Forsyth, T., Lindsay, J.G. & Byron, O. (2010) Solution structure and characterisation of the human pyruvate dehydrogenase complex core assembly. *J Mol Biol.* Vol. 399, No 1, pp. 71-93.

Vélot, C. & Srere, P.A. (2000). Reversible transdominant inhibition of a metabolic pathway. In vivo evidence of interaction between two sequential tricarboxylic acid cycle enzymes in yeast. *J Biol Chem.*, Vol. 275, No. 17, 12926-12933.

Woo, H.A., Yim, S.H., Shin, D.H., Kang, D., Yu, D.Y. & Rhee, S.G. (2010). Inactivation of peroxiredoxin I by phosphorylation allows localized H_2O_2 accumulation for cell signaling. *Cell*, Vol. 140, No. 4, 517-528.

Yang, Y. & Loscalzo, J. (2005). S-nitrosoprotein formation and localization in endothelial cells. *Proc Natl Acad Sci U S A,* Vol. 102, No. 1, 117-122.

Yang, Y., Song, Y. & Loscalzo, J. (2007). Regulation of the protein disulfide proteome by mitochondria in mammalian cells. *Proc Natl Acad Sci U S A,* Vol. 104, No. 26, 10813-10817.

Yue, J., Wei, W., Lam, C.M., Zhao, Y.J., Dong, M., Zhang, L.R., Zhang, L.H. & Lee, H.C. (2009). CD38/cADPR/Ca2+ pathway promotes cell proliferation and delays nerve growth factor-induced differentiation in PC12 cells. *J Biol Chem.,* Vol. 284, No. 43, 29335-29342.

Ziegler, D. M. (1985). Role of Reversible Oxidation-Reduction of Enzyme Thiols-Disulfides in Metabolic Regulation. *Ann.Rev. Biochem.,* Vol. 54, 305-329.